BARRON'S

SAT®

MATH WORKBOOK

4TH EDITION

BARRON'S

SAT®

MATH WORKBOOK

4TH EDITION

Lawrence S. Leff
Former Assistant Principal, Mathematics Supervision
Franklin D. Roosevelt High School
Brooklyn, New York

Dedication

To Rhona:
For the understanding,
for the sacrifices,
and
for the love

All inquiries should be addressed to:
Barron's Educational Series, Inc.
250 Wireless Boulevard
Hauppauge, New York 11788
www.barronseduc.com

Library of Congress Control No.: 2009008921

ISBN-13: 978-0-7641-4196-6
ISBN-10: 0-7641-4196-1

Library of Congress Cataloging-in-Publication Data

Leff, Lawrence S.
Barron's SAT math workbook / Lawrence S. Leff. — 4th ed.
p. cm.
Rev. ed. of: Barron's math workbook for the new SAT. 3rd. ed. c2004.
ISBN-13: 978-0-7641-4196-6
ISBN-10: 0-7641-4196-1
1. Mathematics—Examinations, questions, etc. 2. SAT (Educational test)—Study guides. I. Leff, Lawrence S. Barron's math workbook for the new SAT. II. Title. III. Title: SAT math workbook.

QA43.L4675 2008
510.76—dc22 2009008921

PRINTED IN THE UNITED STATES OF AMERICA

9 8 7 6 5 4

Contents

TAKING PRACTICE TESTS

Preface

This workbook is organized around a simple and easy-to-follow five-step study program:

1. Know what to expect on test day.
2. Become testwise.
3. Learn special math strategies.
4. Review SAT math topics.
5. Take practice exams under test conditions.

As you work your way through this book you will find the following:

- Hundreds of practice SAT-type math questions with detailed solutions
- Special math strategies that will help you get the right answers to different types of problems that you may not know how to solve
- All the math you need to know for the SAT, conveniently organized into sets of easy-to-read lessons. These lessons summarize and illustrate key concepts and formulas. Each lesson is followed by a special set of SAT Tune-Up Exercises that give practice in answering the three different types of SAT questions while reinforcing the math concepts covered in the lesson.
- "Tips for Scoring High," which will help you develop speed and accuracy
- Two SAT math practice tests with complete solutions

If you are pressed for time, you will find at the back of the book a "Quick Review of Key Math Facts," which provides an instant summary of important math skills and concepts.

Lawrence S. Leff

PART 1

LEARNING ABOUT THE SAT MATH

Knowing What You're Up Against

CHAPTER 1

The SAT is designed to assess your reasoning abilities. After reading this chapter, you will know how the mathematics sections of the SAT are organized and what mathematics topics you are expected to know. You will become familiar with the two basic types of SAT math questions and the instructions that accompany them. Learning these instructions before you take the exam will save you valuable time on test day. Knowing the simple test-taking tips and special math strategies that are featured in Chapters 1 and 2 will help you earn additional SAT math points.

LESSONS IN THIS CHAPTER

LESSON

1-1 GETTING ACQUAINTED WITH THE SAT

OVERVIEW

The SAT *is a timed exam lasting 3 hours and 45 minutes. Many colleges require you to take the* SAT *as part of the admissions process. Since the format and mathematical content of the* SAT *are predictable, you can prepare for this important exam. The chapters in this book form a complete and easy-to-follow study program for the math portion of the test. As you move from chapter to chapter, you will acquire the knowledge, skill, and confidence that can lead to large gains in your* SAT *math score.*

WHAT DOES THE SAT MEASURE?

The SAT measures your verbal and mathematical reasoning abilities. It does not test intelligence, creativity, or motivation. The math questions test your reasoning abilities by asking you to solve unfamiliar-looking problems using the basic math facts and skills that every high school student is expected to know and to be able to apply.

WHY DO COLLEGES REQUIRE THE SAT?

College admissions officers know that the students who apply to their colleges come from a wide variety of high schools that may have different grading systems and academic standards. SAT scores make it possible for colleges to compare the course preparation and the performances of applicants by using a common academic yardstick. Your SAT score, together with your high school grades and other information you or your high school may be asked to provide, helps college admission officers to predict your chances of success in the college courses you will take.

HOW ARE THE MATH SECTIONS OF THE SAT ORGANIZED?

The mathematics portion of the SAT is divided into three timed sections as described in Table 1.1.

TABLE 1.1

The Three SAT Math Sections

Mathematics Sections	Question Types	Range of Score
■ Two 25-minute sections ■ One 20-minute section	44 regular multiple-choice plus 10 student-produced response ("grid-in") questions	200–800
Total: 70 minutes	Total: 54 math questions	

ORDER OF TEST SECTIONS

The SAT has a total of ten scored test sections that cover mathematical reasoning, critical reading, and writing:

Don't be concerned about the order of the sections. Different test booklets in the same test administration may have the content sections arranged in different orders.

- The test always begins with a 25-minute essay and always ends with a 10-minute multiple-choice writing section.
- Sections 2 through 7 are 25-minute sections. The order in which these sections appear may not be the same in all test booklets for your exam.
- Sections 8 and 9 are 20-minute sections.

HOW ARE SCORES REPORTED?

After you take the test, you will receive a report that contains a score for the verbal questions and a separate score for the math questions. Raw scores are converted to scores that range from 200 to 800, with 500 representing the average SAT math score.

WHAT MATH DO YOU NEED TO KNOW?

The mathematics sections of the SAT assume that you have studied *three* years of college preparatory high school mathematics that includes concepts from Algebra I, Geometry, and Algebra II. The topics you need to know for the SAT are summarized in Table 1.2.

The SAT will *not* have any questions that require

- the quadratic formula,
- complicated operations with radicals,
- complicated factoring,
- formal geometric proofs, or
- trigonometric ratios.

CAN YOU STUDY FOR THE SAT?

Since the format, the types of questions, and the math background needed to answer SAT math questions do not change from exam to exam, thoughtful preparation before you take the test will boost your math score. To make the best use of your study time, you need a plan. Here is a five-step study plan for the SAT math sections that you can follow simply by reading this book from beginning to end.

A FIVE-STEP STUDY PLAN FOR SAT MATH

Step 1. Know what to expect on test day.
Step 2. Become testwise.
Step 3. Learn special math strategies.
Step 4. Review SAT math topics.
Step 5. Take practice exams under test conditions.

TABLE 1.2

Math Topics You Need to Know

Area of Mathematics	Topics You Need to Know
ARITHMETIC	• Sets, fundamental operations, and arithmetic reasoning • Numbers and divisibility • Number sequences involving exponential growth • Newly defined operations
ALGEBRA I and II	• Basic algebraic equations and inequalities • Integer and fractional exponents • Absolute value • Rational, radical, and exponential equations • Direct and inverse variation • Function notation and evaluation, domain and range, functions as mathematical models • Equations and graphs of linear and quadratic functions • Linear and exponential rates of change • Reflections and translations of function graphs
GEOMETRY	• Geometric notation for length, segment, line, ray, and congruence • 30–60–90 and 45–45–90 right triangle relationships • Properties of tangent lines (e.g., radius of a circle is perpendicular to tangent line at point of contact) • Coordinate geometry (e.g., slope, midpoint, and distance formulas; equation of a line) • Geometric shapes in two and three dimensions
DATA ANALYSIS and PROBABILITY	• Organizing and interpreting data in tables, charts, and graphs • Simple and geometric probability • Mean, median, and mode • Counting methods

Step 1. Know What to Expect on Test Day

Know the format, the types of math questions, and the special directions that appear on every SAT exam. This information will save you valuable testing time when you take the SAT. It will also build your confidence and prevent errors that may arise from not understanding the directions. Chapter 1 gives you this important edge.

Step 2. Become Testwise

Learn the test-taking tips that will help you gain points by increasing your speed and accuracy. Chapter 1 contains many SAT test-taking tips highlighted in a boxed display.

Step 3. Learn Special Math Strategies

Many SAT math questions are not like the standard types of textbook problems you have solved in high school math classes. To better handle these SAT questions, Chapter 2 reveals special math strategies that you may not have learned. These strategies can help you arrive at the correct answer when you get stuck or when traditional methods take too long.

Step 4. Review SAT Math Topics

Although SAT math questions stress problem solving and reasoning skills, you are expected to remember important math concepts taught in arithmetic, Algebra I and II, and beginning geometry courses. Chapters 3–8 serve as brief math refresher courses. In these chapters you will also find a large number of SAT-type math questions organized by lesson topic. If you need a fast review of important SAT math facts, you can also turn to the "Quick Review of Key Math Facts," located at the end of the book.

Step 5. Take Practice Exams Under Test Conditions

Practice makes perfect! Chapter 9 features two full-length SAT math practice tests with Answer Keys and detailed explanations of answers. Each test consists of three math sections that closely resemble the three math sections of an actual SAT exam. These practice tests should be taken under testlike conditions, using the sample SAT answer forms provided at the back of this book. Be sure to use the same calculator that you will bring to the testing room. This activity will provide an opportunity to try out, under test conditions, the tips and strategies that you have learned in this book. It will also give you an idea of how well you can manage your time on the various math sections. Finally, studying the solutions to each of the exam questions you missed or skipped over will help to eliminate any remaining weak spots in your preparation.

LESSON

1-2 REGULAR MULTIPLE-CHOICE QUESTIONS

OVERVIEW

The SAT *groups math problems by question type. More than 80 percent of the math questions that appear on the* SAT *are standard multiple-choice questions, each of which offers five answer choices from which you must pick the correct one.*

THE MOST COMMON TYPE OF SAT MATH QUESTION

Forty-four of the 54 math questions that appear on the SAT are multiple-choice questions that come with five answer choices, labeled from (A) to (E). After you figure out which one of the five answer choices is correct, you must locate the corresponding question number for that section on a separate answer sheet. Using a No. 2 pencil, you then fill in the oval on the answer form that contains the same letter as the correct choice. Since answer forms are scored by a machine, make sure to fill in completely the oval you choose as a correct answer, as shown below.

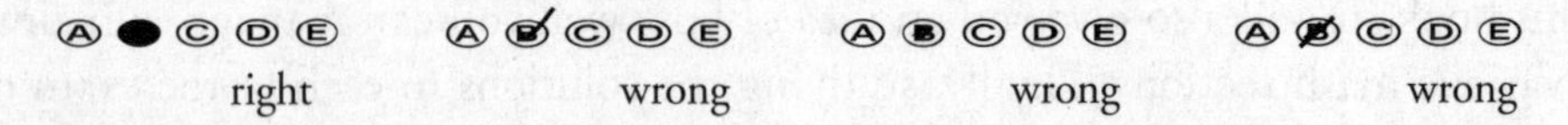

Here is an example of a regular multiple-choice question that involves simple algebra.

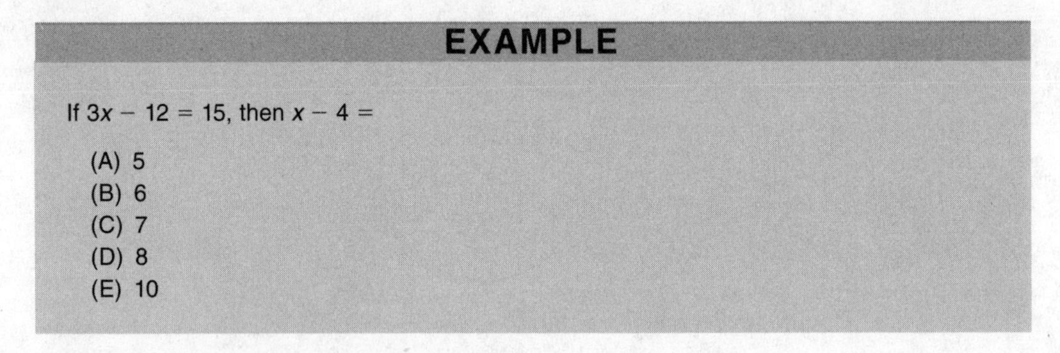

EXAMPLE

If $3x - 12 = 15$, then $x - 4 =$

(A) 5
(B) 6
(C) 7
(D) 8
(E) 10

Solution

Don't waste time by isolating x on one side of the equation. Instead, solve the equation for $x - 4$ by dividing each term of the equation by 3. Thus,

$$\frac{{}^{1}\cancel{3}x}{\cancel{3}} - \frac{12}{3} = \frac{15}{3}$$

$$x - 4 = 5$$

The correct choice is **(A)**.

On your answer sheet you must darken oval A, as shown. 

ROMAN NUMERAL MULTIPLE-CHOICE QUESTIONS

In a special type of multiple-choice question, three statements are numbered as I, II, and III. The facts included in the question make each Roman numeral statement either true or false. Different combinations of these Roman numeral statements are presented as the standard set of five answer choices. You need to figure out which answer choice includes all the Roman numeral statements that are true.

EXAMPLE

b c a

Figure not drawn to scale

If two sides of the above triangle measure 5 and 7, the perimeter of the triangle could be which of the following?

I. 12

II. 19

III. 25

(A) I only

(B) II only

(C) III only

(D) I and II only

(E) II and III only

Solution

For each Roman numeral statement, determine whether it could be true or must be false.

- Consider statement I. Without including the third side, the sum of the lengths of the two given sides is $5 + 7 = 12$. Since the perimeter of the triangle cannot be 12, statement I cannot be true. Mark Roman numeral statement I with an F to indicate that it is false.

- Consider statement II. Assuming that the perimeter was 19, then the third side of the triangle would be $19 - (5 + 7)$, or 7. A triangle can have side lengths of 5, 7, and 7 because each side satisfies the triangle inequality restriction that its length is less than the sum of the lengths of the other two sides. Since the perimeter of the triangle could be 19, mark Roman numeral statement II with a T to indicate that it could be true.

- Consider statement III. Assuming that the perimeter was 25, then the third side of the triangle would be 25 − (5 + 7), or 13. A triangle cannot have side lengths of 5, 7, and 13 since 13 is not less than 5 + 7. Mark Roman numeral statement III with an F to indicate that it must be false.

- Compare your T or F answers for the three Roman numeral statements with the set of answer choices:

I. 12 F
II. 19 T
III. 25 F

✗ (A) I only
✓ (B) II only
✗ (C) III only
✗ (D) I and II only
✗ (E) II and III only

Since only Roman numeral statement II could be true, the correct choice is **(B)**.

TIPS FOR SCORING HIGH

1. Don't keep moving back and forth from the question page to the answer sheet. Record the answer next to each question. After you accumulate a few answers, transfer them to the answer sheet at the same time. This strategy will save you time.
2. After you record a group of answers, check to make sure that you didn't accidentally skip a line and enter the answer to question 3, for example, in the space for question 4. This strategy will save you from a disaster!
3. On the answer sheet, be sure to fill one oval for each question you answer. If you need to change an answer, erase it completely. If the machine that scans your answer sheet "reads" what looks like two marks for the same question, the question will not be scored.
4. Do not try to do all your reasoning and calculations in your head. Freely use the blank areas of the test booklet as a scratch pad and use your calculator as needed.
5. Write a question mark (**?**) to the left of a question that you skip over. If the problem seems too hard or time consuming for you to solve, write a cross-mark (x) instead of a question mark. This will allow you to set priorities for the questions that you need to come back to and retry, if time permits.

LESSON 1-3

GRID-IN QUESTIONS

OVERVIEW

There are only two types of SAT *math questions: regular multiple choice and "grid-in". A grid-in or student-produced response question does not come with any answer choices. You must come up with your own answer and then enter that answer on a special four-column grid following some simple guidelines. The* SAT *includes ten grid-in questions.*

BE FAMILIAR WITH THE ANSWER GRID

When you figure out the answer to a grid-in question, you must record it on a four-column answer grid like the one in Figure 1.1. The answer grid can accommodate whole numbers from 0 to 9999, as well as fractions and decimals.

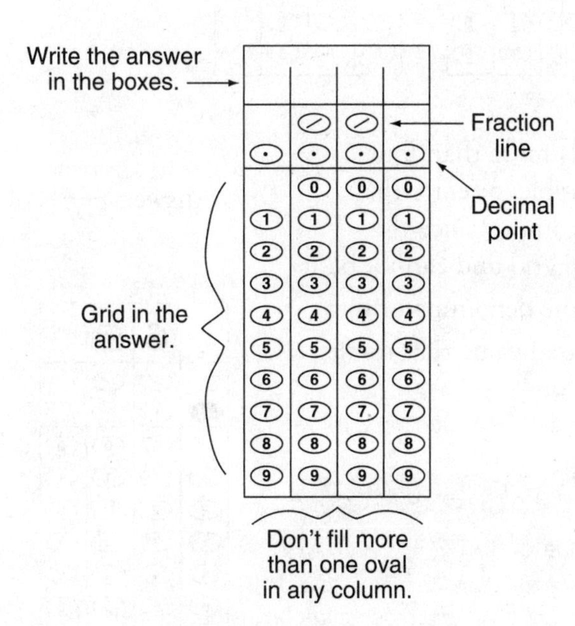

Figure 1.1 An Answer Grid

- The answer grid does *not* contain a negative sign or any letters, so your answer to a grid-in question can *never* be a negative number or contain a letter.
- Since *no* point deduction is made for a wrong answer to a grid-in question, *always* enter an answer for a grid-in question. If you can't make an educated guess, perform some reasonable operation on the numbers given in the problem. For example, multiply, divide, or average them and then use the result as your guess.

To grid in an answer:

- Write the answer in the top row of column boxes of the grid. A decimal point or fraction bar (slash) requires a separate column. Although writing the answer in the column boxes is not required, it will help guide you when you grid the answer in the matching ovals below the column boxes.

- Fill the ovals that match the answer you wrote in the top row of column boxes. Make sure that no more than one oval is filled in any column. Columns that are not needed should be left blank. If you forget to fill in the ovals, the answer that appears in the column boxes will *NOT* be scored.

Here are some examples:

Answer: 0.237 Answer: 23.7 Answer: $\frac{23}{7}$

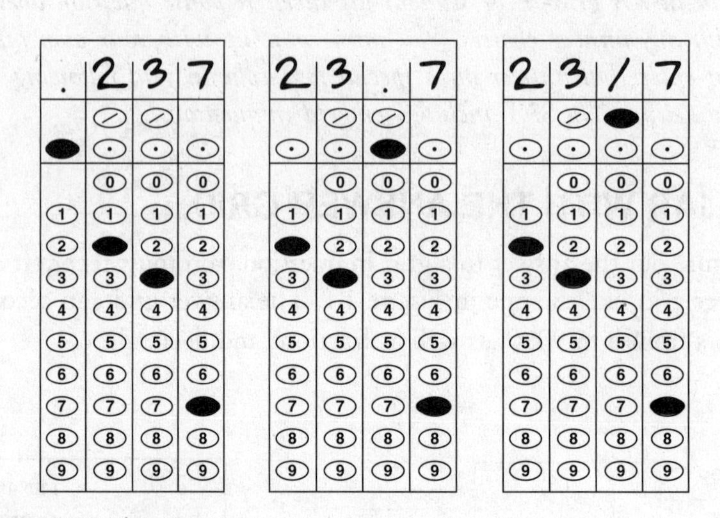

- If the answer is a fraction that needs more than four columns, reduce the fraction, if possible, or enter the decimal form of the fraction. For example, since the fractional answer $\frac{17}{25}$ does not fit the grid and cannot be reduced, use a calculator to divide the denominator into the numerator. Then enter the decimal value that results. Since $17 \div 25 = 0.68$, grid in .68 for $\frac{17}{25}$.

Answer: $\frac{17}{25}$

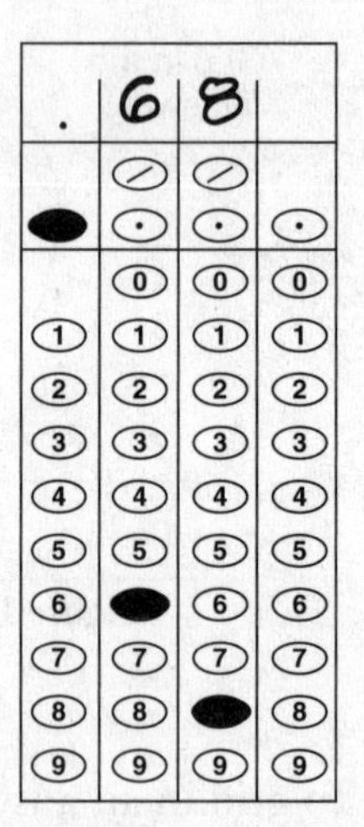

EXAMPLE

If $a \star b = a^3 - b^2 + 1$, what is the value of $3 \star 2$?

Solution

Plug in 3 for *a* and 2 for *b*:

$$3 * 2 = 3^3 - 2^2 + 1$$

Since $3^3 = 3 \times 3 \times 3 = 27$ and $2^2 = 2 \times 2 = 4$,

$$3 * 2 = 27 - 4 + 1 = 24$$

The answer is **24**. Here are one recommended and two acceptable ways of gridding in 24.

Recommended Acceptable

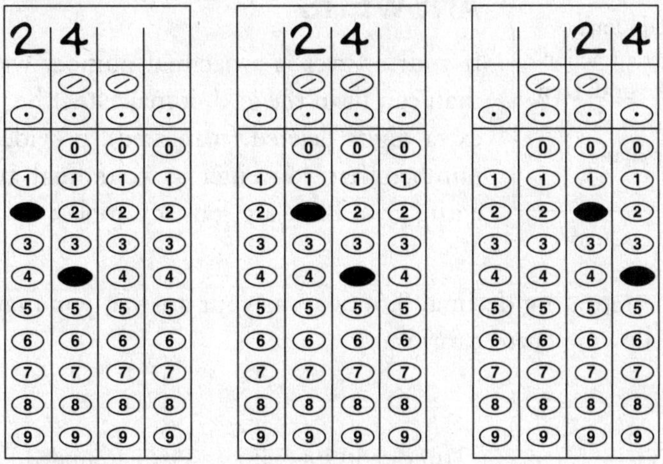

2	4		

	2	4	

		2	4

START ALL ANSWERS, EXCEPT 0, IN THE FIRST COLUMN

If you get into the habit of always starting answers in the first column of the answer grid, you won't waste time thinking about where a particular answer should begin. Note, however, that the first column of the answer grid does not contain 0. Therefore a zero answer can be entered in any column after the first. If your answer is a decimal number less than 1, don't bother writing the answer with a 0 in front of the decimal point. For example, if your answer is 0.126, grid .126 in the four column boxes on the answer grid.

ENTER MIXED NUMBERS AS FRACTIONS OR AS DECIMALS

The answer grid cannot accept a mixed number such as $1\frac{3}{5}$. You *must* change a mixed number into an improper fraction or a decimal before you grid it in. For example, to enter a value of $1\frac{3}{5}$, grid in either 8/5 or 1.6.

Answer: $\frac{8}{5}$

8	/	5	

✓

Answer: 1.6

1	.	6	

✓

Answer: $1\frac{3}{5}$

1	3	/	5

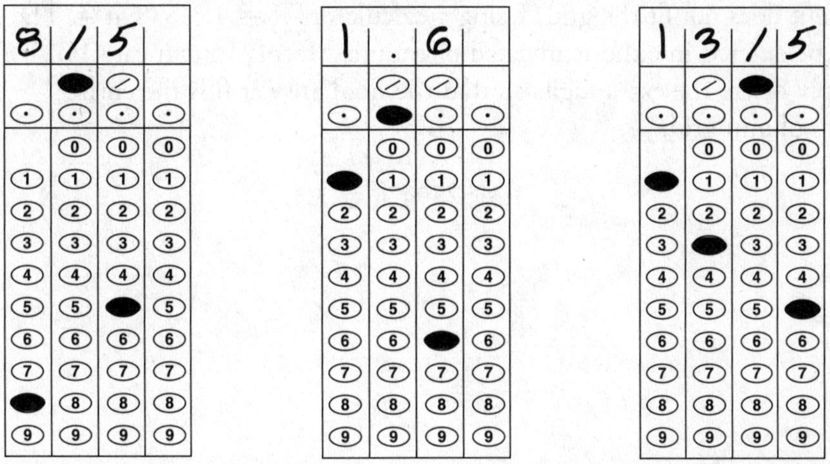

WRONG since the computer interprets the answer as $\frac{13}{5}$

TIME SAVER

If your answer fits the grid, don't change its form. If you get a fraction as an answer and it fits the grid, then do not waste time and risk making a careless error by trying to reduce it or change it into a decimal number.

ENTERING LONG DECIMAL ANSWERS

If your answer is a decimal number with more digits than can fit in the grid, it may either be rounded or the extra digits deleted (truncated), provided the decimal number that you enter as your final answer fills the entire grid. Here are some examples:

- If you get the repeating decimal 0.6666... as your answer, you may enter it in any of the following correct forms:

Fraction $\frac{2}{3}$	Truncated Decimal .666	Rounded Decimal .667
2/3	.666	.667
✓	✓	✓

Entering less accurate answers such as .66 or .67 will be scored as incorrect.

- If your answer is $\frac{25}{19}$, then you must convert the mixed number to a decimal since it does not fit the grid. Using a calculator, $\frac{25}{19} = 1.315789474$. The answer can be entered in either truncated or rounded form. To truncate 1.315789474, simply delete the extra digits so that the final answer fills the entire four-column grid:

$$1.315789474 = 1.31$$

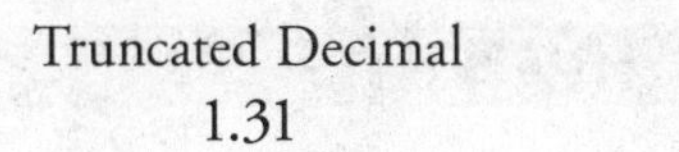

Truncated Decimal	Rounded Decimal	Less Accurate
1.31	1.32	1.3

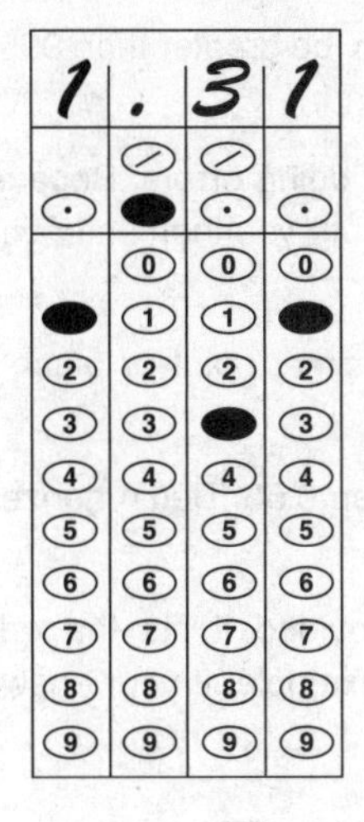

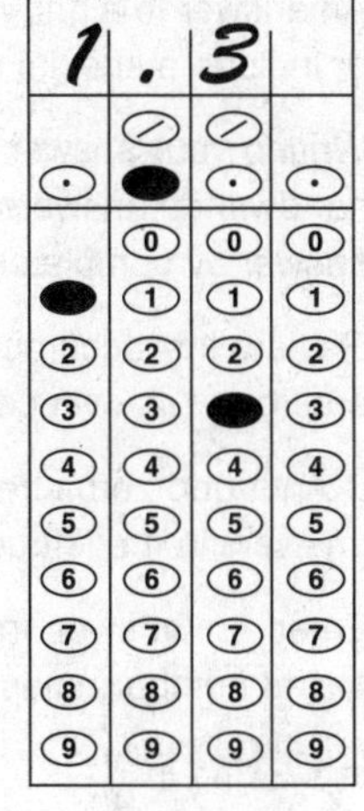

✓	✓	✗
Recommended	Acceptable	Wrong

GRID-IN QUESTIONS MAY HAVE MORE THAN ONE CORRECT ANSWER

When a grid-in question has more than one correct answer, enter only one.

To avoid possible rounding errors, truncate rather than round off long decimal answers. Make sure your final answer fills the entire grid.

EXAMPLE

What is one possible value of x for which $\frac{3}{4} < x < \frac{4}{5}$?

Solution

Since

$$\frac{3}{4} = 0.75 \text{ and } \frac{4}{5} = 0.8$$

any value of x that is between 0.75 and 0.80 satisfies the given inequality. Thus, **0.76** is a correct answer.

TIPS FOR SCORING HIGH

1. An answer to a grid-in question can never contain a negative sign, be greater than 9999, or include a special symbol, as these cannot be entered in the grid.
2. Writing your answer at the top of the grid may help to reduce gridding errors. Because handwritten answers at the top of the grid are *not* scored, make sure you then enter your answer by completely filling in the ovals in the four-column grid.
3. Review the accuracy of your gridding. Make certain that a decimal point or slash is entered in its own column.
4. Do not grid zeros before the decimal point. Grid in .55 rather than 0.55. Begin nonzero answers in the leftmost column of the grid.
5. Enter an answer in its original form, fraction or decimal, provided it fits the grid. Mixed numbers must be changed to a fraction or decimal. For example, if your answer is $1\frac{1}{8}$, grid in $\frac{9}{8}$.
6. If a fraction does not fit the grid, enter it as a decimal. For example, if your answer is $\frac{23}{10}$, grid in 2.3. Truncate long decimal answers that do not fit the grid while making certain that your final answer completely fills the grid.

LESSON 1-4

HOW THE SAT MATH SECTIONS ARE ORGANIZED, TIMED, AND SCORED

OVERVIEW

The 54 SAT math questions that are scored are divided into three timed sections. Two of these sections contain only regular multiple-choice questions. The other math section contains a set of regular multiple-choice questions followed by a set of grid-in questions.

THE THREE TIMED MATH SECTIONS

Table 1.3 shows how the two types of math questions are distributed in the three math sections that are scored. Remember that the math and verbal SAT sections can appear in any order in your test booklet. If you complete a section before the time allowed for it runs out, you may review questions only in that section. You may not go back to an earlier section or start a new one.

TABLE 1.3

Breakdown of the Three Math Sections (RMC = Regular Multiple-Choice)

Question Types	Math Section	Time Allowed
RMC	20 regular multiple-choice	25 minutes
RMC and Grid-in	8 regular multiple-choice plus 10 grid-ins	25 minutes
RMC	16 regular multiple-choice	20 minutes
Total	54 math questions	70 minutes

THE DIFFICULTY LEVELS OF THE QUESTIONS

As you work your way through each math section, questions of the same type (multiple-choice or grid-in) gradually become more difficult. Expect easier questions at the beginning of each section and harder questions at the end. You should, therefore, concentrate on getting as many of the earlier questions right as possible.

When you take the actual SAT, don't panic or become discouraged if you have to omit some questions. Very few students are able to answer all the questions in a section correctly.

TIME SAVER

Only a very small percentage of test takers answer correctly the final couple of questions in a math section. If you are not aiming for a math score over 700, you can make the best use of your time by paying attention to the last couple of questions in each section only after you've done your best to answer all of the earlier questions in that section.

- If a question near the beginning of a math section seems hard, then you are probably not approaching it in the best way. Reread the problem and try solving it again, as problems near the beginning of a math section tend to have easy, straightforward solutions.
- If a question near the end of a math section seems easy, beware—you've probably fallen into a trap!

FIGURING OUT SAT RAW SCORES

Your raw SAT math score is the number of correct answers minus a penalty deduction for any incorrect answers to questions that have answer choices. No deduction is made for unanswered questions.

TIME SAVER

A correct answer to an easy question counts the same as a correct answer to a hard question that may require much more time to figure out. In each test section, you should concentrate on getting as many of the earlier questions right as possible rather than agonizing over the solutions to the last couple of questions in that section.

- One-fourth $(\frac{1}{4})$ of a point is deducted for each incorrect answer to a standard multiple-choice question.
- No penalty deduction is made for an incorrect answer to a grid-in question.

If, after a penalty deduction has been made, the new score includes a fraction, then this score is rounded off to the nearest whole number. The number that results is converted into a score that can range from 200 to 800.

WHEN TO GUESS

Your goal is to answer the maximum number of questions correctly. Keep these points in mind:

- If you do not know the answer to a grid-in question, *always* make an educated guess since there is no penalty deduction for a wrong answer to a grid-in question. If you do not know what to guess, try multiplying, dividing, or averaging the numbers provided in the question, and grid in the result as your answer.
- Since there is a penalty deduction of one-quarter of a point for each incorrect answer to a multiple-choice question, you should *never* guess randomly in a multiple-choice section.
- For a multiple-choice question, you may be able to eliminate an answer choice because the answer is not possible, is unlikely, or its form is very different from that of the other four answer choices. If you can eliminate at least two of the answer choices, then *always* guess from among the remaining choices. If you can eliminate only one answer choice, then it still may be beneficial to guess. See **Strategy 17** on page 43.

LESSON 1-5

CALCULATORS AND THE SAT

OVERVIEW

Plan to bring a calculator that you know how to use to the exam room, although no solution will require a calculator.

WHAT TYPE OF CALCULATOR IS ALLOWED?

Along with a few sharpened pencils, an eraser, and a watch, bring to the exam room a calculator that you know how to use. The exam proctor does not have any calculators to lend, and you will not be allowed to share a calculator. Any one of the following types of calculators may be used:

- A basic four-function calculator
- A scientific calculator
- A graphing calculator

A scientific or graphing calculator is recommended for the SAT.

WHEN SHOULD YOU USE A CALCULATOR?

Although calculators don't tell you *how* to solve problems, they can be helpful when doing routine arithmetic computations. A calculator may come in handy if it is necessary to

- add, subtract, multiply, or divide whole numbers and decimals
- find or estimate a square root
- change a fraction into decimal form

Since SAT math questions never require complicated arithmetic, don't expect to use a calculator on every question. Here is a grid-in question for which using a calculator is appropriate.

EXAMPLE

If the sales tax on a $10.00 scarf is $0.80, what would a $25 shirt cost with tax?

Solution

- Find the sales tax rate by answering this question:

$$\underbrace{\text{What percent}}_{\downarrow\ \frac{p}{100}}\ \underbrace{\text{of}}_{\downarrow\ \times}\ 10\ \underbrace{\text{is}}_{\downarrow\ =}\ 0.80?$$

$$\frac{p}{100} \times 10 = 0.80$$

$$\frac{p}{10} = 0.80$$

$$p = 0.80 \times 10 = 8\%$$

- To add p percent of a given amount (the price of the shirt) to that amount, *use a calculator* to multiply the given amount by the decimal equivalent of $1 + p\%$, as in

$$\$25 \times 1.08 = \mathbf{\$27}$$

EXAMPLE

If $9x = 3^9 + 3^9 + 3^9$, then $x =$

(A) 3^6
(B) 3^8
(C) 3^{10}
(D) 3^{25}
(E) 3^{26}

Solution

You could use a calculator to help you evaluate the right side of the equation, but tedious and routine calculation is not what the SAT is about. Instead, you should recognize that the right side of the equation $9x = 3^9 + 3^9 + 3^9$ can be rewritten as

$$3 \cdot 3^9 = 3^{1+9} = 3^{10}$$

Hence,

$$9x = 3^{10}$$
$$x = \frac{3^{10}}{9}$$
$$= \frac{3^{10}}{3^2}$$
$$= 3^{10-2}$$
$$= 3^8$$

The correct choice is **(B)**.

TIPS FOR SCORING HIGH

1. Before the day of the test, check that your calculator is in good working order. If the calculator uses batteries, insert new ones. If you bring a solar-powered calculator, you will not have to worry about the batteries running down unexpectedly.
2. Approach each problem by first deciding how you will use the given information to obtain the desired answer. Then decide whether your calculator will be helpful.
3. Be aware that a solution involving many steps with complicated arithmetic is probably not the right way to tackle the problem. Look for another method that involves fewer steps and less involved computations.

4. Use a calculator to help avoid careless arithmetic errors, while keeping in mind that you can save time by performing very simple arithmetic mentally or by using mathematical reasoning rather than calculator arithmetic. Calculator icons have been placed next to the solutions to those lesson tune-up exercises that may prompt you to reach for your calculator.
5. Although a calculator can be helpful when answering some test questions, every question can be solved without a calculator.

Math Strategies You Need to Know

CHAPTER 2

This chapter will alert you to critical math strategies that can save you time or help you get the correct answers to test questions that at first glance may seem too difficult. Before you study these strategies, you may wish to read the "Quick Review of Key Math Facts" at the end of the book. Of course, if you know how to solve a problem directly, without first relying on a special strategy, then do so!

LESSONS IN THIS CHAPTER

LESSON

2-1 GENERAL MATH STRATEGIES

OVERVIEW

This lesson deals with math strategies that do not depend on the format of the test question. Knowing different SAT *math strategies can help you solve unfamiliar types of problems. Strategies do not tell you the specific steps to follow when solving a problem. They merely suggest approaches you can try when figuring out the answer to a nonroutine problem.*

STRATEGY

DRAW A DIAGRAM

Drawing a diagram can help you visualize a problem situation and organize the important facts.

EXAMPLE

Amy goes shopping and spends one-third of her money on a new dress. She then goes to another store and spends one-half of the money she has left on shoes. If Amy has $56 left after these two purchases, how much money did she have when she started shopping?

Solution

- Draw a rectangle to represent the amount of money Amy had when she started shopping. Since Amy spends $\frac{1}{3}$ of her money on a dress, divide the rectangle into 3 equal parts:

Dress		

- Two rectangles remain. Amy spends $\frac{1}{2}$ of the *remaining* money on shoes. Hence, the second rectangle represents the amount of money Amy spends on shoes. Since $56 remains, fill in the last rectangle with this amount:

Dress	Shoes	$56

- Since each of the three rectangles is equal, Amy started shopping with $3 \times \$56 = \$168$.

STRATEGY

LOOK AT A SPECIFIC CASE

If a problem does not give specific numbers or the actual dimensions of a figure, make up a simple example using easy numbers.

Grid in **168**. Never attempt to grid in special symbols such as $ and %.

EXAMPLE

The perimeter of a rectangle is 10 times as great as the width of the rectangle. The length of the rectangle is how many times as great as the width of the rectangle?

(A) one-half
(B) two
(C) three
(D) four
(E) the length and width are equal

Solution

- Consider a specific rectangle whose width is 1. If the perimeter of the rectangle is 10 times as great as the width, the perimeter of the rectangle is 10.

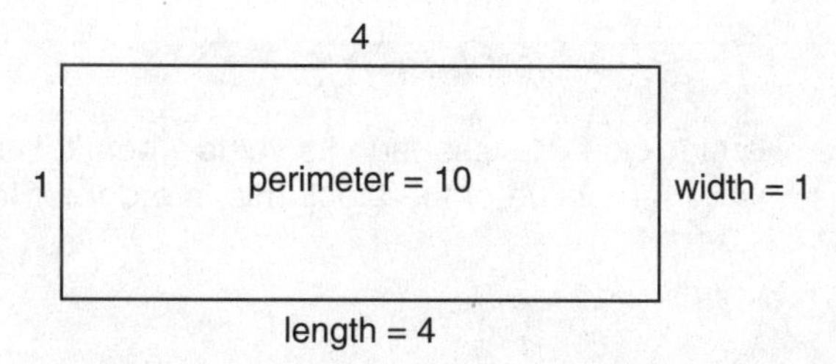

- Because $(2 \times \textit{length}) + 2 \times 1 = 10$, $2 \times \textit{length} = 8$, so the length of the rectangle is 4.
- Since length = 4 and width = 1, the length is 4 times as great as the width.

The correct choice is **(D)**.

STRATEGY 3

PLUG IN NUMBERS TO FIND A PATTERN

You can solve some problems by substituting a few test values for a variable until you discover a pattern.

EXAMPLE

When a positive integer k is divided by 5, the remainder is 3. What is the remainder when $3k$ is divided by 5?

Solution

List a few positive integers that, when divided by 5, give 3 as a remainder. Any positive integer that is the sum of 5 and 3 or of a multiple of 5 (i.e., 10, 15, etc.) and 3 will have this property. For example, when 8, 13, and 18 are each divided by 5, the remainder is 3:

k	$k \div 5$	
8	$8 \div 5 = 1$	remainder 3
13	$13 \div 5 = 2$	remainder 3
18	$18 \div 5 = 3$	remainder 3

Now, using the same values for k, divide $3k$ by 5 and find the remainders.

k	$3k$	$3k \div 5$	
8	24	$24 \div 5 = 4$	remainder 4
13	39	$39 \div 5 = 7$	remainder 4
18	54	$54 \div 5 = 10$	remainder 4

The correct answer is **4**.

CHOOSE A CONVENIENT STARTING VALUE WHEN NONE IS GIVEN

If you need to figure out how a quantity changes without knowing its beginning value, choose any starting value that makes the arithmetic easy. Carry out the computations using this value. Then compare the final answer with the starting value that you chose.

EXAMPLE

The current value of a stock is 20% less than its value when it was purchased. By what percent must the current value of the stock rise in order for the stock to have its original value?

(A) 20%
(B) 25%
(C) 30%
(D) $33\frac{1}{3}$%
(E) 50%

Solution

Choose a convenient starting value of the stock. When working with percents, 100 is usually a good starting value.

- Assume the original value of the stock was $100.
- Find the current value of the stock. Since the current value is 20% less than the original value, the current value is

$$\$100 - 0.20(\$100) = \$100 - \$20 = \$80$$

- Find the amount by which the current value of the stock must increase in order to regain the original value. The value must rise from $80 to $100, which is a change of $20.
- To find the *percent* by which the current value of the stock must rise in order for the stock to have its original value, find what percent $20 is of $80. Since

$$\frac{\$20}{\$80} \times 100\% = \frac{1}{4} \times 100\% = 25\%$$

the current value must increase 25% in order for the stock to regain its original value.

The correct choice is **(B)**.

EXAMPLE

Fred gives $\frac{1}{3}$ of his compact discs to Andy and then gives $\frac{3}{4}$ of the remaining compact discs to Jerry. Fred now has what fraction of the original number of discs?

(A) $\frac{1}{12}$
(B) $\frac{1}{6}$
(C) $\frac{1}{3}$
(D) $\frac{5}{12}$
(E) $\frac{1}{2}$

Solution

Since Fred gives $\frac{1}{3}$ and then $\frac{3}{4}$ of his compact discs away, pick any number that is divisible by both 3 and 4 for the original number of compact discs. Since 12 is the lowest common multiple of 3 and 4, assume that Fred starts with 12 compact discs.

- After Fred gives 4 ($= \frac{1}{3} \times 12$) discs to Andy, he is left with 8 ($= 12 - 4$) compact discs.
- Since $\frac{3}{4}$ of 8 is $\frac{3}{4} \times 8$ or 6, Fred gives 6 of the remaining compact discs to Jerry. This leaves Fred with 2 ($= 8 - 6$) of the original 12 compact discs.
- Since $\frac{2}{12} = \frac{1}{6}$, Fred now has $\frac{1}{6}$ of the original number of discs.

The correct choice is **(B)**.

STRATEGY

MAKE ORGANIZED LISTS

Some problems can be solved by making a list and then discovering a pattern.

EXAMPLE

People enter a room one at a time and are given a name tag in one of five possible colors. The colors are given out in this order: red, blue, white, green, and yellow. What is the color of the name tag that is given to the 93rd person who enters the room?

(A) red
(B) blue
(C) white
(D) green
(E) yellow

Solution

Determine the colors of the name tags that the first ten people who enter the room receive:

person 1: red	person 6: red
person 2: blue	person 7: blue
person 3: white	person 8: white
person 4: green	person 9: green
person 5: yellow	person 10: yellow

The pattern of colors repeats every five people. This means that if the number of the person is divisible by 5, that person will receive a yellow tag. For example, since 10 is divisible by 5, the tenth person will receive a yellow tag. If the number of the person is not divisible by 5, then the remainder tells where the person fits in the cycle of five colors. For example, since $7 \div 5 = 1$ remainder $\underline{2}$, the 7th person receives a blue tag since blue is the second color in the repeating pattern of colors. Since $93 \div 5 = 18$ remainder $\underline{3}$, the 93rd person will receive a white tag which is the third color in the repeating cycle of five colors.

The correct choice is **(C)**.

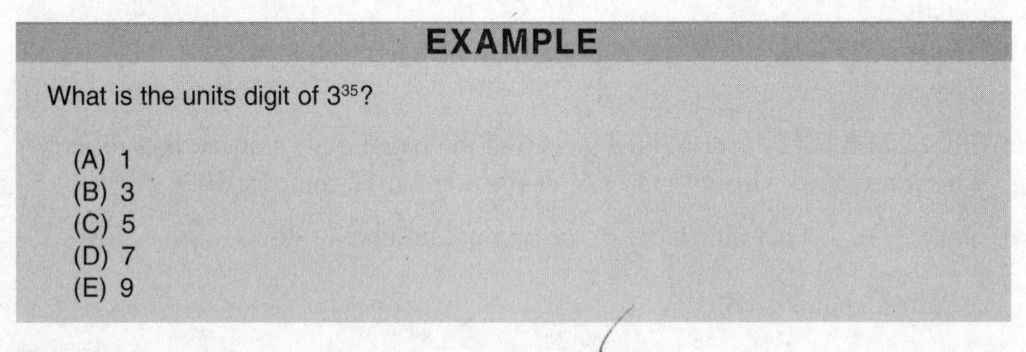

EXAMPLE

What is the units digit of 3^{35}?

(A) 1
(B) 3
(C) 5
(D) 7
(E) 9

Solution

Use your calculator to help make a list of consecutive powers of 3.

- List consecutive powers of 3 beginning with an exponent of 0:

$3^0 = \underline{1}$	$3^4 = 8\underline{1}$	$3^8 = 656\underline{1}$
$3^1 = \underline{3}$	$3^5 = 24\underline{3}$	...
$3^2 = \underline{9}$	$3^6 = 72\underline{9}$	...
$3^3 = 2\underline{7}$	$3^7 = 248\underline{7}$	...

- Do you see a pattern? Notice that as the exponent of 3 increases from 0 to 3, the units digits become 1, 3, 9, and then 7. Similarly, when the exponent of 3 increases from 4 to 7, the sequence of units digits is once again 1, 3, 9, and 7. This repeating pattern of the same four units digits means that the units digit of *any* power of 3 can be obtained by dividing the exponent by 4 and making the remainder the new exponent of 3.

- Since $35 \div 4 = 8$ remainder 3, 3^{35} must have the same units digit as 3^3. The units digit of 3^3 $(= 2\underline{7})$ is 7 so the units digit of 3^{35} is also 7.

Hence, the correct choice is **(D)**.

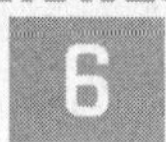

REDRAW FIGURES TO SCALE

If a figure that accompanies a question is labeled "Figure is not drawn to scale," then redrawing the figure to scale may reveal a fact that the test makers are trying to hide.

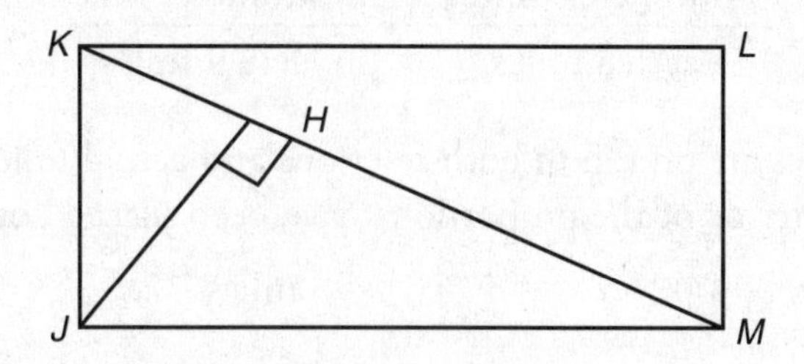

Note: Figure is not drawn to scale.

EXAMPLE

In rectangle *JKLM* above, if $JK = KL$, what is the ratio of *JH* to *KM*?

Solution

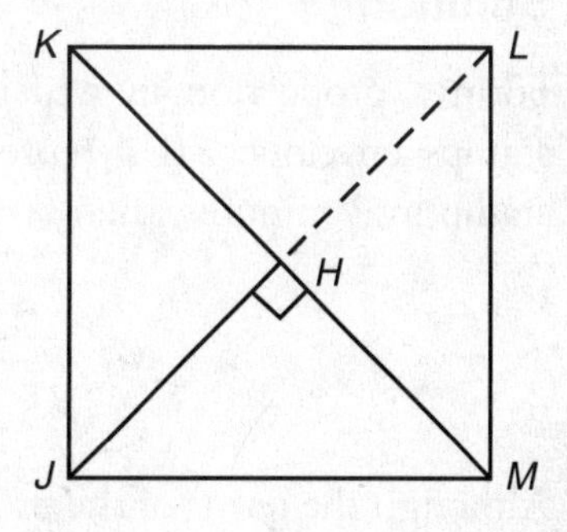

Since the figure is *not* drawn to scale, redraw it so that *JK* looks equal to *KL*. Now it should be clear that since $JK = KL$, the figure is a square. More important, *JH* looks one-half the length of *KM* so the ratio of *JH* to *KM* appears to be 1 : 2.

You could also have arrived at the same conclusion using mathematical reasoning:

- The diagonals of a square bisect each other so $KH = \frac{1}{2}KM$ and $JH = \frac{1}{2}JL$.
- Suppose $KM = 8$. Since the diagonals of a square have the same length, $KM = JL = 8$, which makes $JH = KH = \frac{1}{2} \times 8 = 4$.
- Thus, the ratio of JH $(=KH)$ to KM is 4 to 8, which can be written as the fraction $\frac{1}{2}$ or as 0.5 in decimal form. Grid in as **1/2** or **.5**.

INSERT UNITS OF MEASUREMENT

When writing a proportion to solve a problem involving rates, include the units of measurement to ensure that the terms of the proportion are placed in their correct positions and that the units of measurement are consistent.

EXAMPLE

The trip odometer of an automobile improperly displays only 3 miles for every 4 miles actually driven. If the trip odometer shows 42 miles, how many miles has the automobile actually been driven?

Solution

Form a proportion in which each side represents the rate at which odometer miles translate into actual miles driven. If x represents the number of actual miles driven when the odometer shows 42 miles, then

$$\frac{3 \text{ odometer miles}}{4 \text{ actual miles}} = \frac{42 \text{ odometer miles}}{x \text{ actual miles}}.$$

Since odometer miles are on top in both fractions and actual miles are on bottom in both fractions, the terms of the proportions have been placed correctly. To solve for x, cross-multiply: $3x = 4 \cdot 42$ so $x = \frac{168}{3} = 56$ miles.

EXAMPLE

A machine can stamp 72 envelopes in 45 seconds. How many envelopes can the same machine stamp in 2 minutes?

Solution

Form a proportion in which each side represents the rate at which the machine stamps envelopes. If x represents the number of envelopes that the machine can stamp in 2 minutes, then

$$\frac{72 \text{ envelopes}}{45 \text{ sec}} \stackrel{?}{=} \frac{x \text{ envelopes}}{2 \text{ min}}.$$

Although the terms of the proportion are in their correct positions, it's easy to see that the units of time measurement are *not* consistent. Since 1 minute is equivalent to 60 seconds, 2 minutes = $2 \times 60 = 120$ seconds:

$$\frac{72 \text{ envelopes}}{45 \text{ sec}} = \frac{x \text{ envelopes}}{120 \text{ sec}}.$$

Solve for x by cross-multiplying: $45x = 120 \cdot 72$, $x = \frac{8640}{45} = 192$ envelopes.

STRATEGY

WORK BACKWARDS

When you only know the end result of a computation and want to find the beginning value, reverse the steps that led to that final result.

EXAMPLE

Sara's telephone service cost \$21 per month plus \$0.25 for each local call, and long-distance calls are extra. Last month, Sara's bill was \$36.64, and it included \$6.14 in long-distance charges. How many local calls did she make?

Solution

Work back from the fact that Sara's final bill was \$36.64.

- Since the final bill included \$6.14 in long-distance charges, the part of the bill that did not include any long-distance charges is \$36.64 − \$6.14 = \$30.50.
- The final bill included a fixed monthly charge of \$21. Thus, the remaining part of the bill that includes only charges for local calls is \$30.50 − \$21 = \$9.50.
- Since each local call cost \$0.25, the number of local calls is $\frac{\$9.50}{\$0.25} = \mathbf{38}$.

STRATEGY

USE REFERENCE INFORMATION AS NEEDED

As you work through a test section, you may find a problem that you know how to solve but realize you've forgotten a particular formula needed for the solution. If this happens, look back at the beginning of the test section and see if the formula is included in the reference information box, which looks very similar to the one below:

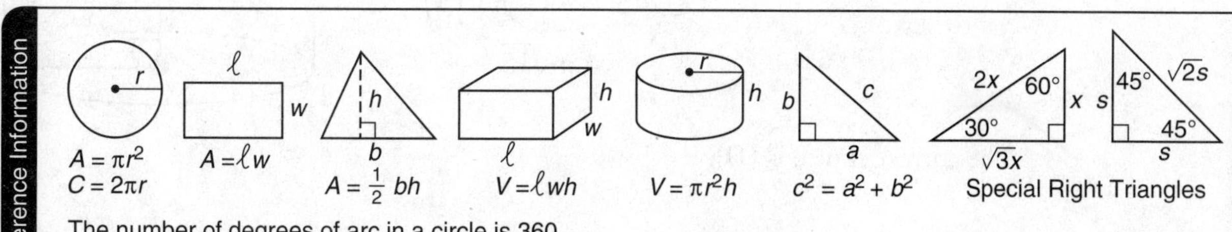

EXAMPLE

The diameter of the base of a right circular cylinder is 2, and the distance from the center of one base to a point on the circumference of the other base is 4. What is the volume of the cylinder in terms of π?

(A) 3π

(B) 5π

(C) $\pi\sqrt{12}$

(D) $\pi\sqrt{15}$

(E) $\pi\sqrt{17}$

Solution

To solve the problem, you need to know the formula for the volume, V, of a right cylinder. The reference section gives this formula as $V = \pi r^2 h$.

You need to know r and h before you can use the volume formula.

- Since the diameter of the base of the cylinder is 2, the radius, r, of the cylinder is $\frac{1}{2}\times 2 = 1$.

- The distance from the center of one base to a point on the circumference of the other base is the hypotenuse of a right triangle. One leg of the right triangle is the height, h, of the cylinder. The other leg is the radius drawn from the given point on the circumference of a base to the center of that base, as shown in the accompanying figure. Use the Pythagorean theorem (see the reference section, if needed) to find h:

$$h^2 + 1^2 = 4^2$$
$$h^2 + 1 = 16$$
$$h^2 = 15$$
$$h = \sqrt{15}$$

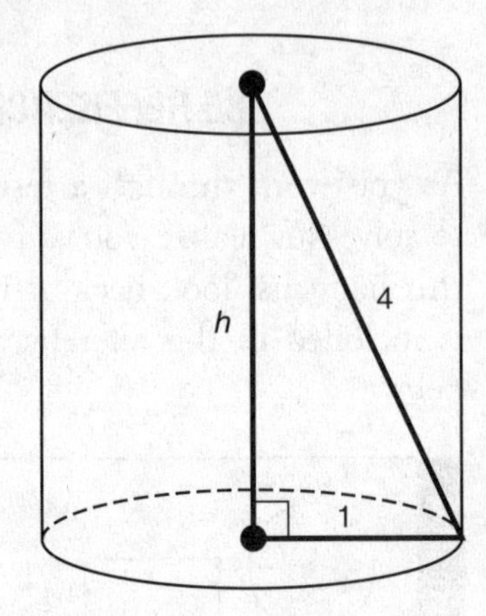

- Use the volume formula:

$$V = \pi r^2 h$$
$$= \pi(1^2)(\sqrt{15})$$
$$= \pi\sqrt{15}$$

The correct choice is **(D)**.

STRATEGY

10

ACCOUNT FOR ALL POSSIBLE CASES

Solving a problem may depend on breaking it down so that all possible cases are considered.

EXAMPLE

In a certain homeroom class, 21 students are enrolled in Math, 17 students are enrolled in Biology, 9 are enrolled in both Math and Biology, and 3 students are *not* enrolled in either course. How many students are in the homeroom class?

(A) 22
(B) 25
(C) 28
(D) 32
(E) 50

Solution

Account for all possible nonoverlapping sets of students that comprise the homeroom class. Students can be enrolled in both courses, in exactly one of the two courses, or in neither of the two courses:

• Students in both courses	$= 9$
• Students in Math but not Biology	$= 21 - 9 = 12$
• Students in Biology but not Math	$= 17 - 9 = 8$
• Students *not* in either course	$= 3$
Total students in homeroom class	$= 32$

By accounting for all possible cases, you know that the total number of students in the homeroom class is $9 + 12 + 8 + 3 = 32$.

The correct choice is **(D)**.

LESSON

2-2 SPECIAL ALGEBRA-BASED STRATEGIES

OVERVIEW

This section looks at a few special math strategies that involve algebra.

STRATEGY 11

WRITE AN ALGEBRAIC EQUATION

Sometimes the most efficient way of solving a problem is by using an algebraic equation. Problems that compare quantities lend themselves to algebraic solutions. In this type of problem, first identify the "base" quantity to which one or more other quantities are being compared. After assigning a variable to the base quantity, represent the other quantities using the same variable. Then translate the condition of the problem that tells how these quantities are related into an equation.

EXAMPLE

Ticket sales receipts for a music concert totaled $2,160. Three times as many tickets were sold for the Saturday night concert as was sold for the Sunday afternoon concert. Two times as many tickets were sold for the Friday night concert as was sold for the Sunday afternoon concert. Tickets for all three concerts sold for $2.00 each. Find the number of tickets sold for the Saturday night concert.

(A) 90
(B) 180
(C) 270
(D) 540
(E) 600

Solution

- Pick out the quantity to which the others are being compared. The numbers of tickets sold for Friday and Saturday night concerts are being compared to the number of tickets sold for the Sunday afternoon concert.
- Assign variables. Since the base quantity is the number of tickets sold for the Sunday afternoon concert:

  Let x = the number of tickets sold for the Sunday afternoon concert.
  Then $3x$ = the number of tickets sold for the Saturday night concert.
  and $2x$ = the number of tickets sold for the Friday night concert.

- Identify the condition that relates the ticket sales for the three concerts. Since each of the tickets costs \$2 and the ticket sales receipts totaled \$2,160:

$$\overbrace{x \cdot \$2 + 3x \cdot \$2 + 2x \cdot \$2}^{\textit{Total ticket sales receipts}} = \$2{,}160$$

$$\$2x + \$6x + \$4x = \$2{,}160$$

$$\$12x = \$2{,}160$$

$$x = \frac{\$2{,}160}{\$12} = 180$$

- Answer the question that is asked. The number of tickets sold for the *Saturday* night concert $= 3x = 3(180) = 540$.

The correct choice is **(D)**.

STRATEGY

12

CREATIVELY MANIPULATE SYSTEMS OF EQUATIONS

SAT problems involving systems of equations may seem complicated; however, they can often be solved simply by combining, multiplying, or dividing the equations.

EXAMPLE

If $2d - e = 32$ and $2e - d = 10$, what is the average of d and e?

Solution

The average of d and e is $\frac{d+e}{2}$. Rather than solving the system of equations for each of the individual variables, think of an easy way of combining the two equations to get a simpler expression. Try adding corresponding sides of the two equations:

$$\begin{array}{r} 2d - e = 32 \\ -d + 2e = 10 \\ \hline d + e = 42 \end{array}$$

This is very close to what you need to find. Now divide each side of $d + e = 42$ by 2:

$$\frac{d+e}{2} = \frac{42}{2} = 21.$$

The average of d and e is **21**.

EXAMPLE

If $\frac{u}{18} = w$ and $3v = 2w$, what is the value of $\frac{u}{v}$?

Solution

The system of equations contains variables u, v, and w. Because you are asked to find the value of $\frac{u}{v}$, look for a way to manipulate the two equations so that w gets eliminated.

- Simplify the first equation by rewriting it as $u = 18w$. The second equation that is given is $3v = 2w$. Look for a clue by writing the second equation underneath the first equation:

$$u = 18w$$
$$3v = 2w$$

- Because $\frac{u}{v}$ means *u divided by v*, try *dividing* corresponding sides of the two equations:

$$\frac{u}{3v} = \frac{18w}{2w}$$

$$\frac{u}{3v} = \frac{18\not{w}}{2\not{w}} = 9$$

- The division process eliminates w, so you can quickly solve for u by multiplying both sides of the new equation by 3:

$$3\left(\frac{u}{3v}\right) = 3(9)$$

$$\not{3}\left(\frac{u}{\not{3}v}\right) = 27$$

$$\frac{u}{v} = 27$$

STRATEGY 13

REWRITE EQUATIONS TO HELP EVALUATE ALGEBRAIC EXPRESSIONS

Given an equation with more than one unknown and an algebraic expression involving the same variables, rewrite the equation in a way that allows you to find the value of the algebraic expression through a substitution using that equation.

EXAMPLE

If $x - 2y = 1$, what is the value of 2^{3x-6y}?

Solution

Do not try to solve the given equation. Instead, think of a simple relationship that connects the equation and the exponent of 2^{3x-6y}. Multiplying both sides of the equation $x - 2y = 1$ by 3 gives $3x - 6y = 3$. Now that you know the exponent is 3, make the substitution and do the math:

$$2^{3x-6y} = 2^3 = 2 \times 2 \times 2 = 8$$

The answer is **8**.

EXAMPLE

If $\frac{m}{p} = \frac{7}{9}$, what is the value of $\frac{2p}{3m}$?

Solution

Solve the proportion for $\frac{p}{m}$ simply by inverting both sides. Since you now know that $\frac{p}{m}$ is $\frac{9}{7}$, make the substitution:

$$\frac{2p}{3m} = \frac{2}{3}\left(\frac{p}{m}\right)$$

$$= \frac{2}{\cancel{3}}\left(\frac{\overset{3}{\cancel{9}}}{7}\right)$$

$$= \frac{6}{7}$$

EXAMPLE

If $(x - y)^2 = 50$ and $xy = 10$, what is the value of $x^2 + y^2$?

Solution

Find a connection between $(x - y)^2 = 50$ and $x^2 + y^2$. Expanding the square of a binomial produces an expression that includes the squares of its two terms:

$$(x-y)^2 = (x-y)(x-y) = x^2 - 2xy + y^2$$

This means that $x^2 - 2xy + y^2 = 50$. You can now find the value of $x^2 + y^2$ by substituting 10 for xy:

$$x^2 - 2(10) + y^2 = 50$$

$$x^2 + y^2 = 50 + 20$$

$$x^2 + y^2 = 70$$

The correct answer is **70**.

EXAMPLE

If $\frac{r}{s} = 2$ and $\frac{s}{t} = 3$, then $\frac{r+s}{s+t} =$

(A) $\frac{2}{3}$

(B) $\frac{7}{9}$

(C) $\frac{5}{6}$

(D) $\frac{9}{4}$

(E) 4

Solution

Since $\frac{r}{s} = 2$ and $\frac{s}{t} = 3$, $r = 2s$ and $s = 3t$. Make these substitutions in the fraction:

$$\frac{r+s}{s+t} = \frac{2s+s}{3t+t} = \frac{3s}{4t} = \frac{3}{4}\left(\frac{s}{t}\right)$$

Going back to the information given, substitute 3 for $\frac{s}{t}$:

$$= \frac{3}{4}(3)$$

$$= \frac{9}{4}$$

The correct choice is **(D)**.

ACCOUNT FOR ALL SOLUTIONS OF A HIGHER-DEGREE EQUATION

Don't forget that a second-degree equation has two solutions, a third-degree equation has three solutions, and so forth.

EXAMPLE

If $x^2 = 4x$, then $x =$

(A) 0 only
(B) 2 only
(C) 4 only
(D) 0 or 2
(E) 0 or 4

Solution

Since the greatest exponent of x in the given equation, $x^2 = 4x$, is 2, the equation has *two* solutions:

- If $x^2 = 4x$, then $x^2 - 4x = 0$.
- Factor out x: $x(x - 4) = 0$. If the product of two quantities is equal to 0, then either quantity may be equal to 0. Hence, $x = 0$ or $x - 4 = 0$, so $x = 4$.
- The two roots of the equation are $x = 0$ or $x = 4$.

The correct choice is **(E)**.

LESSON

2-3 SPECIAL MATH STRATEGIES FOR REGULAR MULTIPLE-CHOICE QUESTIONS

OVERVIEW

The correct answer to a regular multiple-choice question is given to you, along with four other choices that are designed to distract you. Sometimes you can use a "backdoor" strategy that will allow you to identify the correct answer without going through a standard mathematical solution.

STRATEGY 15

TEST NUMERICAL ANSWER CHOICES IN THE QUESTION

When each of the answer choices for a regular multiple-choice question is an easy number, you may be able to find the correct answer by "backsolving." To backsolve, plug each of the possible choices into the question until you find the number that works.

EXAMPLE

When 5 is divided by a number, the result is 3 more than 7 divided by twice the number. What is the number?

(A) $\frac{1}{4}$
(B) $\frac{1}{2}$
(C) 1
(D) $\frac{3}{2}$
(E) 2

Solution

Rather than thinking of an equation and then solving it algebraically, plug in each of the answer choices as the possible correct number until you find the one that works in the problem.

Choice (A): Try $\frac{1}{4}$ as the unknown number:

$$\overbrace{5 \div \frac{1}{4}}^{20} \neq \overbrace{3 + 7 \div \frac{2}{4}}^{3+14} \quad \text{✗}$$

Choice (B): Try $\frac{1}{2}$ as the unknown number:

$$\overbrace{5 \div \frac{1}{2}}^{10} = \overbrace{3 + 7 \div \frac{2}{2}}^{3+7} \quad ✓$$

The correct choice is **(B)**.

EXAMPLE

If $2^{2x-1} = 32$, then $x =$

(A) 2
(B) 3
(C) 4
(D) 5
(E) 6

Solution

Since the answer choices give the possible values of x, replace x with each of these values until you find the one that makes 2^{2x-1} equal to 32. As the value of x increases, so does the value of 2^{2x-1}. Therefore, you may be able to save time by starting with choice (C), the middle x-value. If this value of x makes 2^{2x-1} too large, then the correct answer must be *smaller* than the middle x-value. In this case, you can eliminate choices (C), (D), and (E). Similarly, you can eliminate choices (A), (B), and (C) if the middle x-value makes 2^{2x-1} too *small.*

- Plug in choice (C) for x. If $x = 4$, then

$$\begin{aligned} 2^{2x-1} &= 2^{2(4)-1} \\ &= 2^7 \\ &= 2 \times 2 \times 2 \times 2 \times 2 \times 2 \times 2 \\ &= 128 \end{aligned}$$

Since the correct value of x must be smaller than 4, try choice (B). If choice (B) doesn't work, the correct answer must be choice (A).

- Plug in choice (B) for x. If $x = 3$, then

$$\begin{aligned} 2^{2x-1} &= 2^{2(3)-1} \\ &= 2^5 \\ &= 2 \times 2 \times 2 \times 2 \times 2 \\ &= 32 \end{aligned}$$

Hence, choice **(B)** is correct.

STRATEGY

16

CHANGE VARIABLE ANSWER CHOICES INTO NUMBERS

If you don't know how to find the answer to a regular multiple-choice question in which the answer choices contain letters, substitute simple numbers for the letters.

EXAMPLE

If $m + n$ is an odd number when m and n are positive integers, which expression always represents an even number?

(A) $m - n$
(B) $m^2 + n^2$
(C) $(m + n)^2$
(D) $m^2 - n^2$
(E) $(mn)^2$

Solution

Pick simple numbers for m and n such that $m + n$ is an odd number. Let $m = 2$ and $n = 1$; then $m + n = 3$, and 3 is an odd number. Try the answer choices in turn:

- (A) $m - n = 2 - 1 = 1$
- (B) $m^2 + n^2 = 2^2 + 1^2 = 4 + 1 = 5$
- (C) $(m + n)^2 = (2 + 1)^2 = 3^2 = 9$
- (D) $m^2 - n^2 = 2^2 - 1^2 = 4 - 1 = 3$
- (E) $(mn)^2 = (2 \times 1)^2 = 2^2 = 4$

Since the only even number among the transformed answer choices is 4, the correct choice is **(E)**.

EXAMPLE

If t ties cost d dollars, how many dollars would $t + 1$ ties cost?

(A) $d + 1$

(B) $\frac{dt}{t+1}$

(C) $\frac{d+1}{t+1}$

(D) $\frac{d(t+1)}{t}$

(E) $\frac{t(d+1)}{d}$

Solution

Pick easy numbers for t and d, such as $t = 2$ and $d = 10$. If two ties cost \$10, then one tie costs \$5 and three ties ($t + 1$) cost \$15. Substitute 2 for t and 10 for d in each of the answer choices until you find the one that evaluates to 15.

The correct choice is **(D)**, since

$$\frac{d(t+1)}{t}=\frac{10(2+1)}{2}=\frac{30}{2}=15.$$

Avoid picking 0 and 1, as these numbers tend to produce more than one "correct" answer choice. If more than one answer choice gives the same correct answer, then start over with different numbers.

STRATEGY 17

GUESS AFTER RULING OUT ANSWER CHOICES

If you get stuck on a multiple-choice problem but can eliminate some of the answer choices, guess rather than omit the question. If you skip multiple-choice questions, you have no chance of gaining points for those questions. Each time you guess correctly, however, you add 1 point to your raw score; if you guess incorrectly, you lose only $\frac{1}{4}$ of a point for the wrong answer. You can improve your chances of guessing the correct answer by eliminating answer choices that you know are unlikely or impossible.

- Rule out an answer choice if it stands out as looking very different in form from the other four choices. For example, suppose the five answer choices of a multiple-choice problem are these:

  (A) $\frac{1}{2}\pi r$

  (B) $\pi(r-\sqrt{2})$

  (C) $\pi(r+\sqrt{2})$

  (D) $\pi(r^2+2)$

  (E) $\pi(r^2-2)$

  Choice (A) is the only choice that does not include the product of π and a parenthesized expression. Because this choice looks very different from the other four choices, it is probably not the correct answer, so you can eliminate it.

- Rule out this answer choice:

  (E) It cannot be determined from the information given

  when it belongs to a multiple-choice question that appears late in a test section. This answer choice may be correct when the question appears early in a test section, but it is rarely the correct answer when the question appears near the end of a test section.

- Rule out an answer choice that does not make sense based on the facts of the problem. For instance, consider the example that appears under Strategy 10 on page 32. Since there are 21 students enrolled in Math and 3 students are not enrolled in either Math or Biology, the homeroom class must have at least $21 + 3 = 24$ students. This means that choice (A) can be eliminated. Since it is not reasonable that the answer would simply be the sum $21 + 17 + 9 + 3 = 50$, choice (E) can also be eliminated. This leaves only three possible answer choices. You might also suspect that because the numbers used in the problem are somewhat scattered (21, 17, 9, and 3), it is unlikely that the correct answer is only 1 more than the minimum of 24. This eliminates choice (B) and leaves only two possible answer choices from which you can guess.
- Rule out an answer choice that contradicts arithmetic reasoning, as illustrated in the next example.

EXAMPLE

In June, the price of a CD player that sells for $150 is increased by 10%. In July, the price of the same CD player is decreased by 10% of its current selling price. What is the new selling price of the CD player?

(A) $140
(B) $148.50
(C) $150
(D) $152.50
(E) $160

Analysis

Is the new selling price equal to the original price of $150, less than the original price, or greater than the original price? Rule out choice (C) since "obvious" answers that do not require any work are rarely correct. In June the price of a CD player is increased by 10% of $150. In July the price of the CD player is decreased by 10% of an amount *greater than* $150 (the June selling price). Since the amount of the price decrease was greater than the amount of the price increase, the July price must be *less than* the starting price of $150. You can, therefore, eliminate choices (C), (D), and (E).

This analysis improves your chances of guessing the correct answer. Of course, if you know how to solve the problem without guessing, do so:

STEP 1 June price $= \$150 + (10\% \times \$150) = \$150 + \$15 = \$165$
STEP 2 July price $= \$165 - (10\% \times \$165) = \$165.00 - \$16.50 = \$148.50$
STEP 3 The correct choice is **(B)**.

TIPS FOR SCORING HIGH

1. Before starting your solution, look at the five answer choices. The types of numbers or the forms of the answer choices may help you decide on a solution strategy. For example, if there are whole numbers in the question, and decimal or fractional answers appear as choices, the solution may involve division.

2. If you think you have solved a problem correctly, but do not find your answer among the five choices, try writing your answer in an equivalent form. For example, if you don't find $\frac{3}{2}$, look for 1.5; if you don't find $2a + 4$, look for $2(a + 2)$; if you don't find $\frac{3x-y}{x}$, look for $3 - \frac{y}{x}$ since

$$\frac{3x-y}{x} = \frac{3x}{x} - \frac{y}{x} = 3 - \frac{y}{x}$$

3. Avoid random guessing since one-fourth of a point is deducted for each incorrect answer to a multiple-choice question. If you cannot rule out at least one answer choice as definitely wrong, guessing will probably not help your score. If you can eliminate at least two answer choices as definitely wrong, however, always guess from among the remaining choices rather than omit the question.

4. Guess smartly. Cross out in your test booklet those choices that you know are impossible or unlikely. When deciding which answer choices can be ruled out, ask yourself questions such as these: Should the answer be positive or negative? Greater or less than 1? A whole number? Is there a number that the answer must be greater than or less than? Can an accompanying figure be used to estimate the answer?

5. Before guessing, think through a problem and attempt to solve it mathematically. Guessing should be your last resort.

PART 2

REVIEWING KEY MATH SKILLS AND CONCEPTS

Answer Sheet

Answer Sheet For Tune-Up Exercises

LESSON# ________ :

Before you work through the SAT Tune-Up exercises found at the end of each review lesson, you may wish to reproduce the answer form below. This may contain more ovals and grids than you will need for a particular exercise section. Mark only those you need and leave the rest blank.

Multiple-Choice

1 Ⓐ Ⓑ Ⓒ Ⓓ Ⓔ
2 Ⓐ Ⓑ Ⓒ Ⓓ Ⓔ
3 Ⓐ Ⓑ Ⓒ Ⓓ Ⓔ
4 Ⓐ Ⓑ Ⓒ Ⓓ Ⓔ
5 Ⓐ Ⓑ Ⓒ Ⓓ Ⓔ
6 Ⓐ Ⓑ Ⓒ Ⓓ Ⓔ
7 Ⓐ Ⓑ Ⓒ Ⓓ Ⓔ
8 Ⓐ Ⓑ Ⓒ Ⓓ Ⓔ
9 Ⓐ Ⓑ Ⓒ Ⓓ Ⓔ
10 Ⓐ Ⓑ Ⓒ Ⓓ Ⓔ
11 Ⓐ Ⓑ Ⓒ Ⓓ Ⓔ
12 Ⓐ Ⓑ Ⓒ Ⓓ Ⓔ
13 Ⓐ Ⓑ Ⓒ Ⓓ Ⓔ
14 Ⓐ Ⓑ Ⓒ Ⓓ Ⓔ
15 Ⓐ Ⓑ Ⓒ Ⓓ Ⓔ
16 Ⓐ Ⓑ Ⓒ Ⓓ Ⓔ
17 Ⓐ Ⓑ Ⓒ Ⓓ Ⓔ
18 Ⓐ Ⓑ Ⓒ Ⓓ Ⓔ
19 Ⓐ Ⓑ Ⓒ Ⓓ Ⓔ
20 Ⓐ Ⓑ Ⓒ Ⓓ Ⓔ

Grid-In

1. 2. 3. 4. 5. 6. 7. 8. 9. 10.

CHAPTER 3

Arithmetic Skills and Concepts

This chapter reviews the arithmetic operations and arithmetic reasoning skills that are tested by many of the easy to medium-difficulty questions that appear on the SAT.

LESSONS IN THIS CHAPTER

NUMBERS, SYMBOLS, AND VARIABLES

OVERVIEW

All numbers that appear on the SAT *are real numbers.* ***Real numbers*** *include all the different types of numbers, both positive and negative, you encountered in arithmetic and beginning algebra classes.*

A ***variable*** *is a symbol that serves as a placeholder for an unknown member of a given set of numbers. Letters of the alphabet such as a, b, and x are used as variables to help make general statements about how numbers behave.*

TYPES OF NUMBERS

The set of real numbers is comprised of these sets of numbers:

- *Whole numbers* = 0, 1, 2, 3, . . .
- *Integers* = $\underbrace{\ldots -4,-3,-2,-1}_{\text{Negative integers}},0,\overbrace{1,2,3,4,\ldots}^{\text{Positive integers}}$
- *Rational numbers* =
  - Fractions such as $\frac{2}{3}$ that have integers above and below the fraction bar
  - Terminating decimals such as 0.75
  - Nonending, repeating decimals such as 0.33333. . .
- *Irrational numbers* =
  - Numbers such as π and $\sqrt{2}$ that do not have exact decimal equivalents

COMPARISON SYMBOLS

Table 3.1 summarizes the symbols used to compare two numbers.

TABLE 3.1

Comparison Symbols

Symbol	Translation	Example
$=$	is equal to	$5 = 5$
$\neq$	is *not* equal to	$5 \neq 3$
$>$	is greater than	$5 > 3$
$\geq$	is greater than *or* equal to	$x \geq 5$ means that x can be 5 *or* any number greater than 5.
$<$	is less than	$3 < 5$
$\leq$	is less than *or* equal to	$x \leq 3$ means that x can be 3 *or* any number less than 3.

Since 7 is between 6 and 8, you can write $6 < 7 < 8$. The inequality $6 < 7 < 8$ means that 6 is less than 7 and, at the same time, 7 is less than 8. In general, if x is between a and b with $a < b$, then $a < x < b$. The inequality $a \leq x \leq b$ means that x is between a and b, or may be equal to either a or b, as shown in Figure 3.1.

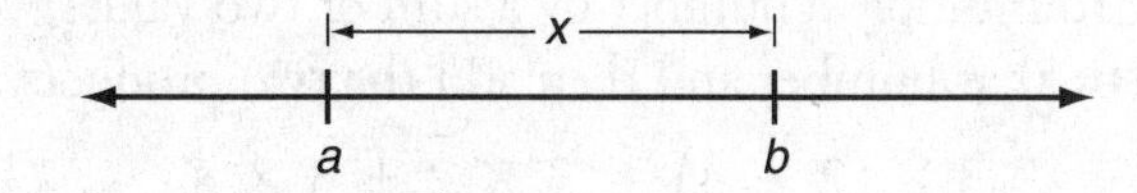

Figure 3.1 $a \leq x \leq b$

INDICATING MULTIPLICATION

When two or more numbers are multiplied together, the answer is called the **product.** Each of the numbers that are being multiplied together is called a **factor** of the product. Since $5 \times 6 = 30$, 30 is the product while 5 and 6 are factors of 30.

Multiplication can be represented in ways that avoid using the symbol $\times$:

- Centering a dot between two quantities indicates multiplication. For example, $4 \cdot y$ means 4 times y.
- Placing parentheses around quantities written next to each other means multiplication. For example, (2)(3) is 2 times 3, $4(y)$ is 4 times y, and $5(x + y)$ is 5 times the sum of x and y.
- Writing a number and one or more variables next to each other means multiplication. For example, ab means a times b, and $5n$ means 5 times n. If $n = 4$, then $5n = 5 \times 4 = 20$. The product $2ab$ means 2 times a times b. If $a = 5$ and $b = 8$, then $2ab = 2 \times 5 \times 8 = 10 \times 8 = 80$.

LIKE TERMS

If a number and a variable are written next to each other, the number is called the **coefficient** of the variable. For $5n$, 5 is the coefficient of n. Since $5n$ means 5 times n, $5n = n + n + n + n + n$. When products such as $3n$ and $2n$ differ only in their numerical coefficients, they are called **like terms.** Like terms may include more than one variable. For example, $4ab$ and $7ab$ are like terms, but $2ab$ and $3ac$ are *not* like terms.

LAWS OF ARITHMETIC

Real numbers behave in predictable ways.

- *Commutative law*: The order in which *two* numbers are added or multiplied does not matter. For example:

$$3 + 4 = 4 + 3 \quad \text{and} \quad 3 \times 4 = 4 \times 3$$

- *Associative law*: The order in which *three* numbers are added or multiplied does not matter. For example:

$$2 \times (3 \times 4) = (2 \times 3) \times 4$$

where the parentheses group the pair of numbers that are multiplied first.

- *Distributive law*: To multiply a number by a sum of two values, you multiply each value in the sum by that number and then add the two products. For example:

$$\begin{aligned} 3 \times (2 + 4) &= 3 \times 2 + 3 \times 4 \\ &= 6 + 12 \\ &= 18 \end{aligned}$$

APPLYING THE LAWS OF ARITHMETIC TO ALGEBRAIC EXPRESSIONS

The commutative, associative, and distributive laws can be applied to expressions that contain variables since these expressions represent real numbers. For example:

- By the commutative law, $x + 3$ and $3 + x$ are equivalent expressions.
- By the distributive law,

$$3(2x + 5) = 3(2x) + 3(5) = 6x + 15$$

- By the reverse of the distributive law,

$$4x + 5x = (4 + 5)x = 9x$$

Thus, like terms are combined by combining their numerical coefficients. Here are three more examples:

EXAMPLE: $3y + 4y = 7y$

EXAMPLE: $9p + 3p + 2p = 14p$

EXAMPLE: $5x - x = 4x$ since $5x - x = 5x - 1x = 4x$

Lesson 3-1 Tune-Up Exercises

Multiple-Choice

1 If $a = 9 \times 23$ and $b = 9 \times 124$, what is the value of $b - a$?

(A) 901
(B) 903
(C) 906
(D) 909
(E) 911

2 By how much does the product of 8 and 25 exceed the product of 15 and 10?

(A) 25
(B) 50
(C) 75
(D) 100
(E) 125

3 How many containers, each holding 16 fluid ounces of milk, are needed to hold 5 quarts of milk? (1 quart = 32 fluid ounces)

(A) 6
(B) 8
(C) 10
(D) 12
(E) 14

4 If the current odometer reading of a car is 31,983 miles, what is the LEAST number of miles that the car must travel before the odometer displays four digits that are the same?

(A) 17
(B) 239
(C) 350
(D) 650
(E) 1350

5 In store *A* a scarf costs $12, and in store *B* the same scarf is on sale for $8. How many scarfs can be bought in store *B* with the amount of money, excluding sales tax, needed to buy 10 scarfs in store *A*?

(A) 4
(B) 12
(C) 15
(D) 18
(E) 21

6 Let * represent one of the four basic arithmetic operations such that, for any nonzero real number *r*,

$$r * 0 = r \text{ and } r * r = 0$$

Which arithmetic operation(s) does the symbol * represent?

(A) + only
(B) − only
(C) + and −
(D) × only
(E) +, −, or ×

7 Kurt has saved \$160 to buy a stereo system that costs \$400 including taxes. If he earns \$8 an hour after all payroll deductions have been made, how many hours will he have to work in order to have exactly enough money to buy the stereo system?

(A) 20
(B) 24
(C) 25
(D) 30
(E) 40

---

8 If the present time is exactly 1:00 P.M., what was the time exactly 39 hours ago?

(A) 4:00 P.M.
(B) 4:00 A.M.
(C) 9:00 P.M.
(D) 9:00 A.M.
(E) 10:00 P.M.

---

9 Let # represent one of the four basic arithmetic operations such that, for any nonzero real numbers r and s,

$$r \# 0 = r \quad \text{and} \quad r \# s = s \# r$$

Which arithmetic operation(s) does the symbol # represent?

(A) + only
(B) × only
(C) − only
(D) − and ×
(E) + and ÷

---

10 If $w = (6)(6)(6)$, $x = (5)(6)(7)$, and $y = (4)(6)(8)$, which inequality statement is true?

(A) $x < y < w$
(B) $w < x < y$
(C) $y < w < x$
(D) $y < x < w$
(E) $w < y < x$

---

11 If x and y are positive integers, $2x + y < 29$, and $y > 4$, what is the greatest possible value of $x - y$?

(A) 5
(B) 6
(C) 7
(D) 8
(E) 9

Grid-In

1 The houses in a certain community are numbered consecutively from 2019 to 2176. How many houses are in the community?

2 If 1 kilobyte of computer memory is equivalent to 1024 bytes and 1 byte is equivalent to 8 bits, how many kilobytes are equivalent to 40,960 bits?

3 A television set will cost, including taxes and finance charges, $495 if the buyer puts $129 down and then pays off the balance in eight equal monthly payments. Under this purchase plan, what will each of the monthly payments be?

4 If $13 \leq k \leq 21, 9 \leq p \leq 19, 2 < m < 6$, and k, p, and m are integers, what is the largest possible value of $\frac{k-p}{m}$?

5 For some fixed value of x, $9(x + 2) = y$. After the value of x is increased by 3, $9(x + 2) = w$. What is the value of $w - y$?

6 If x and y are positive integers, and $3x + 2y = 21$, what is the sum of all possible values of x?

LESSON

3-2 POWERS AND ROOTS

OVERVIEW

The product of identical factors may be indicated by writing the number of times the factor is repeated a half-line above and to the right of the factor. For example:

$$\underbrace{2 \times 2 \times 2 \times 2 \times 2}_{\text{2 is used as a factor 5 times}} = 2^5$$

Exponent: 5; Base: 2

MEANING OF EXPONENT

Repeated multiplication of the same number may be indicated in a more compact form by using an **exponent** that tells the number of times the number, called the **base**, appears as a factor in the product. The notation b^n is read as "b raised to the nth power." In b^n, the number b is the base and n is the exponent.

When a number or variable appears without an exponent, the exponent is understood to be 1. For example, $3 = 3^1$ and $y = y^1$.

EXAMPLE

What is the value of $3^4 - 4^3$?

Solution

Since $3^4 = 3 \times 3 \times 3 \times 3 = 81$ and $4^3 = 4 \times 4 \times 4 = 64$,

$$3^4 - 4^3 = 81 - 64 = 17$$

EXAMPLE

If $2^x = 8$, what is the value of x^3?

Solution

Since $2^x = 8 = 2 \times 2 \times 2 = 2^3$, then $x = 3$. Hence,

$$x^3 = 3^3 = 3 \times 3 \times 3 = 27$$

LAWS OF EXPONENTS

You should know these rules for working with exponents:

- To multiply powers with the *same* nonzero base, *add* their exponents. For example:

$$2^4 \times 2^3 = 2^{4+3} = 2^7 \quad \text{and} \quad b^2 \cdot b^3 = b^{2+3} = b^5$$

Don't try to multiply powers that have different bases. For example:

$$2^5 \times 3^4 \neq (2 \times 3)^{5+4}$$

- To divide powers with the *same* nonzero base, *subtract* their exponents. For example:

$$2^7 \div 2^3 = \frac{2^7}{2^3} = 2^{7-3} = 2^4 \text{ and } \frac{b^5}{b} = \frac{b^5}{b^1} = b^{5-1} = b^4$$

- Don't try to divide powers that have different bases. For example:

$$2^5 \div 3^4 \neq \left(\frac{2}{3}\right)^{5-4}$$

- To raise a power to another power, *multiply* the exponents. For example:

$$(2^3)^7 = 2^{3\times 7} = 2^{21} \quad \text{and} \quad (b^4)^3 = b^{4\times 3} = b^{12}$$

These rules are summarized in Table 3.2, where bases x and y are not 0 and exponents m and n are positive integers.

TABLE 3.2

Rules of Exponents

Product Rule	Quotient Rule	Power Rules
$x^m \cdot x^n = x^{m+n}$	$\frac{x^m}{x^n} = x^{m-n}$	$(x^m)^n = x^{m \cdot n}$ and $(xy)^m = x^m \cdot y^m$

Here are some additional examples:

- $(2a)^3 = 2^3 \cdot a^3 = 8a^3$
- $\left(\frac{x}{2}\right)^3 = \frac{x^3}{2^3} = \frac{x^3}{8}$
- $\frac{x^7 y^2}{x^3 y} = x^{7-3} \cdot y^{2-1} = x^4 y$

ORDER OF OPERATIONS

Arithmetic operations are not necessarily performed in the order in which they are encountered. Instead,

P: First evaluate expressions within *P*arentheses.
E: Then evaluate *E*xponents.
M: Next, *M*ultiply and
D: *D*ivide, working from left to right.
A: Finally, *A*dd and
S: *S*ubtract, again working from left to right.

The first letters of the key words in the sequence of operations spell "PEMDAS," which can be remembered by memorizing the sentence "*P*lease *E*xcuse *M*y *D*ear *A*unt *S*ally."

EXAMPLE

Find the value of $5 + 18 \div (4 - 1)^2$.

Solution

$$\begin{aligned} 5 + 18 \div (4 - 1)^2 &= 5 + 18 \div 3^2 \\ &= 5 + 18 \div 9 \\ &= 5 + 2 \\ &= 7 \end{aligned}$$

SQUARE ROOTS

"Squaring a number" means raising that number to the second power. The square of 5 is 5^2 or 25. Finding the square root of a number reverses the process of squaring a number. The square root of 25 is 5 since $5 \times 5 = 25$, and the square root of 16 is 4 since $4 \times 4 = 16$. The **square root** of a nonnegative number N is one of two identical numbers whose product is N.

- Every positive number has two square roots. The two square roots of 9 are +3 *and* –3, since

$$(+3)(+3) = 9 \quad \text{and} \quad (-3)(-3) = 9$$

The two square roots of 9 taken together may be represented using the shorthand notation ± 3.

- In $\sqrt{9}$, read as the "the square root of 9," the symbol $\sqrt{}$ is called a **radical** or **square root symbol**, and the number underneath it, 9, is called the **radicand**.

MATH REFERENCE FACT

Every positive number N has two square roots: the *principal* or positive square root is $\sqrt{N}$, the negative square root is $-\sqrt{N}$, and the two square roots together are $\pm\sqrt{N}$.

- The symbol $\sqrt{\ }$ always indicates the positive or **principal square root** of the number underneath it. Thus, $\sqrt{9} = 3$, not ± 3.

CUBE ROOTS

The **cube root** of a number N is one of three identical numbers whose product is N. The symbol $\sqrt[3]{\ }$ represents the *cube root* of the number that is written underneath it. For example, $\sqrt[3]{8} = 2$ since $2 \times 2 \times 2 = 8$.

PERFECT SQUARES

The square root of a whole number may not be a whole number. For example, $\sqrt{7}$ does not equal a whole number since it is not possible to find two identical whole numbers whose product is 7. A whole number is a **perfect square** if its square root is also a whole number. The numbers 1, 4, 9, 16, 25, 36, 49, 64, 81, and 100 are perfect squares.

PROPERTIES OF SQUARE ROOT RADICALS

Here are some facts about square root radicals that you should know:

- Because the product of the same two numbers cannot be less than 0, the SAT does not have questions about square roots of negative numbers, such as $\sqrt{-64}$. Whenever you see a square root radical over a variable, assume that the quantity underneath the radical is greater than or equal to 0. Keep in mind, however, that *cube* roots of negative numbers can be evaluated. For example:

$$\sqrt[3]{-64} = -4 \text{ since } (-4) \times (-4) \times (-4) = -64$$

- $\sqrt{a} \times \sqrt{b} = \sqrt{a \times b}$. The product of the square roots of two numbers is the square root of the product. For example:

$$\sqrt{7} \times \sqrt{3} = \sqrt{7 \times 3} = \sqrt{21}$$

- $\dfrac{\sqrt{a}}{\sqrt{b}} = \sqrt{\dfrac{a}{b}}\,(b \neq 0)$. The quotient of the square roots of two numbers is the square root of the quotient. For example:

$$\frac{\sqrt{21}}{\sqrt{7}} = \sqrt{\frac{21}{7}} = \sqrt{3}$$

- $(\sqrt{N})^2 = N$. The square of a square root radical is the radicand. For example:

$$(\sqrt{5})^2 = \sqrt{5} \times \sqrt{5} = 5$$

- $a\sqrt{b} + c\sqrt{b} = (a + c)\sqrt{b}$. To combine square root radicals with the same radicands, combine their rational coefficients and keep the same radical factor, as in

$$5\sqrt{3} + 2\sqrt{3} = 7\sqrt{3}$$

- $\sqrt{a + b} \neq \sqrt{a} + \sqrt{b}$. Radicals cannot be distributed over the operations of addition and subtraction. For example:

$$\sqrt{4 + 9} \neq \sqrt{4} + \sqrt{9}$$

- To simplify a square root radical, write the radicand as the product of two numbers, one of which is the highest perfect square factor of the radicand. Then write the radical over each factor and simplify. For example:

$$\sqrt{20} = \sqrt{4 \times 5} = \sqrt{4} \times \sqrt{5} = 2\sqrt{5}$$

- To eliminate a radical from the denominator of a fraction, multiply the numerator and the denominator of the fraction by that radical, as in

$$\frac{5}{\sqrt{3}} = \frac{5}{\sqrt{3}} \times \frac{\sqrt{3}}{\sqrt{3}} = \frac{5\sqrt{3}}{3}$$

Lesson 3-2 Tune-Up Exercises

Multiple-Choice

If $\dfrac{x^{23}}{x^m} = x^{15}$ and $(x^4)^n = x^{20}$, then mn =

(A) 13
(B) 24
(C) 28
(D) 32
(E) 40

If $x = 32 - 16 \div 2 \times 4$, then x =

(A) 0
(B) 2
(C) 4
(D) 8
(E) 32

What is the greatest number of positive integer values of x for which $1 < x^2 < 50$?

(A) Five
(B) Six
(C) Seven
(D) Eight
(E) Nine

If $2 = p^3$, then $8p$ must equal

(A) p^6
(B) p^8
(C) p^{10}
(D) $8\sqrt{2}$
(E) 16

---

5 If $5 = a^x$, then $\dfrac{5}{a}$ =

(A) a^{x+1}
(B) a^{x-1}
(C) a^{1-x}
(D) $a\dfrac{x}{5}$
(E) $a\dfrac{5}{x}$

---

6 If $b^3 = 4$, then b^6 =

(A) 2
(B) 8
(C) 12
(D) 16
(E) 64

---

7 If w is a positive number and $w^2 = 2$, then w^3 =

(A) $\sqrt{2}$
(B) $2\sqrt{2}$
(C) 4
(D) $3\sqrt{2}$
(E) 6

---

8 If $8^{x+1} = 64$, what is the value of 3^{2x+1}?

(A) 1
(B) 3
(C) 9
(D) 27
(E) 81

9 If $0 < y < x$, which statement MUST be true?

(A) $\sqrt{x} - \sqrt{y} = \sqrt{x - y}$
(B) $\sqrt{x} + \sqrt{x} = \sqrt{2x}$
(C) $x\sqrt{y} = y\sqrt{x}$
(D) $\sqrt{xy} = (\sqrt{x})(\sqrt{y})$
(E) $\sqrt{x + y} = \sqrt{x} + \sqrt{y}$

10 Given $y = wx^2$ and y is not 0. If the values of x and w are each doubled, then the value of y is multiplied by

(A) 1
(B) 2
(C) 4
(D) 6
(E) 8

11 If $\sqrt{n}$ is a positive integer, how many values of n are in the interval $100 < n < 199$?

(A) Three
(B) Four
(C) Five
(D) Six
(E) Seven

12 If $(2^3)^2 = 4^p$, then $3^p =$

(A) 3
(B) 6
(C) 9
(D) 27
(E) 81

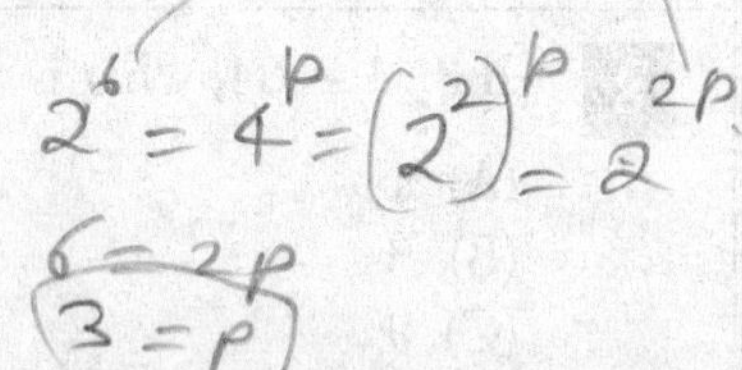

13 If a and b are positive integers, then $3^{a+b} \cdot 6^a =$

(A) 18^{2a+b}
(B) $18^{a(a+b)}$
(C) $3^{6(2a+b)}$
(D) $9^{a+b} \cdot 2^a$
(E) $3^{2a+b} \cdot 2^a$

14 If x is a positive integer, which of the following statements must be true?

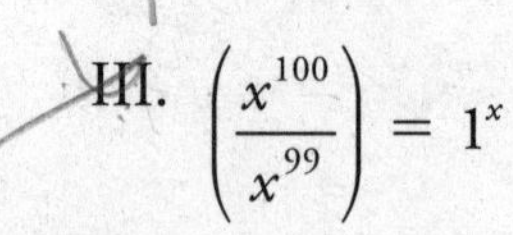

I. $\left(\frac{x}{x}\right)^{99} = \left(\frac{x+1}{x+1}\right)^{100}$

II. $(x^x)^2 = x^{x^2}$

III. $\left(\frac{x^{100}}{x^{99}}\right) = 1^x$

(A) I only
(B) II only
(C) I and III
(D) II and III
(E) I, II, and III

15 If $y = 25 - x^2$ and $1 \le x \le 5$, what is the smallest possible value of y?

(A) 0
(B) 1
(C) 5
(D) 10
(E) 15

16 If $x = \sqrt{6}$ and $y^2 = 12$, then $\frac{4}{xy} =$

(A) $\frac{3}{2\sqrt{2}}$
(B) $\frac{\sqrt{2}}{3}$
(C) $\frac{3}{\sqrt{2}}$
(D) $\frac{2\sqrt{2}}{3}$
(E) $\frac{\sqrt{6}}{3}$

Grid-In

1 If $2^4 \times 4^2 = 16^x$, then $x =$

2 If $a^7 = 7777$ and $\frac{a^6}{b} = 11$, what is the value of ab?

3 If $(y-1)^3 = 8$, what is the value of $(y+1)^2$?

4 If $\frac{p+p+p}{p \cdot p \cdot p} = 12$ and $p > 0$, what is the value of p?

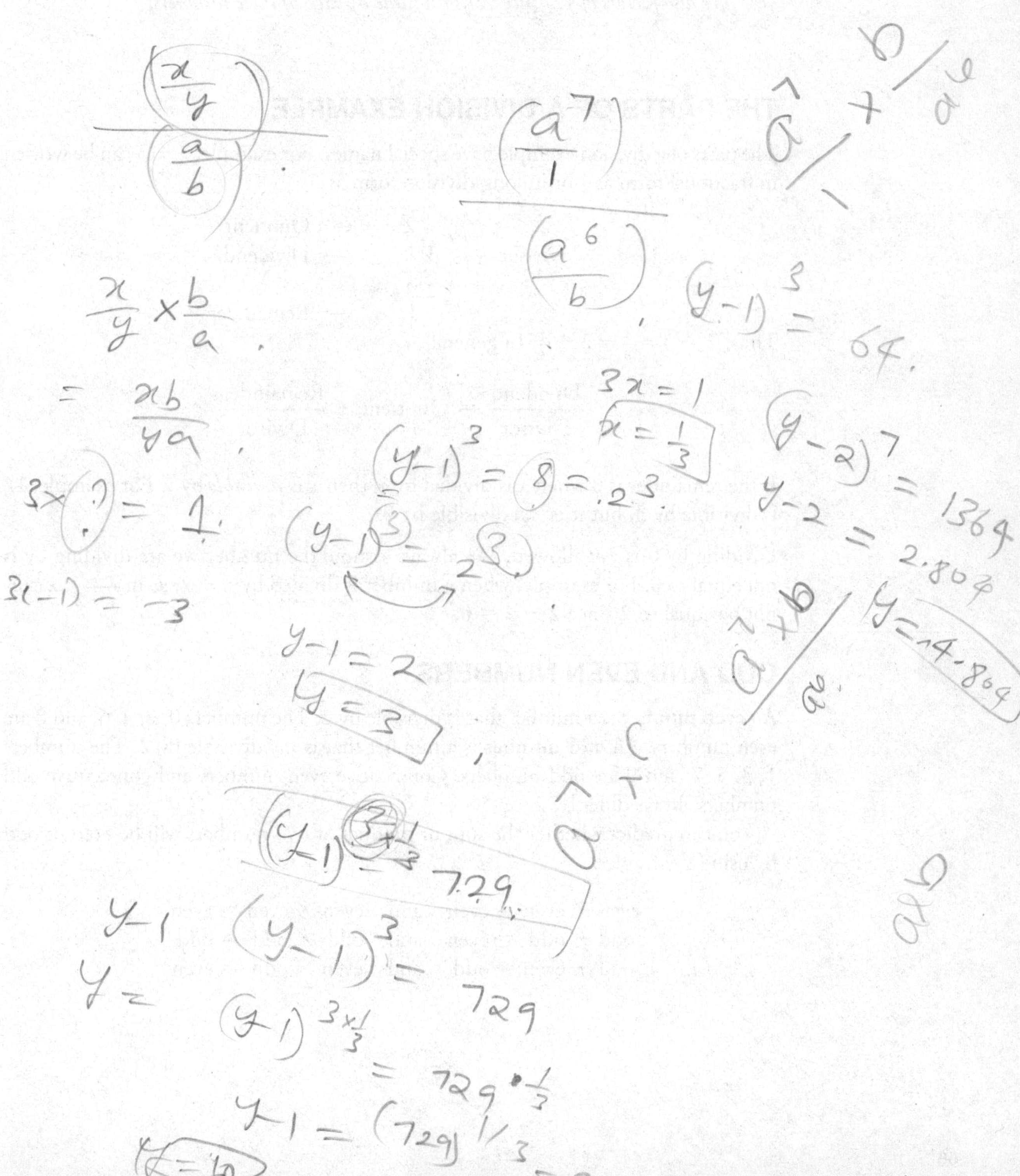

LESSON

3-3 DIVISIBILITY AND FACTORS

OVERVIEW

When 12 *is divided by* 3, *the remainder is* 0. *Thus,* 12 *is* ***divisible by*** 3, *and, as a result,* 3 *is called a factor of* 12. *The numbers* 1, 2, 4, 6, *and* 12 *are also factors of* 12 *since* 12 *is divisible by each of these numbers.*

THE PARTS OF A DIVISION EXAMPLE

The parts of a division example have special names. For example, $7 \div 3$ can be written in fractional form as $\frac{7}{3}$ or in long-division form as

$$\begin{array}{rrl} & 2 & \leftarrow \text{Quotient} \\ \text{Divisor} \rightarrow & 3\overline{)7} & \leftarrow \text{Dividend} \\ & \underline{-6} & \\ & 1 & \leftarrow \text{Remainder} \end{array}$$

Thus, $7 \div 3 = \frac{7}{3} = 2 + \frac{1}{3}$. In general,

$$\frac{\text{Dividend}}{\text{Divisor}} = \text{Quotient} + \frac{\text{Remainder}}{\text{Divisor}}$$

If the remainder is 0 when x is divided by y, then x is *divisible* by y. For example, 12 is divisible by 3, but it is *not* divisible by 7.

Dividing by 0 is *not* allowed. We always assume the number we are dividing by is not equal to 0. For example, when a number is divided by $x - 2$, as in $\frac{5}{x-2}$, x cannot be equal to 2 since $2 - 2 = 0$.

ODD AND EVEN NUMBERS

An **even number** is a number that is divisible by 2. The numbers 0, 2, 4, 6, and 8 are even numbers. An **odd number** is a number that is *not* divisible by 2. The numbers 1, 3, 5, 7, and 9 are odd numbers. Consecutive even numbers and consecutive odd numbers always differ by 2.

You can predict whether the sum or product of two numbers will be even or odd by using these rules:

$$\begin{array}{lcl} \text{even} + \text{even} = \text{even} & \text{and} & \text{even} \times \text{even} = \text{even} \\ \text{odd} + \text{odd} = \text{even} & \text{and} & \text{odd} \times \text{odd} = \text{odd} \\ \text{odd} + \text{even} = \text{odd} & \text{and} & \text{even} \times \text{odd} = \text{even} \end{array}$$

FACTORS AND MULTIPLES

Here are some terms you should know:

- The **factors** of a number *N* are the numbers that can be divided into *N* with a remainder of 0. For example, the factors of 32 are

  1, 2, 4, 8, 16, and 32

  since 32 is divisible by each of these numbers.

- The **common factors** of two numbers are the factors that the numbers have in common. For example:

  Factors of 12: 1, 2, 3, 4, 6, 12

  Factors of 21: 1, 3, 7, 21

  The common factors of 12 and 21 are 1 and 3.

- Any number that can be obtained by multiplying a number *N* by a positive integer is called a **multiple** of *N*. For example, some multiples of 6 are 12, 18, 24, 30, and 36. The multiples of a number are also multiples of any factor of that number. Two factors of 6 are 2 and 3. The numbers 12, 18, 24, 30, and 36 are multiples of 6, and they are also multiples of 2 and 3.

PRIME AND COMPOSITE NUMBERS

A positive integer greater than 1 is either prime or composite, but not both.

- A **prime number** is a number that has exactly two different factors, itself and 1. The set of prime numbers includes

  2, 3, 5, 7, 11, 13, 17, 19, 23, . . .

  The only even number that is prime is 2.

- A **composite number** is a number that has more than two different factors. For example:

  4, 6, 8, 9, 10, 12, 14, 15, 16, . . .

  are composite numbers.

- The number 1 is neither prime nor composite.

PRIME FACTORIZATION

The process of breaking down a number into the product of two or more other numbers is called **factoring.** Factoring reverses multiplication:

$$\text{Multiplying: } 7 \times 3 = 21$$

$$\text{Factoring: } 21 = 7 \times 3$$

Since $21 = 7 \times 3$, 7 and 3 are factors of 21.

The **prime factorization** of a number breaks down the number into the product of prime numbers. For example:

- The prime factorization of 18 is

$$18 = 3 \times 3 \times 2$$

- The prime factorization of 30 is

$$30 = 5 \times 3 \times 2$$

Lesson 3-3 Tune-Up Exercises

Multiple-Choice

1 If the sum of the factors of 18 is S and the sum of prime numbers less than 18 is P, then P exceeds S by what number?

(A) 19
(B) 17
(C) 15
(D) 13
(E) 11

2 When a whole number N is divided by 5, the quotient is 13 and the remainder is 4. What is the value of N?

(A) 55
(B) 59
(C) 65
(D) 69
(E) 79

3 If p is divisible by 3 and q is divisible by 4, then pq must be divisible by each of the following EXCEPT

(A) 3
(B) 4
(C) 6
(D) 9
(E) 12

4 Which number has the most factors?

(A) 12
(B) 18
(C) 25
(D) 70
(E) 100

5 Which number is divisible by 2 and by 3?

(A) 112
(B) 4308
(C) 6122
(D) 23,451
(E) 701,456

6 All numbers that are divisible by both 3 and 10 are also divisible by

(A) 4
(B) 9
(C) 12
(D) 15
(E) 20

7 If x represents any even number and y represents any odd number, which of the following numbers is even?

(A) $y + 2$
(B) $x - 1$
(C) $(x + 1)(y - 1)$
(D) $y(y + 2)$
(E) $x + y$

8 For how many different positive integers p is $\frac{105}{p}$ also an integer?

(A) Five
(B) Six
(C) Seven
(D) Eight
(E) Nine

9 If n is an odd integer, which expression always represents an odd integer?

(A) $(2n - 1)^2$
(B) $n^2 + 2n + 1$
(C) $(n - 1)^2$
(D) $\frac{n + 1}{2}$
(E) $3^n + 1$

10 If $k - 1$ is a multiple of 4, what is the next larger multiple of 4?

(A) $k + 1$
(B) $4k$
(C) $k - 5$
(D) $k + 3$
(E) $4(k - 1)$

11 After m marbles are put into n jars, each jar contains the same number of marbles, with two marbles remaining. In terms of m and n, how many marbles were put into each jar?

(A) $\frac{m}{n} + 2$
(B) $\frac{m}{n} - 2$
(C) $\frac{m + 2}{n}$
(D) $\frac{m - 2}{n}$
(E) $\frac{mn}{n + 2}$

12 When p is divided by 4, the remainder is 3; and when p is divided by 3, the remainder is 0. What is a possible value of p?

(A) 8
(B) 11
(C) 15
(D) 18
(E) 21

13 If n is an integer and $n^2 + 5$ is an odd integer, then which statement(s) must be true?

I. $n^2 - 1$ is even.
II. n is even.
III. $5n$ is even.

(A) I only
(B) I and II only
(C) I and III only
(D) II and III only
(E) I, II, and III

14 When the number of people who contribute equally to a gift decreases from four to three, each person must pay an additional $10. What is the cost of the gift?

(A) $30
(B) $60
(C) $90
(D) $120
(E) $180

15 If n is any even integer, what is the remainder when $(n + 1)^2$ is divided by 4?

(A) 0
(B) 1
(C) 2
(D) 3
(E) 4

16 A jar contains between 40 and 50 marbles. If the marbles are taken out of the jar three at a time, two marbles will be left in the jar. If the marbles are taken out of the jar five at a time, four marbles will be left in the jar. How many marbles are in the jar?

(A) 41
(B) 43
(C) 44
(D) 47
(E) 49

Grid-In

1 When a is divided by 7, the remainder is 5; and when b is divided by 7, the remainder is 4. What is the remainder when $a + b$ is divided by 7?

2 How many integers from -3000 to 3000, inclusive, are divisible by 3?

3 When a positive integer k is divided by 6, the remainder is 1. What is the remainder when $5k$ is divided by 3?

4 What is the smallest positive integer p for which $2^{2p} + 1$ is not a prime number?

LESSON

3-4 NUMBER LINES AND SIGNED NUMBERS

OVERVIEW

This lesson reviews the rules for working with positive and negative numbers.

ORDERING NUMBERS ON A NUMBER LINE

The size order of real numbers can be pictured by drawing a rulerlike line called a **number line**. As shown in Figure 3.2, positive numbers are located in increasing size order to the right of 0, and their opposites appear in descending size order to the left of 0.

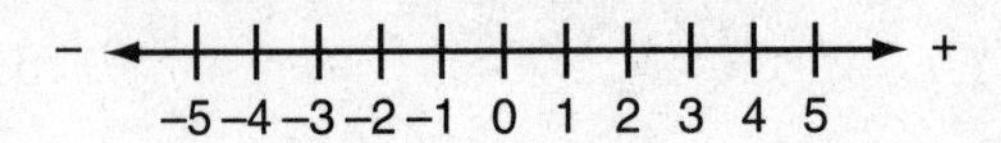

Figure 3.2 The Real Number Line

On the number line:

- Each number is *less than* the number that is located to its right. For example, $-3 < -2$.
- The larger of two negative numbers is the number that is closer to 0. For example, $-1 > -100$, since -1 is closer to 0 than is -100.
- If $a > b$, then $-b > -a$. For example, since $7 > 5$, $-5 > -7$.

MULTIPLYING AND DIVIDING SIGNED NUMBERS

The product or quotient of two numbers with the same sign is positive, and the product or quotient of two numbers with different signs is negative.

$$(+) \times (+) = + \qquad (+) \div (+) = +$$

$$(-) \times (-) = + \qquad (-) \div (-) = +$$

$$(+) \times (-) = - \qquad (+) \div (-) = -$$

$$(-) \times (+) = - \qquad (-) \div (+) = -$$

Thus, the square of either a positive or a negative number is positive. If $x^2 = 9$, x may be equal to either 3 *or* -3 since

$$(3)^2 = (-3)^2 = 9$$

MULTIPLYING MORE THAN TWO SIGNED NUMBERS

When more than two signed numbers are multiplied together, the sign of the product can be determined by using these rules:

- The product of an *even* number of negative factors is *positive.* For example:

$$(-2)^4 = \underbrace{(-2) \times (-2) \times (-2) \times (-2)}_{\text{4 factors}} = 16$$

- The product of an *odd* number of negative factors is *negative.* For example:

$$(-2)^3 = \underbrace{(-2) \times (-2) \times (-2)}_{\text{3 factors}} = -8$$

ADDING AND SUBTRACTING SIGNED NUMBERS

The rules for adding and subtracting signed numbers are summarized in Table 3.3.

TABLE 3.3

Rules for Adding and Subtracting Signed Numbers

Operation	Sign of Numbers	Procedure
Addition	Same $(+) + (+) = +$ $(-) + (-) = -$	Add numbers while ignoring their signs. Write the sum using the *common* sign. • $(+5) + (+8) = \mathbf{+13}$ • $(-5) + (-8) = \mathbf{-13}$
	Different	Subtract numbers while ignoring their signs. The answer has the same sign as the number having the *larger value when disregarding the signs.* • $(+5) + (-8) = \mathbf{-3}$ • $(-5) + (+8) = \mathbf{+3}$
Subtraction	Same	$(+5) - (+8) = (+5) + (-8) = \mathbf{-3}$ Take the opposite and *add.*
	Different	$(+5) - (-8) = (+5) + (+8) = \mathbf{+13}$ Take the opposite and *add.*

Lesson 3-4 Tune-Up Exercises

Multiple-Choice

1 If $2b = -3$, what is the value of $1 - 4b$?

(A) −7
(B) −5
(C) 5
(D) 6
(E) 7

2 If a is a negative integer and b is a positive integer, which of the following statements must be true?

I. $b + a > 0$

II. $\frac{b - a}{a} < 0$

III. $a^b < 0$

(A) None
(B) I only
(C) II only
(D) III only
(E) I and II only

3 If the product of five numbers is positive, then, at most, how many of the five numbers could be negative?

(A) One
(B) Two
(C) Three
(D) Four
(E) Five

4 For which value of k is the value of $k(k - 2)(k + 1)$ negative?

(A) −2
(B) −1
(C) 0
(D) 2
(E) 3

5 $p^2(2 - 5) + (-p)^2 =$

(A) $-4p^2$
(B) $-p^2$
(C) $-2p^2$
(D) $2p^2$
(E) $4p^2$

6 If $n + 5$ is an odd integer, then n could be which of the following?

(A) 1
(B) −1
(C) −2
(D) −3
(E) −7

7 Which of the following statements must be true when $a < 0$ and $b > 0$?

I. $a + b > 0$
II. $b - a > 0$
III. $a\left(\frac{a}{b}\right) > 0$

(A) I only
(B) II only
(C) I and III only
(D) II and III only
(E) I, II, and III

8 $(-2)^3 + (-3)^2 =$

(A) −12
(B) −1
(C) 0
(D) 1
(E) 2

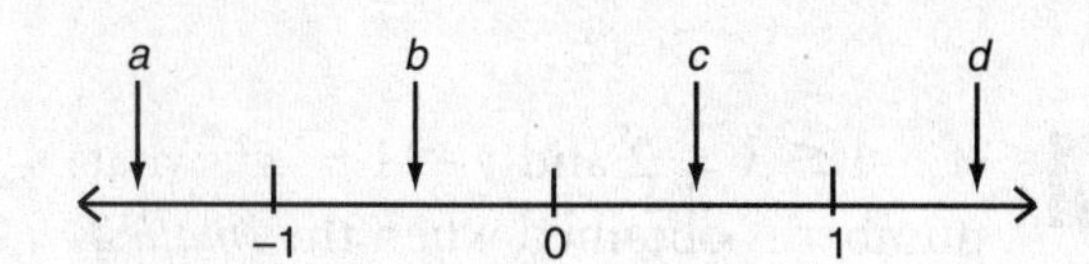

Questions 9 and 10 refer to the diagram above.

9 Which of the following statements must be true?

I. $c^2 < c$
II. $a^2 > c$
III. $b < \frac{1}{b}$

(A) I only
(B) I and II only
(C) I and III only
(D) I, II, and III
(E) None

10 Which of the following statements must be true?

I. $ad > b$
II. $ab > ad$
III. $\frac{1}{a} > \frac{1}{b}$

(A) II only
(B) I and II only
(C) II and III only
(D) I, II, and III
(E) None

11 If $(y - 3)^2 = 16$, what is the smallest possible value of y^2?

(A) −4
(B) 1
(C) 7
(D) 16
(E) 49

12 If X represents the sum of the 10 greatest negative integers and Y represents the sum of the 10 least positive integers, which of the following must be true?

I. $X + Y < 0$
II. $Y - X = 2Y$
III. $X^2 = Y^2$

(A) None
(B) I only
(C) III only
(D) I and II only
(E) II and III only

13 If $a^2b^3c > 0$, which of the following statements must be true?

I. $bc > 0$
II. $ac > 0$
III. $ab > 0$

(A) I only
(B) I and II only
(C) I and III only
(D) II and III only
(E) I, II, and III

14 If $a^2b^3c^5$ is negative, which product is always negative?

(A) bc
(B) b^2c
(C) ac
(D) ab
(E) bc^2

15 If $a \neq 0$, which of the following statements must be true?

I. $(-a)^2 = a^2 - 2a^2$
II. $a - b = -(b - a)$
III. $a > -a$

(A) I only
(B) II only
(C) III only
(D) I and III only
(E) II and III only

Grid-In

1 If $p^2 = 16$ and $q^2 = 36$, what is the largest possible value of $q - p$?

2 If $-4 \leq x \leq 2$ and $y = 1 - x^2$, what number is obtained when the *smallest* possible value of y is subtracted from the *largest* possible value of y?

LESSON

3-5

FRACTIONS AND DECIMALS

OVERVIEW

*A **fraction** represents a specific number of the equal parts in a whole. Thus, since there are 24 equal hours in a day, 3 hours represents $\frac{3}{24}$ of a day.*

*The number above the fraction bar is the **numerator** of the fraction, and the number below the fraction bar is the **denominator.** In the fraction $\frac{3}{24}$, 3 is the numerator and 24 is the denominator. Since the fraction bar means division,*

$$\frac{3}{24} = 3 \div 24 = 0.125$$

*The **decimal number** 0.125 is another name for the fraction $\frac{3}{24}$.*

TYPES OF FRACTIONS

The numerator of a fraction may be less than, greater than, or equal to the denominator.

- A **proper fraction** is a fraction such as $\frac{2}{3}$, in which the numerator is less than the denominator. The value of a proper fraction is always less than 1.
- An **improper fraction** is a fraction such as $\frac{3}{2}$, in which the numerator is greater than or equal to the denominator. The value of an improper fraction is always greater than or equal to 1.

EQUIVALENT FRACTIONS

Equivalent fractions are fractions that have the same value. The fractions $\frac{1}{2}$ and $\frac{5}{10}$ are equivalent because they name the same part of a whole:

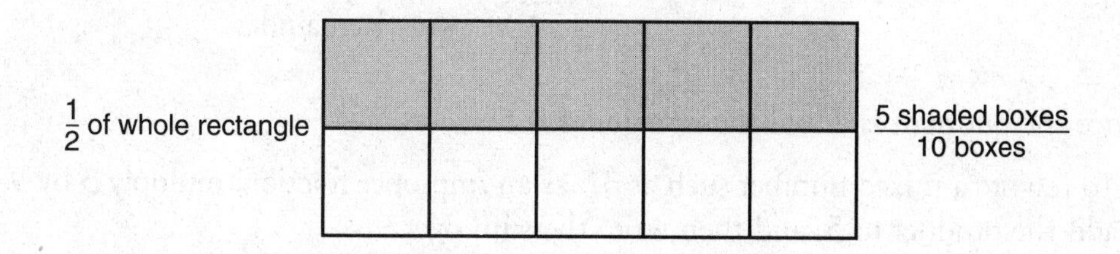

Multiplying or dividing the numerator and denominator of a fraction by the same nonzero number always produces an equivalent fraction:

$$\frac{1}{2} = \frac{1 \times 5}{2 \times 5} = \frac{5}{10} \quad \text{and} \quad \frac{5}{10} = \frac{5 \div 5}{10 \div 5} = \frac{1}{2}$$

REDUCING FRACTIONS TO LOWEST TERMS

A fraction is in **lowest terms** when its numerator and denominator do not have any common factors other than 1. To write a fraction in lowest terms, divide the numerator and denominator by their largest common factor.

EXAMPLE

Write $\frac{16}{24}$ in lowest terms.

Solution

The largest number by which 16 and 24 are both divisible is 8.

$$\frac{16}{24} = \frac{16 \div 8}{24 \div 8} = \frac{2}{3}$$

You may find it easier to perform the division mentally by thinking "How many times does 8 go into 16?" and "How many times does 8 go into 24?" Write the answers above the numerator and denominator of the fraction as shown below:

$$\frac{\overset{2}{\cancel{16}}}{\underset{3}{\cancel{24}}} = \frac{2}{3}$$

MIXED NUMBERS

A **mixed number** is a number such as $2\frac{3}{8}$, which represents the sum of a whole number and a proper fraction. Thus, $2\frac{3}{8}$ means $2 + \frac{3}{8}$.

- To change an improper fraction into a mixed number, divide the denominator of the improper fraction into the numerator. Write the quotient with the remainder expressed as a fraction in lowest terms. For example:

$$\frac{19}{7} = 19 \div 7 = 7\overline{)\,19\,}\quad \text{quotient } 2$$
$$\begin{array}{r} 19 \\ -14 \\ \hline 5 \end{array} \quad \leftarrow \text{Remainder}$$

Since the quotient is 2 and the remainder is 5, $\frac{19}{7} = 2\frac{5}{7}$

- To rewrite a mixed number such as $3\frac{5}{9}$ as an improper fraction, multiply 3 by 9, add the product to 5, and then write the sum over 9:

$$3\frac{5}{9} = \frac{(3 \times 9) + 5}{9} = \frac{27 + 5}{9} = \frac{32}{9}$$

PLACE VALUE IN DECIMAL NUMBERS

The **place value** of each digit of a decimal number is 10 times as great as the place value of the digit to its right.

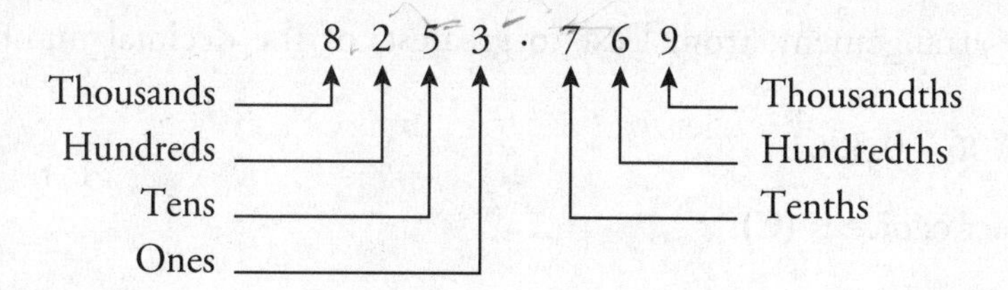

Thus, the decimal number 8253.769 is equivalent to

$$8253.769 = 8 \times 1000 + 2 \times 100 + 5 \times 10 + 3 \times 1 + 7 \times 0.1 + 6 \times 0.01 + 9 \times 0.001$$

COMPARING DECIMALS

To compare two decimal numbers:

- Write one decimal underneath the other so that the decimal points and digits with the same place value are aligned.
- If one decimal is shorter than the other, add zeros to the right of the last digit of the shorter decimal until the decimal numbers have the same number of digits.
- Start from the left and compare the digits that are in the same column. Stop when you find unlike digits. The greater digit indicates the decimal with the greater value.

EXAMPLE

In which of the following lists are the numbers written in order from least to greatest?

(A) 0.0361, 0.3061, 0.306
(B) 0.3061, 0.306, 0.0361
(C) 0.0361, 0.306, 0.3061
(D) 0.306, 0.3061, 0.0301
(E) 0.306, 0.0361, 0.3061

Solution

Write the three decimal numbers that are being compared, one underneath the other, so that digits with the same place value are aligned in the same vertical columns:

0.0361
0.3061
0.3060 ← Add 0.

Then compare the digits in the same columns. The least decimal is 0.0361. Also, $0.306\underline{0} < 0.306\underline{1}$ since $0 < 1$ in the last decimal position.

The correct arrangement, from least to greatest, of the decimal numbers is as follows:

0.0361, 0.306, 0.3061

Thus, the correct choice is **(C)**.

REWRITING FRACTIONS AS DECIMALS

To rewrite a fraction as a decimal number, use a calculator to divide the denominator of the fraction into the numerator of the fraction.

EXAMPLE

Rewrite $\frac{11}{40}$ as a decimal.

Solution

Using a calculator, divide 40 into 11:

$$\frac{11}{40} = 11 \div 40 = 0.275$$

COMPARING FRACTIONS

To compare two fractions, use these facts:

- If two fractions have the same denominator but different numerators, then the larger fraction is the fraction with the greater numerator. For example, $\frac{5}{9} > \frac{4}{9}$.
- If two fractions have the same numerator but different denominators, then the larger fraction is the fraction with the smaller denominator. For example, $\frac{4}{5} > \frac{4}{7}$.
- If two fractions have different numerators and different denominators, then the larger fraction can be determined by comparing the decimal values of the two fractions. For example, $\frac{11}{23} > \frac{7}{15}$ since

$$\frac{11}{23} = 0.47826\ldots \text{ and } \frac{7}{15} = 0.46666\ldots$$

MULTIPLYING AND DIVIDING DECIMALS BY POWERS OF 10

Multiplying or dividing a decimal number by a power of 10 (10, 100, 1000, etc.) affects only the position of the decimal point of the number.

- To multiply a decimal number by a power of 10, move the decimal point one place to the *right* for each 0 in the power of 10. For example,

$$6.25 \times 1000 = 6.250 \times 1000 = 6250.$$

- To divide a decimal number by a power of 10, move the decimal point one place to the *left* for each 0 in the power of 10. For example,

$$8.24 \div 100 = 08.24 \div 100 = 0.0824.$$

Lesson 3-5 Tune-Up Exercises

Multiple-Choice

1 What part of an hour elapses from 4:56 P.M. to 5:32 P.M.?

(A) $\frac{1}{4}$
(B) $\frac{1}{2}$
(C) $\frac{3}{5}$
(D) $\frac{2}{3}$
(E) $\frac{3}{4}$

2 If each of the fractions $\frac{3}{k}, \frac{4}{k}, \frac{5}{k}$ is in lowest terms, which of the following could be the value of k?

(A) 48
(B) 49
(C) 50
(D) 51
(E) 52

3 Which number has the greatest value?

(A) 0.2093
(B) 0.2908
(C) 0.2893
(D) 0.2938
(E) 0.2909

4 The elapsed time from 11:00 A.M. to 3:00 P.M. on Wednesday of the same day is what fraction of the elapsed time from 11:00 A.M. on Wednesday to 3:00 P.M. on Friday of the same week?

(A) $\frac{1}{15}$
(B) $\frac{1}{14}$
(C) $\frac{1}{13}$
(D) $\frac{4}{51}$
(E) $\frac{1}{12}$

5 In which number is the digit 3 in the hundredths place?

(A) 300.000
(B) 30.000
(C) 0.300
(D) 0.030
(E) 0.003

6 In which arrangement are the fractions listed from least to greatest?

(A) $\frac{9}{19}, \frac{1}{2}, \frac{8}{15}$
(B) $\frac{1}{2}, \frac{8}{15}, \frac{9}{19}$
(C) $\frac{9}{19}, \frac{8}{15}, \frac{1}{2}$
(D) $\frac{1}{2}, \frac{9}{19}, \frac{8}{15}$
(E) $\frac{8}{15}, \frac{1}{2}, \frac{9}{19}$

7 After the formula $V = \frac{4}{3}\pi r^3$ has been evaluated for some positive value of r, the formula is again evaluated using one-half of the original value of r. The new value of V is what fractional part of the original value of V?

(A) $\frac{1}{16}$
(B) $\frac{1}{9}$
(C) $\frac{1}{8}$
(D) $\frac{1}{4}$
(E) $\frac{1}{2}$

8 Each inch on ruler A is marked in equal $\frac{1}{8}$-inch units, and each inch on ruler B is marked in equal $\frac{1}{12}$-inch units. When ruler A is used, a side of a triangle measures 12 of the $\frac{1}{8}$-inch units. When ruler B is used, how many $\frac{1}{12}$-inch units will the same side measure?

(A) 8
(B) 12
(C) 18
(D) 20
(E) 24

9 $60 + 2 + \frac{4}{8} + \frac{3}{500} =$

(A) 60.256
(B) 62.43
(C) 62.506
(D) 62.53
(E) 62.560

10 If $N \times \frac{7}{12} = \frac{7}{12} \times \frac{3}{14}$, then $\frac{1}{N} =$

(A) 8
(B) $\frac{14}{3}$
(C) $\frac{12}{7}$
(D) $\frac{3}{14}$
(E) $\frac{1}{6}$

11 If $n = 2.5 \times 10^{25}$, then $\sqrt{n} =$

(A) 0.5×10^5
(B) 0.5×10^{12}
(C) $5 \times 10^{\sqrt{24}}$
(D) 5×10^5
(E) 5×10^{12}

12 If y is a real number and $y = \frac{x-2}{x+3}$, then x CANNOT equal which of the following numbers?

(A) -3
(B) -2
(C) 0
(D) 2
(E) 3

13 If eight pencils cost $0.42, how many pencils can be purchased with $2.10?

(A) 16
(B) 24
(C) 30
(D) 36
(E) 40

14 A store sells 8-ounce containers of orange juice at $0.69 each and 12-ounce containers of orange juice at $0.95 each. How much money will be saved by purchasing a total of 48 ounces of orange juice in 12-ounce rather than 8-ounce containers?

(A) $0.24
(B) $0.32
(C) $0.34
(D) $0.48
(E) $0.56

15 In the repeating decimal

$$0.\overline{31752} = 0.3175231752\ldots$$

the set of digits 31752 repeats endlessly. Which digit is in the 968th place to the right of the decimal point?

(A) 1
(B) 2
(C) 3
(D) 5
(E) 7

Grid-In

1 If $\frac{4\Delta}{6\Delta} + \frac{5}{17} = 1$, what digit does Δ represent?

2 On a certain map, 1.5 inches represents a distance of 45 miles. If two points on the map are 0.8 inch apart, how many miles apart are these two points?

3 Four lemons cost $0.68. At the same rate, 1 pound of lemons costs $1.19. How many lemons typically weigh 1 pound?

4 If the charge for a taxi ride is $2.50 for the first $\frac{1}{2}$ mile and $0.75 for each additional $\frac{1}{8}$ mile, how many miles did the taxi travel for a ride that cost $10.75?

5 One cubic foot of a certain metal weighs 8 pounds and costs $4.20 per pound. If 1 cubic foot is equivalent to 1728 cubic inches, what is the cost of 288 cubic inches of the same metal?

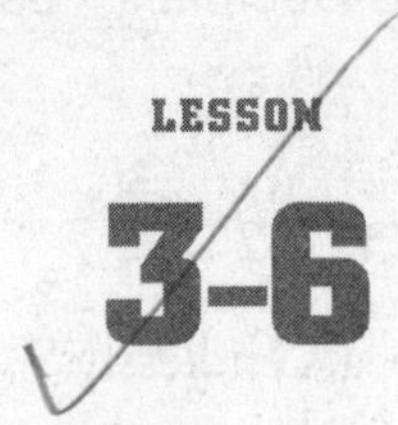

OPERATIONS WITH FRACTIONS

OVERVIEW

Fractions are multiplied, divided, and combined (added or subtracted) according to the following rules, where a, b, c, and d stand for real numbers with b, d ≠ 0:

- $\frac{a}{b} \times \frac{c}{d} = \frac{a \times c}{b \times d}$
- $\frac{a}{b} \pm \frac{c}{b} = \frac{a \pm c}{b}$
- $\frac{a}{b} \div \frac{c}{d} = \frac{a}{b} \times \frac{d}{c} = \frac{a \times d}{b \times c}$
- $\frac{a}{b} \pm \frac{c}{d} = \frac{ad \pm bc}{bd}$

MULTIPLYING FRACTIONS

To multiply fractions, write the product of the numerators over the product of the denominators. Then write the result in lowest terms. For example:

$$\frac{4}{9} \times \frac{5}{8} = \frac{20}{72} = \frac{20 \div 4}{72 \div 4} = \frac{5}{18}$$

Sometimes it is easier to divide out pairs of common factors of any numerator and denominator *before* multiplying. For example:

$$\frac{\overset{2}{\cancel{8}}}{\underset{3}{\cancel{21}}} \times \frac{\overset{1}{\cancel{7}}}{\underset{5}{\cancel{20}}} = \frac{2}{15}$$

FINDING A FRACTION OF A NUMBER

To find a fraction of another number, replace the word *of* with the times symbol. Then multiply.

EXAMPLE

What is $\frac{3}{5}$ of 20?

Solution

$\frac{3}{5}$ of $20 = \frac{3}{5} \times 20 = 3 \times 4 = 12$

EXAMPLE

After John spends $\frac{1}{4}$ of his salary on food and $\frac{2}{3}$ of what remains on clothes, what part of his original salary does he have left?

Solution 1

- After John spends $\frac{1}{4}$ of his original salary on food, $\frac{3}{4}\left(= 1 - \frac{1}{4}\right)$ of his salary remains.
- John spends $\frac{2}{3}$ of what remains on clothes. Since

$$\frac{2}{3} \text{ of } \frac{3}{4} - \frac{2}{3} \times \frac{3}{4} = \frac{1}{2}$$

John spends $\frac{1}{2}$ of his original salary on clothes.

- After John spends $\frac{1}{4}$ of his original salary on food and $\frac{1}{2}$ of his original salary on clothes, he has $\frac{1}{4}\left(= 1 - \frac{1}{4} - \frac{1}{2}\right)$ of his original salary left.

Solution 2

Pick a convenient amount as John's original salary. Choose a number that is divisible by 4, for example, $100.

- John spends $\frac{1}{4}$ of his original salary on food. Since $\frac{1}{4} \times \$100 = \$25$, he spends $25 on food, so he has $75 left.
- John spends $\frac{2}{3}$ of $75 on clothes. Since $\frac{2}{3} \times \$75 = \$50$, he spends $50 on clothes, so $25 remains from his original salary.
- Since $\frac{\$25}{\$100} = \frac{1}{4}$, John has $\frac{1}{4}$ of his original salary left.

DIVIDING FRACTIONS

To divide a fraction by another fraction, invert the second fraction by interchanging its numerator and denominator. Then multiply. For example:

$$\frac{12}{35} \div \frac{3}{14} = \frac{\overset{4}{\cancel{12}}}{\underset{5}{\cancel{35}}} \times \frac{\overset{2}{\cancel{14}}}{\underset{1}{\cancel{3}}} = \frac{4 \times 2}{5 \times 1} = \frac{8}{5}$$

EXAMPLE

Divide: $3 \div \frac{12}{7}$

Solution

$$3 \div \frac{12}{7} = \frac{3}{1} \times \frac{7}{12} = \frac{7}{4}$$

RECIPROCALS

The **reciprocal** of any nonzero number x is $\frac{1}{x}$. For example, the reciprocal of 3 is $\frac{1}{3}$. The reciprocal of any nonzero fraction $\frac{a}{b}(b \neq 0)$ is $\frac{b}{a}$. For example, the reciprocal of $\frac{2}{5}$ is $\frac{5}{2}$.

The product of any nonzero number and its reciprocal is 1.

POWERS, SQUARE ROOTS, AND RECIPROCALS OF PROPER FRACTIONS

If a number is between 0 and 1, then:

- The square of the number is *less than* the original number. For example:

$$\left(\frac{1}{3}\right)^2 < \frac{1}{3} \text{ since } \frac{1}{9} < \frac{1}{3}$$

  In general, as the exponent of a proper fraction increases, the value of that expression decreases.

- The square root of the number is *greater than* the original number;

$$\sqrt{0.25} > 0.25 \text{ since } \sqrt{0.25} = 0.5 \text{ and } 0.5 > 0.25$$

- The reciprocal of the number is *greater than* the original number. For example, the reciprocal of $\frac{1}{2}$ is 2, which is greater than $\frac{1}{2}$.

In each of the above cases, taking powers, roots, or reciprocals of numbers that are *greater than* 1 gives results that are exactly opposite, as shown in Table 3.4.

TABLE 3.4

Rules of Powers, Square Roots, and Reciprocals

x is between 0 and 1.	$x^2 < x$	$\sqrt{x} > x$	$\frac{1}{x} > x$
x is greater than 1.	$x^2 > x$	$\sqrt{x} < x$	$\frac{1}{x} < x$

ADDING AND SUBTRACTING FRACTIONS

To add or subtract fractions with like denominators:

- Write the sum or difference of the numerators over the common denominator.
- Express the result in lowest terms, if possible.

EXAMPLE

Add: $\frac{7}{12} + \frac{1}{12}$

Solution

$$\frac{7}{12} + \frac{1}{12} = \frac{7+1}{12} = \frac{8}{12} = \frac{2}{3}$$

EXAMPLE

Write $\frac{3x}{8} - \frac{x}{8}$ as a single fraction.

Solution

$$\begin{aligned} \frac{3x}{8} - \frac{x}{8} &= \frac{3x - x}{8} \\ &= \frac{2x}{8} \\ &= \frac{x}{4} \end{aligned}$$

To combine fractions with unlike denominators:

- Find the *L*east *C*ommon *D*enominator (LCD) of the fractions. For example, the LCD of $\frac{3}{8}$ and $\frac{5}{12}$ is 24 since 24 is the smallest number that is divisible by both 8 and 12.
- Change each fraction into an equivalent fraction that has the LCD as its denominator.
- Write the sum or difference of the numerators over the common denominator.
- Express the result in lowest terms, if possible.

EXAMPLE

Add: $\frac{1}{3} + \frac{2}{5}$

Solution

The LCD is 15. Change each fraction into an equivalent fraction that has 15 as its denominator. Multiply the first fraction by 1 in the form of $\frac{5}{5}$, and multiply the second fraction by 1 in the form of $\frac{3}{3}$.

$$\begin{aligned} \frac{1}{3} + \frac{2}{5} &= \frac{5}{5}\left(\frac{1}{3}\right) + \frac{3}{3}\left(\frac{2}{5}\right) \\ &= \frac{5}{15} + \frac{6}{15} \\ &= \frac{11}{15} \end{aligned}$$

OPERATIONS WITH MIXED NUMBERS

Mixed numbers have a whole number and a fractional part.

- To multiply or divide mixed numbers, first change the mixed numbers to improper fractions. Then follow the rules for multiplying and dividing fractions.

EXAMPLE

Multiply: $1\frac{7}{8} \times 3\frac{1}{3}$

Solution

$$1\frac{7}{8} \times 3\frac{1}{3} = \frac{15}{8} \times \frac{10}{3} = \frac{25}{4}$$

EXAMPLE

Divide: $5\frac{1}{2} \div 1\frac{3}{8}$

Solution

$$5\frac{1}{2} \div 1\frac{3}{8} = \frac{11}{2} \div \frac{11}{8} = \frac{11}{2} \times \frac{8}{11} = 4$$

- To add or subtract mixed numbers, write the second mixed number under the first mixed number. If the denominators of the fractions are different, find the LCD and change the fractions into equivalent fractions with the LCD as denominator. Add or subtract the fractions. Then add or subtract the whole numbers.

EXAMPLE

Add: $2\frac{1}{4} + 5\frac{2}{3}$

Solution

$$\begin{array}{r} 2\frac{1}{4} = 2\frac{3}{12} \\ +\,5\frac{2}{3} = 5\frac{8}{12} \\ \hline = 7\frac{11}{12} \end{array}$$

EXAMPLE

Subtract: $6\frac{2}{9} - 4\frac{4}{9}$

Solution

$$\begin{array}{rcl} 6\frac{2}{9} = 5\frac{9+2}{9} & = & 5\frac{11}{9} \\ -\ 4\frac{4}{9} & = & -4\frac{4}{9} \\ \hline & = & 1\frac{7}{9} \end{array}$$

SIMPLIFYING COMPLEX FRACTIONS

A fraction that has another fraction in its numerator or denominator is called a **complex fraction**. To simplify a complex fraction:

- Find the LCD of the denominators of the fractions that are contained in the numerator or denominator of the complex fraction.
- Multiply the numerator and denominator of the complex fraction by the LCD.
- Simplify the resulting fraction.

EXAMPLE

Simplify: $\frac{4}{\frac{3}{2}}$

Solution 1

Multiply the numerator and the denominator by 2:

$$\frac{2}{2} \times \frac{4}{\frac{3}{2}} = \frac{2 \times 4}{2 \times \frac{3}{2}} = \frac{8}{3}$$

Solution 2

Divide the numerator by the denominator:

$$\frac{4}{\frac{3}{2}} = 4 \div \frac{3}{2} = 4 \times \frac{2}{3} = \frac{8}{3}$$

Lesson 3-6 Tune-Up Exercises

Multiple-Choice

1 John completes a race in $9\frac{1}{3}$ minutes, and Steve finishes the same race in $7\frac{3}{4}$ minutes. How many seconds after Steve finishes the race does John complete the race?

(A) 72
(B) 90
(C) 92
(D) 95
(E) 100

2 Multiplying a number by $\frac{1}{2}$ and then dividing the result by $\frac{3}{4}$ is equivalent to performing which of the following operations on the number?

(A) Multiplying by $\frac{2}{3}$
(B) Dividing by $\frac{2}{3}$
(C) Multiplying by $\frac{3}{8}$
(D) Dividing by $\frac{3}{8}$
(E) Multiplying by 3

3 $\left(\frac{1}{10}\right)^3 + \left(\frac{2}{10}\right)^3 + \left(\frac{3}{10}\right)^3 =$

(A) 0.018
(B) 0.025
(C) 0.032
(D) 0.036
(E) 0.040

4 What is $\frac{3}{4}$ of $\frac{2}{3}$ of 12?

(A) 2
(B) 3
(C) 4
(D) 6
(E) 8

5 What is $\frac{2}{21}$ of $3 \times 5 \times 7$?

(A) 6
(B) 9
(C) 10
(D) 12
(E) 15

6 For which value of n is $\frac{1}{4} < n < \frac{1}{3}$?

(A) $\frac{5}{24}$
(B) $\frac{6}{24}$
(C) $\frac{7}{24}$
(D) $\frac{8}{24}$
(E) $\frac{9}{24}$

7 If $0 < x < 1$, then which of the following statements must be true?

I. $\frac{1}{x} - x < 0$

II. $x^2 + x^2 > x$

III. $\frac{x}{\sqrt{x}} > x^2$

(A) I only
(B) II only
(C) III only
(D) I and II only
(E) I and III only

8 If $k = \frac{c}{\frac{a}{b}}$, which expression equals $\frac{1}{k}$?

(A) $\frac{ac}{b}$
(B) $\frac{b}{ac}$
(C) $\frac{a}{bc}$
(D) $\frac{bc}{a}$
(E) $\frac{1}{abc}$

9 An item is on sale on Monday at $\frac{1}{3}$ off the list price of d dollars. On Friday, the sale price of the item is reduced by $\frac{1}{4}$ of the current price. What is the number of dollars in the final sale price of the item?

(A) $\frac{d}{2}$
(B) $\frac{d}{3}$
(C) $\frac{d}{4}$
(D) $\frac{d}{6}$
(E) $\frac{d}{8}$

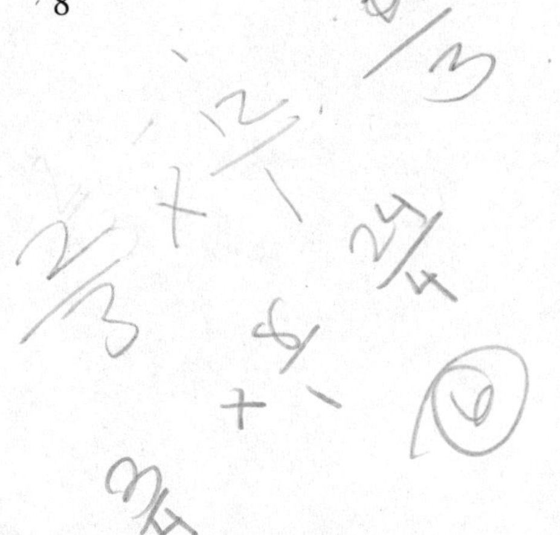

10 If $\frac{a}{b}$ is a positive fraction less than 1, which of the following fractions MUST be greater than 1?

(A) $\left(\frac{a}{b}\right)^2$
(B) $2\left(\frac{a}{b}\right)$
(C) $\frac{a}{2b}$
(D) $\frac{a+2}{b+2}$
(E) $\frac{2}{\frac{a}{b}}$

11 Juan gives $\frac{2}{3}$ of his p pencils to Roger and then gives $\frac{1}{4}$ of the pencils that he has left to Maria. In terms of p, how many pencils does Juan now have?

(A) $\frac{1}{6}p$
(B) $\frac{1}{4}p$
(C) $\frac{1}{3}p$
(D) $\frac{1}{2}p$
(E) $\frac{5}{6}p$

12 The statement $a^3 < a^2 < a$ is true when

I. $a > 1$
II. $0 < a < 1$
III. $-1 < a < 0$

(A) I and III only
(B) II and III only
(C) II only
(D) III only
(E) None

Grid-In

1 If $y = \sqrt{\dfrac{x+4}{2}}$, what is the value of y when $x = \dfrac{1}{2}$?

2 What fraction of $\dfrac{10}{9}$ is $\dfrac{5}{6}$?

3 If $25\left(\dfrac{x}{y}\right) = 4$, what is the value of $100\left(\dfrac{y}{x}\right)$?

4 $P = \left(1 + \dfrac{1}{2}\right)\left(1 + \dfrac{1}{3}\right)\left(1 + \dfrac{1}{4}\right)\cdots\left(1 + \dfrac{1}{17}\right)$

If each factor in the above product has the form $\left(1 + \dfrac{1}{k}\right)$, where k represents all the consecutive integers from 2 to 17, what is the value of P?

LESSON 3-7

Next Class

FRACTION WORD PROBLEMS

OVERVIEW

Many types of fraction problems can be solved either by:

- *forming a fraction in which the number in the numerator is being compared with the number in the denominator or*
- *finding the whole when a fractional part of it is given.*

FINDING WHAT FRACTION ONE NUMBER IS OF ANOTHER

To find what fraction a given number "$\frac{\text{is}}{\text{of}}$" another number, form the fraction

$$\frac{\text{Part } is}{of \text{ WHOLE}}$$

In the fraction above, the "part" is the given number and the "whole" is the number to which the part is being compared.

EXAMPLE

There are 15 boys and 12 girls in a class. The number of boys is what fraction of the total number of students in the class?

Solution

Since the number of boys is being compared to the total number of students in the class, form the fraction in which the number of boys is the numerator and the total number of students in the class, boys plus girls, is the denominator:

$$\frac{\text{Boys}}{\text{Whole class}} = \frac{15}{15 + 12} = \frac{15}{27} = \frac{5}{9}$$

FINDING THE WHOLE WHEN A PART OF IT IS KNOWN

If $\frac{4}{7}$ of some unknown number is 12, then multiplying 12 (the "part") by the reciprocal of the given fraction gives the unknown number (the "whole"). Thus,

$$\text{Whole} = \overset{3}{\cancel{12}} \times \frac{7}{\underset{1}{\cancel{4}}} = 3 \times 7 = 21$$

You can also find the unknown number by following these steps:

- Find $\frac{1}{7}$ of the number. Since $\frac{4}{7}$ of the unknown number is 12, $\frac{1}{7}$ of the unknown number is $\frac{1}{4}$ of 12, which is 3.
- Multiply to find the whole number. Since $\frac{1}{7}$ of the unknown number is 3, $\frac{7}{7}$ of the unknown number is $7 \times 3 = 21$.

  Hence, the unknown number is 21.

The above procedure can be quickly applied using the following shorthand notation:

$\frac{4}{7} \Rightarrow 12$ $\left[\text{You are told that } \frac{4}{7} \text{ of the unknown number is 12.}\right]$

$\frac{1}{7} \Rightarrow \frac{1}{4} \times 12 = 3$ $\left[\frac{1}{7} \text{ of the unknown number is 3.}\right]$

$\frac{7}{7} \Rightarrow 7 \times 3 = 21$ $\left[\frac{7}{7} \text{ of the unknown number is 21.}\right]$

EXAMPLE

Of the books that are on a shelf, $\frac{1}{3}$ are math books, $\frac{1}{4}$ are science books, and the remaining books are history books. If the shelf contains 10 history books, how many books are on the shelf?

Solution 1

- Math and science books account for $\frac{7}{12}$ of the books on the shelf since

$$\frac{1}{3} + \frac{1}{4} = \frac{4}{12} + \frac{3}{12} = \frac{7}{12}$$

- History books make up $\frac{5}{12}\left(= 1 - \frac{7}{12}\right)$ of the total number of books on the shelf.
- Since there are 10 history books, the problem reduces to answering the question "$\frac{5}{12}$ of what number is 10?"

$\frac{5}{12} \Rightarrow 10$ $\left[\frac{5}{12} \text{ of the unknown number is 10.}\right]$

$\frac{1}{12} \Rightarrow \frac{1}{5} \times 10 = 2$ $\left[\frac{1}{12} \text{ of the unknown number is 2.}\right]$

$\frac{12}{12} \Rightarrow 12 \times 2 = 24$ $\left[\frac{12}{12} \text{ of the whole number is 24.}\right]$

The shelf contains a total of 24 books.

Solution 2

- Use algebra by letting *x* represent the total number of books on the shelf.

- Write an equation that states that the sum of the numbers of math, science, and history books is x:

$$\frac{x}{3} + \frac{x}{4} + 10 = x$$

- To eliminate the fractions from this equation, multiply each member of the equation by 12, which is the LCD of 3 and 4:

$$\begin{aligned}
12\left(\frac{x}{3}\right) + 12\left(\frac{x}{4}\right) + 12(10) &= 12x \\
4x + 3x + 120 &= 12x \\
120 &= 12x - 7x \\
120 &= 5x \\
\frac{120}{5} &= x \\
24 &= x
\end{aligned}$$

Lesson 3-7 Tune-Up Exercises

Multiple-Choice

1 If $\frac{3}{8}$ of a number is 6, what is $\frac{7}{8}$ of the same number?

(A) 8
(B) 12
(C) 14
(D) 16
(E) 24

2 After Claire has read the first $\frac{5}{8}$ of a book, there are 120 pages left to read. How many pages of the book has Claire read?

(A) 160
(B) 200
(C) 240
(D) 300
(E) 320

3 $\frac{4}{9}$ of 27 is $\frac{6}{5}$ of what number?

(A) 5
(B) 10
(C) 12
(D) 15
(E) 20

4 If $\frac{3}{5}$ of a class that includes 10 girls are boys, how many students are in the class?

(A) 15
(B) 20
(C) 21
(D) 25
(E) 30

5 What is the sum of all two-digit whole numbers in which one digit is $\frac{3}{4}$ of the other digit?

(A) 77
(B) 102
(C) 129
(D) 154
(E) 231

6 Boris has D dollars. If he lends $\frac{1}{4}$ of this amount of money and then spends $\frac{1}{3}$ of the money that he has left, how many dollars, in terms of D, does Boris now have?

(A) $\frac{D}{2}$
(B) $\frac{D}{3}$
(C) $\frac{D}{4} - \frac{1}{3}$
(D) $\frac{3}{4}D - \frac{1}{3}$
(E) $\frac{11}{12}D$

7 The value obtained by increasing a by $\frac{1}{5}$ of its value is numerically equal to the value obtained by decreasing b by $\frac{1}{2}$ of its value. Which equation expresses this fact?

(A) $1.2a = 0.5b$
(B) $0.2a = 0.5b$
(C) $0.8a = 1.5b$
(D) $1.2a = 1.5b$
(E) $a + 0.2 = b - 0.5$

8 How many times must a jogger run around a circular $\frac{1}{4}$-mile track in order to have run $3\frac{1}{2}$ miles?

(A) 9
(B) 12
(C) 14
(D) 15
(E) 16

9 A chocolate bar that weighs $\frac{9}{16}$ of a pound is cut into seven equal parts. How much do three parts weigh?

(A) $\frac{21}{112}$ pound

(B) $\frac{27}{112}$ pound

(C) $\frac{16}{63}$ pound

(D) $\frac{47}{63}$ pound

(E) $\frac{85}{112}$ pound

10 At a high school basketball game, $\frac{3}{5}$ of the students who attended were seniors, $\frac{1}{3}$ of the other students who attended were juniors, and the remaining 80 students who attended were all sophomores. How many seniors attended this game?

(A) 120
(B) 175
(C) 180
(D) 210
(E) 300

11 Of the 75 people in a room, $\frac{2}{5}$ are college graduates. If $\frac{4}{9}$ of the students who are not college graduates are seniors in high school, how many people in the room are neither college graduates nor high school seniors?

(A) 15
(B) 20
(C) 25
(D) 36
(E) 40

12 If $\frac{2}{3}$ of $\frac{3}{4}$ of a number is 24, what is $\frac{1}{4}$ of the same number?

(A) 8
(B) 12
(C) 16
(D) 20
(E) 24

13 A man paints $\frac{3}{4}$ of a house in 2 days. If he continues to work at the same rate, how much more time will he need to paint the rest of the house?

(A) $\frac{1}{4}$ day

(B) $\frac{1}{2}$ day

(C) $\frac{2}{3}$ day

(D) 1 day

(E) $\frac{4}{3}$ days

14 A water tank is $\frac{3}{5}$ full. After 12 gallons are poured out, the tank is $\frac{1}{3}$ full. When the tank is full, how many gallons of water does it hold?

(A) 25
(B) 32
(C) 35
(D) 42
(E) 45

15 In a school election, Susan received $\frac{2}{3}$ of the ballots cast, Mary received $\frac{1}{5}$ of the remaining ballots, and Bill received all of the other votes. If Bill received 48 votes, how many votes did Susan receive?

(A) 75
(B) 90
(C) 120
(D) 150
(E) 180

16 After Arlene pumps gas into the gas tank of her car, the gas gauge moves from exactly $\frac{1}{8}$ full to exactly $\frac{7}{8}$ full. If the gas costs $1.50 per gallon and Arlene is charged $18.00 for the gas, what is the capacity, in gallons, of the gas tank?

(A) 24
(B) 20
(C) 18
(D) 16
(E) 15

17 At the beginning of the day, the prices of stocks A and B are the same. At the end of the day, the price of stock A has increased by $\frac{1}{10}$ of its original price and the price of stock B has decreased by $\frac{1}{10}$ of its original price. The new price of stock A is what fraction of the new price of stock B?

(A) $\frac{2}{10}$
(B) $\frac{9}{11}$
(C) $\frac{9}{10}$
(D) $\frac{11}{10}$
(E) $\frac{11}{9}$

18 After $\frac{3}{4}$ of the people in a room leave, three people enter the same room. The number of people who are now in the room, assuming no other people enter or leave, is $\frac{1}{3}$ of the original number of people who were in the room. How many people left the room?

(A) 9
(B) 18
(C) 24
(D) 27
(E) 36

Grid-In

1 After a number is increased by $\frac{1}{3}$ of its value, the result is 24. What was the original number?

2 In an election, $\frac{1}{2}$ of the male voters and $\frac{2}{3}$ of the female voters cast their ballots for candidate *A*. If the number of female voters was $1\frac{1}{2}$ times the number of male voters, what fraction of the total number of votes cast did candidate *A* receive?

LESSON

3-8 PERCENT

OVERVIEW

Since there are 100 *cents in* 1 *dollar,* 15 *cents represents* 15 *percent of a dollar.* ***Percent*** *means the number of* hundred*ths or the number of parts out of* 100. *Instead of writing the word* percent, *you can use the symbol* %. *Thus,*

$$P\% = \frac{P}{100}$$

REWRITING FRACTIONS AND DECIMALS AS PERCENTS

Since fractions, decimals, and percents all represent parts of a whole, you can change from one of these types of numbers to another. Fractions and decimals greater than 1 are equivalent to percents greater than 100.

- To rewrite a decimal as a percent, move the decimal point two places to the right and add the percent sign (%). For example:

$$0.45 = 45\% \qquad 0.08 = 8\% \qquad 1.5 = 1.50 = 150\%$$

- To rewrite a fraction as a percent, first change the fraction to a decimal. Then change the decimal to a percent. For example:

$$\frac{3}{4} = 0.75 = 75\% \qquad \frac{5}{2} = 2.50 = 250\% \qquad 4 = 4.00 = 400\%$$

REWRITING PERCENTS AS DECIMALS AND AS FRACTIONS

You may need to express a percent as a decimal or as a fraction.

- To rewrite a percent as a decimal, drop the percent symbol and divide the number that remains by 100. For example:

$$85\% = \frac{85}{100} = 0.85 \qquad \frac{3}{4}\% = \frac{3}{4} \div 100 = \frac{0.75}{100} = 0.0075$$

- To rewrite a percent as a fraction, drop the percent symbol and make the number that remains the numerator of a fraction whose denominator is 100. For example:

$$27\% = \frac{27}{100} \qquad 7\frac{1}{2}\% = \frac{7\frac{1}{2}}{100} = \frac{7.5}{100} = \frac{7.5 \times 10}{100 \times 10} = \frac{75}{1000}$$

FINDING THE PERCENT OF A NUMBER

To find the percent of a given number:

- Rewrite the percent as a decimal.
- Use your calculator to multiply the given number by the decimal form of the percent.

EXAMPLE

What is 15% of 80?

Solution

Rewrite 15% as 0.15. Then multiply:

$$15\% \text{ of } 80 = 0.15 \times 80 = 12$$

FINDING AN AMOUNT AFTER A PERCENT CHANGE

If a 20% tip is left on a restaurant bill of $80, then to find the total amount of the bill including the tip, do the following:

- STEP 1: Find the amount of the tip: $80 × 0.20 = $16
- STEP 2: Add the tip to the bill: $80 + $16 = $96

Time Saver

Find the final result of increasing or decreasing a number by a percent in one step by multiplying the number by the *total* percentage:

- If 80 is *increased* by 20%, the total percentage is 100% + 20% = 120%, so the final amount is 80 × 1.20 = 96.
- If 50 is decreased by 30%, the total percentage is 100% – 30% = 70%, so the final amount is 50 × 0.70 = 35.

EXAMPLE

If the length of a rectangle is increased by 30% and its width is increased by 10%, by what percentage will the area of the rectangle be increased?

(A) 33%
(B) 37%
(C) 40%
(D) 43%
(E) 45%

Solution

Pick easy numbers for the length and width of the rectangle. Assume the length and width of the rectangle are each 10 units, so the area of the original rectangle is 10 × 10 = 100 square units.

- The total percent increase of the length is 100% + 30% = 130%. The length of the new rectangle is 10 × 1.3 = 13.
- The total percent increase of the width is 100% + 10% = 110%. The width of the new rectangle is 10 × 1.1 = 11.
- The area of the new rectangle is 13 × 11 = 143 square units. Compared to the original area of 100 square units, this is a 43% increase.

The correct choice is **(D)**.

FINDING THE PERCENT OF INCREASE OR DECREASE

When a quantity goes up or down in value, the percent of change can be calculated by comparing the amount of the change to the original amount.

$$\text{Percent of change} = \frac{\text{Amount of change}}{\text{Original amount}} \times 100\%$$

EXAMPLE

If the price of an item increases from $70 to $84, what is the percent of increase in price?

Solution

The original amount is $70, and the amount of increase is $84 – $70 = $14.

$$\begin{aligned}\text{Percent of change} &= \frac{\text{Amount of increase}}{\text{Original amount}} \times 100\% \\ &= \frac{14}{70} \times 100\% \\ &= \frac{1}{5} \times 100\% \\ &= 20\%\end{aligned}$$

FINDING AN ORIGINAL AMOUNT AFTER A PERCENT CHANGE

If you know the number that results after a given number is increased by P%, you can find the original number by dividing the new amount by the total percentage.

$$\text{Original amount} = \frac{\text{New amount after an increase of } P\%}{100\% + P\%}$$

Similarly, if you know the number that results after a given number is decreased by $P\%$, you can find the original number by dividing the new amount by the total percentage.

$$\text{Original amount} = \frac{\text{New amount after a decrease of } P\%}{100\% - P\%}$$

EXAMPLE

A pair of tennis shoes cost $48.60 including sales tax. If the sales tax rate is 8%, what is the cost of the tennis shoes before the tax is added?

Solution

The total percentage is $100\% + 8\% = 108\%$.

$$\text{Cost of tennis shoes} = \frac{\text{Cost with tax included}}{\text{Total percentage}}$$

$$= \frac{48.60}{108\%}$$

$$= \frac{48.60}{1.08}$$

Use a calculator to divide: $= 45$

The tennis shoes cost $45 without tax.

Lesson 3-8 Tune-Up Exercises

Multiple-Choice

1 What is 20% of $\frac{2}{3}$ of 15?

(A) 1
(B) 2
(C) 3
(D) 4
(E) 5

2 Which expression is equivalent to $\frac{2}{5}\%$?

(A) 0.40
(B) 0.04
(C) 0.004
(D) 0.0004
(E) 0.00004

3 The fraction $\frac{0.25 + 0.15}{0.50}$ is equivalent to what percent?

(A) 20%
(B) 25%
(C) 40%
(D) 60%
(E) 80%

4 In a movie theater, 480 of the 500 seats were occupied. What percent of the seats were NOT occupied?

(A) 0.4%
(B) 2%
(C) 4%
(D) 20%
(E) 40%

5 After 2 months on a diet, John's weight dropped from 168 pounds to 147 pounds. By what percent did John's weight drop?

(A) $12\frac{1}{2}\%$
(B) $14\frac{2}{7}\%$
(C) 21%
(D) 25%
(E) $28\frac{4}{7}\%$

6 Which expression is equivalent to 0.01%?

(A) $\sqrt{\frac{1}{100}}$
(B) $\frac{0.01}{10}$
(C) $\frac{1}{100}$
(D) $\frac{\frac{1}{100}}{\frac{1}{10}}$
(E) $\frac{1}{10^4}$

7 If the result of increasing a by 300% of a is b, then a is what percent of b?

(A) 20%
(B) 25%
(C) $33\frac{1}{3}\%$
(D) 40%
(E) $66\frac{2}{3}\%$

8 After a 20% increase, the new price of a radio is \$78.00. What was the original price of the radio?

(A) \$15.60
(B) \$60.00
(C) \$62.40
(D) \$65.00
(E) \$97.50

9 After a discount of 15%, the price of a shirt is \$51. What was the original price of the shirt?

(A) \$44.35
(B) \$58.65
(C) \$60.00
(D) \$64.00
(E) \$65.00

10 Three students use a computer for a total of 3 hours. If the first student uses the computer 28% of the total time, and the second student uses the computer 52% of the total time, how many minutes does the third student use the computer?

(A) 24
(B) 30
(C) 36
(D) 42
(E) 50

11 What is 20% of 25% of $\frac{4}{5}$?

(A) 0.0025
(B) 0.004
(C) 0.005
(D) 0.04
(E) 0.05

12 In a factory that manufactures light bulbs, 0.04% of the bulbs manufactured are defective. It is expected that there will be one defective light bulb in what number of bulbs that are manufactured?

(A) 2500
(B) 1250
(C) 1000
(D) 500
(E) 250

13 A discount of 25% on the price of a pair of shoes, followed by another discount of 8% on the new price of the shoes, is equivalent to a single discount of what percent of the original price?

(A) 17%
(B) 29%
(C) 31%
(D) 33%
(E) 35%

14 If 8 kilograms of alcohol are added to 17 kilograms of pure water, what percent by weight of the resulting solution is alcohol?

(A) 8%
(B) 17%
(C) 32%
(D) 68%
(E) 75%

15 If a is 40% of b, then b exceeds a by what percent of a?

(A) 60%
(B) 100%
(C) 140%
(D) 150%
(E) 250%

Grid-In

1 A store offers a 4% discount if a consumer pays cash rather than paying by credit card. If the cash price of an item is $84.00, what is the credit-card purchase price of the same item?

2 During course registration, 28 students enroll in a certain college class. After three boys are dropped from the class, 44% of the class consists of boys. What percent of the original class did girls comprise?

ANSWERS TO CHAPTER 3 TUNE-UP EXERCISES

Lesson 3-1 (*Multiple-Choice*)

1. **(D)** If $a = 9 \times 23$ and $b = 9 \times 124$, then
$$b - a = 9 \times 124 - 9 \times 23$$
You can use your calculator to do the arithmetic on the right side of the equation, but it's easier to use the reverse of the distributive law:
$$\begin{aligned} b - a &= 9(124 - 23) \\ &= 9(101) \\ &= 909 \end{aligned}$$
2. **(B)** To figure out by what amount quantity *A* exceeds quantity *B*, calculate $A - B$:
$$(8 \times 25) - (15 \times 10) = 200 - 150 = 50$$
3. **(C)** Since 1 quart = 32 fluid ounces, 5 quarts = $5 \times 32 = 160$ fluid ounces. If each container holds 16 fluid ounces, then $\frac{160}{16} = 10$ containers are needed to hold 5 quarts of milk.
4. **(B)** *Solution 1*: If the current odometer reading of a car is 31,983 miles, then the next mileage reading in which at least four digits are the same will be 32,222. Hence, the least number of miles that the car must travel before the odometer displays four digits that are the same is 32,222 − 31,983 = 239.
Solution 2: Find the sum of 31,983 and the number given in each answer choice beginning with (A). Stop when you get a sum (31,983 + 239) in which at least four of the digits are the same. Hence, the correct choice is (B).
5. **(C)** In store *A* 10 scarfs cost $10 \times \$12 = \$120$. Since the same scarf costs \$8 in store *B*, $\frac{120}{8} = 15$ scarfs can be bought in store *B* with \$120.
6. **(B)** If $r * 0 = r$, then $*$ represents either addition or subtraction or both. It is also given that $r * r = 0$. Since $r - r = 0$ and $r + r \neq 0$ when r is not 0, $*$ can represent only subtraction.
7. **(D)** Since Kurt has saved \$160 to buy a stereo system that costs \$400, he needs to earn an additional \$400 − \$160 = \$240. Earning \$8 an hour, he will need to work $\frac{240}{8} = 30$ hours to have enough money to buy the stereo system.
8. **(E)** Since 24 hours + 12 hours + 3 hours = 39 hours, break down the problem by figuring out the time 24 hours ago, 12 hours before that time, and then 3 hours earlier. If the present time is exactly 1:00 P.M., then 24 hours ago it was also 1:00 P.M., and 12 hours before that time it was 1:00 A.M. Three hours before 1:00 A.M. the time was 10:00 P.M.
9. **(A)** If $r \# 0 = r$, then the symbol # can represent either addition or subtraction, but not multiplication or division. Since it is also given that $r \# s = s \# r$, the operation # is commutative. Since addition is commutative and subtraction is not commutative, the symbol # represents only addition.
10. **(D)** Each product contains 6, so 6 can be ignored. Then $w = 36$, $x = 35$, and $y = 32$, so $y < x < w$.
11. **(B)** If $2x + y < 29$, the expression $x - y$ will have its greatest possible value when y has its least possible value and x has its maximum value.
    - Since y is an integer and $y > 4$, the least possible value for y is 5.
    - If $y = 5$, then $2x + y < 29$ becomes $2x + 5 < 29$ so $2x < 24$ and $x < 12$.
    - Since x is an integer and $x < 12$, the maximum value of x is 11.
    - Hence, $11 - 5 = 6$ is the greatest possible value of $x - y$.

(*Grid-In*)

1. **158** In general, if *A* and *B* are positive integers, then the number of integers from *A* to *B* is $(B - A) + 1$. If the number of houses in a certain community are numbered consecutively from 2019 to 2176, there are $(2176 - 2019) + 1 = 157 + 1 = 158$ houses in the community.

2. **5** Since 1 kilobyte is equivalent to 1024×8 or 8192 bits, 40,960 bits are equivalent to $\frac{40,960}{8192}$ or 5 kilobytes.

3. **45.8** The balance that needs to be paid off is \$495 − \$129 or \$366. Since eight equal monthly payments will be made, each monthly payment is $\frac{\$366}{8}$ or \$45.75. Since 45.75 will not fit the grid, grid in 45.8.

4. **4** The fraction will have its largest value when $\frac{k-p}{m}$ has its greatest value and m has its least value. The largest value of $k - p$ is $21 - 9$ or 12. The inequality $2 < m < 6$ means that m is greater than 2 but less than 6. Since m is an integer, the least value of m is 3. Hence, the largest possible value of $\frac{k-p}{m}$ is $\frac{12}{3}$ or 4.

5. **27** For some fixed value of x, $9(x + 2) = y$. If the value of x is increased by 3, then the value of y is increased by $9 \times 3 = 27$. Since after x is increased by 3, $9(x + 2) = w$, the value of $w - y$ is 27.

6. **9** To find the sum of all possible values of x given $3x + 2y = 21$, where x and y are positive integers, plug successive integer values for x starting with 1 into the given equation and note which values of x produce positive integer values for y.

x	$y = \frac{21 - 3x}{2}$	
1	$y = \frac{21 - 3}{2} = \frac{18}{2} = 9$	
2	$y = \frac{21 - 3(2)}{2} = \frac{15}{2}$	
3	$y = \frac{21 - 3(3)}{2} = \frac{12}{2} = 6$	
4	$y = \frac{21 - 3(4)}{2} = \frac{9}{2}$	
5	$y = \frac{21 - 3(5)}{2} = \frac{6}{2} = 3$	
6	$y = \frac{21 - 3(6)}{2} = \frac{3}{2}$	
7	$y = \frac{21 - 3(7)}{2} = \frac{0}{2}$	Stop!

Any value of x greater than 7 will produce a negative value for y. Since $x = 1$, 3, and 5 produce positive integer values for y, the sum of all possible values of x is $1 + 3 + 5 = 9$.

LESSON 3-2 (*Multiple-Choice*)

1. **(E)** Since $\frac{x^{23}}{x^m} = x^{15}$, $x^m = \frac{x^{23}}{x^{15}} = x^{23-15} = x^8$, so $m = 8$. If $\left(x^4\right)^n = x^{20}$, then $4n = 20$, so $n = \frac{20}{4} = 5$. Thus, $mn = 8 \times 5 = 40$.

2. **(A)** Given $x = 32 - 16 \div 2 \times 4$, find the value of x by working from left to right, doing any divisions and multiplications before any additions or subtractions:
$$\begin{aligned} x &= 32 - 16 \div 2 \times 4 \\ &= 32 - 8 \times 4 \\ &= 32 - 32 \\ &= 0 \end{aligned}$$

3. **(B)** If $1 < x^2 < 50$, then $\sqrt{1} < \sqrt{x^2} < \sqrt{50}$, which can be written as $1 < x < \sqrt{50}$. Since $\sqrt{50}$ is between 7 and 8, there are six integer values of x that are greater than 1 but less than $\sqrt{50}$: 2, 3, 4, 5, 6, and 7.

4. **(C)** If $2 = p^3$, then
$$\begin{aligned} (2)^3 &= (p^3)^3 \\ 8 &= p^{3\times3} \\ &= p^9 \end{aligned}$$
Hence, $8p = p^9 \times p = p^{10}$.

5. **(B)** To divide powers with the *same* base, keep the base and *subtract* the exponents. If $5 = a^x$, then
$$\begin{aligned} \frac{5}{a} &= \frac{a^x}{a} \\ &= a^{x-1} \end{aligned}$$

6. **(D)** If $b^3 = 4$, then
$$b^6 = (b^3)^2 = (4)^2 = 16$$

7. **(B)** If w is a positive number and $w^2 = 2$, then $w = \sqrt{2}$, so
$$w^3 = w^2 \bullet w = 2\sqrt{2}$$

8. **(D)** If powers of the same base are equal, then their exponents must be equal. Since $8^{x+1} = 64 = 8^2$, then $x + 1 = 2$, so $x = 1$.

To find the value of 3^{2x+1}, replace x with 1:

$$3^{2x+1} = 3^{2(1)+1} = 3^3 = 27$$

9. **(D)** Test each choice in turn, using particular values for x and y. Since it is given that $0 < y < x$, let $y = 1$ and $x = 4$. The only statement that is true is (D), $\sqrt{xy} = (\sqrt{x})(\sqrt{y})$, since

$$\sqrt{xy} = \sqrt{4 \cdot 1} = \sqrt{4} = 2$$

and

$$(\sqrt{x})(\sqrt{y}) = (\sqrt{4})(\sqrt{1}) = (2)(1) = 2$$

10. **(E)** Given $y = wx^2$ and y is not 0. Since the values of x and w are each doubled, replace w with $2w$ and x with $2x$ in the original equation:

$$y_{new} = (2w)(2x)^2 = (2w)(4x^2) = 8(wx^2) = 8y$$

Hence, the original value of y is multiplied by 8.

11. **(B)** If $\sqrt{n}$ is a positive integer, then n must be a perfect square integer. The perfect square integers in the interval $100 < n < 199$ are as follows:

$121 (= 11^2)$, $144 (= 12^2)$, $169 (= 13^2)$, $196 (= 14^2)$

Hence, there are four perfect square integers in the given interval.

12. **(D)** If $(2^3)^2 = 4^p$, you can find p by expressing each side of the equation as a power of the same base and then setting the exponents of the two bases equal:

$$(2^3)^2 = 4^p$$
$$2^6 = 2^{2p}$$
$$6 = 2p$$
$$3 = p$$

Since $p = 3$, then

$$3^p = 3^3 = 3 \times 3 \times 3 = 27$$

13. **(E)** Rewrite 6^a as $(3 \cdot 2)^a$. Then use the laws of exponents:

$$3^{a+b} \cdot 6^a = 3^{a+b} \cdot (3 \cdot 2)^a = 3^{a+b} \cdot 3^a \cdot 2^a = (3^{a+b} \cdot 3^a) \cdot 2^a = 3^{2a+b} \cdot 2^a$$

14. **(A)** Determine whether each Roman numeral statement is true or false when x is a positive integer.

- I. Since $\left(\frac{x}{x}\right)^{99} = (1)^{99} = 1$ and $\left(\frac{x+1}{x+1}\right)^{100} = (1)^{100} = 1$, statement I is true.
- II. Since $(x^x)^2 = x^{2x}$ and x^{x^2} is not equal to x^{2x} for all positive integer values of x, statement II is false.
- III. Since $\frac{x^{100}}{x^{99}} = x^{100-99} = x^1 = x$ and $1^x = 1$, statement III is false.

Since only Roman numeral statement I is true, the correct choice is (A).

15. **(A)** If $y = 25 - x^2$, the smallest possible value of y is obtained by subtracting the largest possible value of x^2 from 25. Since $1 \leq x \leq 5$, the largest possible value of x^2 is $5^2 = 25$. When $x^2 = 25$, then $y = 25 - 25 = 0$.

16. **(B)** If $x = \frac{\$0.68}{4}$ and $y^2 = 12$, then $y = \sqrt{12}$, so

$$\frac{4}{xy} = \frac{4}{\sqrt{6}\sqrt{12}} = \frac{4}{\sqrt{72}} = \frac{4}{\sqrt{36}\sqrt{2}} = \frac{4}{6\sqrt{2}} = \frac{2}{3\sqrt{2}} \cdot \frac{\sqrt{2}}{\sqrt{2}} = \frac{2\sqrt{2}}{3 \cdot 2} = \frac{\sqrt{2}}{3}$$

(Grid-In)

1. **2** Given $2^4 \times 4^2 = 16^x$, find the value of x by expressing each side of the equation as a power of the same base.

$$2^4 \times (2^2)^2 = (2^4)^x$$
$$2^4 \times 2^4 = 2^{4x}$$
$$2^{4+4} = 2^{4x}$$
$$2^8 = 2^{4x}$$
$$8 = 4x, \text{ so } x = 2$$

2. **707** Since $\frac{a^6}{b} = 11$, $a^6 = 11b$. Thus,

$$\begin{aligned} a^7 &= a \times a^6 = 7777 \\ a \times (11b) &= 7777 \\ 11ab &= 7777 \\ \frac{11ab}{11} &= \frac{7777}{11} \\ ab &= 707 \end{aligned}$$

3. **16** Since $2^3 = 2 \times 2 \times 2 = 8$ and $(y-1)^3 = 8$, then $y - 1 = 2$, so $y = 3$. Hence, $(y+1)^2 = (3+1)^2 = 4^2 = 4 \times 4 = 16$
4. **1/2** Since

$$\frac{p+p+p}{p \cdot p \cdot p} = \frac{3p}{p \cdot p \cdot p} = \frac{3}{p^2} = 12$$

then $\frac{p^2}{3} = \frac{1}{12}$, so $p^2 = \frac{3}{12} = \frac{1}{4}$. Hence, $p = \frac{1}{2}$ since $\frac{1}{2} \times \frac{1}{2} = \frac{1}{4}$. Grid in as 1/2.

Lesson 3-3 (*Multiple-Choice*)

1. (**A**) Find the value of $P - S$:
   - If S represents the sum of the factors of 18, then
     $S = 1 + 2 + 3 + 6 + 9 + 18 = 39$
   - If P represents the sum of the prime numbers less than 18, then
     $P = 2 + 3 + 5 + 7 + 11 + 13 + 17 = 58$
   - Hence, $P - S = 58 - 39 = 19$.
2. (**D**) In any division example, the divisor times the quotient plus the remainder should equal the dividend. If the quotient of N divided by 5 is 13 and the remainder is 4, then $\frac{N}{5} = 13 + \frac{4}{5}$, so

   $N = (5 \times 13) + 4 = 65 + 4 = 69$
3. (**D**) *Solution 1*: If p is divisible by 3 and q is divisible by 4, then pq must be divisible by any combination of prime factors of 3 and 4. Since $3 = 3 \times 1$ and $4 = 2 \times 2$, pq is divisible by each of the following: 3, 4, 3×2 or 6, and $3 \times 2 \times 2$ or 12. Since no product of prime factors of 3 and 4 equals 9, pq cannot be divisible by 9.

   Solution 2: Pick numbers for p and q. Then test each choice until you find a number that does not divide pq evenly. For example, if $p = 6$ and $q = 8$, then $pq = 48$. Testing each answer choice, you find that 48 is divisible by 3, 4, and 6, but not by 9.
4. (**E**) When figuring out how many factors a number has, be sure to include the number itself and 1. Try each choice in turn:
   - (A) There are 5 factors of 12: 1, 3, 4, 6, and 12.
   - (B) There are 6 factors of 18: 1, 2, 3, 6, 9, and 18.
   - (C) There are 3 factors of 25: 1, 5, and 25.
   - (D) There are 8 factors of 70: 1, 2, 5, 7, 10, 14, 35, and 70.
   - (E) There are 9 factors of 100: 1, 2, 4, 5, 10, 20, 25, 50, and 100.

   Hence, 100 has the most factors. The correct choice is (E).
5. (**B**) *Solution 1*: A number is divisible by 3 if the sum of its digits is divisible by 3. In choice (B), the sum of the digits of 4308 is $4 + 3 + 0 + 8 = 15$, which is divisible by 3. In choice (E), the sum of the digits of 23,451 is $2 + 3 + 4 + 5 + 1 = 15$, so (E) is also divisible by 3. A number is divisible by 2 only if its last digit is even. Since the last digit of 4308 is even but the last digit of 23,451 is odd, the correct choice must be (B).

   Solution 2: Using a calculator, test each choice in turn to find a number that gives a 0 remainder when divided by 2 and by 3.
6. (**D**) *Solution 1*: Since 3 and 10 do not have any common factors other than 1, any number that is divisible by both 3 and 10 must be divisible by their product, 30. Any number that is divisible by 30 must also be divisible by any factor of 30. Since 15 is the only answer choice that is a factor of 30, any number that is divisible by both 3 and 10 must also be divisible by 15.

   Solution 2: Pick an easy number that is divisible by both 3 and 10, say 30. Then divide 30 by each of the answer choices in turn. Stop when you find a number (15) that divides evenly into 30. The correct choice is (D).

7. **(C)** Since x represents any even number and y represents any odd number, let $x = 2$ and $y = 3$. Evaluate the expression in each of the answer choices until you find one that produces an even number.
   - (A) $y + 2 = 3 + 2 = 5$
   - (B) $x - 1 = 2 - 1 = 1$
   - (C) $(x + 1)(y - 1) = (2 + 1)(3 - 1) = (3)(2) = 6$. There is no need to go further. The correct choice is (C).
8. **(D)** Break down 105 into its prime factors:
   $$\frac{105}{p} = \frac{1 \times 5 \times 21}{p} = \frac{1 \times 3 \times 5 \times 7}{p}$$
   Thus, when p equals any of the eight positive integers 1, 3, 5, 7, 15 (3×5), 21 (3×7), 35 (5×7), or 105, $\frac{105}{p}$ is an integer.
9. **(A)** Since n is an odd integer, let $n = 3$. Evaluate each answer choice in turn until you find an odd number. For choice (A),
   $$(2n - 1)^2 = (2(3) - 1)^2 = (6-1)^2 = 25 = 2(3) - 1 = 5$$
   There is no need to go further. The correct choice is (A).
10. **(D)** Consecutive multiples of 4, such as 4, 8, and 12, always differ by 4. If $k - 1$ is a multiple of 4, then the next larger multiple of 4 is obtained by adding 4 to $k - 1$, which gives $k - 1 + 4$ or $k + 3$.
11. **(D)** You are told that, after m marbles are put into n jars, each jar contains the same number of marbles, with two marbles remaining. If x represents the number of marbles put into each jar, m divided by n equals x with a remainder of 2. This statement can be written as
   $$\frac{m}{n} = x + \frac{2}{n}$$
   Since
   $$x = \frac{m}{n} - \frac{2}{n} = \frac{m - 2}{n},$$
   $\frac{m-2}{n}$ marbles were put into each jar.
12. **(C)** You are given that, when p is divided by 4, the remainder is 3, and when p is divided by 3, the remainder is 0. Identify the answer choices that are divisible by 3. Then substitute each of these choices for p until you find the choice that produces a remainder of 3 when divided by 4.
   Choices (C), (D), and (E) are each divisible by 3. For choice (C), let $p = 15$. When 15 is divided by 4, the remainder is 3.
   There is no need to go further. The correct choice is (C).
13. **(D)** Determine whether each Roman numeral statement is true or false.
   - I. Subtracting any multiple of 2 from an odd integer always produces another odd integer. For example, $7 - 4 = 3$. Since $n^2 + 5$ is an odd integer and $(n^2 + 5) - 6 = n^2 - 1$, $n^2 - 1$ is also an odd integer. Hence, statement I is false.
   - II. The sum of an even integer and an odd integer is an odd integer. Since $n^2 + 5$ is an odd integer, n^2 must be even, so n must also be even. Hence, statement II is true.
   - III. The product of an odd integer and an even integer is an even integer. Since you have determined that n is even, $5n$ is also even. Hence, statement III is true.

   Since only Roman numeral statements II and III are true, the correct choice is (D).
14. **(D)** You are told that, if the number of people who contribute equally to a gift decreases from four to three, each person must pay an additional \$10. You can eliminate choices (A) and (C), which are not divisible by both 3 and 4. Check each of the remaining choices until you find the right one.
   - (B) 60 divided by 3 is 20, and 60 divided by 4 is 15. The difference between 20 and 15 is 5.
   - (D) 120 divided by 3 is 40, and 120 divided by 4 is 30. The difference between 40 and 30 is 10.

   There is no need to test the last choice. The correct choice is (D).
15. **(B)** Since n is any even integer, pick a simple number for n. If $n = 4$, then $(n+1)^2 = (4+1)^2 = 25$. When 25 is divided by 4, the remainder is 1.
16. **(C)** You need to find the answer choice that produces remainders of 2 and 4 when divided by 3 and 5, respectively.

- (A) The remainder when 41 is divided by 3 is 2, and the remainder when 41 is divided by 5 is 1.
- (B) The remainder when 43 is divided by 3 is 1.
- (C) The remainder when 44 is divided by 3 is 2, and the remainder when 44 is divided by 5 is 4.

There is no need to continue. The correct choice is (C).

(*Grid-In*)

1. **2** When a is divided by 7, the remainder is 5, so let $a = 12$. When b is divided by 7, the remainder is 4, so let $b = 11$. Then $a + b = 23$. When 23 is divided by 7, the remainder is 2.
2. **2001** All multiples of 3 from -3000 to 3000 are divisible by 3. Since
$$3 = 1 \times 3, 6 = 2 \times 3, 9 = 3 \times 3, \ldots,$$
$$3000 = 1000 \times 3$$
there are 1000 multiples of 3 from 3 to 3000, inclusive. Similarly, there are 1000 multiples of 3 from -3000 to -3. Since 0 is also divisible by 3, there are $1000 + 1000 + 1$ or 2001 integers from -3000 to 3000, inclusive, that are divisible by 3.
3. **2** When a positive integer k is divided by 6, the remainder is 1, so let $k = 7$. Then $5k = 35$. When 35 is divided by 3, the remainder is 2.
4. **3** To find the least positive integer p for which $2^{2p} + 1$ is not a prime number, plug in consecutive integer values for p starting with 1 until you find one that does not make $2^{2p} + 1$ a prime number.
   - For $p = 1, 2^{2p} + 1 = 2^2 + 1 = 4 + 1 = 5$, which is prime.
   - For $p = 2, 2^{2p} + 1 = 2^4 + 1 = 16 + 1 = 17$, which is prime.
   - For $p = 3, 2^{2p} + 1 = 2^6 + 1 = 64 + 1 = 65$, which is not prime since 65 is divisible by 5.

Lesson 3-4 (*Multiple-Choice*)

1. **(E)** Multiplying both sides of the given equation, $2b = -3$, by -2:
$$-2(2b) = -2(-3)$$
Since $-4b = 6$, then $1 - 4b = 1 + 6 = 7$.
2. **(C)** Using the given facts that a is a negative integer and b is a positive integer, determine whether each Roman numeral statement is true or false.
   - I. $a + b$ may be negative, 0, or positive, depending on the particular values of a and b. For example, if $a = -2$ and $b = 2$, $a + b = 0$. If, however, $a = -3$, and $b = 4$, $a + b > 0$. Statement I is false.
   - II. $\frac{b-a}{a} = \frac{b}{a} - \frac{a}{a} = \frac{b}{a} - 1$. Since a positive integer divided by a negative integer is negative, $\frac{b}{a}$ is negative, making $\frac{b}{a} - 1$ negative. Hence $\frac{b-a}{a} < 0$. Statement II is true.
   - III. When b is an even integer, $a^b > 0$. Statement III, $a^b < 0$, is not always true.

   Hence, only statement II must be true.
3. **(D)** The product of five negative numbers is negative. Four negative numbers yield
$$\underbrace{(-) \times (-)}_{(+)} \times \underbrace{(-) \times (-)}_{(+)} \times (+) = (+)$$
Thus, four numbers in the product, at most, could be negative.
4. **(A)** If k equals 0, 2, or -1, $k(k-2)(k+1)$ evaluates to 0. Hence, you can eliminate answer choices (B), (C), and (D). Plug each of the two remaining answer choices into $k(k-2)(k+1)$ to see which results in a negative value. For answer choice (A), $k = -2$:
$$\begin{aligned} k(k-2)(k+1) &= -2(-2-2)(-2+1) \\ &= -2(-4)(-1) \\ &= -8 \end{aligned}$$
The correct choice is (A).
5. **(C)** $p^2(2-5) + (-p)^2 = -3p^2 + p^2 = -2p^2$
6. **(C)** If $n + 5$ is an odd integer, then n could be -2 since $-2 + 5 = 3$.
7. **(D)** Determine whether each Roman numeral statement is true or false when $a < 0$ and $b > 0$.
   - I. $a + b$ may be negative, 0, or positive, depending on the particular values of a

and b. For example, if $a = -2$ and $b = 2$, then $a + b = 0$. If $a = -3$ and $b = 2$, then $a + b < 0$. Hence, statement I is false.

- II. Pick numbers for a and b. If $a = -3$ and $b = 2$, then $b - a = 2 - (-3) = 2 + 3 = 5$. Since $b - a > 0$, statement II is true.
- III. Since the square of a negative number is positive, $a\left(\frac{a}{b}\right) = \frac{a^2}{b} > 0$, so statement III is true.

Only Roman numeral statements II and III are true.

8. **(D)** Since $(-2)^3 = (-2)\times(-2)\times(-2) = -8$ and $(-3)^2 = (-3)\times(-3) = 9$, then

$$(-2)^3 + (-3)^2 = -8 + 9 = 1$$

Questions 9 and 10.

From the diagram, $a < -1$, $-1 < b < 0$, $0 < c < 1$, and $d > 1$.

9. **(B)** Determine whether each Roman numeral statement is true or false.
   - I. Pick a number for c. If $c = \frac{1}{2}$, then $c^2 = \frac{1}{2} \times \frac{1}{2} = \frac{1}{4}$. Since $\frac{1}{4} < \frac{1}{2}$, $c^2 < c$. Hence, statement I is true.
   - II. Since $a < -1$, then $a^2 > 1$. For example, if $a = -3$, then $a^2 = 9$. Since $0 < c < 1$, $a^2 > c$. Hence, statement II is true.
   - III. Pick a number for b. If $b = -\frac{1}{2}$, then $\frac{1}{b} = -\frac{2}{1} = -2$. Since $-\frac{1}{2} > -2$, then $b > \frac{1}{b}$. Hence, statement III is false.

   Only Roman numeral statements I and II are true.

10. **(C)** Determine whether each Roman numeral statement is true or false.
    - I. On the basis of the diagram, pick numbers for a and d. If $a = -2$ and $d = 2$, then $ad = -4 < b$, so statement I is false.
    - II. Since ab is the product of two negative numbers, $ab > 0$. Since ad is the product of numbers with opposite signs, $ad < 0$. Since $ab > ad$, statement II is true.
    - III. Since $a < b$, the reciprocals of a and b have the opposite size relationship. Since $\frac{1}{a} > \frac{1}{b}$, statement III is true.

    Hence, only Roman numeral statements II and III are true.

11. **(B)** If $(y-3)^2 = 16$, then the number inside the parentheses must be either 4 or -4. If $y - 3 = 4$, then $y = 7$ and $y^2 = 49$. If $y - 3 = -4$, then $y = 3-4 = -1$, so $y^2 = 1$. The smallest possible value of y^2 is 1.
12. **(E)** It is given that $Y = 1 + 2 + 3 + \ldots + 10$ and $X = (-1) + (-2) + (-3) + \ldots + (-10) = -(1 + 2 + 3 \ldots + 10) = -Y$. Since $X = -Y$:
    - $X + Y = -Y + Y = 0$, which makes Roman numeral choice I false.
    - $Y - X = Y-(-Y) = Y + Y = 2Y$, so Roman numeral choice II must be true.
    - $(X)^2 = (-Y)^2$ or, equivalently, $X^2 = Y^2$, so Roman numeral choice III must be true.
    - Since only Roman numeral choices II and III must be true, the correct choice is (E).
13. **(A)** Rewrite a^2b^3c as $(a^2b^2)bc > 0$. Then determine whether each Roman numeral statement is true or false.
    - I. Since $(a^2b^2)bc > 0$, $a^2 > 0$, and $b^2 > 0$, it must be the case that $bc > 0$. Hence, statement I is true.
    - II. Using $(a^2b^2)bc > 0$, you cannot tell whether ac is positive or negative. Hence, statement II is false.
    - III. In the product $(a^2b^2)bc > 0$, there is no restriction on the signs of a and b, so their product can be positive or negative. Hence, statement III is false.

    Only Roman numeral statement I must be true.
14. **(A)** Since $a^2b^3c^5 = (a^2b^2c^4)bc < 0$ and $a^2b^2c^4$ is always positive, it must be the case that bc is negative.
15. **(B)** Determine whether each Roman numeral statement is true or false when $a \neq 0$.
    - I. $(-a)^2 = (-a) \times (-a) = a^2$ and $a^2 - 2a^2 = -a^2$. Since $(-a)^2 \neq -a^2$, statement I is false.
    - II. Since $-(b-a) = -b-(-a) - b + a = a - b$, statement II is true.
    - III. If $a > 0$, then $a > -a$. However, if $a < 0$, then $a < -a$, so statement III is false.

    Hence, only Roman numeral statement II is always true.

(*Grid-In*)

1. **10** Since $p^2 = 16$, $p = 4$ or -4. Also, $q^2 = 36$, so $q = 6$ or -6. Hence, the largest possible value of $q - p$ is $6 - (-4) = 6 + 4 = 10$.
2. **16** Follow these steps:
   - Find the largest possible value of y. Since x^2 is always nonnegative, the largest possible value of $y = 1 - x^2$ occurs when x^2 has its smallest value. Since $-4 \leq x \leq 2$, the smallest value of x^2 is 0. The *largest* possible value of y is $y = 1 - x^2 = 1 - 0 = 1$.
   - Find the smallest possible value of y. The smallest possible value of $y = 1 - x^2$ occurs when x^2 has its largest value. The largest value of x^2 is $(-4)^2 = 16$. The *smallest* possible value of y is $y = 1 - x^2 = 1 - 16 = -15$.
   - Subtract. The number obtained when the *smallest* possible value of y is subtracted from the *largest* possible value of y is $1 - (-15) = 1 + 15 = 16$.

Lesson 3-5 (*Multiple-Choice*)

1. **(C)** To find the number of minutes that elapse from 4:56 P.M. to 5:32 P.M., subtract 4 hours and 56 minutes from 5 hours and 32 minutes:

$$\begin{array}{r} 5\text{ hours } 32\text{ min} \\ -\ \underline{4\text{ hours } 56\text{ min}} \end{array} \Rightarrow \begin{array}{r} 4\text{ hours } 92\text{ min} \\ -\ \underline{4\text{ hours } 56\text{ min}} \\ 36\text{ min} \end{array}$$

Since

$$\frac{36\text{ minutes}}{1\text{ hour}} = \frac{36}{60} = \frac{12 \times 3}{12 \times 5} = \frac{3}{5}$$

the number of minutes that elapse from 4:56 P.M. to 5:32 P.M. is $\frac{3}{5}$ of an hour.

2. **(B)** If each of the fractions $\frac{3}{k}, \frac{4}{k}, \frac{5}{k}$ is in lowest terms, then k cannot be divisible by 3, 4, and 5. Of the answer choices, only 49 is not divisible by 3, 4, and 5.
3. **(D)** To find the number that has the greatest value, compare each digit, reading from left to right. The numbers in choices (A) and (C) can be eliminated since they are less than any of the other choices. Since the thousandths digit of 0.2938 is greater than the thousandths digit of each of the remaining answer choices, 0.2938 is the largest number.
4. **(C)** Four hours elapse from 11:00 A.M. to 3:00 P.M. on Wednesday of the same day, and $52(= 24 + 24 + 4)$ hours elapse from 11:00 A.M. on Wednesday to 3:00 P.M. on Friday of the same week. To compare these elapsed times, form the fraction

$$\frac{4}{52} = \frac{\overset{1}{\cancel{4}}}{\cancel{4} \times 13} = \frac{1}{13}$$

5. **(D)** In 0.030, the digit 3 is located two places to the *right* of the decimal point, so it is in the hundredths place.
6. **(A)** Since $\frac{9.5}{19} = \frac{1}{2}$, then $\frac{9}{19} < \frac{1}{2}$. Similarly, $\frac{7.5}{15} = \frac{1}{2}$, so $\frac{8}{15} > \frac{1}{2}$. Hence, you can eliminate answer choices (B), (C), (D), and (E). Choice (A) is the correct answer.
7. **(C)** Since $\left(\frac{1}{2}\right)^3 = \frac{1}{8}$, the new value of V is $\frac{1}{8}$ of the original value of V.
8. **(C)** Since each inch on ruler A is marked in equal $\frac{1}{8}$-inch units, a side that measures 12 of these $\frac{1}{8}$-inch units is

$$12 \times \frac{1}{8} = \frac{3}{2} = 1\frac{1}{2} \text{ inches long}$$

If the same side is measured with ruler B, which is marked in equal $\frac{1}{12}$-inch units, the side will measure 18 of these $\frac{1}{12}$-inch units since

$$1\frac{1}{2} = \frac{12}{12} + \frac{6}{12}$$

and $12 + 6 = 18$.

9. **(C)** $60 + 2 + \frac{4}{8} + \frac{3}{500} = 62 + 0.5 + 0.006 = 62.506$.
10. **(B)** Since $N \times \frac{7}{12} = \frac{7}{12} \times \frac{3}{14}$, you can cancel $\frac{7}{12}$ on each side to get $N = \frac{3}{14}$, so $\frac{1}{N} = \frac{14}{3}$.
11. **(E)** If $n = 2.5 \times 10^{25}$, then

$$\begin{aligned} \sqrt{n} &= \sqrt{2.5 \times 10^{25}} \\ &= \sqrt{25 \times 10^{24}} \\ &= \sqrt{25} \times \sqrt{10^{24}} \\ &= 5 \times \sqrt{10^{12} \cdot 10^{12}} \\ &= 5 \times 10^{12} \end{aligned}$$

12. **(A)** Since division by 0 is not allowed, a variable in the denominator of a fraction cannot be equal to a number that makes the denominator evaluate to 0. If $y = \frac{x-2}{x+3}$, then x cannot be equal to -3 since $-3 + 3 = 0$.
13. **(E)** Eight pencils cost \$0.42, and \$2.10 $\div$ 0.42 = 5. Hence, $8 \times 5 = 40$ pencils can be purchased.
14. **(C)** For a total of 48 ounces of orange juice, six 8-ounce containers must be purchased since 6×8 ounces = 48 ounces. If each 8-ounce container costs \$0.69, six of these containers will cost $6 \times \$0.69 = \$4.14$.
For a total of 48 ounces of orange juice, four 12-ounce containers must be purchased since 4×12 ounces = 48 ounces. If each 12-ounce container costs \$0.95, four of these containers will cost $4 \times \$0.95 = \$3.80$. The amount of money that will be saved by purchasing the 12-ounce containers is $\$4.14 - \$3.80 = \$0.34$.
15. **(E)** There are five digits that repeat in order. The remainder obtained by dividing the number of positions to the right of the decimal point by 5 produces the following cyclical pattern.

Nth position to right of decimal point	Remainder when N is divided by 5	Repeating digit in Nth position
1	1	3
2	2	1
3	3	7
4	4	5
5	0	2
6	1	3
7	2	1
8	3	7
9	4	5
10	0	2
...	...	...

Hence, the repeating digit (3, 1, 7, 5, or 2) that is in the *N*th decimal position is the repeating digit from the above table that corresponds to the remainder when *N* is divided by 5. For example, a remainder of 1 corresponds to the repeating digit 3, a remainder of 2 corresponds to the repeating digit 1, and so forth.
Because $968 \div 5 = 193$ remainder 3, the repeating digit 7, obtained from the above table, is in the 968th place to the right of the decimal point.

(*Grid-In*)

1. **8** If $\frac{4\Delta}{6\Delta} + \frac{5}{17} = 1$, then $\frac{4\Delta}{6\Delta} = \frac{12}{17}$ since $\frac{12}{17} + \frac{5}{17} = 1$.
Since $\frac{4\Delta}{6\Delta} = \frac{12}{17}$, 4Δ must be a multiple of 12. Hence, $4\Delta = 48$, so $\Delta = 8$.
2. **24** Since 1.5 inches represents 45 miles, 1 inch represents $\frac{45}{1.5}$ or 30 miles. Hence, 0.8 inch represents 0.8×30 or 24 miles.
3. **7** If four lemons cost \$0.68, one lemon costs $\frac{\$0.68}{4}$ or \$0.17. Since $\frac{\$1.19}{\$0.17} = 7$, there are typically seven lemons in 1 pound of lemons.

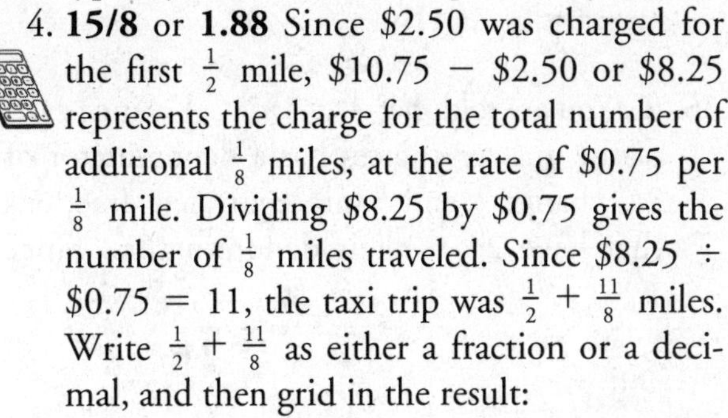

4. **15/8** or **1.88** Since \$2.50 was charged for the first $\frac{1}{2}$ mile, \$10.75 − \$2.50 or \$8.25 represents the charge for the total number of additional $\frac{1}{8}$ miles, at the rate of \$0.75 per $\frac{1}{8}$ mile. Dividing \$8.25 by \$0.75 gives the number of $\frac{1}{8}$ miles traveled. Since \$8.25 $\div$ \$0.75 = 11, the taxi trip was $\frac{1}{2} + \frac{11}{8}$ miles. Write $\frac{1}{2} + \frac{11}{8}$ as either a fraction or a decimal, and then grid in the result:
   - since $\frac{1}{2} + \frac{11}{8} + \frac{4}{8} + \frac{11}{8} = \frac{15}{8}$, grid 15/8;

   or
   - since $\frac{1}{2} + \frac{11}{8} = 0.5 + 1.375 = 1.875$, grid 1.88.
5. **5.6** Since $\frac{288}{1728} = \frac{1}{6}$, 288 cubic inches are equivalent to $\frac{1}{6}$ of a cubic foot. If 1 cubic foot of the metal weighs 8 pounds, then $\frac{1}{6}$ of a cubic foot will weigh

$$\frac{1}{\cancel{6}} \times \overset{4}{\cancel{8}} \text{ or } \frac{4}{3} \text{ pounds}$$

Since each pound of the metal costs \$4.20, the cost of $\frac{4}{3}$ pounds equals

$$\frac{4}{3} \times \$4.20 = 4 \times \$1.40 = \$5.60$$

Omit the dollar sign and grid 5.6

Lesson 3-6 (*Multiple-Choice*)

1. **(D)** To subtract $7\frac{3}{4}$ minutes from $9\frac{1}{3}$ minutes, first change $\frac{3}{4}$ to $\frac{9}{12}$ and $\frac{1}{3}$ to $\frac{4}{12}$:

$$9\frac{4}{12} - 7\frac{9}{12} = 8\frac{16}{12} - 7\frac{9}{12} = 1\frac{7}{12} \text{ minutes}$$

Since 1 minute = 60 seconds, $1\frac{7}{12}$ minutes = 60 + 35 = 95 seconds.

2. **(A)** Since $\frac{1}{2} \div \frac{3}{4} = \frac{1}{2} \times \frac{4}{3} = \frac{2}{3}$, multiplying a number by $\frac{1}{2}$ and then dividing the result by $\frac{3}{4}$ is equivalent to multiplying the original number by $\frac{2}{3}$.

3. **(D)** $\left(\frac{1}{10}\right)^3 + \left(\frac{2}{10}\right)^3 + \left(\frac{3}{10}\right)^3$
$= 0.001 + 0.008 + 0.027$
$= 0.036$

4. **(D)** $\frac{3}{4}$ of $\frac{2}{3}$ of $12 = \frac{\cancel{3}^{1}}{\cancel{4}_{1}} \times \frac{2}{\cancel{3}_{1}} \times \frac{\cancel{12}^{3}}{1} = 6$

5. **(C)** $\frac{2}{21}$ of $3 \times 5 \times 7 =$

$$\frac{\cancel{2}}{\cancel{21}_{1}} \times \cancel{3} \times \cancel{5} \times \cancel{7} = 10$$

6. **(C)** You are told that $\frac{1}{4} < n < \frac{1}{3}$. Since each of the answer choices has a denominator of 24, change $\frac{1}{4}$ and $\frac{1}{3}$ into equivalent fractions that have 24 as their denominators. Since

$$\frac{1}{4} = \frac{1}{4} \cdot \frac{6}{6} = \frac{6}{24}$$

and

$$\frac{1}{3} = \frac{1}{3} \cdot \frac{6}{6} = \frac{6}{18}$$

then $\frac{6}{24} < n < \frac{8}{24}$. A possible value of n is $\frac{7}{24}$.

7. **(C)** It is given that $0 < x < 1$.
   - If $x = \frac{1}{2}, \frac{1}{x} - x = 2 - \frac{1}{2} > 0$ so Roman numeral choice I is false.
   - If $x = \frac{1}{2}, x^2 + x^2 = \left(\frac{1}{2}\right)^2 + \left(\frac{1}{2}\right)^2 = \frac{1}{4} + \frac{1}{4} = \frac{1}{2} = x$ which means Roman numeral choice II is false.
   - Since $\frac{x}{\sqrt{x}} = \frac{\sqrt{x} \cdot \sqrt{x}}{\sqrt{x}} = \sqrt{x}$, Roman numeral choice III is equivalent to $\sqrt{x} > x^2$, which, after squaring both sides of the inequality, is equivalent to $x > x^4$. When $0 < x < 1$, as the exponent of x increases, the power of x decreases. Since it is true that $x > x^4$, Roman numeral choice III must be true, which makes (C) the correct choice.

8. **(C)** If $k = \frac{c}{\frac{a}{b}}$, then

$$k = \frac{c}{\frac{a}{b}} = c \div \frac{a}{b} = c \times \frac{b}{a} = \frac{bc}{a}$$

so $\frac{1}{k} = \frac{a}{bc}$.

9. **(A)** If an item is on sale at $\frac{1}{3}$ off the list price of d dollars, its sale price is $d - \frac{1}{3}d = \frac{2}{3}d$. If the sale price of $\frac{2}{3}d$ is reduced by $\frac{1}{4}$ of that price, then the final sale price is $\frac{2}{3}d - \frac{1}{4}\left(\frac{2}{3}d\right)$. Simplify:

$$\begin{aligned}\frac{2}{3}d - \frac{1}{4}\left(\frac{2}{3}d\right) &= \frac{2}{3}d - \frac{1}{6}d \\ &= \frac{4}{6}d - \frac{1}{6}d \\ &= \frac{3}{6}d \\ &= \frac{d}{2}\end{aligned}$$

10. **(E)** *Solution 1*: If $\frac{a}{b}$ is a fraction less than 1, then its reciprocal is greater than 1. For example, $\frac{3}{4} < 1$ and $\frac{4}{3} > 1$. Since the reciprocal of $\frac{a}{b}$ may be written as

$$\frac{1}{\frac{a}{b}}, \frac{1}{\frac{a}{b}} > 1$$

Two times this value, or $\frac{2}{\frac{a}{b}}$ must also be greater than 1.

Solution 2: Pick numbers for a and b that make $\frac{a}{b} < 1$. For example, let $a = 3$ and $b = 4$. Plug these numbers into each answer choice until you find the one (E) that produces a number greater than 1.

11. **(B)** After Juan gives $\frac{2}{3}$ of his p pencils to Roger, Juan is left with $\frac{p}{3}$ pencils. When he gives $\frac{1}{4}$ of the pencils that he has left to Maria, Juan has $\frac{p}{3} - \frac{1}{4}\left(\frac{p}{3}\right)$ pencils left. Simplify:

$$\begin{aligned}\frac{p}{3} - \frac{1}{4}\left(\frac{p}{3}\right) &= \frac{p}{3} - \frac{p}{12} \\ &= \frac{4p}{12} - \frac{p}{12} \\ &= \frac{3p}{12} \\ &= \frac{1}{4}p\end{aligned}$$

12. **(C)** Determine whether each Roman numeral inequality makes the statement true or false.
   - I and II. If a number that is greater than 1 is raised to a power, then, as the exponent gets larger, the value of that expression also becomes larger. Thus $a^3 < a^2 < a$ is false when $a > 1$, and true when $0 < a < 1$. Hence, inequality I is false, and inequality II is true.
   - III. If $-1 < a < 0$, then $a^2 > 0$ and $a < 0$, so $a^2 > a$. Therefore, $a^3 < a^2 < a$ is not true. Thus, inequality III is false.

   Only Roman numeral inequality II is true.

(*Grid-In*)

1. **3/2** When $x = \frac{1}{2}$,

$$y = \sqrt{\frac{x+4}{2}} = \sqrt{\frac{1}{2}\left(\frac{1}{2}+4\right)}$$
$$= \sqrt{\frac{1}{2}\left(\frac{9}{2}\right)} = \sqrt{\frac{9}{4}} = \frac{3}{2}$$

Grid in as 3/2.

2. **3/4** To find what fraction of a is b, divide b by a, as in

$$\frac{5}{6} \div \frac{10}{9} = \frac{5}{6} \times \frac{9}{10} = \frac{3}{4}$$

Grid in as 3/4.

3. **625** If $25\left(\frac{x}{y}\right) = 4, \frac{x}{y} = \frac{4}{25}$ so $\frac{x}{y} = \frac{25}{4}$.

Hence, $100\left(\frac{y}{x}\right) = 100\left(\frac{25}{4}\right) = 25(25) = 625$.

4. **9** Express the sum inside each set of parentheses as an improper fraction. Then multiply by matching pairs of numerators and denominators that cancel out to 1, except for the denominator of the first fraction and the numerator of the last fraction:

$$p = \left(1+\frac{1}{2}\right)\left(1+\frac{1}{3}\right)\left(1+\frac{1}{4}\right)\cdots\left(1+\frac{1}{16}\right)\left(1+\frac{1}{17}\right)$$
$$= \left(\frac{\not{3}}{2}\right)\left(\frac{\not{4}}{\not{3}}\right)\left(\frac{5}{\not{4}}\right)\cdots\left(\frac{\not{17}}{16}\right)\left(\frac{18}{\not{17}}\right)$$
$$= \left(\frac{1}{2}\right)\left(\frac{1}{1}\right)\left(\frac{1}{1}\right)\cdots\left(\frac{1}{1}\right)\left(\frac{18}{1}\right)$$
$$= \frac{18}{2}$$
$$= 9$$

Lesson 3-7 (*Multiple-Choice*)

1. **(C)** If $\frac{3}{8}$ of a number is 6, then $\frac{1}{8}$ of the same number is $\frac{1}{3}$ of 6, which is 2. Since $\frac{1}{8}$ of the number is 2, $\frac{7}{8}$ of the same number is 7×2 or 14.
2. **(B)** If Claire has read $\frac{5}{8}$ of a book, $\frac{3}{8}$ of the book remains to be read. Since there are 120 pages left to read, $\frac{3}{8}$ of the total number of pages in the book equals 120. Hence, $\frac{1}{8}$ of the total number of pages in the book is $\frac{1}{3}$ of 120 or 40. Since there are 40 pages in $\frac{1}{8}$ of the book, there are 5×40 or 200 pages in the first $\frac{5}{8}$ of the book that Claire has read.
3. **(B)** $\frac{4}{9}$ of $27 = \frac{4}{\not{9}_1} \times \frac{\not{27}^3}{1} = 12$

   If $\frac{6}{5}$ of a number is 12, then $\frac{1}{5}$ of that number is $\frac{1}{6}$ of 12 or 2. Since $\frac{1}{5}$ of a number is 2, the number must be 5×2 or 10.
4. **(D)** If $\frac{3}{5}$ of a class are boys, then $\frac{2}{5}$ are girls. Since there are 10 girls in the class, $\frac{2}{5}$ of the class represents 10 students, so in $\frac{1}{5}$ of the class there are $\frac{1}{2}$ of 10 or 5 students. Since there are 5 students in $\frac{1}{5}$ of the class, there is a total of 5×5 or 25 students in the whole class.
5. **(E)** The two-digit whole numbers in which either digit is $\frac{3}{4}$ of the other digit are 34, 43, 68, and 86. Use a calculator to find the sum of these four numbers: $34 + 43 + 68 + 86 = 231$.
6. **(A)** After Boris lends $\frac{1}{4}$ of D dollars, he has $\frac{3}{4}D$ dollars left. After he spends $\frac{1}{3}$ of the money that he has left, the number of dollars he now has is $\frac{3}{4}D - \frac{1}{3}\left(\frac{3}{4}D\right)$. Simplify:

$$\frac{3}{4}D - \frac{1}{3}\left(\frac{3}{4}D\right) = \frac{3}{4}D - \frac{1}{4}D$$
$$= \frac{2}{4}D$$
$$= \frac{1}{2}D \text{ or } \frac{D}{2}$$

7. **(A)** The value obtained by increasing a by $\frac{1}{5}$ of its value is

$$a + \frac{1}{5}a = a + 0.2a = 1.2a$$

The value obtained by decreasing b by $\frac{1}{2}$ of its value is

$$b - \frac{1}{2}b = b - 0.5b = 0.5b$$

Since it is given that these two values are equal, $1.2a = 0.5b$.

8. **(C)** $3\frac{1}{2} \div \frac{1}{4} = \frac{7}{\cancel{2}} \times \frac{\overset{2}{\cancel{4}}}{1} = 14$

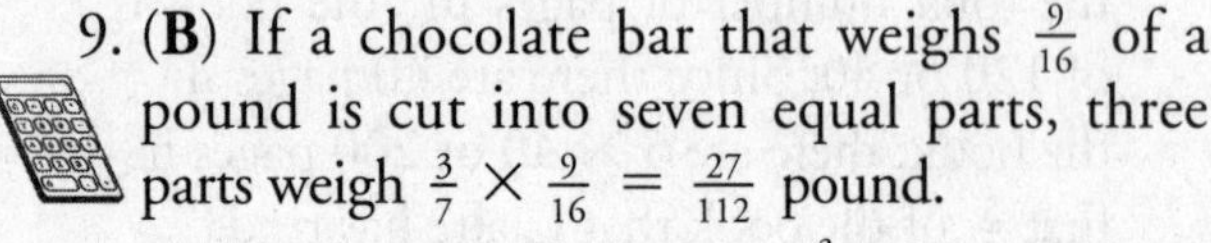

9. **(B)** If a chocolate bar that weighs $\frac{9}{16}$ of a pound is cut into seven equal parts, three parts weigh $\frac{3}{7} \times \frac{9}{16} = \frac{27}{112}$ pound.
10. **(C)** At the basketball game, $\frac{3}{5}$ of the students who attended were seniors, so $\frac{2}{5}$ were not seniors. Of the students who were not seniors, $\frac{1}{3}$ were juniors.
    - The number of juniors who attended the game comprised $\frac{1}{3}$ of $\frac{2}{5}$ or $\frac{2}{15}$ of the total number of students.
    - The number of seniors and juniors who attended represented

      $$\frac{3}{5} + \frac{2}{15} = \frac{9}{15} + \frac{2}{15} = \frac{11}{15}$$

      of the total number of students.
    - The remaining 80 students were all sophomores. Since $1 - \frac{11}{15} = \frac{4}{15}$, $\frac{4}{15}$ of the students in attendance were sophomores. Thus,

      $$\frac{4}{15} \rightarrow 80, \text{ so } \frac{1}{15} \rightarrow \frac{1}{4} \times 80 = 20$$

    - Since $\frac{1}{15}$ of the total number of students is 20 and $15 \times 20 = 300$, a total of 300 students attended the game.
    - Since the number of seniors comprised $\frac{3}{5}$ of the total number of students, the number of seniors who attended the game was

      $$\frac{3}{5} \text{ of } 300 = \frac{3}{5} \times 300 = 180$$

    Hence, 180 seniors attended the basketball game.
11. **(C)** If, of the 75 people in a room, $\frac{2}{5}$ are college graduates, then $\frac{3}{5}$ are *not* college graduates. If $\frac{4}{9}$ of the students who are *not* college graduates are seniors in high school, then $\frac{5}{9}$ of the students who are *not* college graduates are *not* seniors in high school. Thus, $\frac{5}{9} \times \frac{3}{5}$ or $\frac{1}{3}$ of the number of people in the room are neither college graduates nor high school seniors. Since there are 75 people in the room and $\frac{1}{3} \times 75 = 25$, 25 people are neither college graduates nor high school seniors.
12. **(B)** You are given that $\frac{2}{3}$ of $\frac{3}{4}$ of a number is 24. Since

    $$\frac{\overset{1}{\cancel{2}}}{\cancel{3}} \times \frac{\overset{1}{\cancel{3}}}{\underset{2}{\cancel{4}}} = \frac{1}{2}$$

    $\frac{1}{2}$ of the number is 24, so the number is 2×24 or 48. Hence, $\frac{1}{4}$ of the number is $\frac{1}{4} \times 48$ or 12.
13. **(C)** If a man paints $\frac{3}{4}$ of a house in 2 days, then he painted $\frac{1}{4}$ of the house in $\frac{1}{3} \times 2$ or $\frac{2}{3}$ of a day. Since $\frac{1}{4}$ of the house remains to be painted, it will take him $\frac{2}{3}$ of a day to complete the job.
14. **(E)** If a water tank drops from being $\frac{3}{5}$ full to $\frac{1}{3}$ full, then the level of the water tank drops $\frac{4}{15}$ of a full tank since

    $$\frac{3}{5} - \frac{1}{3} = \frac{9}{15} - \frac{5}{15} = \frac{4}{15}$$

    The water tank drops $\frac{4}{15}$ of a full tank as a result of 12 gallons being poured out. Thus,

    $$\frac{4}{15} \rightarrow 12, \text{ so } \frac{1}{15} \rightarrow \frac{1}{4} \times 12 = 3$$

    Since $\frac{1}{15}$ of a full tank holds 3 gallons, $\frac{15}{15}$ or a full tank holds 15×3 or 45 gallons.
15. **(C)** Since Susan received $\frac{2}{3}$ of the ballots cast, $\frac{1}{3}$ of the ballots were cast for the other candidates. Bill received $\frac{5}{5} - \frac{1}{5} = \frac{4}{15}$ of the remaining ballots, which amounted to 48 votes. Bill's votes represent $\frac{4}{5} \times \frac{1}{3}$ or $\frac{4}{15}$ of the total number of ballots. Hence,

    $$\frac{4}{15} \rightarrow 48, \text{ so } \frac{1}{15} \rightarrow \frac{1}{4} \times 48 = 12$$

    Since 12 votes are $\frac{1}{15}$ of the total number cast, the total number of votes cast was $12 \times 15 = 180$. Since Susan received $\frac{2}{3}$ of the 180 ballots cast, she received $\frac{2}{3} \times 180 = 120$ votes.

16. **(D)** You can find the capacity of the gas tank by reasoning as follows:
   - If a gas gauge moves from exactly $\frac{1}{8}$ full to exactly $\frac{7}{8}$ full, then the amount of gas put into the tank represents $\frac{3}{4}\left(= \frac{7}{8} - \frac{1}{8}\right)$ of the capacity of the tank.
   - If the gas costs \$1.50 per gallon and Arlene is charged \$18.00 for the gas, then the number of gallons put into the tank equals $18 \div 1.5 = 12$.

   Since $\frac{3}{4} \rightarrow 12, \frac{1}{4} \rightarrow \frac{1}{3} \times 12 = 4$.

   $\frac{4}{4}$(full tank) $\rightarrow 4 \times 4 = 16$ gallons

   The capacity of the gas tank is 16 gallons.
17. **(E)** Suppose the prices of stocks A and B were each \$100 at the beginning of the day. Then:
   - If the price of stock A increases by $\frac{1}{10}$ of its original value, the new price of stock A is

   $$\$100 + \left[\frac{1}{10} \times \$100\right] = \$100 + \$10 = \$110$$

   - If the price of stock B decreases by $\frac{1}{10}$ of its original value, the new price of stock B is

   $$\$100 - \left(\frac{1}{10} \times \$100\right) = \$100 - \$10 = \$90$$

   - The relationship between the new price of stock A and the new price of stock B can be found by forming a fraction in which the numerator is the new price of stock A and the denominator is the new price of stock B: $\frac{110}{90} = \frac{11}{9}$.

   The new price of stock A is $\frac{11}{9}$ of the new price of stock B.
18. **(D)** Suppose n represents the original number of people in the room. When $\frac{3}{4}$ of these n people leave the room, $\frac{1}{4}n$ people remain. After three people enter the same room, there are $\frac{1}{4}n + 3$ people in the room. Since the number of people now in the room is $\frac{1}{3}$ of the original number of people in the room, $\frac{1}{4}n + 3 = \frac{1}{3}n$. Hence:

   $$3 = \frac{1}{3}n - \frac{1}{4}n$$

   Simplify the right side of this equation by subtracting fractions after changing them to their lowest common denominator:

   $$\begin{aligned} 3 &= \frac{4}{12}n - \frac{3}{12}n \\ &= \frac{1}{12}n \\ 3 \times 12 &= n \\ 36 &= n \end{aligned}$$

   You now know that the original number of people in the room was 36. The question asks for the number of people who left the room: $\frac{3}{4} \times 36 = 27$.

(Grid-In)

1. **18** After a number is increased by $\frac{1}{3}$ of its value, the new number is $\frac{4}{3}$ of its original value. Since the result of the increase is 24,

   $$\frac{4}{3} \rightarrow 24, \text{ so } \frac{1}{3} \rightarrow \frac{1}{4} \times 24 = 6$$

   Since $\frac{1}{3}$ of the original number is 6, multiplying by 3:

   $$\frac{3}{3} \rightarrow 3 \times 6 = 18$$

   gives 18 as the original number.
2. **3/5** Pick a convenient number, say 100, for the number of male voters. You are told that the number of female voters was $1\frac{1}{2}$ times the number of male voters. Then there were $100 \times 1\frac{1}{2}$ or 150 female voters.
   - Since $\frac{1}{2}$ of the male voters cast their ballots for candidate A, this candidate received $\frac{1}{2} \times 100$ or 50 votes from the males.
   - Since $\frac{2}{3}$ of the female voters cast their ballots for candidate A, this candidate received $\frac{2}{3} \times 150$ or 100 votes from the females.
   - A total of 100 + 150 or 250 ballots were cast. Candidate A received 50 + 100 or 150 votes. Since

   $$\frac{150}{250} = \frac{3 \times \overset{1}{\cancel{50}}}{5 \times \cancel{50}} = \frac{3}{5},$$

   candidate A received $\frac{3}{5}$ of the total number of votes cast.

   Grid in as 3/5.

Lesson 3-8 (*Multiple-Choice*)

1. **(B)** 20% of $\frac{2}{3}$ of 15 $= \frac{1}{5} \times \frac{2}{3} \times 15 = 2$

2. **(C)** To change from a percent to a decimal, divide the percent by 100 and drop the percent sign:

$$\frac{2}{5}\% = \frac{\frac{2}{5}}{100} = \frac{0.4}{100} = 0.004$$

3. **(E)** $\frac{0.25 + 0.15}{0.50} = \frac{0.40}{0.50} = \frac{4}{5} = 0.80$
$= 80\%$

4. **(C)** If 480 of 500 seats were occupied, then 20 seats were *not* occupied. Hence, the percent of the seats that were *not* occupied is

$$\frac{20}{500} \times 100\% = \frac{1}{25} \times 100\% = 4\%$$

5. **(A)**

$$\text{Percent decrease} = \frac{\text{Amount of decrease}}{\text{Original amount}} \times 100\%$$

100%

$$= \frac{168 - 147}{168} \times 100\%$$

$$= \frac{21}{168} \times 100\%$$

$$= 0.125 \times 100\%$$

$$= 12.5\% \text{ or } 12\frac{1}{2}\%$$

6. **(E)** $0.01\% = \frac{0.01}{100} = \frac{1}{10{,}000} = \frac{1}{10^4}$

7. **(B)** If the result of increasing a by 300% of a is b, then $a + 3a = 4a = b$. Dividing both sides of $4a = b$ by b gives $\frac{4a}{b} = 1$, so $\frac{a}{b} = \frac{1}{4}$. Since $\frac{1}{4} = 25\%$, a is 25% of b.

8. **(D)** After a 20% price increase, the new price of a radio is $78.00. Hence,

$$\text{original price} = \frac{\$78.00}{1 + 20\%}$$

$$= \frac{\$78.00}{1.2}$$

$$= \$78.00 \div 1.2$$

$$= \$65.00$$

9. **(C)** After a discount of 15%, the price of a shirt is $51. Hence,

$$\text{original price} = \frac{\$51}{1 - 15\%}$$

$$= \frac{\$51}{1 - 0.15}$$

$$= \$51 \div 0.85$$

$$= \$60.00$$

10. **(C)** Two students use the computer a total of 28% + 52% or 80% of the total time of 3 hours or 180 minutes. The third student uses the computer for 100% − 80% or 20% of 180 minutes, which equals $0.2 \times 180 = 36$ minutes.

11. **(D)** 20% of 25% of $\frac{4}{5} = \frac{1}{5} \times \frac{1}{\not{4}} \times \frac{\not{4}^{1}}{5} =$

$$= \frac{1}{25} = 0.04$$

12. **(A)** Since

$$0.04\% = \frac{0.04}{100} = \frac{4}{10{,}000} = \frac{1}{2{,}500}$$

it can be expected that one of every 2500 light bulbs manufactured will be defective.

13. **(C)** Suppose the pair of shoes cost $100. After a discount of 25%, the price of the shoes is $75. If this new price is then discounted 8%, the final price is calculated by subtracting 8% of $75 from $75:

$$\text{final price} = \$75 - (8\% \text{ of } \$75)$$

$$= \$75 - (0.08 \times \$75)$$

$$= \$75 - \$6$$

$$= \$69$$

Since the final selling price of $69 is $31 less than the original selling price of $100 and $\frac{\$31}{\$100} = 31\%$, successive discounts of 25% and 8% are equivalent to a single discount of 31%.

14. **(C)**

$$\text{Percent of alcohol} = \frac{\text{Amount of alcohol by weight}}{\text{Total weight of solution}} \times 100\%$$

$$= \frac{8}{8+17} \times 100\%$$

$$= \frac{8}{25} \times 100\%$$

$$= 8 \times 4\%$$

$$= 32\%$$

15. **(D)** Since 40% of $\frac{2}{5}$ and a is 40% of b, then $a = \frac{2}{5}b$. Thus:

$$b = \frac{5}{2}a = a + \frac{3}{2}a$$

Since b exceeds a by $\frac{3}{2}a$ and $\frac{3}{2} = 1.5 \times 100\% = 150\%$, then b exceeds a by 150% of a.

(Grid-In)

1. **87.5** The cash price of $84.00 reflects the new amount after a discount of 4%. Hence,

$$\text{Original amount} = \frac{\text{New amount after decrease of } P\%}{1 - P\%}$$

$$= \frac{\$84}{1 - 4\%}$$

$$= \frac{\$84}{1 - 0.04}$$

$$= \frac{\$84}{0.96}$$

$$= \$87.5$$

The credit-card purchase price of the same item is $8.75.

2. **50** After three boys are dropped from the class, 25 students remain. Of the 25 students, 44% are boys, so 56% are girls. Since 56% of $25 = 0.56 \times 25 = 14$, 14 girls are enrolled in the class. Hence, 14 of the 28 students in the original class were girls. Thus, the number of girls in the original class comprised $\frac{1}{2}$ or 50% of that class.

Algebraic Methods

CHAPTER 4

This chapter reviews equations and inequalities, as well as some related algebraic methods that you need to know for the SAT.

LESSONS IN THIS CHAPTER

LESSON

4-1 SOLVING EQUATIONS

OVERVIEW

*A **linear** or **first-degree equation** is an equation such as* $2y + 1 = 13$ *in which the exponent of the variable is* 1. *To solve a linear equation, do the same things to both sides of the equation until the resulting equation has the form*

$$\textit{letter} = \text{number}$$

SOLVING LINEAR EQUATIONS BY USING INVERSE OPERATIONS

To isolate the letter (variable) in a linear equation, you must perform the same arithmetic operation on each side of the equation.

- In the equation $x - 7 = -3$, x can be isolated by adding 7 on each side of the equation:

$$\begin{aligned} x - 7 &= -3 \\ +7 &= +7 \\ \hline x + 0 &= 4 \\ x &= 4 \end{aligned}$$

- In the equation $x + 5 = 3$, x can be isolated by subtracting 5 on each side of the equation:

$$\begin{aligned} x + 5 &= 3 \\ -5 &= -5 \\ \hline x + 0 &= -2 \\ x &= -2 \end{aligned}$$

- In the equation $3x = 21$, x can be isolated by dividing each side of the equation by 3:

$$\begin{aligned} 3x &= 21 \\ \frac{3x}{3} &= \frac{21}{3} \\ x &= 7 \end{aligned}$$

- In the equation $\frac{2}{3}x = -8$, x can be isolated by multiplying each side of the equation by the reciprocal of the fractional coefficient of x:

$$\frac{2}{3}x = -8$$

$$\frac{3}{2}\left(\frac{2}{3}x\right) = \frac{3}{\not{2}}(\not{-8}^{-4})$$

$$x = -12$$

SOLVING LINEAR EQUATIONS BY USING TWO OPERATIONS

To isolate a letter in a linear equation, it may be necessary to add or subtract and then to multiply or divide. For example, to solve $2y + 1 = 13$, begin by subtracting 1 on each side of the equation to get $2y = 12$. Dividing each side of this equation by 2 gives $y = 6$.

SOLVING EQUATIONS WITH PARENTHESES

If an equation contains parentheses, remove them by applying the distributive law. Then solve the resulting equation.

EXAMPLE

Solve for b: $3(b + 2) + 2b = 21$

Solution

$$3(b + 2) + 2b = 21$$

- Remove the parentheses by multiplying each term inside the parentheses by 3: $3b + 6 + 2b = 21$
- Combine like terms: $5b + 6 = 21$
- Subtract 6 from each side of the equation: $5b + 6 - 6 = 21 - 6$
- Simplify: $5b = 15$
- Divide each side of the equation by 5: $\frac{5b}{5} = \frac{15}{5}$

$$b = 3$$

SOLVING EQUATIONS THAT CONTAIN FRACTIONS

If an equation contains fractions, eliminate them by multiplying each member of *both* sides of the equation by the LCD of all the denominators. Then solve the resulting equation.

EXAMPLE

Solve for m: $\frac{m}{3} - \frac{m}{4} = 2$

Solution

Since the LCD of 3 and 4 is 12, multiply each member of the given equation by 12:

$$12\left(\frac{m}{3}\right) - 12\left(\frac{m}{4}\right) = 12(2)$$
$$4m - 3m = 24$$
$$m = 24$$

SOLVING EQUATIONS BY CROSS-MULTIPLYING

Since $\frac{4}{8} = \frac{1}{2}$, $4 \times 2 = 8 \times 1$. This operation is sometimes referred to as **cross-multiplying**. If an equation sets one fraction equal to another fraction, then cross-multiply and solve the equation that results.

EXAMPLE

If $\frac{x}{3} = \frac{x+4}{5}$, what is the value of x?

Solution

Cross-multiply by setting the products of opposite pairs of terms equal:

$$\frac{x}{3} \times \frac{x+4}{5}$$

$$5x = 3(x + 4)$$
$$5x = 3x + 12$$
$$2x = 12$$
$$x = \frac{12}{2} = 6$$

USING A ROOT OF AN EQUATION TO ANSWER A QUESTION

To find the value of an algebraic expression that involves a particular letter, you may need to solve an equation.

EXAMPLE

If $2x + 5 = 11$, what is the value of $2x - 5$?

Solution

Instead of solving for x, solve for $2x$. Then subtract 5 from both sides of that equation. Since $2x + 5 = 11$, $2x = 6$, so $2x - 5 = 6 - 5 = 1$.

The value of $2x - 5$ is 1.

In your regular math class, solving for x typically gives you the final answer to the problem. But x is not always the final answer on the SAT. When reading a question, circle or underline the quantity that the question asks you to find. After you solve the problem, check that your answer matches what you were required to find by looking back at what you circled or underlined.

TIPS FOR SCORING HIGH

1. If time permits, plug your solution of an equation back into the original equation to make sure that it works.
2. If you get stuck on a multiple-choice question that asks for the value of a variable in an equation, you may be able to find the solution by testing each of the numerical answer choices in the equation until you find the one that works.
3. Answer the question asked. For example, if the question is "If $3y - 2 = 4$, what is the value of $y - 2$?" make sure your answer is the value of $y - 2$, not the value of y.

Lesson 4-1 Tune-Up Exercises

Multiple-Choice

1 If $x + 4x - 3x + 8 = 0$, then $x =$

(A) -4
(B) -2
(C) 0
(D) 2
(E) 6

2 If $(9 - 4)(x + 4) = 30$, then $x =$

(A) 2
(B) 4
(C) 6
(D) 8
(E) 12

3 If $2(1 + 5) = 3(w - 4)$, then $w =$

(A) 0
(B) 2
(C) 4
(D) 6
(E) 8

4 If $\frac{x}{3} - 1 = 1 - 3$, then $x =$

(A) -6
(B) -3
(C) -1
(D) 3
(E) 6

5 If $3(x - 2) - 2x = 7$, then $x =$

(A) 1
(B) 4
(C) 8
(D) 13
(E) 14

6 If $3x - 6 = 18$, then $x - 2 =$

(A) 2
(B) 4
(C) 6
(D) 8
(E) 10

7 If $2x + y = y + 14$, then $x =$

(A) $\frac{7}{2}$
(B) 7
(C) 14
(D) 21
(E) It cannot be determined.

8 If $\frac{4}{7}k = 36$, then $\frac{3}{7}k =$

(A) 21
(B) 27
(C) 32
(D) 35
(E) 42

9 If $\frac{1}{2}x + \frac{1}{4}x + \frac{1}{8}x = 14$, then $x =$

(A) 4
(B) 8
(C) 12
(D) 16
(E) 24

10 If $2s + 3t = 12$ and $4s = 36$, then $t =$

(A) -3
(B) -2
(C) 2
(D) 3
(E) 9

11 If $2x + 5 = -25$ and $-3y - 6 = 48$, then $xy =$

(A) -270
(B) -90
(C) 45
(D) 90
(E) 270

12 If $\frac{2}{x} = 2$, then $x + 2 =$

(A) $\frac{3}{2}$
(B) $\frac{5}{2}$
(C) 3
(D) 4
(E) 6

13 If $\frac{y-2}{2} = y + 2$, then $y =$

(A) -6
(B) -4
(C) -2
(D) 4
(E) 6

14 If $\frac{2y}{7} = \frac{y+3}{4}$, then $y =$

(A) 5
(B) 9
(C) 13
(D) 17
(E) 21

15 If $\frac{y}{3} = 4$, then $3y =$

(A) 4
(B) 12
(C) 24
(D) 30
(E) 36

16 If $\frac{1}{q} + \frac{3}{q} = 12$, then $q =$

(A) $\frac{1}{4}$
(B) $\frac{1}{3}$
(C) $\frac{1}{2}$
(D) 3
(E) 4

17 If $8 - 3x = 2x + 13$, then $x =$

(A) -4
(B) -3
(C) -2
(D) -1
(E) 1

18 If $(1-2-3-4)\,w = (-1)(-2)(-3)(-4)$, then $w =$

(A) -3
(B) -2
(C) -1
(D) 2
(E) 3

19 If $\frac{1}{3} + \frac{3}{p} = 1$, then $p =$

(A) $\frac{3}{2}$
(B) 2
(C) $\frac{5}{2}$
(D) $\frac{9}{2}$
(E) 6

20 If $\frac{3}{4} = \frac{6}{x} = \frac{9}{y}$, then $x + y =$

(A) 4
(B) 8
(C) 12
(D) 16
(E) 20

Grid-In

1 If $2x - 1 = 11$ and $3y = 12$, what is the value of $\frac{x}{y}$?

2 If $7(a + b) - 4(a + b) = 24$, what is the value of $a + b$?

3 If $1 - x - 2x - 3x = 6x - 1$, what is the value of x?

4 In the equation $p = \frac{5b}{c^2}$, what is a value of c when $p = 9$ and $b = 20$?

5 If $\frac{3(y + 8)}{2y} = 9$, what is the value of y?

LESSON

4-2

EQUATIONS WITH MORE THAN ONE VARIABLE

OVERVIEW

An equation that has more than one letter can have many different numerical solutions. For example, in the equation $x + y = 8$, if $x = 1$, then $y = 7$; if $x = 2$, then $y = 6$. Since the value of y depends on the value of x, and x may be any real number, the equation $x + y = 8$ has infinitely many solutions.

The SAT *may include questions that ask you to:*

- *Find the value of an expression that is a multiple of one side of an equation. For example, if $x + 2y = 9$, then the value of $2x + 4y$ is* 18 *since*

$$2x + 4y = 2(x + 2y) = 2(9) = 18$$

- *Solve for one letter in terms of the other letter(s) of an equation. For example, if $x + y = 8$, then x in terms of y is $x = 8 - y$.*

WORKING WITH AN EQUATION IN TWO UNKNOWNS

Some SAT questions can be answered by multiplying or dividing an equation by a suitable number.

EXAMPLE

If $3x + 3y = 12$, what is the value of $x + y - 6$?

Solution

- Solve the given equation for $x + y$ by dividing each member by 3:

$$\frac{3x}{3} + \frac{3y}{3} = \frac{12}{3}$$

$$x + y = 4$$

- Evaluate the required expression by replacing $x + y$ with 4:

$$\begin{aligned} x + y - 6 &= 4 - 6 \\ &= -2 \end{aligned}$$

The value of $x + y - 6$ is -2.

EXAMPLE

If $\frac{t}{s} = \frac{3}{4}$, what is the value of $\frac{3s}{4t}$?

Solution

Since $\frac{t}{s} = \frac{3}{4}$, then $\frac{s}{t} = \frac{4}{3}$. Substitute $\frac{4}{3}$ for $\frac{s}{t}$ in the required expression:

$$\frac{3s}{4t} = \frac{3}{4} \cdot \frac{s}{t} = \frac{3}{4} \cdot \frac{4}{3} = 1$$

EXAMPLE

If $x + 5 = t$, then $2x + 9 =$

(A) $t - 1$

(B) $t + 1$

(C) $2t$

(D) $2t - 1$

(E) $2t + 1$

Solution

Method 1: Use algebraic reasoning:

- Since the coefficient of x in $2x + 9$ is 2, multiply the given equation by 2:

$$2(x + 5) = 2t, \text{ so } 2x + 10 = 2t$$

- Comparing the left side of $2x + 10 = 2t$ with $2x + 9$ suggests that you subtract 1 from both sides of the equation:

$$(2x + 10) - 1 = 2t - 1$$

$$2x + 9 = 2t - 1$$

The correct choice is (**D**).

Method 2: Pick easy numbers for x and t that satisfy $x + 5 = t$. Then use these values for x and t to compare $2x + 9$ with each of the answer choices. For example, when $x = 1$, $1 + 5 = t = 6$. Using $x = 1$, you know that $2x + 9 = 2(1) + 9 = 11$. Hence, the correct answer choice is the one that evaluates to 11 when $t = 6$:

Choice (A):	$t - 1 = 6 - 1$	$= 5$	✗
Choice (B):	$t + 1 = 6 + 1$	$= 7$	✗
Choice (C):	$2t = 2(6)$	$= 12$	✗
Choice (D):	$2t - 1 = 2(6) - 1$	$= 11$	✓
Choice (E):	$2t + 1 = 2(6) + 1$	$= 13$	✗

If an equation involving two or more variables appears early in a test section, try making a simple substitution. For example, if it is given that $3(a - b)(a + b) = 60$ and $a - b = 2$, then you can find the value of $a + b$ simply by replacing $a - b$ with 2 in the equation:

$$3(2)(a + b) = 60, \text{ so } a + b = \frac{60}{6} = 10$$

SOLVING FOR ONE LETTER IN TERMS OF ANOTHER

To solve an equation that has different letters, isolate in the usual way the letter that is being solved for.

EXAMPLE

If $2x - z = y$, what is x in terms of y and z?

Solution

Treat y and z as constants, and isolate x on one side of the equation.

- Add z on both sides of $2x - z = y$ to get $2x = y + z$.
- Divide both sides of $2x = y + z$ by 2:

$$x = \frac{y + z}{2}$$

Lesson 4-2 Tune-Up Exercises

Multiple-Choice

1 If $6 = 2x + 4y$, what is the value of $x + 2y$?

(A) 2
(B) 3
(C) 6
(D) 8
(E) 12

2 If $\frac{a}{2} + \frac{b}{2} = 3$, what is the value of $2a + 2b$?

(A) 6
(B) 8
(C) 12
(D) 16
(E) 24

3 If $2s - 3t = 3t - s$, what is s in terms of t?

(A) $\frac{t}{2}$
(B) $2t$
(C) $t + 2$
(D) $\frac{t}{2} + 1$
(E) $3t$

4 If $a + b = 5$ and $\frac{c}{2} = 3$, what is the value of $2a + 2b + 2c$?

(A) 12
(B) 14
(C) 16
(D) 20
(E) 22

5 If $xy + z = y$, what is x in terms of y and z?

(A) $\frac{y + z}{y}$
(B) $\frac{y - z}{z}$
(C) $\frac{y - z}{y}$
(D) $1 - z$
(E) $\frac{z - y}{y}$

6 If $b(x + 2y) = 60$ and $by = 15$, what is the value of bx?

(A) 15
(B) 20
(C) 25
(D) 30
(E) 45

7 If $2x^2 + 3y^2 = 0$, what is the value of $3x + 2y$?

(A) -1
(B) 0
(C) 1
(D) 3
(E) 5

8 If $\frac{a - b}{b} = \frac{2}{3}$, what is the value of $\frac{a}{b}$?

(A) $\frac{1}{2}$
(B) $\frac{3}{5}$
(C) $\frac{3}{2}$
(D) $\frac{5}{3}$
(E) 2

9 If $\frac{10^x}{10^y} = 100$, then $x =$

(A) $y + 2$

(B) $y - 2$

(C) $2 - y$

(D) $2y$

(E) $\frac{y}{2}$

10 If $\frac{1}{p+q} = r$ and $p \neq -q$, what is p in terms of r and q?

(A) $\frac{rq-1}{q}$

(B) $\frac{1+rq}{q}$

(C) $\frac{r}{1+rq}$

(D) $\frac{1-rq}{r}$

(E) $\frac{1-q}{rq}$

11 If $a = 2b = 5c$, then $4a$ is equal to which of the following expressions?

I. $8c$

II. $4b + 10c$

III. $2b + 15c$

(A) I, II, and III

(B) II and III only

(C) II only

(D) III only

(E) None

12 If $\frac{a+b+c}{3} = \frac{a+b}{2}$, then $c =$

(A) $\frac{a-b}{2}$

(B) $\frac{a+b}{2}$

(C) $5a + 5b$

(D) $\frac{a+b}{5}$

(E) $-a - b$

13 If $wx = z$, which of the following expressions is equal to xz?

(A) $\frac{w}{z^2}$

(B) $\frac{w^2}{z}$

(C) wz^2

(D) w^2z

(E) $\frac{z^2}{w}$

14 If the value of n nickels plus d dimes is c cents, what is n in terms of d and e?

(A) $\frac{c}{5} - 2d$

(B) $5c - 2d$

(C) $\frac{c-d}{10}$

(D) $\frac{cd}{10}$

(E) $\frac{c+10d}{5}$

15 If $\frac{c}{d} - \frac{a}{b} = x$, $a = 2c$, and $b = 5d$, what is the value of $\frac{c}{d}$ in terms of x?

(A) $\frac{2}{3}x$

(B) $\frac{3}{4}x$

(C) $\frac{4}{3}x$

(D) $\frac{5}{3}x$

(E) $\frac{7}{2}x$

16 If $kx - 4 = (k-1)x$, which of the following must be true?

(A) $x = -5$

(B) $x = -4$

(C) $x = -3$

(D) $x = 4$

(E) $x = 5$

17 If $c = b + 1$ and $p = 4b + 5$, which of the following is an expression for p, in terms of c?

(A) $4c$
(B) $4c - 4$
(C) $\frac{c+1}{4}$
(D) $4c + 1$
(E) $4c + 9$

18 If p and r are positive integers and $2p + r + 1 = 2r + p + 1$, which of the following must be true?

I. p and r are consecutive integers.
II. p is even.
III. r is odd.

(A) None
(B) I only
(C) II only
(D) III only
(E) I, II, and III

Grid-In

1 If $16 \times a^2 \times 64 = (4 \times b)^2$ and a and b are positive integers, then b is how many times greater than a?

2 If $3a - c = 5b$ and $3a + 3b - c = 40$, what is the value of b?

3 If $a = 2x + 3$ and $b = 4x - 7$, for what value of x is $3b = 5a$?

4 $\frac{x}{8} + \frac{y}{5} = \frac{31}{40}$

In the equation above, if x and y are positive integers, what is the value of $x + y$?

LESSON

4-3

POLYNOMIALS AND ALGEBRAIC FRACTIONS

OVERVIEW

A ***polynomial*** *is a single term or the sum or difference of two or more* unlike *terms. For example, the polynomial $a + 2b + 3c$ represents the sum of the three* unlike *terms a, $2b$, and $3c$. Since polynomials represent real numbers, they can be added, subtracted, multiplied, and divided using the laws of arithmetic.*

Whenever a letter appears in the denominator of a fraction, you may assume that it cannot represent a number that makes the denominator of the fraction equal to 0.

CLASSIFYING POLYNOMIALS

A polynomial can be classified according to the number of terms it contains.

- A polynomial with one term, as in $3x^2$, is called a **monomial**.
- A polynomial with two unlike terms, as in $2x + 3y$, is called a **binomial**.
- A polynomial with three unlike terms, as in $x^2 + 3x - 5$, is called a **trinomial**.

OPERATIONS WITH POLYNOMIALS

Polynomials may be added, subtracted, multiplied, and divided.

- To add polynomials, write one polynomial on top of the other one so that like terms are aligned in the same vertical columns. Then combine like terms. For example:

$$\begin{array}{r} 2x^2 - 3x + 7 \\ +\quad x^2 + 5x - 9 \\ \hline 3x^2 + 2x - 2 \end{array}$$

- To subtract polynomials, take the opposite of each term of the polynomial that is being subtracted. Then add the two polynomials. For instance, the difference

$$(7x - 3y - 9z) - (5x + y - 4z)$$

can be changed into a sum by adding the opposite of each term of the second polynomial to the first polynomial:

$$\begin{array}{r} 7x - 3y - 9z \\ + -5x - y + 4z \\ \hline 2x - 4y - 5z \end{array}$$

- To multiply monomials, multiply their numerical coefficients and multiply *like* variable factors by *adding* their exponents. For example:

$$\begin{aligned}(-2a^2b)(4a^3b^2) &= (-2)(4)(a^2 \cdot a^3)(b \cdot b^2)\\ &= -8(a^{2+3})(b^{1+2})\\ &= -8a^5b^3\end{aligned}$$

- To divide monomials, divide their numerical coefficients and divide *like* variable factors by *subtracting* their exponents. For example:

$$\begin{aligned}\frac{14x^5y^2}{21x^2y^2} &= \left(\frac{14}{21}\right)\left(\frac{x^5}{x^2}\right)\left(\frac{y^2}{y^2}\right)\\ &= \left(\frac{2}{3}\right)(x^{5-2})(1)\\ &= \frac{2}{3}x^3\end{aligned}$$

- To multiply a polynomial by a monomial, multiply each term of the polynomial by the monomial and add the resulting products. For example:

$$\begin{aligned}5x(3x - y + 2) &= 5x(3x) + 5x(-y) + 5x(2)\\ &= 15x^2 - 5xy + 10x\end{aligned}$$

The expression $-(a - b)$ can be interpreted as "take the opposite of whatever is inside the parentheses." The result is $b - a$ since

$$\begin{aligned}-(a - b) &= -1(a - b)\\ &= (-1)a + (-1)(-b)\\ &= -a + b \text{ or } b - a\end{aligned}$$

- To divide a polynomial by a monomial, divide each term of the polynomial by the monomial and add the resulting quotients. For example:

$$\frac{6x + 15}{3} = \frac{6x}{3} + \frac{15}{3} = 2x + 5$$

MULTIPLYING BINOMIALS USING FOIL

To find the product of two binomials, write them next to each other and then add the products of their *F*irst, *O*uter, *I*nner, and *L*ast pairs of terms.

EXAMPLE

Multiply $(2x + 1)$ by $(x - 5)$.

Solution

For the product $(2x + 1)(x - 5)$, $2x$ and x are the *F*irst pair of terms; $2x$ and -5 are the *O*utermost pairs of terms; 1 and x are the *I*nnermost terms; 1 and -5 are the *L*ast terms of the binomials. Thus:

$$(2x + 1)(x - 5) = \overbrace{(2x)(x)}^{F} + \overbrace{(2x)(-5)}^{O} + \overbrace{(1)(x)}^{I} + \overbrace{(1)(-5)}^{L}$$
$$= 2x^2 + [-10x + x] - 5$$
$$= 2x^2 - 9x - 5$$

PRODUCTS OF SPECIAL PAIRS OF BINOMIALS

SAT problems may involve these special products: $(a - b)(a + b)$, $(a + b)^2$, and $(a - b)^2$.

TIME SAVER

You can save some time if you memorize and learn to recognize when these special multiplication rules can be applied.

- $(a - b)(a + b) = a^2 - b^2$
- $(a + b)^2 = (a + b)(a + b) = a^2 + 2ab + b^2$
- $(a - b)^2 = (a - b)(a - b) = a^2 - 2ab + b^2$

EXAMPLE

Express $(x + 3)(x - 3)$ as a binomial.

Solution

$$(x + 3)(x - 3) = (x)^2 - (3)^2$$
$$= x^2 - 9$$

EXAMPLE

Express $(2y - 1)(2y + 1)$ as a binomial.

Solution

$$(2y - 1)(2y + 1) = (2y)^2 - (1)^2$$
$$= 4y^2 - 1$$

EXAMPLE

If $(x + y)^2 - (x - y)^2 = 28$, what is the value of xy?

Solution

- Square each binomial:

$$(x + y)^2 - (x - y)^2 = 28$$
$$(x^2 + 2xy + y^2) - (x^2 - 2xy + y^2) = 28$$

- Write the first squared binomial without the parentheses. Then remove the second set of parentheses by changing the sign of each term inside the parentheses to its opposite.

$$x^2 + 2xy + y^2 - x^2 + 2xy - y^2 = 28$$

- Combine like terms. Adding x^2 and $-x^2$ gives 0, as does adding y^2 and $-y^2$. The sum of $2xy$ and $2xy$ is $4xy$. The result is

$$4xy = 28$$

$$xy = \frac{28}{4} = 7$$

COMBINING ALGEBRAIC FRACTIONS

Algebraic fractions are combined in much the same way as fractions in arithmetic.

EXAMPLE

Write $\frac{w}{2} - \frac{w}{3}$ as a single fraction.

Solution

The LCD of 2 and 3 is 6. Change each fraction into an equivalent fraction that has 6 as its denominator.

$$\begin{aligned} \frac{w}{2} - \frac{w}{3} &= \frac{3}{3}\left(\frac{w}{2}\right) - \frac{2}{2}\left(\frac{w}{3}\right) \\ &= \frac{3w}{6} - \frac{2w}{6} \\ &= \frac{3w - 2w}{6} \\ &= \frac{w}{6} \end{aligned}$$

EXAMPLE

If $h = \frac{y}{x-y}$, what is $h + 1$ in terms of x and y?

Solution

- Add 1 on both sides of the given equation:

$$h + 1 = \frac{y}{x - y} + 1$$

- On the right side of the equation, replace 1 with $\frac{x-y}{x-y}$:

$$h + 1 = \frac{y}{x - y} + \frac{x - y}{x - y}$$

- Write the sum of the numerators over the common denominator:

$$h + 1 = \frac{y + x - y}{x - y} = \frac{x}{x - y}$$

THE SUM AND DIFFERENCE OF TWO RECIPROCALS

The formulas for the sum and the difference of the reciprocals of two nonzero numbers, x and y, are worth remembering:

TIME SAVER

Reciprocal Rules

If x and y are not 0, then

$$\frac{1}{x}+\frac{1}{y}=\frac{x+y}{xy} \text{ and } \frac{1}{x}-\frac{1}{y}=\frac{y-x}{xy}.$$

EXAMPLE

If $t = \frac{1}{r}+\frac{1}{s}$, then what is $\frac{1}{t}$ in terms of r and s?

Solution

Since $t = \frac{1}{r} + \frac{1}{s} = \frac{r+s}{rs}$, then $\frac{1}{t} = \frac{rs}{r+s}$.

Lesson 4-3 Tune-Up Exercises

Multiple-Choice

1 $\frac{20b^3 - 8b}{4b} =$

(A) $5b^2 - 2b$
(B) $5b^3 - 2$
(C) $5b^2 - 8b$
(D) $5b^2 - 2$
(E) $5b - 2$

2 $(39)^2 + 2(39)(61) + (61)^2 =$

(A) 9,099
(B) 9,909
(C) 9,990
(D) 10,000
(E) 10,001

3 $(4a + a - 3)(4b - 2b - b) =$

(A) $2a - b$
(B) $2ab - 2b$
(C) $5ab - b$
(D) $5ab - 3b$
(E) $3ab$

4 If $(x - y)^2 = 50$ and $xy = 7$, what is the value of $x^2 + y^2$?

(A) 8
(B) 36
(C) 43
(D) 57
(E) 64

5 If $p = \frac{a}{a-b}$ and $a \neq b$, then, in terms of a and b, $1 - p =$

(A) $\frac{a}{b-a}$
(B) $\frac{b}{b-a}$
(C) $\frac{a}{a-b}$
(D) $\frac{b}{a-b}$
(E) $\frac{a+b}{a-b}$

6 If $(p - q)^2 = 25$ and $pq = 14$, what is the value of $(p + q)^2$?

(A) 25
(B) 36
(C) 53
(D) 64
(E) 81

7 If $a - b = p$ and $a + b = k$, then $a^2 - b^2 =$

(A) pk
(B) $p^2 - k^2$
(C) $p + k$
(D) $\frac{p^2}{k^2}$
(E) $k^2 - p^2$

8 If $(2y + k)^2 = 4y^2 - 12y + k^2$, what is the value of k?

(A) 3
(B) 1
(C) -1
(D) -2
(E) -3

9 Which statement is true for all real values of x and y?

(A) $(x + y)^2 = x^2 + y^2$

(B) $x^2 + x^2 = x^4$

(C) $\frac{2^{x+2}}{2^x} = 4$

(D) $(3x)^2 = 6x^2$

(E) $x^5 - x^3 = x^2$

10 If $(px + q)^2 = 9x^2 + kx + 16$, what is the value of k?

(A) 3
(B) 4
(C) 12
(D) 24
(E) 30

11 If $\left(\frac{x}{100} - 1\right)\left(\frac{x}{100} + 1\right) =$ $kx^2 - 1$, then $k =$

(A) 0.1
(B) 0.01
(C) 0.001
(D) 0.0001
(E) 0.000001

12 If $(x + 5)(x + p) = x^2 + 2x + k$, then

(A) $p = 3$ and $k = 5$
(B) $p = -3$ and $k = 15$
(C) $p = 3$ and $k = -15$
(D) $p = 3$ and $k = 15$
(E) $p = -3$ and $k = -15$

13 For what value of p is $(x - 2)(x + 2) = x(x - p)$?

(A) -4

(B) 0

(C) $\frac{2}{x}$

(D) $\frac{4}{x}$

(E) $-\frac{4}{x}$

14 For $x, y > 0$, which expression is equal to $\frac{1}{\frac{1}{x} + \frac{1}{y}}$?

(A) $x + y$

(B) $\frac{x + y}{xy}$

(C) $\frac{xy}{x + y}$

(D) $\frac{x}{y} + \frac{y}{x}$

(E) $\frac{1}{xy}$

15 If $\left(k + \frac{1}{k}\right)^2 = 16$, then $k^2 + \frac{1}{k^2} =$

(A) 4

(B) 8

(C) 12

(D) 14

(E) 18

16 If $\frac{a}{b} = 1 - \frac{x}{y}$, then $\frac{b}{a} =$

(A) $\frac{x}{y - x}$

(B) $\frac{y}{x} - 1$

(C) $\frac{y}{x - y}$

(D) $\frac{x}{y} + 1$

(E) $\frac{y}{y - x}$

Grid-In

1 If $(3y - 1)(2y + k) = ay^2 + by - 5$ for all values of y, what is the value of $a + b$?

2 If $4x^2 + 20x + r = (2x + s)^2$ for all values of x, what is the value of $r - s$?

LESSON

4-4

FACTORING

OVERVIEW

***Factoring** reverses multiplication.*

Operation	*Example*
Multiplication	$2(x + 3y) = 2x + 6y$
Factoring	$2x + 6y = 2(x + 3y)$

There are three basic types of factoring that you need to know for the SAT:

- *Factoring out a common monomial factor, as in*

$$4x^2 - 6x = 2x(x - 3) \quad \text{and} \quad ay - by = y(a - b)$$

- *Factoring a quadratic trinomial using the reverse of FOIL, as in*

$$x^2 - x - 6 = (x - 3)(x + 2) \quad \text{and}$$
$$x^2 - 2x + 1 = (x - 1)(x - 1) = (x - 1)^2$$

- *Factoring the difference between two squares using the rule*

$$a^2 - b^2 = (a + b)(a - b)$$

FACTORING A POLYNOMIAL BY REMOVING A COMMON FACTOR

If all the terms of a polynomial have factors in common, the polynomial can be factored by using the reverse of the distributive law to remove these common factors. For example, in

$$24x^3 + 16x = 8x(3x^2 + 2)$$

$8x$ is the *G*reatest *C*ommon *F*actor (GCF) of $24x^3$ and $16x$ since 8 is the GCF of 24 and 16, and x is the greatest power of that variable that is contained in both $24x^3$ and $16x$. The factor that corresponds to $8x$ can be obtained by dividing $24x^3 + 16x$ by $8x$:

$$\frac{24x^3 + 16x}{8x} = \frac{24x^3}{8x} + \frac{16x}{8x} = 3x^2 + 2$$

You can check that the factorization is correct by multiplying $3x^2 + 2$ by $8x$ and verifying that the product is $24x^3 + 16x$.

USING FACTORING TO ISOLATE VARIABLES IN EQUATIONS

It may be necessary to use factoring to help isolate a variable in an equation in which terms involving the variable cannot be combined into a single term.

EXAMPLE

If $ax - c = bx + d$, what is x in terms of a, b, c, and d?

Solution

Isolate terms involving x on the same side of the equation.

- On each side of the equation add c and subtract bx:

$$ax - bx = c + d$$

- Factor out x from the left side of the equation:

$$x(a - b) = c + d$$

- Divide both sides of the new equation by the coefficient of x:

$$\frac{x(a-b)}{a-b} = \frac{c+d}{a-b}$$

$$x = \frac{c+d}{a-b}$$

FACTORING A QUADRATIC TRINOMIAL

A quadratic trinomial like $x^2 - 7x + 12$ contains x^2 as well as x. Quadratic trinomials that appear on the SAT can be factored as the product of two binomials by reversing the FOIL multiplication process.

- Think: "What two integers when multiplied together give $+12$ and when added together give -7?"
- Recall that, since the product of these integers is *positive* 12, the two integers must have the same sign. Hence, the integers are limited to the following pairs of factors of $+12$:

$$1 \text{ and } 12; \quad -1 \text{ and } -12$$

$$2 \text{ and } 6; \quad -2 \text{ and } -6$$

$$3 \text{ and } 4; \quad -3 \text{ and } -4$$

- Choose -3 and -4 as the factors of 12 since they add up to -7. Thus,

$$x^2 - 7x + 12 = (x - 3)(x - 4)$$

- Use FOIL to check that the product $(x - 3)(x - 4)$ is $x^2 - 7x + 12$.

EXAMPLE

Factor $n^2 - 5n - 14$.

Solution

Find two integers that when multiplied together give −14 and when added together give −5. The two factors of −14 must have different signs since their product is negative. Since (+2)(−7) = −14 and 2 + (−7) = −5, the factors of −14 you are looking for are +2 and −7. Thus:

$$n^2 - 5n - 14 = (n + 2)(n - 7)$$

FACTORING THE DIFFERENCE BETWEEN TWO SQUARES

Since $(a + b)(a - b) = a^2 - b^2$, any binomial of the form $a^2 - b^2$ can be rewritten as $(a + b)$ times $(a - b)$.

$$a^2 - b^2 = (a + b)(a - b)$$

This means that the difference between two squares can be factored as the product of the sum and difference of the quantities that are being squared. Here are some examples in which this factoring rule is used:

- $y^2 - 16 = (y + 4)(y - 4)$
- $x^2 - 0.25 = (x + 0.5)(x - 0.5)$
- $100 - x^4 = (10 - x^2)(10 + x^2)$

FACTORING COMPLETELY

To factor a polynomial into factors that cannot be further factored, it may be necessary to use more than one factoring technique. For example:

- $3t^2 - 75 = 3(t^2 - 25) = 3(t + 5)(t - 5)$
- $t^3 - 6t^2 + 9t = t(t^2 - 6t + 9) = t(t - 3)(t - 3)$ or $t(t - 3)^2$

USING FACTORING TO SIMPLIFY ALGEBRAIC FRACTIONS

To simplify an algebraic fraction, factor the numerator and the denominator. Then cancel any factor that is in both the numerator and the denominator of the fraction since any nonzero quantity divided by itself is 1.

EXAMPLE

Simplify $\dfrac{2b-2a}{a^2-b^2}$.

Solution

- Factor the numerator and the denominator: $\dfrac{2b - 2a}{a^2 - b^2} = \dfrac{2(b - a)}{(a + b)(a - b)}$
- Rewrite $b - a$ as $-(a - b)$: $= \dfrac{2[-(a - b)]}{(a + b)(a - b)}$
- Cancel $a - b$ since it is a factor of the numerator and a factor of the denominator: $= \dfrac{-2\overset{1}{\cancel{(a - b)}}}{(a + b)\cancel{(a - b)}}$

$$= \frac{-2}{a + b}$$

Lesson 4-4 Tune-Up Exercises

Multiple-Choice

1 $\dfrac{4x + 4y}{2x^2 - 2y^2} =$

(A) $\dfrac{2}{x + y}$

(B) $2(x - y)$

(C) $\dfrac{xy}{2(x - y)}$

(D) $2\left(\dfrac{1}{x + y}\right)$

(E) $\dfrac{2}{x - y}$

2 The sum of $\dfrac{b}{a^2 - b^2}$ and $\dfrac{b}{a^2 - b^2}$ is

(A) $\dfrac{1}{a - b}$

(B) $\dfrac{a}{a - b}$

(C) $\dfrac{b}{a - b}$

(D) $\dfrac{a + b}{a - b}$

(E) $\dfrac{ab}{a - b}$

3 If $ax + x^2 = y^2 - ay$, what is a in terms of x and y?

(A) $y - x$

(B) $x - y$

(C) $x + y$

(D) $\dfrac{x^2 + y^2}{x + y}$

(E) $\dfrac{x^2 + y^2}{x - y}$

4 If $\dfrac{xy}{x + y} = 1$ and $x \neq -y$, what is x in terms of y?

(A) $\dfrac{y + 1}{y - 1}$

(B) $\dfrac{y + 1}{y}$

(C) $\dfrac{y}{y - 1}$

(D) $\dfrac{y}{y + 1}$

(E) $1 - \dfrac{1}{y}$

5 If $x^2 = r^2 + 2rs + s^2$, $y^2 = r^2 - s^2$, $x > 0$, and $y > 0$, then $\dfrac{x}{y} =$

(A) $\dfrac{r + s}{r - s}$

(B) $\sqrt{\dfrac{r + s}{r - s}}$

(C) $\dfrac{r - s}{rs}$

(D) $\sqrt{\dfrac{r^2 + s^2}{r^2 - s^2}}$

(E) $\sqrt{r + s}$

6 If $h = \dfrac{x^2 - 1}{x + 1} + \dfrac{x^2 - 1}{x - 1}$, what is x in terms of h?

(A) $\dfrac{h}{2}$

(B) $2h + 1$

(C) $2h - 1$

(D) $\sqrt{\dfrac{h}{2}}$

(E) $\sqrt{2h}$

7 If $ax^2 - bx = ay^2 + by$, then $\frac{a}{b} =$

(A) $\frac{1}{x-y}$

(B) $\frac{1}{x+y}$

(C) $\frac{x-y}{x+y}$

(D) $\frac{x+y}{x-y}$

(E) $\frac{x}{y}$

8 If $a \neq b$ and $\frac{a^2 - b^2}{9} = a + b$, then what is the value of $a - b$?

(A) $\frac{1}{3}$

(B) 3

(C) 9

(D) 12

(E) It cannot be determined from the information given.

LESSON

4-5 QUADRATIC EQUATIONS

OVERVIEW

*A **quadratic equation** is an equation in which the greatest exponent of the variable is 2, as in $x^2 + 3x - 10 = 0$. A quadratic equation has two roots, which can be found by breaking down the quadratic equation into two first-degree equations.*

ZERO-PRODUCT RULE

If the product of two or more numbers is 0, at least one of these numbers is 0.

EXAMPLE

For what values of x is $(x - 1)(x + 3) = 0$?

Solution

Since $(x - 1)(x + 3) = 0$, either $x - 1 = 0$ or $x + 3 = 0$.

- If $x - 1 = 0$, then $x = 1$.
- If $x + 3 = 0$, then $x = -3$.

The possible values of x are 1 and -3.

SOLVING A QUADRATIC EQUATION BY FACTORING

The two roots of a quadratic equation may or may not be equal. For example, the equation $(x - 3)^2 = 0$ has a double root of $x = 3$. The quadratic equation $x^2 = 9$, however, has two unequal roots, $x = 3$ and $x = -3$. More complicated quadratic equations on the SAT can be solved by factoring the quadratic expression.

EXAMPLE

Solve $x^2 + 2x = 0$ for x.

Solution

- Factor the left side of the quadratic equation: $x^2 + 2x = 0$
  $x(x + 2) = 0$
- Form two first-degree equations by setting each factor equal to 0: $x = 0$ or $x + 2 = 0$
- Solve each first-degree equation: $x = 0$ or $x = -2$

The two roots are 0 and -2.

If a quadratic equation does not have all its nonzero terms on the same side of the equation, you must put the equation into this form before factoring.

EXAMPLE

Solve $x^2 + 3x = 10$ for x.

Solution

To rewrite the quadratic equation so that all the nonzero terms are on the same side:

- Subtract 10 from both sides of $x^2 + 3x = 10$: $x^2 + 3x - 10 = 0$
- Factor the quadratic polynomial: $(x + 5)(x - 2) = 0$
- Set each factor equal to 0: $x + 5 = 0$ or $x - 2 = 0$
- Solve each equation: $x = -5$ or $x = 2$

The two roots are -5 and 2.

You can check that $x = -5$ and $x = 2$ are the roots by plugging each value into $x^2 + 3x = 10$ and verifying that the left side then equals 10, the right side.

EXAMPLE

If 4 is a root of $x^2 - x - w = 0$, what is the value of w?

Solution

Since 4 is a root of the given equation, replacing x with 4 in that equation gives an equation that can be used to solve for w:

$$4^2 - 4 - w = 0$$
$$16 - 4 - w = 0$$
$$12 - w = 0$$
$$12 = w$$

Hence, $w = 12$.

QUADRATIC EQUATIONS WITH EQUAL OR NO ROOTS

A quadratic equation may have equal roots. If $x^2 - 2x + 1 = 0$, then

$$(x - 1)(x - 1) = 0$$
$$x - 1 = 0 \quad or \quad x - 1 = 0$$
$$x = 1 \qquad\qquad x = 1$$

Thus, the equation $x^2 - 2x + 1 = 0$ has a double root of 1.

Although a quadratic equation may have no real roots, as in $x^2 + 1 = 0$, on the SAT, all quadratic equations have real roots.

Every quadratic equation on the SAT has two real roots, but the roots may be equal.

Lesson 4-5 Tune-Up Exercises

Multiple-Choice

1 If $(x + 7)(x - 3) = 0$, then $x =$

(A) 7 or 3
(B) 7 or -3
(C) -7 or 3
(D) -7 or -3
(E) -4 or -3

2 If $\frac{a^2}{2} = 2a$, then a equals

(A) 0 or -2
(B) 0 or 2
(C) 0 or -4
(D) 0 or 4
(E) 2 or -2

3 If x is a positive integer and $(x - 6)^2 = 9$, then $x^2 =$

(A) 3
(B) 9
(C) 16
(D) 36
(E) 49

4 If $a(x - y) = 0$, then which of the following statements is (are) always true?

I. $a = 0$
II. $y = 0$
III. $x = y$

(A) I and II only
(B) I and III only
(C) II and III only
(D) I, II, and III
(E) None

5 If $(x - 1)^2 - (x - 1) = 0$, then,

(A) $x = 0$ or $x = 2$
(B) $x = 0$ or $x = 1$
(C) $x = -1$ or $x = 2$
(D) $x = 1$ or $x = 2$
(E) $x = 0$ or $x = -1$

6 If w and y are positive numbers, $w^2 - 2w = 0$, and $2y^2 - y = 0$, $\frac{w}{y} =$

(A) $\frac{1}{2}$
(B) $\frac{2}{3}$
(C) 1
(D) 2
(E) 4

7 For which of the following equations is $x = -2$ a root?

I. $\frac{2}{x} - x = 0$
II. $x^2 + 4 = 0$
III. $x^2 + 4x + 4 = 0$

(A) I only
(B) II only
(C) III only
(D) I and III
(E) II and III

8 If $\frac{x^2}{3} = x$, then $x =$

(A) 0 or -3
(B) 3 or -3
(C) 3 only
(D) 0 only
(E) 0 or 3

9 By how much does the sum of the roots of the equation $(x + 1)(x - 3) = 0$ exceed the product of its roots ?

(A) 1
(B) 2
(C) 3
(D) 4
(E) 5

10 If $x^2 - 63x - 64 = 0$ and p and n are integers such that $p^n = x$, which of the following CANNOT be a value for p?

(A) -8
(B) -4
(C) -1
(D) 4
(E) 64

11 If $r > 0$ and $r^t = 6.25r^{t+2}$, then $r =$

(A) $\frac{2}{5}$
(B) $\frac{4}{9}$
(C) $\frac{5}{8}$
(D) $\frac{3}{4}$
(E) $\frac{5}{4}$

Grid-In

1 If $(4p + 1)^2 = 81$ and $p > 0$, what is a possible value of p?

2 If $(x - 1)(x - 3) = -1$, what is a possible value of x?

3 By what amount does the sum of the roots exceed the product of the roots of the equation $(x - 5)(x + 2) = 0$?

LESSON

4-6 SYSTEMS OF EQUATIONS

OVERVIEW

*A **system of equations** is a set of equations whose solution makes each of the equations true at the same time.* SAT *questions involving systems of two equations with two different letters can usually be solved by:*

- *substituting the solution of one equation into the other equation to eliminate one of the letters in that equation; or*
- *adding or subtracting corresponding sides of the two equations so that an equation with only one letter results.*

SOLVING A SYSTEM OF TWO EQUATIONS BY SUBSTITUTION

If one of the equations in a system of two equations has the form

$$\textit{letter} = \text{math expression}$$

the same letter in the other equation can be replaced with that math expression. For example, if

$$3x = 2 \quad \text{and} \quad 5y + 3x = 7$$

we can find the value of y by replacing $3x$ with 2 in the second equation to obtain $5y + 2 = 7$. Hence, $5y = 5$, so $y = 1$.

EXAMPLE

If $y = 2x - 3$ and $x + y = 18$, what is the value of $y - x$?

Solution

- Substitute $2x - 3$ for y in the equation $x + y = 18$. Then solve for x.

$$x + y = 18$$

$$x + \overbrace{2x - 3}^{y} = 18$$

$$3x - 3 = 18$$

$$3x = 21$$

$$x = \frac{21}{3} = 7$$

- Find the corresponding value of y by substituting 7 for x in either of the original equations.

$$x + y = 18$$
$$7 + y = 18$$
$$y = 11$$

- The value of $y - x$ is $11 - 7 = 4$.

EXAMPLE

If $8b = 40$ and $bc = 1$, what is the value of c?

Solution

Use the first equation to help eliminate b in the second equation.

- Since $8b = 40$, $b = \dfrac{40}{8} = 5$.
- Substitute 5 for b in the second equation:

$$5c = 1, \quad \text{so} \quad c = \frac{1}{5}$$

EXAMPLE

If $2x - 3y = 6$ and $y - 5 = 3 - y$, what is the value of x?

Solution

- Since the second equation does not depend on x, solve it for y:

$$y - 5 = 3 - y$$
$$2y = 8$$
$$y = \frac{8}{2} = 4$$

- Substitute 4 for y in the first equation:

$$2x - 3y = 6$$
$$2x - 3(4) = 6$$
$$2x - 12 = 6$$
$$2x = 18$$
$$x = \frac{18}{2} = 9$$

SOLVING A SYSTEM OF EQUATIONS BY COMBINING CORRESPONDING SIDES

It may be possible to solve a system of two equations by writing one equation above the other equation and then adding or subtracting the like terms in each column so that one letter is eliminated.

EXAMPLE

If $x - 2y = 5$ and $x + 2y = 11$, what is the value of x?

Solution

Since the numerical coefficients of y in the two equations are opposites, adding the corresponding sides of the equations will eliminate y.

$$\begin{array}{r} x - 2y = 5 \\ + \quad x + 2y = 11 \\ \hline 2x + 0 = 16 \\ x = \dfrac{16}{2} = 8 \end{array}$$

The value of x is 8.

Before the equations are combined in a system of equations, it may be necessary to multiply one or both equations by a number that will eliminate one of the two letters when the two equations are added together.

EXAMPLE

If $2a = b + 7$ and $5a = 2b + 15$, what is the value of a?

Solution

- Rewrite each equation so that all letters are on the same side:

$$2a = b + 7 \quad \rightarrow \quad 2a - b = 7$$

$$5a = 2b + 15 \quad \rightarrow \quad 5a - 2b = 15$$

- Multiply the first equation by -2, so that the coefficient of b becomes $+2$. Then add the two equations to eliminate b:

$$\begin{array}{rcr} 2a - b = 7 & \rightarrow & -4a + 2b = -14 \\ 5a - 2b = 15 & \rightarrow & 5a - 2b = 15 \\ \hline & & a = 1 \end{array}$$

The value of a is 1.

SOLVING OTHER TYPES OF SYSTEMS OF EQUATIONS

If a system of equations has more letters than equations, you may be asked to solve for some combination of letters.

EXAMPLE

If $2r = s$ and $24t = 3s$, what is r in terms of t?

Solution

Since the question asks for r in terms of t, work toward eliminating s.

- Substitute $2r$ for s in the second equation:

$$24t = 3s = 3(2r) = 6r$$

- Solve for r in the equation:

$$24t = 6r$$
$$\frac{24t}{6} = \frac{6r}{6}$$
$$4t = r$$

Hence, $r = 4t$.

EXAMPLE

If $ab - 3 = 12$ and $2bc = 5$, what is the value of $\frac{a}{c}$?

Solution

Since the question asks for $\frac{a}{c}$, you must eliminate b.

- Find the value of ab in the first equation. Since $ab - 3 = 12$, then $ab = 15$.
- To eliminate b, divide corresponding sides of $ab = 15$ and $2bc = 5$:

$$\frac{ab}{2bc} = \frac{15}{5}$$
$$\frac{ab}{2bc} = 3$$
$$\frac{a}{2c} = 3$$

Solve the resulting equation for $\frac{a}{c}$:

$$2\left(\frac{a}{2c}\right) = 2(3)$$
$$\frac{a}{c} = 6$$

The value of $\frac{a}{c}$ is 6.

Look back at strategies 12 and 13 on pages 35 and 36 for more examples.

Lesson 4-6 Tune-Up Exercises

Multiple-Choice

1 If $2x - 3y = 11$ and $3x + 15 = 0$, what is the value of y?

(A) -7
(B) -5
(C) $\frac{1}{3}$
(D) 3
(E) 10

2 If $2a = 3b$ and $4a + b = 21$, then $b =$

(A) 1
(B) 3
(C) 4
(D) 7
(E) 8

3 If $2p + q = 11$ and $p + 2q = 13$, then $p + q =$

(A) 6
(B) 8
(C) 9
(D) 12
(E) 18

4 If $m + p + k = 70$, $p = 2m$, and $k = 2p$, then $m =$

(A) 2
(B) 5
(C) 7
(D) 10
(E) 14

5 If $x - y = 3$ and $x + y = 5$, what is the value of y?

(A) -4
(B) -2
(C) -1
(D) 1
(E) 2

6 If $5x + y = 19$ and $x - 3y = 7$, then $x + y =$

(A) -4
(B) -1
(C) 3
(D) 4
(E) It cannot be determined from the information given.

7 If $x - 9 = 2y$ and $x + 3 = 5y$, what is the value of x?

(A) -2
(B) 4
(C) 11
(D) 15
(E) 17

8 If $\frac{1}{x} + \frac{1}{y} = \frac{1}{4}$ and $\frac{1}{x} - \frac{1}{y} = \frac{3}{4}$, then $x =$

(A) $\frac{1}{4}$
(B) $\frac{1}{2}$
(C) 1
(D) 2
(E) 4

9 If $5a + 3b = 35$ and $\frac{a}{b} = \frac{2}{5}$, what is the value of a?

(A) $\frac{14}{5}$
(B) $\frac{7}{2}$
(C) 5
(D) 7
(E) 9

10 If $\frac{x}{y} = 6, \frac{y}{w} = 4$ and $x = 36$, what is the value of w?

(A) $\frac{1}{2}$
(B) $\frac{3}{2}$
(C) 2
(D) 4
(E) 6

11 If $4r + 7s = 23$ and $r - 2s = 17$ then $3r + 3s =$

(A) 8
(B) 24
(C) 32
(D) 40
(E) 48

12 If $\frac{p-q}{2} = 3$ and $rp - rq = 12$, then $r =$

(A) -1
(B) 1
(C) 2
(D) 4
(E) It cannot be determined from the information given.

13 If $(a + b)^2 = 9$ and $(a - b)^2 = 49$, what is the value of $a^2 + b^2$?

(A) 17
(B) 20
(C) 29
(D) 58
(E) 116

14 If $\frac{r+s}{r} = 3$ and $\frac{t+r}{t} = 5$, what is the value of $\frac{s}{t}$?

(A) $\frac{1}{2}$
(B) $\frac{3}{5}$
(C) 4
(D) 8
(E) 16

15 If $3x + y = c$ and $x + y = b$, what is the value of x in terms of c and b?

(A) $\frac{c-b}{3}$
(B) $\frac{c-b}{2}$
(C) $\frac{b-c}{3}$
(D) $\frac{b-c}{2}$
(E) $\frac{c-b}{4}$

16 If $a + b = 11$ and $a - b = 7$, then $ab =$

(A) 6
(B) 8
(C) 10
(D) 12
(E) 18

$$x - z = 7$$
$$x + y = 3$$
$$z - y = 6$$

17 For the above system of three equations, $x =$

(A) 5
(B) 6
(C) 7
(D) 8
(E) 9

18 In a drama club, x students each contributed y dollars to buy a $60 gift for their adviser. If four more students had contributed, each pupil could have contributed 3 dollars less to buy the same gift. Which of the following pairs of equations expresses this relationship?

(A) $xy = 60$
$(x + 4)(y - 3) = 60$
(B) $xy = 60$
$(x - 4)(y + 3) = 60$
(C) $xy = 60$
$(x + 3)(y - 4) = 60$
(D) $xy = 60$
$xy = 60 - (4 \times 3)$
(E) $x + y = 60$
$(4x)(3y) = 60$

19 If $r^8 = 5$ and $r^7 = \frac{3}{t}$, what is the value of r in terms of t?

(A) $\frac{5}{3}t$
(B) $\frac{3}{5}t$
(C) $5 - \frac{3}{t}$
(D) $3 + \frac{5}{t}$
(E) $\frac{t}{15}$

20 If $\frac{a}{b} = \frac{6}{7}$ and $\frac{a}{c} = \frac{2}{5}$, what is the value of $3b + c$ in terms of a?

(A) $12a$
(B) $9a$
(C) $8a$
(D) $6a$
(E) $4a$

Grid-In

1 If 5 sips + 4 gulps = 1 glass and 13 sips + 7 gulps = 2 glasses, how many sips equal a gulp?

2 If $2a = 9 - b$ and $4a = 3b - 12$, what is the value of a?

3 John and Sara each bought the same type of pen and notebook in the school bookstore, which does not charge sales tax. John paid $5.55 for two pens and three notebooks, and Sara paid $3.50 for one pen and two notebooks. How much does the school bookstore charge for one notebook?

LESSON **4-7**

ALGEBRAIC INEQUALITIES

OVERVIEW

Linear inequalities *such as* $2x - 3 \leqslant 7$ *are solved by isolating the letter in much the same way that linear (first-degree) equations are solved. Multiplying or dividing both sides of an inequality by a negative number reverses the direction of the inequality sign. For example:*

$$5 > 3, \text{ so } 5 \times (-2) < 3 \times (-2)$$

Any general property of inequalities that is stated in this lesson for the "is less than" (<) relation is also true for each of the other (≤, >, and ≥) inequality relations.

SOME PROPERTIES OF INEQUALITIES

An equivalent inequality results when

- The same number, positive or negative, is added or subtracted on both sides of the inequality. For example:

$$6 > 4, \quad \text{so} \quad 6 + 2 > 4 + 2$$

$$6 > 4, \quad \text{so} \quad 6 - 2 > 4 - 2$$

> **HELPFUL HINT**
> The direction of the inequality sign also gets reversed when comparing the reciprocals of two positive numbers x and y:
>
> $$\text{If } x > y, \text{ then } \frac{1}{x} < \frac{1}{y}.$$

- Both sides of the inequality are multiplied or divided by the same *positive* number. For example:

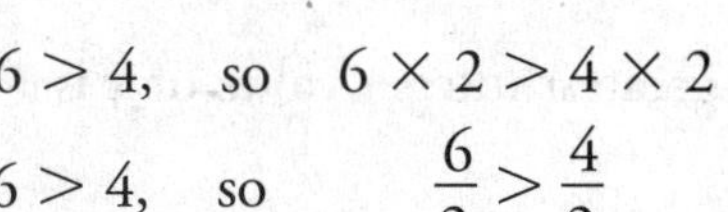

$$6 > 4, \quad \text{so} \quad 6 \times 2 > 4 \times 2$$

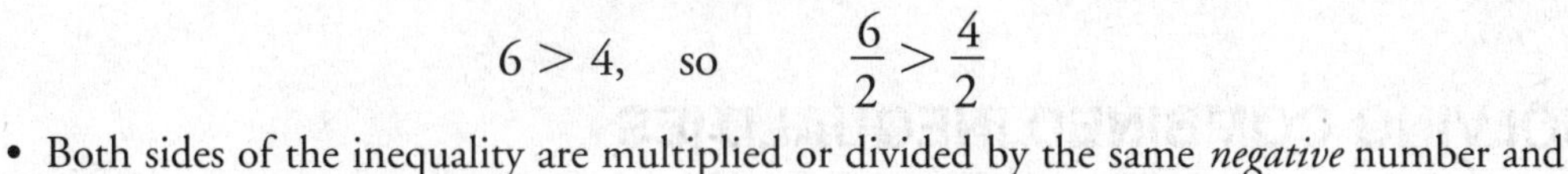

$$6 > 4, \quad \text{so} \quad \frac{6}{2} > \frac{4}{2}$$

- Both sides of the inequality are multiplied or divided by the same *negative* number and the direction of the inequality is reversed. For example:

$$6 > 4, \quad \text{so} \quad 6 \times (-2) < 4 \times (-2)$$

$$6 > 4, \quad \text{so} \quad \frac{6}{-2} < \frac{4}{-2}$$

SOLVING LINEAR INEQUALITIES

To solve a linear inequality, isolate the letter by performing the same arithmetic operation on both sides of the inequality. Remember to reverse the direction of the inequality sign whenever multiplying or dividing the inequality by a negative number.

EXAMPLE

If $3x - 2 < 10$, find x.

Solution

$$3x - 2 < 10$$
$$3x < 12$$
$$x < \frac{12}{3}$$
$$x < 4$$

The solution consists of all real numbers less than 4.

EXAMPLE

What is the *greatest* integer value of x such that $1 - 2x > 6$?

Solution

$$1 - 2x > 6$$
$$-2x > 5$$

Reverse the inequality: $x < \frac{5}{-2}$

Since x is less than $\frac{-5}{2}$, the greatest integer value of x is -3.

SOLVING COMBINED INEQUALITIES

To solve an inequality that has the form $c \leq ax + b \leq d$, where a, b, c, and d stand for numbers, isolate the letter by performing the same operation on each member of the inequality.

EXAMPLE

If $-2 \leq 3x - 7 \leq 8$, find x.

Solution

- Add 7 to each member of the inequality: $-2 \leq 3x - 7 \leq 8$

$$-2 + 7 \leq 3x - 7 + 7 \leq 8 + 7$$
$$5 \leq 3x \leq 15$$

- Divide each member of the inequality by 3: $\frac{5}{3} \leqslant \frac{3x}{3} \leqslant \frac{15}{3}$

$$\frac{5}{3} \leqslant x \leqslant 5$$

The solution consists of all real numbers greater than or equal to $\frac{5}{3}$ and less than or equal to 5.

ORDERING PROPERTIES OF INEQUALITIES

- If $a < b$ and $b < c$, then $a < c$. For example:

$$2 < 3 \text{ and } 3 < 4, \quad \text{so} \quad 2 < 4$$

- If $a < b$ and $x < y$, then $a + x < b + y$. For example:

$$2 < 3 \text{ and } 4 < 5, \quad \text{so} \quad 2 + 4 < 3 + 5$$

- If $a < b$ and $x > y$, then the relationship between $a + x$ and $b + y$ cannot be determined until each of the letters is replaced by a specific number.

Lesson 4-7 Tune-Up Exercises

Multiple-Choice

1 What is the largest integer value of p that satisfies the inequality $4 + 3p < p + 1$?

(A) -2
(B) -1
(C) 0
(D) 1
(E) 2

2 If $-3 < 2x + 5 < 9$, which of the following CANNOT be a possible value of x?

(A) -2
(B) -1
(C) 0
(D) 1
(E) 2

3 If the sum of a number and the original number increased by 5 is greater than 11, which could be a possible value of the number?

(A) -5
(B) -1
(C) 1
(D) 3
(E) 4

4 If $0 < a^2 < b$, which of the following statements is (are) always true?

I. $a < \dfrac{b}{a}$
II. $a^4 < a^2 b$
III. $\dfrac{a^2}{b} < 1$

(A) I only
(B) I and II
(C) II and III
(D) I and III
(E) I, II, and III

5 What is the smallest integer value of x that satisfies the inequality $4 - 3x < 11$?

(A) -3
(B) -2
(C) -1
(D) 0
(E) 1

6 If $a > b > c > 0$, which of the following statements must be true?

I. $\dfrac{a-c}{b-a} > \dfrac{b-c}{b-a}$
II. $ab > ac$
III. $\dfrac{b}{a} > \dfrac{b}{c}$

(A) I only
(B) II only
(C) III only
(D) I and II
(E) II and III

7 If n is an integer, how many different values of n satisfy the inequality $-4 \leq 3n \leq 87$?

(A) 32
(B) 31
(C) 30
(D) 29
(E) 28

8 Which of the following statements must be true when $a^2 < b^2$ and a and b are not 0?

I. $\dfrac{a^2}{a} < \dfrac{b^2}{a}$
II. $\dfrac{1}{a^2} > \dfrac{1}{b^2}$
III. $(a + b)(a - b) < 0$

(A) I only
(B) II only
(C) III only
(D) I and II
(E) II and III

9 For how many integer values of b is $b + 3 > 0$ and $1 > 2b - 9$?

(A) Four
(B) Five
(C) Six
(D) Seven
(E) Eight

10 If $xy > 1$ and $z < 0$, which of the following statements must be true?

I. $x > z$
II. $xyz < -1$
III. $\frac{xy}{z} < \frac{1}{z}$

(A) I only
(B) II only
(C) III only
(D) II and III
(E) None

Grid-In

1 For what integer value of y is $y + 5 > 8$ and $2y - 3 < 7$?

2 If 2 times an integer x is increased by 5, the result is always greater than 16 and less than 29. What is the least value of x?

3 If $2 < 20x - 13 < 3$, what is one possible value for x?

4 $\frac{1}{7} + \frac{1}{8} - \frac{1}{9} + \frac{1}{10} < \frac{1}{8} - \frac{1}{9} + \frac{1}{10} + \frac{1}{n}$

For the above inequality, what is the greatest possible positive integer value of n?

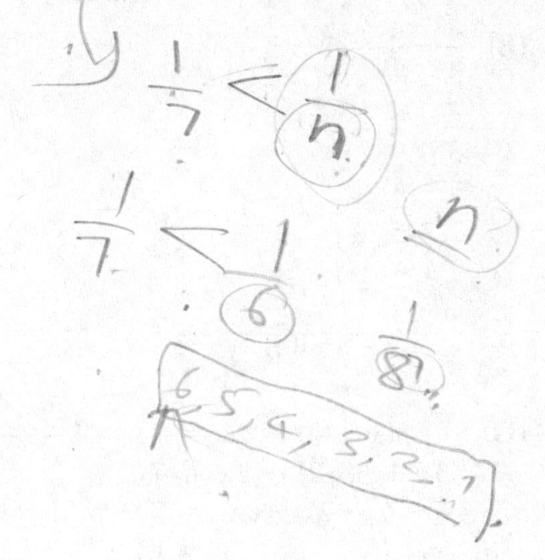

ANSWERS TO CHAPTER 4 TUNE-UP EXERCISES

Lesson 4-1 (*Multiple-Choice*)

1. **(A)**
$$\begin{aligned} x + 4x - 3x + 8 &= 0 \\ 5x - 3x + 8 &= 0 \\ 2x + 8 &= 0 \\ 2x &= -8 \\ x &= \frac{-8}{2} = -4 \end{aligned}$$

2. **(A)**
$$\begin{aligned} (9 - 4)(x + 4) &= 30 \\ 5(x + 4) &= 30 \\ x + 4 &= \frac{30}{5} = 6 \\ x &= 6 - 4 \\ &= 2 \end{aligned}$$

3. **(E)**
$$\begin{aligned} 2(1 + 5) &= 3(w - 4) \\ 2(6) &= 3(w) + 3(-4) \\ 12 &= 3w - 12 \\ 12 + 12 &= 3w \\ 24 &= 3w \\ \frac{24}{3} &= w \\ 8 &= w \end{aligned}$$

4. **(B)**
$$\begin{aligned} \frac{x}{3} - 1 &= 1 - 3 \\ \frac{x}{3} - 1 &= -2 \\ \frac{x}{3} &= -2 + 1 \\ \frac{x}{3} &= -1 \\ x &= -1(3) = -3 \end{aligned}$$

5. **(D)**
$$\begin{aligned} 3(x - 2) - 2x &= 7 \\ 3x + 3(-2) - 2x &= 7 \\ 3x - 6 - 2x + 6 &= 7 + 6 \\ x &= 13 \end{aligned}$$

6. **(C)** *Solution 1*: Solve $3x - 6 = 18$ for $x - 2$ by dividing each member of the equation by 2:
$$\begin{aligned} \frac{3x}{3} - \frac{6}{3} &= \frac{18}{3} \\ x - 2 &= 6 \end{aligned}$$

Solution 2: The answer choices represent possible values for $x - 2$. Add 2 to each answer choice to obtain the possible values for x. Then substitute each of these values into the given equation until you find one that works:
$$\begin{aligned} 3x - 6 &= 18 \\ 3(6 + 2) - 6 &= 18 \\ 24 - 6 &= 18 \end{aligned}$$

7. **(B)** If $2x + y = y + 14$, subtracting y from each side of the equation gives $2x = 14$, so
$$x = \frac{14}{2} = 7$$

8. **(B)** If $\frac{4}{7}k = 36$, then
$$\frac{1}{7}k = \frac{1}{4}(36) = 9$$
Since $\frac{1}{7}k = 9$, then
$$\frac{3}{7}k = 3(9) = 27$$

9. **(D)**
$$\begin{aligned} \frac{1}{2}x + \frac{1}{4}x + \frac{1}{8}x &= 14 \\ \frac{4}{8}x + \frac{2}{8}x + \frac{1}{8}x &= 14 \\ \frac{7}{8}x &= 14 \\ \frac{8}{7}\left(\frac{7}{8}x\right) &= \frac{8}{7}(14) \\ x &= 16 \end{aligned}$$

10. **(B)** If $2s + 3t = 12$ and $4s = 36$, the second equation can be used to eliminate s in the first equation. Since $\frac{4s}{2} = 2s = \frac{36}{2} = 18$, replace $2s$ with 18 in the first equation. Then solve for t:
$$\begin{aligned} 18 + 3t &= 12 \\ 3t &= 12 - 18 \\ 3t &= -6 \\ t &= \frac{-6}{3} = -2 \end{aligned}$$

11. **(E)**
    - If $2x + 5 = -25$, then $2x = -30$, so

$$x = \frac{-30}{2} = -15$$

    - If $-3y - 6 = 48$, then $-3y = 54$, so

$$y = \frac{-54}{3} = -18$$

    - Hence, $xy = (-15)(-18) = 270$.

12. **(C)** If $\frac{2}{x} = 2$, then $x = 1$ since $\frac{2}{1} = 2$. Hence,

$$x + 2 = 1 + 2 = 3$$

13. **(A)** If $\frac{y-2}{2} = y + 2$, then $y - 2 = 2(y + 2)$. Eliminate the parentheses, and then collect all the terms involving y on the same side of the equation.

$$\begin{aligned} y - 2 &= 2(y + 2) \\ &= 2y + 4 \\ y - 2y &= 4 + 2 \\ -y &= 6, \text{ so } y = -6 \end{aligned}$$

14. **(E)** If $\frac{2y}{7} = \frac{y+3}{4}$, set the cross-products equal and then solve the resulting equation.

$$\begin{aligned} \frac{2y}{7} &= \frac{y+3}{4} \\ 4(2y) &= 7(y + 3) \\ 8y &= 7y + 21 \\ 8y - 7y &= 21 \\ y &= 21 \end{aligned}$$

15. **(E)** If $\frac{y}{3} = 4$, then $y = 3(4) = 12$, so

$$3y = 3(12) = 36$$

16. **(B)** To add fractions that have the same denominator, write the sum of the numerators over the common denominator. Since $\frac{1}{q} + \frac{3}{q} = 12$, then $\frac{4}{q} = 12$, so $4 = 12q$. Dividing each side of the equation $4 = 12q$ by 12 gives

$$q = \frac{4}{12} = \frac{1}{3}$$

17. **(D)** To solve $8 - 3x = 2x + 13$, collect the numbers on the left side of the equation and the letters on the right side.

$$\begin{aligned} 8 - 3x &= 2x + 13 \\ 8 - 13 &= 2x + 3x \\ -5 &= 5x \\ \frac{-5}{5} &= \frac{5x}{5} \\ -1 &= x \end{aligned}$$

18. **(A)** To solve $(1-2-3-4)w = (-1)(-2)(-3)(-4)$, first simplify the left side of the equation. Since $1 - 2 - 3 - 4 = 1 - 9 = -8$, then $-8w = (-1)(-2)(-3)(-4)$. The product of an even number of negative numbers is positive, so $(-1)(-2)(-3)(-4) = 24$. The original equation simplifies to

$$\begin{aligned} -8w &= 24 \\ w &= \frac{24}{-8} = -3 \end{aligned}$$

19. **(D)** Since $\frac{1}{3} + \frac{3}{p} = 1$, isolate the letter by subtracting $\frac{1}{3}$ from each side of the equation:

$$\frac{3}{p} = 1 - \frac{1}{3} = \frac{2}{3}$$

Eliminate the fractions in this equation by cross-multiplying:

$$\begin{aligned} \frac{3}{p} &= \frac{2}{3} \\ 2p &= 9 \\ p &= \frac{9}{2} \end{aligned}$$

20. **(E)** Write the equation $\frac{3}{4} = \frac{6}{x} = \frac{9}{y}$ as two equations:

$$\frac{3}{4} = \frac{6}{x} \text{ and } \frac{3}{4} = \frac{9}{y}$$

Solve the first equation for x:

$$\begin{aligned} \frac{3}{4} &= \frac{6}{x} \\ 3x &= 24 \\ x &= \frac{24}{3} = 8 \end{aligned}$$

Solve the second equation for y:

$$\frac{3}{4} = \frac{9}{y}$$
$$3y = 36$$
$$y = \frac{36}{3} = 12$$

Hence,

$$x + y = 8 + 12 = 20$$

(*Grid-In*)

1. **6/4** If $2x - 1 = 11$, then $2x = 12$, so $x = \frac{12}{2} = 6$. Since $3y = 12$, then $y = \frac{12}{3} = 4$. Hence, $\frac{x}{y} = \frac{6}{4}$. Grid in as 6/4 or 1.5.

2. **8** If $7(a + b) - 4(a + b) = 24$, then $3(a + b) = 24$, so

$$a + b = \frac{24}{3} = 8$$

3. **2/12** If $1 - x - 2x - 3x = 6x - 1$, then $1 - 6x = 6x - 1$, so $1 + 1 = 6x + 6x$ and $2 = 12x$. Thus, $x = \frac{2}{12}$. Grid in as 2/12.

4. **10/3** In the equation $p = \frac{5b}{c^2}$, if $p = 9$ and $b = 20$, then $9 = \frac{5(20)}{c^2}$ or $9c^2 = 100$, so $c^2 = \frac{100}{9}$. Taking the positive square root of both sides of the equation gives

$$c = \frac{\sqrt{100}}{\sqrt{9}} = \frac{10}{3}$$

Grid in as 10/3.

5. **8/5** If $\frac{3(y + 8)}{2y} = 9$, then $3(y + 8) = 9 \times 2y$, so $3 \cdot y + 3 \cdot 8 = 18y$. Collecting like terms on the same side of the equation makes $24 = 15y$, so $y = \frac{24}{15} = \frac{8}{5} = 1.6$. Grid in as 8/5 or 1.6.

Lesson 4-2 (*Multiple-Choice*)

1. **(B)** Dividing each member of the given equation, $6 = 2x + 4y$, by 2 will leave $x + 2y$ on the right side of the equation:

$$\frac{6}{2} = \frac{2x}{2} + 4y$$
$$3 = x + 2y$$

The value of $x + 2y$ is 3.

2. **(C)** Multiplying each member of the given equation, $\frac{a}{2} + \frac{b}{2} = 3$, by 4 will make the left side of the equation equal to $2a + 2b$:

$$4\left(\frac{a}{2}\right) + 4\left(\frac{b}{2}\right) = 4(3)$$
$$2a + 2b = 12$$

The value of $2a + 2b$ is 12.

3. **(B)** For the given equation, $2s - 3t = 3t - s$, finding s in terms of t means solving the equation for s by treating t as a constant. Work toward isolating s by first adding $3t$ on each side of the equation:

$$2s = 3t + 3t - s$$
$$= 6t - s$$

Next, add s on each side of the equation:

$$2s + s = 6t$$
$$3s = 6t$$
$$s = \frac{6t}{3} = 2t$$

4. **(E)** • If $a + b = 5$, then $2(a + b) = 2(5)$, so $2a + 2b = 10$.
   • If $\frac{c}{2} = 3$, then $4\left(\frac{c}{2}\right) = 4(3)$, so $2c = 12$.

   Hence, $2a + 2b + 2c = 10 + 12 = 22$.

5. **(C)** If $xy + z = y$, then $xy = y - z$, so $x = \frac{y - z}{y}$.

6. **(D)** If $b(x + 2y) = 60$, then $bx + 2by = 60$. Since it is also given that $by = 15$:

$$bx + 2by = 60$$
$$bx + 2(15) = 60$$
$$bx + 30 = 60$$
$$bx = 60 - 30 = 30$$

7. **(B)** Since x^2 and y^2 are both greater than or equal to 0, the only values of x^2 and y^2 for which $2x^2 + 3y^2 = 0$ are $x^2 = y^2 = 0$. If $x^2 = y^2 = 0$, then $x = y = 0$, so
$$3x + 2y = 3(0) + 2(0) = 0$$

8. **(D)** If $\frac{a-b}{b} = \frac{2}{3}$, then
$$\frac{a}{b} - \frac{b}{b} = \frac{2}{3}$$
$$\frac{a}{b} - 1 = \frac{2}{3}$$
$$\frac{a}{b} = 1 + \frac{2}{3}$$
$$= \frac{5}{3}$$
The value of $\frac{a}{b}$ is $\frac{5}{3}$.

9. **(A)** To divide powers that have the same base, *subtract* their exponents. If $\frac{10^x}{10^y} = 100$, then $100^{x-y} = 100 = 10^2$, so $x - y = 2$. Hence, $x = y + 2$.

10. **(D)** Since $\frac{1}{p+q} = r = \frac{r}{1}$, eliminate the fractions by cross-multiplying:
$$r(p + q) = 1(1)$$
$$rp + rq = 1$$
$$rp = 1 - rq$$
$$p = \frac{1 - rq}{r}$$

11. **(B)** Determine whether each Roman numeral expression is equal to $4a$.
- I. Since $a = 2b = 5c$, then $a = 5c$, so
$$4a = 4(5c) = 20c$$
Expression I is not equal to $4a$.
- II. Rewrite $4a$ as $2a + 2a$. Since $a = 2b = 5c$, then $2a = 2(2b) = 4b$ and $2a = 2(5c) = 10c$, so
$$4a = 2a + 2a = 4b + 10c$$
Expression II is equal to $4a$.
- III. Rewrite $4a$ as $a + 3a$. Since $a = 2b$ and $a = 5c$, then $3a = 3(5c) = 15c$. Hence,
$$4a = a + 3a = 2b + 15c$$
Expression III is equal to $4a$.

Only Roman numeral expressions II and III are equal to $4a$.

12. **(B)** Since $\frac{a+b+c}{3} = \frac{a+b}{2}$, eliminate the fractions by cross-multiplying:
$$2(a + b + c) = 3(a + b)$$
$$2a + 2b + 2c = 3a + 3b$$
$$2c = (3a - 2a) + (3b - 2b)$$
$$= a + b$$
$$c = \frac{a+b}{2}$$

13. **(E)** If $wx = z$, then $x = \frac{z}{w}$, so
$$xz = z\left(\frac{z}{w}\right) = \frac{z^2}{w}$$

14. **(A)** The value in cents of n nickels plus d dimes is $5n + 10d$, which you are told is equal to c cents. Hence, $5n + 10d = c$ or $5n = c - 10d$, so
$$n = \frac{c}{5} - \frac{10d}{5} = \frac{c}{5} - 2d$$

15. **(D)** Since $a = 2c$ and $b = 5d$, replace a in the equation $\frac{c}{d} - \frac{a}{b} = x$, with $2c$ and replace b with $5d$:
$$\frac{c}{d} - \frac{2c}{5d} = x$$
$$\frac{5c}{5d} - \frac{2c}{5d} = x$$
$$\frac{5c - 2c}{5d} = x$$
$$\frac{3c}{5d} = x$$
To find the value of $\frac{c}{d}$ in terms of x, multiply both sides of the equation $\frac{3c}{5d} = x$ by the reciprocal of $\frac{3}{5}$:
$$\frac{5}{3}\left(\frac{3c}{5d}\right) = \frac{5}{3}x$$
$$\frac{c}{d} = \frac{5}{3}x$$

16. **(D)** If $kx - 4 = (k - 1)x$, removing parentheses makes $kx - 4 = kx - x$ so $-4 = -x$ or $x = 4$

17. **(D)** If $c = b + 1$, then $b = c - 1$. Hence, $p = 4b + 5 = 4(c - 1) + 5 = 4c - 4 + 5 = 4c + 1$.

18. **(A)** If $2p + r - 1 = 2r + p + 1$, then, after like terms are collected on the same side of the equation, $p = r + 2$ where p and r are given as positive integers.
    - I. Since p is 2 more than r, p and r cannot be consecutive integers. Hence, Roman numeral choice I is false.
    - II. Since $p = r + 2$, p can be either odd (if r is odd) or even (if r is even). Hence, Roman numeral choice II is false.
    - III. Roman numeral choice III is also false since r can be either even or odd. Since none of the Roman numeral choices must be true, the correct choice is (A).

(*Grid-In*)

1. **8** Since $(4 \times b)^2 = 4^2 \times b^2 = 16 \times b^2$,

$$\begin{aligned} 16 \times a^2 \times 64 &= 16 \times b^2 \\ \not{16} \times a^2 \times 64 &= \not{16} \times b^2 \\ a^2 \times 64 &= b^2 \\ \sqrt{a^2 \times 64} &= \sqrt{b^2} \\ a \times 8 &= b \end{aligned}$$

Hence, b is 8 times as great as a.

2. **5** Rearrange the terms of $3a + 3b - c = 40$ to get $(3a - c) + 3b = 40$. Because $3a - c = 5b$, $5b + 3b = 40$; so, $8b = 40$ and $b = \frac{40}{8} = 5$.

3. **18** If $a = 2x + 3$ and $b = 4x - 7$, when $3b = 5a$, x must satisfy the equation $3(4x - 7) = 5(2x + 3)$. Removing parentheses makes $12x - 21 = 10x + 15$. Collecting like terms gives $12x - 10x = 15 + 21$ or $2x = 36$ so $x = \frac{36}{2} = 18$.

4. **5** Multiplying both sides of $\frac{x}{8} = \frac{y}{5} = \frac{31}{40}$ by 40 produces the eqivalent equation $5x + 8y = 31$. Substitute consecutive positive integer values for x until you find one that makes y have a positive integer value.

x	$5x$	$y = \frac{31 - 5x}{8}$	
1	5	$y = \frac{31 - 5}{8} = \frac{26}{8}$	
2	10	$y = \frac{31 - 10}{8} = \frac{21}{8}$	
3	15	$y = \frac{31 - 15}{8} = \frac{16}{8} = 2$	Stop!

Hence, $x + y = 3 + 2 = 5$.

LESSON 4-3 (*Multiple-Choice*)

1. **(D)** Write each term of the polynomial numerator separately over the monomial denominator. Then divide powers of the same base by subtracting their exponents.

$$\begin{aligned} \frac{20b^3 - 8b}{4b} &= \frac{20b^3}{4b} - \frac{8b}{4b} \\ &= 5b^{3-1} - 2 \\ &= 5b^2 - 2 \end{aligned}$$

2. **(D)** The given expression $(39)^2 + 2(39)(61) + (61)^2$ has the form

$$x^2 + 2xy + y^2 = (x + y)^2$$

where $x = 39$ and $y = 61$. Hence,

$$\begin{aligned} (39)^2 + 2(39)(61) + (61)^2 &= (39 + 61)^2 \\ &= (100)^2 \\ &= 10{,}000 \end{aligned}$$

3. **(D)** Before multiplying, simplify each expression inside the parentheses:

$$\begin{aligned} (4a + a - 3)(4b - 2b - b) &= (5a - 3)(4b - 3b) \\ &= (5a - 3)(b) \\ &= 5ab - 3b \end{aligned}$$

4. **(E)** If $(x - y)^2 = 50$ and $xy = 7$, then

$$\begin{aligned} (x - y) = x^2 - 2xy + y^2 &= 50 \\ x^2 - 2(7) + y^2 &= 50 \\ x^2 - 14 + y^2 &= 50 \\ x^2 + y^2 &= 50 + 14 \\ &= 64 \end{aligned}$$

5. **(B)** *Solution 1:* Do the algebra. If $p = \dfrac{a}{a-b}$, then

$$1 - p = 1 - \frac{a}{a-b}$$
$$= \frac{a-b}{a-b} - \frac{a}{a-b}$$
$$= \frac{a-b-a}{a-b}$$
$$= \frac{-b}{a-b}$$

Since $\dfrac{-b}{a-b}$ is not one of the answer choices, eliminate the negative sign in the numerator by multiplying the numerator and the denominator by -1:

$$1 - p = \left(\frac{-1}{-1}\right)\frac{-b}{a-b}$$
$$= \frac{(-1)(-b)}{(-1)(a-b)} = \frac{b}{-a+b} = \frac{b}{b-a}$$

Solution 2: Substitute numbers for the letters. Let $a = 3$ and $b = 2$; then

$$p = \frac{b}{a-b} = \frac{3}{3-2} = 3$$

so $1 - p = 1 - 3 = -2$. When you plug in 3 for a and 2 for b in each of the answer choices, you find that only choice (B) produces -2.

6. **(E)** You are given that $(p - q)^2 = 25$ and $pq = 14$. Use the formula for the square of a binomial to expand $(p - q)^2$:

$$(p-q)^2 = p^2 - 2pq + q^2 = 25$$
$$p^2 - 2(14) + q^2 = 25$$
$$p^2 + q^2 = 25 + 28 = 53$$

Now use the formula for $(p + q)^2$:

$$(p+q)^2 = p^2 + 2pq + q^2$$
$$= 53 + 2(14)$$
$$= 53 + 28 = 81$$

7. **(A)** If $a - b = p$ and $a + b = k$, then

$$a^2 - b^2 = (a-b)(a+b)$$
$$= (p)(k)$$
$$= pk$$

8. **(E)** You are told that $(2y + k)^2 = 4y^2 - 12y + k^2$. Use the formula for expanding the square of a binomial:

$$(2y+k)^2 = (2y)^2 + 2(2yk) + k^2$$
$$= 4y^2 + 4yk + k^2$$

Compare $4y^2 - 12y + k^2$ with $4y^2 + 4yk + k^2$. Since the two trinomials must be the same, then $-12y = 4yk$, so

$$k = \frac{-12y}{4y} = -3$$

9. **(C)** Examine each choice in turn:

- (A) $(x + y)^2 = x^2 + y^2$ is false since $(x + y)^2 = x^2 + 2xy + y^2$.
- (B) $x^2 + x^2 = x^4$ is false since $x^2 + x^2 = 1x^2 + 1x^2 = 2x^2$.
- (C) $\dfrac{2^{x+2}}{2^x} = 4$ is true since

$$\frac{2^{x+2}}{2^x} = 2^{(x+2)-x}$$
$$= 2^2$$
$$= 4$$

- (D) $(3x)^2 = 6x^2$ is false since $(3x)^2 = 3^2 \cdot x^2 = 9x^2$.
- (E) $x^5 - x^3 = x^2$ is false since only like terms can be subtracted.

10. **(D)** Since

$$(px + q)^2 = p^2x^2 + 2pq + q^2$$

and

$$(px + q)^2 = 9x^2 + kx + 16,$$

then $p^2 = 9$ and $q^2 = 16$, so $p = \pm 3$ and $q = \pm 4$. Since the answer choices contain only positive numbers, use $p = 3$ and $q = 4$. The middle terms of $p^2x^2 + 2pq + q^2$ and $9x^2 + kx + 16$ must be equal, so

$$k = 2pq = 2(3)(4) = 24.$$

11. **(D)** $\left(\dfrac{x}{100} - 1\right)\left(\dfrac{x}{100} + 1\right) = kx^2 - 1$

$$(0.01x - 1)(0.01x + 1) = kx^2 - 1$$
$$0.0001x^2 - 1 = kx^2 - 1$$

Since the coefficients of x on both sides of the equation must be the same, $k = 0.0001$.

12. **(E)** Use FOIL to multiply the left side of the given equation:
$$(x + 5)(x + p) = x^2 + 2x + k$$
$$x^2 + 5x + px + 5p = x^2 + 2x + k$$
Since $5x + px$ must be equal to $2x$, $p = -3$. Hence,
$$k = 5p = 5(-3) = -15$$

13. **(D)** Multiply on each side of the given equation, $(x - 2)(x + 2) = x(x - p)$. The result is
$$x^2 - 4 = x^2 - xp$$
so $4 = xp$ and $p = \frac{4}{x}$.

14. **(C)** Use the formula $\frac{1}{x} + \frac{1}{y} = \frac{x + y}{xy}$. Since
$$\frac{1}{\frac{1}{x} + \frac{1}{y}}$$
is the reciprocal of $\frac{x + y}{xy}$, then
$$\frac{1}{\frac{1}{x} + \frac{1}{y}} = \frac{xy}{x + y}$$

15. **(D)** Use the formula for the square of a binomial to expand the left side of the given equation:
$$\left(k + \frac{1}{k}\right)^2 = 16$$
$$k^2 + 2(k)\left(\frac{1}{k}\right) + \left(\frac{1}{k}\right)^2 = 16$$
$$k^2 + 2 + \frac{1}{k^2} = 16$$
$$k^2 + \frac{1}{k^2} = 16 - 2 = 14$$

16. **(E)** Change the right side of the given equation, $\frac{a}{b} = 1 - \frac{x}{y}$, into a single fraction:
$$\frac{a}{b} = 1 - \frac{x}{y}$$
$$= \frac{y}{y} - \frac{x}{y}$$
$$= \frac{y - x}{y}$$
Since $\frac{b}{a}$ is the reciprocal of $\frac{a}{b}$,
$$\frac{b}{a} = \frac{y}{y - x}$$

(Grid-In)

1. **19** Since
$$(3y - 1)(2y + k) = ay^2 + by - 5$$
and the product of the last terms of the two binomial factors is equal to the constant term, $(-1)(k) = -5$, so $k = 5$. Now multiply the two binomials together:
$$(3y - 1)(2y + 5) = (3y)(2y) + (3y)(5) + (-1)(2y) + (-1)(5)$$
$$= 6y^2 + 15y - 2y - 5$$
$$= 6y^2 + 13y - 5$$
Since
$$(3y - 1)(2y + 5) = 6y^2 + 13y - 5 = ay^2 + by - 5$$
equating the coefficients makes $a = 6$ and $b = 13$, so
$$a + b = 6 + 13 = 19$$

2. **20**
$$4x^2 + 20x + r = (2x + s)^2$$
$$= (2x + s)(2x + s)$$
$$= (2x)(2x) + (2x)(s) + (s)(2x) + (s)(s)$$
$$= 4x^2 + 2sx + 2sx + s^2$$
$$= 4x^2 + 4sx + s^2$$
Since the coefficients of x on each side of the equation must be the same, $20 = 4s$, so $s = 5$. Comparing the last terms of the polynomials on the two sides of the equation makes $r = s^2 = 5^2 = 25$. Hence,
$$r - s = 25 - 5 = 20$$

LESSON 4-4 *(Multiple-Choice)*

1. **(E)** Factor the numerator and the denominator. Then divide out any factor that is common to both the numerator and the denominator:
$$\frac{4x + 4y}{2x^2 - 2y^2} = \frac{4(x + y)}{2(x^2 - y^2)}$$
$$= \frac{4(x + y)}{2(x + y)(x - y)}$$
$$= \frac{2}{x - y}$$

2. **(A)** Combine the fractions and then simplify:

$$\frac{a}{a^2-b^2}+\frac{b}{a^2-b^2}=\frac{a+b}{a^2-b^2}$$
$$=\frac{\overset{1}{\cancel{(a+b)}}}{\cancel{(a+b)}(a-b)}$$
$$=\frac{1}{a-b}$$

3. **(A)** To solve $ax + x^2 = y^2 - ay$ for a in terms of x and y, isolate a on the left side of the equation:

$$ax + x^2 = y^2 - ay$$
$$ax + ay = y^2 - x^2$$
$$a(x + y) = (y - x)(y + x)$$
$$\frac{a(x+y)}{(x+y)}=\frac{(y-x)\overset{1}{\cancel{(y+x)}}}{\cancel{(x+y)}}$$
$$a = y - x$$

4. **(C)** If $\frac{xy}{x+y} = 1$, then multiplying both sides of the equation by $x + y$ gives $xy = x + y$, so $xy - x = y$. Hence, $x(y - 1) = y$, so

$$x=\frac{y}{y-1}$$

5. **(B)** Since

$$x^2 = r^2 + 2rs + s^2 = (r + s)^2$$

and

$$y^2 = r^2 - s^2 = (r + s)(r - s)$$

then

$$\frac{x^2}{y^2}=\frac{(r+s)(r+s)}{(r+s)(r-s)}=\frac{r+s}{r-s}$$

You are told that x and $y > 0$, so

$$\frac{x}{y}=\sqrt{\frac{r+s}{r-s}}$$

6. **(A)** Simplify each fraction, then add:

$$h=\frac{x^2-1}{x+1}+\frac{x^2-1}{x-1}$$
$$=\frac{(x+1)(x-1)}{x+1}+\frac{(x+1)(x-1)}{x-1}$$
$$=(x-1)+(x+1)$$
$$=2x$$
$$\frac{h}{2}=x$$

7. **(A)** Collect the terms involving a on one side of the given equation and the terms involving b on the opposite side of the equation:

$$ax^2 - bx = ay^2 + by$$
$$ax^2 - ay^2 = bx + by$$

Factor each side of the equation:

$$ax^2 - ay^2 = bx + by$$
$$a(x^2 - y^2) = b(x + y)$$

Divide each side of the equation by b and $x^2 - y^2$:

$$\frac{a}{b}=\frac{x+y}{x^2-y^2}$$
$$=\frac{x+y}{(x+y)(x-y)}$$
$$=\frac{1}{x-y}$$

8. **(C)** Solve the given equation, $\frac{a^2-b^2}{9} = a + b$, for $a^2 - b^2$. Then solve the equation that results for $a - b$.

$$\frac{a^2-b^2}{9}=a+b$$
$$a^2 - b^2 = 9(a + b)$$
$$(a + b)(a - b) = 9(a + b)$$
$$a-b=\frac{9(a+b)}{a+b}=9$$

LESSON 4-5 (*Multiple-Choice*)

1. **(C)** If $(x + 7)(x - 3) = 0$, then either or both factors may be equal to 0. If $x + 7 = 0$, then $x = -7$. Also, if $x - 3 = 0$, then $x = 3$. Hence, x may be equal to -7 or 3.

2. **(D)** Multiplying both sides of the given equation, $\frac{a^2}{2} = 2a$, by 2 gives $a^2 = 4a$. To apply the zero-product rule, one side of the equation must be 0. After $4a$ is subtracted from both sides of the equation, $a^2 - 4a = 0$, which can be factored as $a(a - 4) = 0$. Thus, either $a = 0$ or $a - 4 = 0$. Hence, a equals 0 or 4.
3. **(B)** If x is a positive integer and $(x - 6)^2 = 9$, the expression inside the parentheses must be equal to 3 or -3.
   - If $x - 6 = 3$, then $x = 9$ and $x^2 = 81$. But 81 is not one of the choices.
   - If $x - 6 = -3$, then $x = -3 + 6 = 3$, so $x^2 = 9$.
4. **(E)** If $a(x - y) = 0$, then either $a = 0$ *or* $x - y = 0$ *or* both factors are 0. Determine whether each Roman numeral statement is always true.
   - I. It may be the case that $x - y = 0$ and $a \neq 0$, so statement I is not always true.
   - II. Since y can be any number, provided that $a = 0$ or $x = y$, statement II is not necessarily true.
   - III. It may be the case that $a = 0$ and $x - y \neq 0$, so statement III is not always true.

   None of the Roman numeral statements is always true.
5. **(D)** *Solution 1*: If $(x - 1)^2 - (x - 1) = 0$, then factoring the left side of the equation gives
   $$(x - 1)\,[(x - 1) - 1] = 0$$
   which simplifies to $(x - 1)(x - 2) = 0$. Hence, x may be equal to 1 or 2.
   Solution 2: Substitute the pair of values in each of the answer choices into the given equation, $(x - 1)^2 - (x - 1) = 0$, until you find a pair, $x = 1$ or $x = 2$, that works.
6. **(E)** If $w^2 - 2w = 0$, then $w(w - 2) = 0$, so $w = 0$ or $w = 2$.
   - If $2y^2 - y = 0$, then $y(2y - 1) = 0$, so $y = 0$ or $2y - 1 = 0$.
   - If $2y - 1 = 0$, then $y = \frac{1}{2}$.

   Since you are told that w and y are positive, $w = 2$ and $y = \frac{1}{2}$. Hence:
   $$\frac{w}{y} = 2 \div \frac{1}{2} = 2 \times 2 = 4$$
7. **(C)** For each Roman numeral equation, check whether -2 is a root.
   - I. If $\frac{2}{x} - x = 0$, then $\frac{2}{x} = x$. If $x = -2$, then
   $$\frac{2}{x} = \frac{2}{-2} = -1 \neq x$$

   Hence, -2 is not a root of equation I.
   - II. If $x^2 + 4 = 0$, then $(-2)^2 + 4 = 4 + 4 \neq 0$

   Hence, -2 is not a root of equation II.
   - III. If $x^2 + 4x + 4 = 0$, then $(-2)^2 + 4(-2) + 4 = 4 - 8 + 2 = 0$

   Hence, -2 is a root of equation III.
   Only equation III has -2 as one of its roots.
8. **(E)** If $\frac{x^2}{3} = x$, then $x^2 = 3x$, so $x^2 - 3x = 0$. Factoring $x^2 - 3x = 0$ gives $x(x - 3) = 0$. Thus, $x = 0$ or $x = 3$.
9. **(E)** If $(x + 1)(x - 3) = 0$, then $x + 1 = 0$ or $x - 3 = 0$. Hence, $x = -1$ or $x = 3$. The sum of these roots is $-1 + 3$ or 2, and their product is $(-1) \times (3) = -3$. Since
   $$2 - (-3) = 2 + 3 = 5$$
   the sum of the roots of the equation exceeds the product of its roots by 5.
10. **(B)** If $x^2 - 63x - 64 = 0$, then
    $$(x - 64)(x + 1) = 0$$
    so $x = 64$ or $x = -1$. If p and n are integers such that $p^n = x$, then either $p^n = 64$ or $p^n = -1$. Examine each answer choice in turn until you find a number that cannot be the value of p in either $p^n = 64$ or $p^n = -1$.
    - (A) If $p = -8$, then $(-8)^n = 64$, so $n = 2$.
    - (B) If $p = -4$, then there is no integer value of n for which $(-4)^n = 64$ or $(-4)^n = -1$.

11. **(A)** Isolate variable r by dividing both sides of the equation by r^t:

$$\begin{aligned} r^t &= 6.25r^{t+2} \\ 1 &= \frac{6.25r^{t+2}}{r^t} \\ &= 6.25r^{(t+2)-t} \\ &= 6.25r^2 \end{aligned}$$

Using a calculator, divide both sides of the equation by 6.25. Since $\frac{1}{6.25} = 0.16 = r^2$,

$$r = \sqrt{0.16} = 0.4 \text{ or } \frac{2}{5}$$

(*Grid-In*)

1. **2** If $(4p + 1)^2 = 81$ and $p > 0$, the expression inside the parentheses is either 9 or -9. Since $p > 0$, let $4p + 1 = 9$; then $4p = 8$ and $p = 2$. A possible value of p is 2.
2. **2** If $(x - 1)(x - 3) = -1$, then
$$x^2 - 4x + 3 = -1$$
so $x^2 - 4x + 4 = 0$. Factoring this equation gives $(x - 2)(x - 2) = 0$. Hence, a possible value of x is 2.
3. **13** The roots of the equation $(x - 5)(x + 2) = 0$ are the values of x that make the equation a true statement: $x = 5$ or $x = -2$. The sum of the roots is $5 + (-2)$ or 3, and the product of the roots is $(5)(-2)$ or -10. Hence, the sum, 3, exceeds the product, -10, by $3 - (-10)$ or 13.

LESSON 4-6 (*Multiple-Choice*)

1. **(A)** First solve the equation that contains one variable. Since $3x + 15 = 0$, then $3x = -15$, so
$$x = \frac{-15}{3} = -5$$
Substituting -5 for x in the other equation, $2x - 3y = 11$, gives $2(-5) - 3y = 11$ or $-10 - 3y = 11$. Adding 10 to both sides of the equation makes $-3y = 21$, so
$$y = \frac{-21}{3} = -7$$
2. **(B)** If $2a = 3b$, then $4a = 6b$. Substituting $6b$ for $4a$ in $4a + b = 21$ gives $6b + b = 21$ or $7b = 21$, so
$$b = \frac{21}{7} = 3$$
3. **(B)** Add corresponding sides of the two given equations:
$$\begin{aligned} 2p + q &= 11 \\ + \ p + 2q &= 13 \\ \hline 3p + 3q &= 24 \end{aligned}$$
Dividing each member of $3p + 3q = 24$ by 3 gives $p + q = 8$.
4. **(D)** Since $p = 2m$ and $k = 2p$, then
$$k = 2(2m) = 4m$$
Substituting for p and k in $m + p + k = 70$ gives
$$\begin{aligned} m + 2m + 4m &= 70 \\ 7m &= 70 \\ m &= \frac{70}{7} = 10 \end{aligned}$$
5. **(D)** Eliminate y by adding corresponding sides of the two equations:
$$\begin{aligned} x - y &= 3 \\ + \ x + y &= 5 \\ \hline 2x + 0 &= 8, \quad \text{so} \quad x = \frac{8}{2} = 4 \end{aligned}$$
Since $x = 4$ and $x - y = 3$, then $4 - y = 3$, so $y = 1$.
6. **(C)** Subtract corresponding sides of the two given equations:
$$\begin{aligned} 5x + y &= 19 & \qquad 5x + y &= 19 \\ -(x - 3y &= 7) \ \rightarrow & + \ -x + 3y &= -7 \\ \hline & & 4x + 4y &= 12 \end{aligned}$$
Dividing each member of the equation $4x + 4y = 12$ by 4 gives $x + y = 3$.
7. **(E)** Subtract corresponding sides of the two given equations:
$$\begin{aligned} x + 3 &= 5y & \qquad x + 3 &= 5y \\ -(x - 9 &= 2y) \ \rightarrow & + \ -x + 9 &= -2y \\ \hline & & 0 + 12 &= 3y \\ & & \frac{12}{3} &= y \text{ or } y = 4 \end{aligned}$$

Since $y = 4$ and $x + 3 = 5y$, then $x + 3 = 5(4) = 20$, so

$$x = 20 - 3 = 17$$

8. **(D)** Eliminate y by adding corresponding sides of the given equations:

$$\begin{array}{r} \frac{1}{x} + \frac{1}{y} = \frac{1}{4} \\ + \frac{1}{x} + \frac{1}{y} = \frac{3}{4} \\ \hline \frac{2}{x} + 0 = \frac{4}{4} = 1 \end{array}$$

Since $\frac{2}{x} = 1$, then $x = 2$.

9. **(A)** In the equation $\frac{a}{b} = \frac{2}{5}$, cross-multiplying gives $5a = 2b$. Since $5a + 3b = 35$ and $5a = 2b$, then

$$2b + 3b = 35$$
$$5b = 35$$
$$b = 7$$

Since $5a = 2b = 2(7) = 14$,

$$a = \frac{14}{5}$$

10. **(B)** If $\frac{x}{y} = 6$ and $x = 36$, then $\frac{36}{y} = 6$, so $y = 6$. Since $\frac{y}{w} = 4$ and $y = 6$,

$$\frac{6}{w} = 4$$
$$4w = 6$$
$$w = \frac{6}{4} = \frac{3}{2}$$

11. **(B)** Add corresponding sides of the given equations:

$$\begin{array}{r} 4r + 7s = 23 \\ + \; r - 2s = 17 \\ \hline 5r + 5s = 40 \end{array}$$

Dividing each member of $5r + 5s = 40$ by 5 gives $r + s = 8$. Since $r + s = 8$, then

$$3r + 3s = 3(8) = 24$$

12. **(C)** If $\frac{p - q}{2} = 3$ and $rp - rq = 12$, then

$$p - q = 2(3) = 6$$

and

$$r(p - q) = 12$$

so $r(6) = 12$ or $6r = 12$. Hence,

$$r = \frac{12}{6} = 2$$

13. **(C)** If $(a + b)^2 = 9$, then

$$a^2 + 2ab + b^2 = 9$$

If $(a - b)^2 = 49$, then

$$a^2 - 2ab + b^2 = 49$$

Add corresponding sides of the two equations:

$$\begin{array}{r} a^2 + 2ab + b^2 = 9 \\ + \; a^2 - 2ab + b^2 = 49 \\ \hline 2a^2 + 0 \quad + 2b^2 = 58 \end{array}$$

Dividing each member of $2a^2 + 2b^2 = 58$ by 2 gives

$$a^2 + b^2 = 29$$

14. **(D)** Proceed as follows:

- Find the value of $\frac{s}{r}$. If $\frac{r + s}{r} = 3$, then

$$\frac{r}{r} + \frac{s}{r} = 3 \text{ or } 1 + \frac{s}{r} = 3$$

so $\frac{s}{r} = 2$.

- Find the value of $\frac{r}{t} = 4$. If $\frac{t + r}{t} = 5$, then

$$\frac{t}{t} + \frac{r}{t} = 5 \text{ or } 1 + \frac{r}{t} = 5$$

so $\frac{r}{t} = 4$.

- Multiply corresponding sides of the equations $\frac{s}{r} = 2$ and $\frac{r}{t} = 4$:

$$\frac{s}{r} \times \frac{r}{t} = 2 \times 4$$
$$\frac{s}{t} = 8$$

15. **(B)** Eliminate y by subtracting corresponding sides of the given equations:

$$\begin{array}{r} 3x + y = c \\ - \; x + y = b \\ \hline 2x + 0 = c - b \end{array}$$

so $x = \frac{c - b}{2}$

16. **(E)** If $a + b = 11$ and $a - b = 7$, then adding corresponding sides of the two equations gives $2a = 18$, so

$$a = \frac{18}{2} = 9$$

If $a + b = 11$ and $a = 9$, then $9 + b = 11$, so

$$b = 11 - 9 = 2$$

Hence,

$$ab = (9)(2) = 18$$

17. **(D)** For the given system of three equations,

$$x - z = 7$$
$$x + y = 3$$
$$z - y = 6$$

add the equations two at a time to eliminate the variables y and z.

- Eliminate y by adding corresponding sides of the second and third equations. The result is $x + z = 9$.
- Eliminate z by adding $x + z = 9$ to the first equation. The result is $2x = 16$.

Hence, $x = \frac{16}{2} = 8$.

18. **(A)** If x students each contributed y dollars to buy a $60 gift for their advisor, then $xy = 60$ (equation 1). If four more, or $x + 4$ students in total, had contributed, each pupil could have contributed 3 dollars less, or $y - 3$ dollars each, to buy the same gift. Hence, $(x + 4)(y - 3) = 60$ (equation 2). Choice (A) gives these two equations.

19. **(A)** Divide corresponding sides of the given equations:

$$\frac{r^8}{r^7} = 5 \div \frac{3}{t}$$
$$r = 5 \times \frac{t}{3}$$
$$= \frac{5}{3}t$$

20. **(D)** • If $\frac{a}{b} = \frac{6}{7}$, then $6b = 7a$, so

$$3b = \frac{7}{2}a$$

• If $\frac{a}{c} = \frac{2}{5}$, then $2c = 5a$, so

$$c = \frac{5}{2}a$$

• Hence,

$$3b + c = \frac{7}{2}a + \frac{5}{2}a = \frac{12}{2}a = 6a$$

(Grid-In)

1. **3** Since

$$5 \text{ sips} + 4 \text{ gulps} = 1 \text{ glass}$$

and

$$13 \text{ sips} + 7 \text{ gulps} = 2 \text{ glasses}$$

then

$$13 \text{ sips} + 7 \text{ gulps} = 2\overbrace{(5 \text{ sips} + 4 \text{ gulps})}^{1 \text{ glass}}$$
$$= 10 \text{ sips} + 8 \text{ gulps}$$
$$13 - 10 \text{ sips} = 8 - 7 \text{ gulps}$$
$$3 \text{ sips} = 1 \text{ gulp}$$

2. **1.5** Write one equation underneath the other, aligning like terms in the same vertical column. Eliminate b by multiplying the first equation by 3 and then adding the result to the second equation:

$$\begin{array}{l} 2a = 9 - b \\ \underline{4a = 3b - 12} \end{array} \quad \xrightarrow{3\times} \quad \begin{array}{rcl} 6a &=& 27 - 3b \\ +\ 4b &=& -12 + 3b \\ \hline 10a &=& 15 + 0 \\ a &=& \frac{15}{10} = 1.5 \end{array}$$

3. **1.45** If p = the cost of a pen and n = the cost of a notebook then

$$\begin{array}{l} 1p + 2n = 3.50 \\ \underline{2p + 3n = 5.55} \end{array} \quad \xrightarrow{2\times} \quad \begin{array}{rcl} 2p + 4n &=& 7.00 \\ -\ 2p + 3n &=& 5.55 \\ \hline n &=& 1.45 \end{array}$$

The charge for one notebook is $1.45.

LESSON 4-7 (*Multiple-Choice*)

1. **(A)** *Solution 1*: Since $4 + 3p < p + 1$, then
$$3p - p < 1 - 4 \quad \text{or} \quad 2p < -3$$
so $p < -\frac{3}{2}$. Hence, the largest integer value for p is -2.
Solution 2: Plug each of the answer choices for p into $4 + 3p < p + 1$ until you find one that makes the inequality a true statement. Since choice (A) gives
$$4 + 3(-2) < (-2) + 1$$
there is no need to continue.
2. **(E)** *Solution 1*: Solve $-3 < 2x + 5 < 9$ by first subtracting 5 from each member. The result is $-8 < 2x < 4$. Now divide each member of this inequality by 2, obtaining $-4 < x < 2$. Examine each of the answer choices until you find one (E) that is not between -4 and 2. Since x is less than 2, 2 is not a possible value of x.
Solution 2: Plug each of the answer choices for x into $-3 < 2x + 5 < 9$ until you find one (E) that does not make the inequality a true statement.
3. **(E)** If the sum of a number, x, and the original number increased by 5, $x + 5$, is greater than 11, then $x + (x + 5) > 11$, so $2x + 5 > 11$. Then $2x > 6$, so $x > 3$. The only answer choice that is greater than 3 is (E).
4. **(C)** Determine whether each Roman numeral statement is always true when $0 < a^2 < b$.
   - I. From the given inequality, you know that $a^2 < b$. Although a^2 is positive, a may or may not be positive. If $a > 0$, then $a < \frac{b}{a}$. If $a < 0$, then dividing each side of $a^2 < b$ by a reverses the inequality sign, so $a > \frac{b}{a}$. Hence, statement I is not always true.
   - II. Multiplying both sides of $a^2 < b$ by a^2 gives $a^4 < a^2b$, so statement II is always true.
   - III. Since $b > 0$, dividing both sides of $a^2 < b$ by b gives $\frac{a^2}{b} < 1$, so statement III is always true.

   Only Roman numeral statements II and III are always true.
5. **(B)** *Solution 1*: If $4 - 3x < 11$, then $-3x < 7$, so $x > -\frac{7}{3}$. Since $-\frac{7}{3}$ is between -2 and -3, the smallest integer value of x that satisfies this inequality is -2.
Solution 2: Plug each of the answer choices for x, starting with (A), into $4 - 3x < 11$ until you find one that makes the inequality a true statement. Choice (A) gives
$$4 - 3(-3) < 11$$
$$13 < 11$$
which is not a true statement.
Choice (B) gives
$$4 - 3(-2) < 11$$
$$10 < 11$$
which is true, so there is no need to continue.
6. **(B)** Determine whether each Roman numeral statement is always true when $a > b > c > 0$.
   - I. Since $a > b$, then $a - c > b - c$, and $b - a$ represents a negative number. Hence, when both sides of the inequality $a - c > b - c$ are divided by $b - a$, a true inequality results only if the direction of the inequality is reversed. Thus,
$$\frac{a-c}{b-a} < \frac{b-c}{b-a}$$
so statement I is not always true.
   - II. Since $b > c$ and $a > 0$, multiplying both sides of the inequality $b > c$ by a results in the true inequality $ab > ac$, so statement II is always true.
   - III. Since $a > c$, then $\frac{1}{a} < \frac{1}{c}$. Since $b > 0$, multiplying both sides of the inequality $\frac{1}{a} < \frac{1}{c}$ by b produces the true inequality $\frac{b}{a} > \frac{b}{c}$, so statement II is not always true.

   Hence, only Roman numeral statement II is always true.
7. **(B)** If $-4 \le 3n \le 87$, then
$$-\frac{4}{3} \le n \le \frac{87}{3}$$
or, equivalently,
$$-1\frac{1}{3} \le n \le 29$$
Since n is an integer, n can be any integer from -1 to 29, inclusive. Including 0, there are 31 integers in this interval.

8. **(E)** If $a^2 < b^2$ and a and b are not 0, then a and b may be either positive or negative numbers. Determine whether each Roman numeral statement must be true.
   - I. If $a > 0$, then $\frac{a^2}{a} < \frac{b^2}{a}$. Since dividing both sides of an inequality by a negative number reverses the inequality sign, if $a < 0$, then $\frac{a^2}{a} < \frac{b^2}{a}$. Hence, statement I is not always true.
   - II. Since $a^2 < b^2$, their reciprocals have the opposite size relationship, so $\frac{1}{a^2} > \frac{1}{b^2}$. Statement II is always true.
   - III. Since $a^2 < b^2$, then $a^2 - b^2 < 0$. Factoring the left side of this inequality gives $(a + b)(a - b) < 0$, so statement III is always true.

   Only Roman numeral statements II and III must be true.
9. **(D)** If $b + 3 > 0$, then $b > -3$. Since $1 > 2b - 9$, then $10 > 2b$, so $5 > b$ or $b < 5$. Since b is an integer, b may be equal to any of these seven integers: $-2, -1, 0, 1, 2, 3$, or 4.
10. **(C)** Determine whether each Roman numeral statement is always true when $xy > 1$ and $z < 0$.
    - I. If $x > 0$, then $x > z$. However, the fact that x may be a negative number could mean that $x < z$, so statement I is not always true.
    - II. Multiplying an inequality by a negative quantity ($z < 0$) reverses the direction of the inequality, so $(xy)z < (1)z$, or $xyz < z$. Since z may or may not be greater than or equal to -1, the inequality $xyz < -1$ may or may not be true. Hence, statement II is not always true.
    - III. Dividing $xy > 1$ by a negative quantity reverses the direction of the inequality, so $\frac{xy}{z} < \frac{1}{z}$. Statement III is always true.

    Only Roman numeral statement III is always true.

(Grid-In)

1. **4** If $2y - 3 < 7$, then $2y < 10$, so $y < 5$. Since
   $$y + 5 > 8 \quad \text{and} \quad 2y - 3 < 7$$
   then $y > 3$ and at the same time $y < 5$. The integer for which the question asks must be 4.
2. **6** When 2 times an integer x is increased by 5, the result is always greater than 16 and less than 29, so $16 < 2x + 5 < 29$. Subtracting 5 from each member of this inequality gives $11 < 2x < 24$. Then
   $$\frac{11}{2} < \frac{2x}{2} < \frac{24}{2}$$
   so $5\frac{1}{2} < x < 12$. According to this inequality, x is greater than $5\frac{1}{2}$, so the least integer value of x is 6.
3. **.76** If $2 < 20x - 13 < 3$, adding 13 to each member of the combined inequality makes $15 < 20x < 16$ or $\frac{15}{20} < x < \frac{16}{20}$, which can also be written as $0.75 < x < 0.80$. Hence, one possible value for x is 0.76. Grid in as .76.
4. **6** Canceling identical terms on either side of the given inequality, $\frac{1}{7} + \frac{1}{8} - \frac{1}{9} + \frac{1}{10} < \frac{1}{8} - \frac{1}{9} + \frac{1}{10} + \frac{1}{n}$, results in $\frac{1}{7} < \frac{1}{n}$ or, equivalently, $n < 7$. Hence, the greatest possible integer value for n is 6.

Word Problems

CHAPTER 5

This chapter uses algebraic methods and reasoning to help prepare you to solve the different types of word problems that may appear on the SAT.

LESSONS IN THIS CHAPTER

LESSON

5-1 TRANSLATING FROM ENGLISH TO ALGEBRA

OVERVIEW

Some SAT *algebra questions require that you translate an English sentence into an algebraic sentence.*

ASSOCIATING ENGLISH PHRASES WITH MATHEMATICAL OPERATIONS

When translating an English phrase into a mathematical expression, look for key words that tell you which arithmetic operation is needed. For example:

- "A number n *increased by* 3" is translated as $n + 3$.
- "Three *less than* a number y" is translated as $y - 3$.
- "Two *times* the *sum* of a number x and 3" is translated as $2(x + 3)$.
- "Twelve *divided by* a number p" is translated as $\frac{12}{p}$.
- "A number y exceeds two times a number x by 5" is translated as $y = 2x + 5$.
- "The *sum* of a number x and 3 is *at least* 7" is translated as $x + 3 \geq 7$.
- "When 2 is *subtracted from* a number n, the difference is *at most* 5" is translated as $n - 2 \leq 5$.

TRANSLATING AN ENGLISH SENTENCE INTO AN EQUATION

To translate an English sentence into an equation, identify key phrases that can be translated directly into mathematical terms. For example:

- $\underbrace{\text{Two times}}_{2\times}\ \underbrace{\text{a number } n}_{n}\ \underbrace{\text{increased by 5}}_{+5}\ \underbrace{\text{is 17}}_{=17}.$

  The equation is $2n + 5 = 17$.

- $\underbrace{\text{One-half of}}_{\frac{1}{2}\ \times}\ \underbrace{\text{a number } n}_{n}\ \underbrace{\text{diminished by 5}}_{-5}\ \underbrace{\text{is 17}}_{=17}.$

  The equation is $\frac{1}{2}n - 5 = 17$.

- $\underbrace{\text{Five times a number } n}_{5n}\ \underbrace{\text{exceeds}}_{=}\ \underbrace{\text{2 times that number}}_{2n}\ \underbrace{\text{by 21}}_{+21}.$

  The equation is $5n = 2n + 21$.

REPRESENTING ONE QUANTITY IN TERMS OF ANOTHER

Sometimes it is necessary to compare two quantities by representing one quantity in terms of another. For example:

- Tim weighs 13 pounds more than Sue. If x represents Sue's weight, then Tim's weight is $x + 13$.
- The number of dimes exceeds 3 times the number of pennies by 2. If x represents the number of pennies, then $3x + 2$ represents the number of dimes.
- Bill has 7 fewer dollars than twice the number Kim has. If x represents the number of dollars Kim has, then the number of dollars Bill has is $2x - 7$.
- The sum of two integers is 25. If x represents one of these integers, then $25 - x$ represents the other integer.

LESS, LESS THAN, AND *IS LESS THAN*

A common mistake is to think of *less* and *less than* as interchangeable. Although both imply subtraction, *less than* reverses the order in which the given numbers are subtracted: "5 less 3" is translated as $5 - 3$, while "5 *less than* 3" becomes $3 - 5$. The phrase *is less than* refers to a comparison of two numbers rather than an arithmetic operation. Thus, "3 is less than 5" is translated as $3 < 5$.

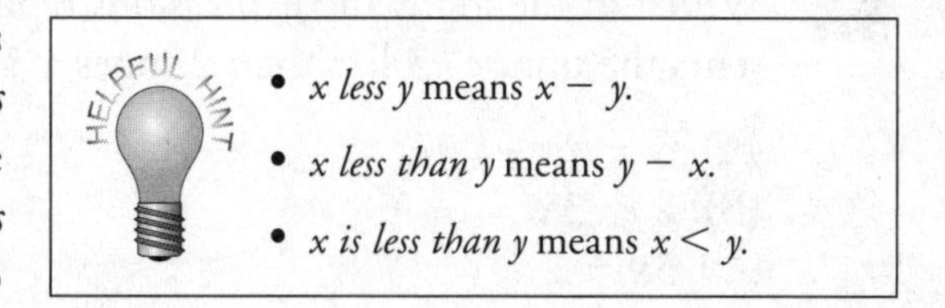

- x *less* y means $x - y$.
- x *less than* y means $y - x$.
- x *is less than* y means $x < y$.

EXAMPLE

If 11 less than 7 times a certain number is 7 more than 4 times the number, what is the number?

Solution

Assume that x represents the unknown number. The expression "11 less than 7 times a certain number" means $7x$ decreased by 11. Also, remember the rule that *less than* reverses the order of subtraction, so 11 gets subtracted from $7x$ as in $7x - 11$. The expression "7 more than 4 times the number" means $4x$ increased by 7, which is translated as $4x + 7$. Thus,

$$\overbrace{7x}^{\text{7 times the number}}\ \overbrace{-11}^{\text{decreased by 11}}\ \overbrace{=}^{\text{is}}\ \overbrace{4x}^{\text{4 times the number}}\ \overbrace{+7}^{\text{increased by 7}}$$

Solve $7x - 11 = 4x + 7$ by collecting like terms on the same side of the equation:

$$\begin{aligned} 7x - 4x &= 11 + 7 \\ 3x &= 18 \\ \frac{3x}{3} &= \frac{18}{3} \\ x &= 6 \end{aligned}$$

The number is **6**.

Lesson 5-1 Tune-Up Exercises

Multiple-Choice

1 Carl has 7 fewer than twice the number of course credits that Steve has. Steve has 5 more course credits than Gary. If Gary has 8 course credits, how many does Carl have?

(A) 6
(B) 19
(C) 20
(D) 27
(E) 33

2 Which of the following expressions represents the phrase "3 less than 2 times x"?

(A) $3 - 2x$
(B) $2 - 3x$
(C) $3x - 2$
(D) $2x - 3$
(E) $2(3 - x)$

3 When 4 times a number n is increased by 9, the result is 21.

Which of the following equations represents the statement above?

(A) $4(n + 9) = 21$
(B) $4n = 9 + 21$
(C) $n + 4 \times 9 = 21$
(D) $4n + 9 = 21$
(E) $4 + 9n = 21$

4 If $x - 4$ is 2 greater than $y + 1$, then by how much is $x + 6$ greater than y?

(A) 7
(B) 8
(C) 13
(D) 14
(E) 15

5 When 3 is subtracted from 5 times a number n, the result is 27.

Which of the following equations represents the statement above?

(A) $5n - 3 = 27$
(B) $3n - 5 = 27$
(C) $5(n - 3) = 27$
(D) $3(n - 5) = 27$
(E) $3 - 5n = 27$

6 If $\frac{1}{3}$ of a number is 4 less than $\frac{1}{2}$ of the number, the number is

(A) 12
(B) 18
(C) 24
(D) 30
(E) 36

7 Susan weighs p pounds. If Susan gains 17 pounds, she will weigh as much as Carol, who weighs 8 pounds less than Judy. If Judy weighs x pounds, then Susan's weight, p, in terms of x is

(A) $x - 25$
(B) $x - 9$
(C) $x + 9$
(D) $17x - 8$
(E) $x + 25$

8 When 2 times a number n is decreased by 5, the result is at least 11.

Which of the following expressions represents the sentence above?

(A) $2(n - 5) \geq 11$
(B) $2n - 5 \leq 11$
(C) $2n - 5 \geq 11$
(D) $2n - 5 = 11$
(E) $2(n - 5) \leq 11$

9 When 3 times a number n is increased by 7, the result is at most 4 times the number decreased by 1.

Which of the following expressions represents the sentence above?

(A) $3n + 7 \leq 4n - 1$
(B) $3n + 7 > 4n - 1$
(C) $3n + 7 \leq 4(n - 1)$
(D) $3n + 7 \geq 4(n - 1)$
(E) $3(n + 7) \leq 4(n - 1)$

10 If t multiplied by the sum of x and $2y$ is divided by $4y$, the result is

(A) $\frac{xt}{2} + \frac{yt}{4}$
(B) $\frac{xt}{4} + \frac{yt}{2}$
(C) $\frac{xt}{4y} + \frac{t}{2}$
(D) $\frac{xt}{2y} + \frac{t}{4y}$
(E) $\frac{xt + yt}{2y}$

11 If 4 less than x is 1 more than y, what is x in terms of y?

(A) $y - 3$
(B) $y + 1$
(C) $y + 3$
(D) $y + 4$
(E) $y + 5$

12 If the number $a - 5$ is 3 less than b, which expression has the same value as 1 more than b?

(A) $a - 2$
(B) $a - 1$
(C) a
(D) $a + 1$
(E) $a + 2$

13 If $x + y$ is 4 more than $x - y$, which of the following statements must be true?

I. $x = 2$
II. $y = 2$
III. xy has more than one value.

(A) I only
(B) II only
(C) III only
(D) I and III only
(E) II and III only

14 Jill's present weight is 14 pounds less than her weight a year ago. If her weight at that time was $\frac{9}{8}$ of her present weight, what is her present weight in pounds?

(A) 98
(B) 104
(C) 112
(D) 118
(E) 120

15 A number x is 3 less than 4 times the number y. Two times the sum of x and y is 9. Which of the following pairs of equations could be used to find the values of x and y?

(A) $x = 4y - 3$
$2(x + y) = 9$
(B) $y = 4x - 3$
$2(x + y) = 9$
(C) $x = 4(y - 3)$
$2(x + y) = 9$
(D) $y = 4(x - 3)$
$2x + y = 9$
(E) $x = 4y - 3$
$2(xy) = 9$

16 Arthur has 3 times as many marbles as Vladimir. If Arthur gives Vladimir 6 marbles, Arthur will be left with 4 more marbles than Vladimir. What is the total number of marbles that Arthur and Vladimir have?

(A) 36
(B) 32
(C) 30
(D) 24
(E) 21

17 A video store rents each DVD at the rate of x dollars for the first day and y dollars for each additional day the DVD is out. When Sara returns a DVD to this store, she is charged c dollars. In terms of x, y, and c, what is the number of rental days for which Sara is charged?

(A) $x + cy$
(B) $\frac{c - x}{y}$
(C) $\frac{x + y}{c}$
(D) $x + y(c - 1)$
(E) $1 + \frac{c - x}{y}$

18 A computer program is designed so that, when a number is entered, the computer output is obtained by multiplying the number by 3 and then subtracting 4 from the product. If the output that results from entering a number x is then entered, which expression represents, in terms of x, the final output?

(A) $3x - 8$
(B) $3x - 12$
(C) $9x - 8$
(D) $9x - 16$
(E) $6x + 9$

19 There are 36 identical cartons, $\frac{3}{4}$ of which must be taped. If every 3 cartons require 2 rolls of tape, what is the total number of rolls of tape that will be needed?

(A) 18
(B) 21
(C) 24
(D) 27
(E) 30

20 After a certain number of people enter an empty room, $\frac{2}{3}$ of the people who entered the room leave. After 2 more people leave, $\frac{1}{4}$ of the original number of people who entered the room remain. What was the original number of people who entered the empty room?

(A) 9
(B) 12
(C) 18
(D) 24
(E) 30

21 In a high school that has a total of 950 students, the number of seniors is $\frac{3}{4}$ of the number of juniors, and the number of juniors is $\frac{2}{3}$ of the number of sophomores. If this school has the same number of freshmen as sophomores, how many students are seniors?

(A) 120
(B) 150
(C) 180
(D) 200
(E) 300

Grid-In

1 If $\frac{5}{4}$ of x is 20, what number is x decreased by 1?

2 Half the difference of two positive numbers is 10. If the smaller of the two numbers is 3, what is the sum of the two numbers?

3 In a certain college class, each student received a grade of A, B, C, or D or an "Incomplete." In this class, $\frac{1}{6}$ of the students received As, $\frac{1}{4}$ received Bs, $\frac{1}{3}$ received Cs, and $\frac{1}{6}$ received Ds. If 3 students received grades of "Incomplete," how many students were in the class?

LESSON

5-2 PERCENT PROBLEMS

OVERVIEW

Simple algebraic equations can be used to help solve different types of percent problems.

THE THREE TYPES OF PERCENT PROBLEMS

You can solve each of the three basic types of percent problems by writing and solving an equation.

- **Type 1:** Finding a percent of a given number.

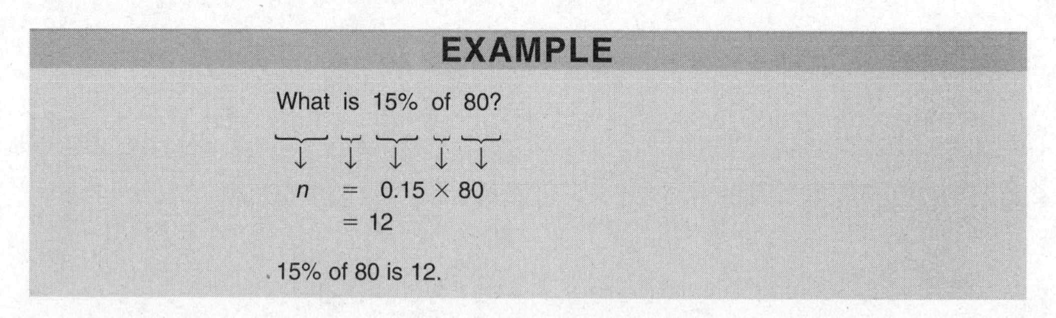

EXAMPLE

What is 15% of 80?

$n = 0.15 \times 80$

$= 12$

15% of 80 is 12.

- **Type 2:** Finding a number when a percent of it is given.

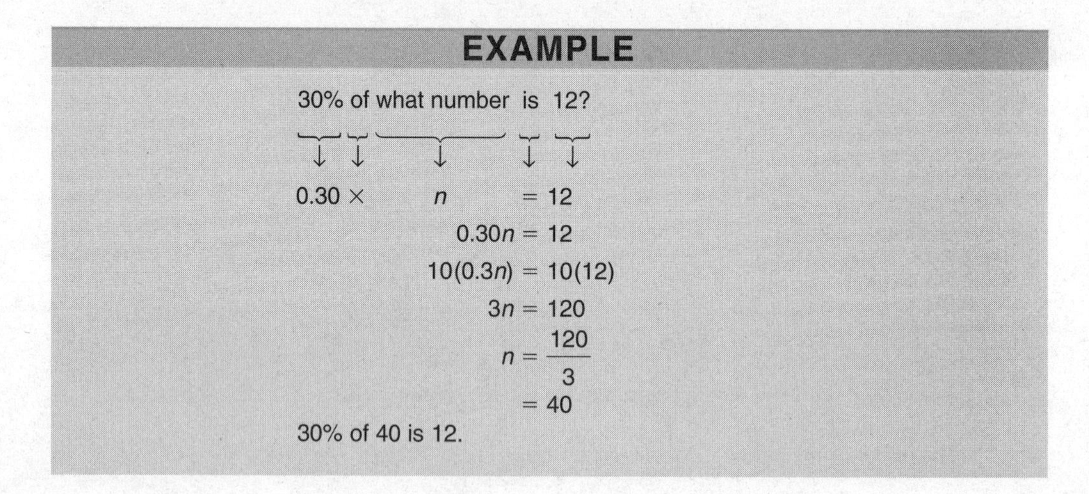

EXAMPLE

30% of what number is 12?

$0.30 \times n = 12$

$0.30n = 12$

$10(0.3n) = 10(12)$

$3n = 120$

$n = \frac{120}{3}$

$= 40$

30% of 40 is 12.

- **Type 3:** Finding what percent one number is of another.

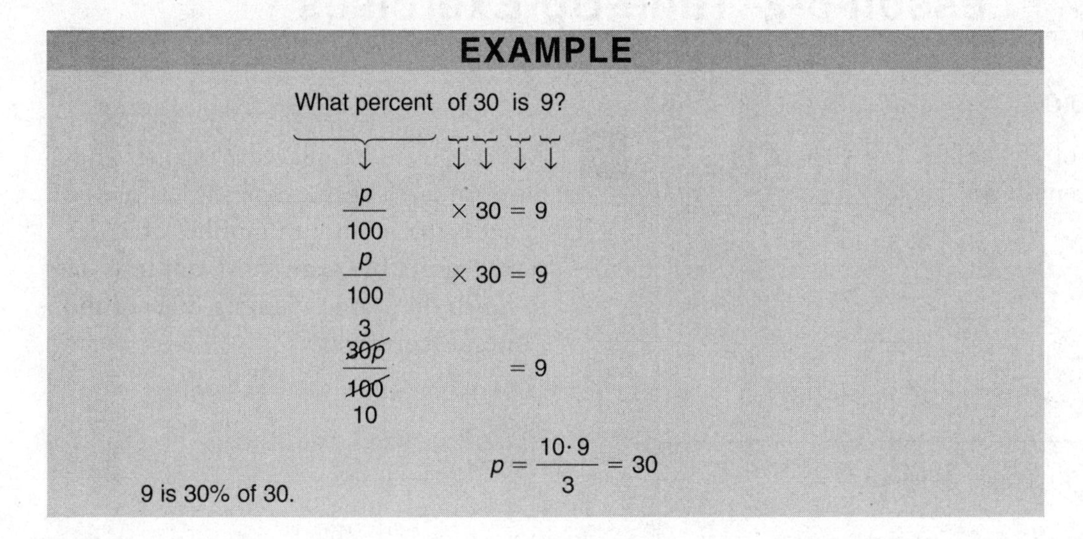

EXAMPLE

What percent of 30 is 9?

$$\frac{p}{100} \times 30 = 9$$

$$\frac{p}{100} \times 30 = 9$$

$$\frac{\overset{3}{\cancel{30}p}}{\underset{10}{\cancel{100}}} = 9$$

$$p = \frac{10 \cdot 9}{3} = 30$$

9 is 30% of 30.

Lesson 5-2 Tune-Up Exercises

Multiple-Choice

1 The sum of 25% of 32 and 40% of 15 is what percent of 35?

(A) 20%
(B) 25%
(C) 30%
(D) 40%
(E) 50%

2 30% of 150 equals 4.5% of

(A) 10
(B) 100
(C) 250
(D) 1000
(E) 10,000

3 If A is 125% of B, then B is what percent of A?

(A) 60%
(B) 75%
(C) 80%
(D) 88%
(E) 90%

4 By the end of the school year, Terry had passed 80% of his science tests. If Terry failed 4 science tests, how many science tests did Terry pass?

(A) 12
(B) 15
(C) 16
(D) 18
(E) 20

5 If 30% of x is 21, what is 80% of x?

(A) 28
(B) 35
(C) 42
(D) 48
(E) 56

6 A soccer team has played 25 games and has won 60% of the games it has played. What is the minimum number of additional games the team must win in order to finish the season winning 80% of the games it has played?

(A) 28
(B) 25
(C) 21
(D) 18
(E) 15

7 What percent of 800 is 5?

(A) $\frac{1}{160}\%$
(B) $\frac{5}{8}\%$
(C) 1.6%
(D) 62.5%
(E) $\frac{800}{5}\%$

8 $\frac{1}{2}$ is what percent of $\frac{1}{5}$?

(A) 250%
(B) 210%
(C) 140%
(D) 40%
(E) 20%

9 In an opinion poll of 50 men and 40 women, 70% of the men and 25% of the women said that they preferred fiction to nonfiction books. What percent of the number of people polled preferred to read fiction?

(A) 40%
(B) 45%
(C) 50%
(D) 60%
(E) 75%

10 If 25% of x is 12.5, what is 12.5% of $2x$?

(A) 6.25
(B) 12.5
(C) 25
(D) 37.5
(E) 50

11 If $\frac{1}{8}$ of a number is 9, what is 75% of the same number?

(A) 36
(B) 48
(C) 54
(D) 72
(E) 81

12 300% of 6 is what percent of 24?

(A) 40%
(B) 50%
(C) 60%
(D) 75%
(E) 80%

13 The regular price of software at a computer superstore is 12% off the retail price. During an annual sale, the same software is 25% off the regular price. If the retail price is p, which expression represents the sale price?

(A) $0.34p$
(B) $0.37p$
(C) $0.63p$
(D) $0.64p$
(E) $0.66p$

14 The price of a stock falls 25%. By what percent of the new price must the stock price rise in order to reach its original value?

(A) 25%
(B) 30%
(C) $33\frac{1}{3}$%
(D) 40%
(E) 75%

15 If a is 30% greater than A and b is 20% greater than B, then ab is what percent greater than AB?

(A) 25%
(B) 50%
(C) 56%
(D) 60%
(E) 75%

16 A number a increased by 20% of a results in a number b. When b is decreased by $33\frac{1}{3}$% of b, the result is c. The number c is what percent of a?

(A) 40%
(B) 60%
(C) 80%
(D) 120%
(E) 150%

17

VOTING POLL	
Candidate A	30%
Candidate B	50%
Undecided	20%

The table above summarizes the results of an election poll in which 4000 voters participated. In the actual election, all 4000 of these people voted and those people who chose a candidate in the poll voted for that candidate. People who were undecided voted for candidate A in the same proportion as the people who cast votes for candidates in the poll. Of the people polled, how many voted for candidate A in the actual election?

(A) 1420
(B) 1500
(C) 1640
(D) 1680
(E) 1800

Grid-In

1 A high school tennis team is scheduled to play 28 matches. If the team wins 60% of the first 15 matches, how many additional matches must the team win in order to finish the season winning 75% of its scheduled matches?

2 In a club of 35 boys and 28 girls, 80% of the boys and 25% of the girls have been members for more than 2 years. If n percent of the club have been members for more than 2 years, what is the value of n?

LESSON

5-3

SOME SPECIAL TYPES OF WORD PROBLEMS

OVERVIEW

This lesson applies algebra translation skills to solving consecutive integer problems and age problems.

CONSECUTIVE INTEGER PROBLEMS

Consecutive integers differ by 1, and consecutive even or odd integers differ by 2.

- A list of consecutive integers that begins with x is

$$x, x + 1, x + 2, x + 3, \ldots$$

- A list of consecutive even or odd integers that begins with x is

$$x, x + 2, x + 4, x + 6, \ldots$$

EXAMPLE

In a set of 3 consecutive odd integers, twice the sum of the second and the third integers is 43 more than 3 times the first integer. What is the smallest of the 3 integers?

Solution

- Represent the unknown quantities in mathematical terms:

Let $x =$ the first of the three consecutive odd integers.

Then $x + 2 =$ the second of the three consecutive odd integers,

and $x + 4 =$ the third of the three consecutive odd integers.

- Write an equation by directly translating the key words of the English sentence into mathematical terms:

$$\underbrace{\text{Twice the sum of the second and the third}}_{2[(x+2)+(x+4)]}\ \underbrace{\text{is}}_{=}\ \underbrace{\text{43 more than}}_{43\,+}\ \underbrace{\text{3 times the first}}_{3x}$$

- Solve the equation:

$$2[(x + 2) + (x + 4)] = 43 + 3x$$
$$2[2x + 6] = 43 + 3x$$
$$4x + 12 = 43 + 3x$$
$$4x - 3x = 43 - 12$$
$$x = 31$$

Hence, the smallest of the three consecutive odd integers is 31.

AGE PROBLEMS

If x represents a person's present age, then subtracting a number of years from x gives the person's age in the past. Similarly, adding a number of years to x gives the person's age in the future. For example:

- $x - 5$ represents a person's age 5 years ago.
- $x + 5$ represents a person's age 5 years from now.

EXAMPLE

If 7 years from now Susan will be 2 times as old as she was 3 years ago, what is Susan's present age?

Solution

Let $x =$ Susan's present age.

Then $x + 7 =$ Susan's age 7 years from now,

and $x - 3 =$ Susan's age 3 years ago.

$$\underbrace{\text{7 years from now Susan}}_{x+7}\ \underbrace{\text{will be}}_{=}\ \underbrace{\text{2 times as old as she was 3 years ago}}_{2(x-3)}$$

$$x + 7 = 2(x - 3)$$
$$x + 7 = 2x - 6$$
$$13 = x$$

Susan is now 13 years old.

Lesson 5-3 Tune-Up Exercises

Multiple-Choice

1 In an ordered set of 4 consecutive odd integers, the sum of 3 times the second integer and the greatest integer is 104. Which is the least integer in the set?

(A) 11
(B) 17
(C) 19
(D) 23
(E) 27

2 Four years ago Jim was one-half the age he will be 7 years from now.

If x represents Jim's present age, which of the following equations represents the above statement?

(A) $\frac{1}{2}(x - 4) = x + 7$
(B) $x - 4 = \frac{1}{2}(x + 7)$
(C) $\frac{1}{2}(x + 4) = x + 7$
(D) $\frac{1}{2}(x + 4) = x - 7$
(E) $(x - 4) = 2(x + 7)$

3 The sum of n consecutive integers is 11. If the least of these integers is -10, what is the value of n?

(A) 11
(B) 12
(C) 20
(D) 21
(E) 22

4 How old is David if his age 6 years from now will be twice his age 7 years ago?

(A) 13
(B) 15
(C) 17
(D) 20
(E) 23

5 If the sum of two consecutive integers is k, what is the larger of these two integers expressed in terms of k?

(A) $\frac{k}{2} - 1$
(B) $\frac{k}{2} + 1$
(C) $2k - 1$
(D) $\frac{k - 1}{2}$
(E) $\frac{k + 1}{2}$

6 The greatest of 4 consecutive odd integers is 11 more than twice the sum of the first and the second odd integers.

If x is the least of the 4 consecutive odd integers, which of the following equations represents the above statement?

(A) $x + 4 = 2[(x + 1) + (x + 2)] + 11$
(B) $x + 8 = 2[(x + 2) + (x + 4)] + 11$
(C) $(x + 6) + 11 = 2[x + (x + 2)]$
(D) $x + 5 = 2[x + (x + 1)] + 11$
(E) $x + 6 = 2[x + (x + 2)] + 11$

7 If x is the least of 4 consecutive even integers whose sum is S, what is x in terms of S?

(A) $\frac{S}{4} - 3$
(B) $\frac{3(S + 1)}{4}$
(C) $\frac{S + 3}{4}$
(D) $\frac{4S}{3}$
(E) $3S - 4$

8 If John's age is increased by Mary's age, the result is 2 times John's age 3 years ago. If Mary is now M years old, what is John's present age in terms of M?

(A) $M + 3$
(B) $M + 6$
(C) $2M - 3$
(D) $2M - 6$
(E) $2M + 3$

9 If $3x$ years from today Reyna will be $(3y + 4)$ times her present age, what is Reyna's present age in terms of x and y?

(A) $\frac{3x}{4(y-1)}$
(B) $\frac{3x}{3y+4}$
(C) $\frac{x}{y+1}$
(D) $\frac{3x}{4y}$
(E) $\frac{x}{4y-1}$

LESSON

5-4

RATIO AND VARIATION

OVERVIEW

*A **ratio** is a comparison by division of two quantities that are measured in the same units. For example, if Mary is 16 years old and her brother Gary is 8 years old, then Mary is 2 times as old as Gary. The ratio of Mary's age to Gary's age is 2 : 1 (read as "2 to 1") since*

$$\frac{\text{Mary's age}}{\text{Gary's age}} = \frac{16 \text{ years}}{8 \text{ years}} = \frac{2}{1} \text{ or } 2:1$$

One quantity may be related to another quantity so that either the ratio or product of these quantities always remains the same.

RATIO OF *a* TO *b*

The ratio of a to b ($b \neq 0$) is the fraction $\frac{a}{b}$, which can be written as $a : b$ (read as "a is to b").

EXAMPLE

The ratio of the number of girls to the number of boys in a certain class is 3 : 5. If there is a total of 32 students in the class, how many girls are in the class?

Solution

Since the number of girls is a multiple of 3 and the number of boys is the same multiple of 5, let

$$3x = \text{the number of girls in the class}$$
$$\text{and } 5x = \text{the number of boys in the class}$$

Then

$$3x + 5x = 32$$
$$8x = 32$$
$$x = \frac{32}{8} = 4$$

The number of girls $= 3x = 3(4) = 12$.

RATIO OF *A* TO *B* TO *C*

If $A : B$ represents the ratio of A to B and $B : C$ represents the ratio of B to C, then the ratio of A to C is $A : C$, provided that B stands for the same number in both ratios. For example, if the ratio of A to B is $3 : 5$ and the ratio of B to C is $5 : 7$, then the ratio of A to C is $3 : 7$. In this case B represents the number 5 in both ratios.

EXAMPLE

If the ratio of A to B is $3 : 5$ and the ratio of B to C is $2 : 7$, what is the ratio of A to C?

Solution

Change each ratio into an equivalent ratio in which the term that corresponds to B is the same number.

- The ratio of A to B is $3 : 5$, so the term corresponding to B in this ratio is 5. The ratio of B to C is $2 : 7$, so the term corresponding to B in this ratio is 2.
- The least common multiple of 5 and 2 is 10. You need to change each ratio into an equivalent ratio in which the term corresponding to B is 10.
- Multiplying each term of the ratio $3 : 5$ by 2 gives the equivalent ratio $6 : 10$. Multiplying each term of the ratio $2 : 7$ by 5 gives $10 : 35$.
- Since the ratio of A to B is equivalent to $6 : 10$ and the ratio of B to C is equivalent to $10 : 35$, the ratio of A to C is $6 : 35$.

DIRECT VARIATION

If two quantities change in value so that their ratio always remains the same, then one quantity is said to vary **directly** with the other. When one quantity varies directly with another quantity, a change in one causes a change in the other in the same direction—both increase or both decrease.

EXAMPLE

If 28 pennies weigh 42 grams, what is the weight in grams of 50 pennies?

Solution

The number of pennies and their weight vary directly since multiplying one of the two quantities of pennies by a constant causes the other to be multiplied by the same constant. If x represents the weight in grams of 50 pennies, then

$$\frac{\text{Pennies}}{\text{Grams}} = \frac{28}{42} = \frac{50}{x}$$

Cross-multiply:

$$28x = 42(50)$$

$$x = \frac{2100}{28} = 75$$

The weight of 50 pennies is 75 grams.

INVERSE VARIATION

If two quantities change in opposite directions, so that their product always remains the same, then one quantity is said to vary **inversely** with the other.

EXAMPLE

Four workers can build a house in 9 days. How many days would it take 3 workers to build the same house?

Solution

As the number of people working on the house *decreases*, the number of days needed to build the house *increases*. Since this is an inverse variation, the number of workers times the number of days needed to build the house stays constant.

If d represents the number of days that 3 workers take to build the house, then

$$3 \times d = 4 \times 9$$

$$3d = 36$$

$$d = \frac{36}{3} = 12$$

Three people working together would take 12 days to build the house.

Lesson 5-4 Tune-Up Exercises

Multiple-Choice

1 A recipe for 4 servings requires salt and pepper to be added in the ratio of 2 : 3. If the recipe is adjusted from 4 to 8 servings, what is the ratio of the salt and pepper that must now be added?

(A) 4 : 3
(B) 2 : 6
(C) 2 : 3
(D) 3 : 2
(E) 8 : 4

2 On a certain map, $\frac{3}{8}$ of an inch represents 120 miles. How many miles does $1\frac{3}{4}$ inches represent?

(A) 300
(B) 360
(C) 400
(D) 480
(E) 560

3 The population of a bacteria culture doubles in number every 12 minutes. The ratio of the number of bacteria at the end of 1 hour to the number of bacteria at the beginning of that hour is

(A) 64 : 1
(B) 60 : 1
(C) 32 : 1
(D) 16 : 1
(E) 8 : 1

4 At the end of the season, the ratio of the number of games a team has won to the number of games it lost is 4 : 3. If the team won 12 games and each game played ended in either a win or a loss, how many games did the team play during the season?

(A) 9
(B) 15
(C) 18
(D) 21
(E) 24

5 If s and t are integers, $8 < t < 40$, and $\frac{s}{t} = \frac{4}{7}$, how many possible values are there for t?

(A) Two
(B) Three
(C) Four
(D) Five
(E) Six

6 A school club includes only sophomores, juniors, and seniors, in the ratio of 1 : 3 : 2. If the club has 42 members, how many seniors are in the club?

(A) 6
(B) 7
(C) 12
(D) 14
(E) 21

7 If $\frac{c - 3d}{4} = \frac{d}{2}$, what is the ratio of c to d?

(A) 5 : 1
(B) 3 : 2
(C) 4 : 3
(D) 3 : 4
(E) 2 : 3

8 If 4 pairs of socks costs $10.00, how many pairs of socks can be purchased for $22.50?

(A) 5
(B) 7
(C) 8
(D) 9
(E) 10

9 Two boys can paint a fence in 5 hours. How many hours would it take 3 boys to paint the same fence?

(A) $\frac{3}{2}$
(B) 3
(C) $3\frac{1}{3}$
(D) $4\frac{2}{3}$
(E) $7\frac{1}{2}$

10 A car moving at a constant rate travels 96 miles in 2 hours. If the car maintains this rate, how many miles will the car travel in 5 hours?

(A) 480
(B) 240
(C) 210
(D) 192
(E) 144

11 The number of kilograms of corn needed to feed 5000 chickens is 30 less than twice the number of kilograms needed to feed 2800 chickens. How many kilograms of corn are needed to feed 2800 chickens?

(A) 70
(B) 110
(C) 140
(D) 190
(E) 250

12 In an ordered list of five consecutive positive even integers, the ratio of the greatest integer to the least integer is 2 to 1. Which of the following is the middle integer in the list?

(A) 10
(B) 12
(C) 14
(D) 16
(E) 18

13 If the ratio of p to q is 3 : 2, what is the ratio of $2p$ to q?

(A) 1 : 3
(B) 2 : 3
(C) 3 : 3
(D) 3 : 1
(E) 3 : 4

14

$$\frac{x}{z} = \frac{1}{3}$$

If in the equation above x and z are integers, which are possible values of $\frac{x^2}{z}$?

I. $\frac{1}{9}$
II. $\frac{1}{3}$
III. 3

(A) II only
(B) III only
(C) I and III only
(D) II and III only
(E) None

15 If $a - 3b = 9b - 7a$, then the ratio of a to b is

(A) 3 : 2
(B) 2 : 3
(C) 3 : 4
(D) 4 : 3
(E) 1 : 2

16 The ratio of A to B is $a : 8$, and the ratio of B to C is $12 : c$. If the ratio of A to C is $2 : 1$, what is the ratio of a to c?

(A) 2 : 3
(B) 3 : 2
(C) 4 : 3
(D) 3 : 4
(E) 1 : 3

17 If $8^r = 4^t$, the ratio of r to t is

(A) 2 : 3
(B) 3 : 2
(C) 4 : 3
(D) 3 : 4
(E) 1 : 2

18 If $\frac{a+b}{b} = 4$ and $\frac{a+c}{c} = 3$, what is the ratio of c to b?

(A) 2 : 3
(B) 3 : 2
(C) 2 : 1
(D) 3 : 1
(E) 6 : 1

19 In a certain college, the ratio of mathematics majors to English majors is 3 : 8. If in the following school year the number of mathematics majors increases 20% and the number of English majors decreases 15%, what is the new ratio of mathematics majors to English majors?

(A) 4 : 9
(B) 1 : 2
(C) 9 : 17
(D) 17 : 32
(E) 7 : 12

20 At a college basketball game, the ratio of the number of freshmen who attended to the number of juniors who attended is 3 : 4. The ratio of the number of juniors who attended to the number of seniors who attended is 7 : 6. What is the ratio of the number of freshmen to the number of seniors who attended the basketball game?

(A) 7 : 8
(B) 3 : 4
(C) 2 : 3
(D) 1 : 2
(E) 1 : 3

21 Given that y varies inversely as x and x varies directly as z, if z is doubled, then y is

(A) divided by 4
(B) divided by 2
(C) multiplied by 2
(D) multiplied by 4
(E) unchanged

22 It took 12 men 5 hours to build an airstrip. Working at the same rate, how many additional men could have been hired in order for the job to have taken 1 hour less?

(A) two
(B) three
(C) four
(D) six
(E) eight

23 The price per person to rent a limousine for a prom varies inversely with the number of passengers. If nine people rent the limousine, the cost is \$70 each. How many people are renting the limousine when the cost *per couple* is \$105?

(A) 4
(B) 6
(C) 8
(D) 12
(E) 16

Grid-In

1 A string is cut into 2 pieces that have lengths in the ratio of 2 : 9. If the difference between the lengths of the 2 pieces of string is 42 inches, what is the length in inches of the shorter piece?

2 The ratio of a to b is 5 : 9, and the ratio of x to y is 10 : 3. The ratio of ay to bx is equivalent to the ratio of 1 to what number?

3 The ratio of dimes to pennies in a purse is 3 : 4. If 3 pennies are taken out of the purse, the ratio of dimes to pennies becomes 1 : 1. How many dimes are in the purse?

4 For integer values of a and b, $b^a = 8$. The ratio of a to b is equivalent to the ratio of c to d, where c and d are integers. What is the value of c when $d = 10$?

5 If $6a - 8b = 0$ and $c = 12b$, the ratio of a to c is equivalent to the ratio of 1 to what number?

6 Jars A, B, and C each contain 8 marbles. What is the minimum number of marbles that must be transferred among the jars so that the ratio of the number of marbles in jar A to the number in jar B to the number in jar C is 1 : 2 : 3?

LESSON

5-5 RATE PROBLEMS

OVERVIEW

*A ratio of two quantities that have different units of measurement is called a **rate**. For example, if a car travels a total distance of 150 miles in 3 hours, the average rate of speed is the distance traveled divided by the amount of time required to travel that distance:*

$$\text{Rate} = \frac{\text{Distance}}{\text{Time}} = \frac{150 \text{ miles}}{3 \text{ hours}} = 50 \text{ miles per hour}$$

Rate problems are usually solved using the general relationship

$$\text{Rate (of } A \text{ per unit } B) \times B = A$$

You are using the correct rate relationship if the units check, as in

$$\underbrace{\text{Rate}}_{\downarrow} \times \underbrace{\text{Time}}_{\downarrow} = \underbrace{\text{Distance}}_{\downarrow}$$

$$\frac{\text{Miles}}{\cancel{\text{Hour}}} \times \cancel{\text{Hours}} = \text{Miles}$$

UNIT COST PROBLEMS

Unit cost problems require you to figure out a rate by calculating the cost per item. Multiplying this rate by a given number of items gives the total cost of those items.

EXAMPLE

If 5 cans of soup cost $1.95, how much do 3 cans of soup cost?

Solution 1

Since 5 cans of soup cost $1.95, the cost of 1 can is

$$\frac{\$1.95}{5 \text{ cans}} = 0.39 \text{ dollar per can}$$

To find the cost of 3 cans, multiply the rate of 0.39 dollar per can by 3:

$$\text{Cost of 3 cans} = 3 \text{ cans} \times 0.39 \frac{\$}{\text{can}} = \$1.17$$

Solution 2

The number of cans of soup varies directly with the cost of the cans. Form a proportion in which x represents the cost of 3 cans of soup:

$$\frac{\text{Cost}}{\text{Number of cans}} = \frac{x}{3} = \frac{1.95}{5}$$

$$5x = 3(1.95)$$

$$x = 5.85$$

$$x = \frac{5.85}{5} = 1.17$$

The cost of 3 cans of soup is \$1.17.

MOTION PROBLEMS

The solution of motion problems depends on the relationship

$$\text{Rate} \times \text{Time} = \text{Distance}$$

EXAMPLE

John rode his bicycle to town at the rate of 15 miles per hour. He left the bicycle in town for minor repairs and walked home along the same route at the rate of 3 miles per hour. Excluding the time John spent in taking the bike into the repair shop, the trip took 3 hours. How many hours did John take to walk back?

Solution

Since the trip took a total of 3 hours,

let $x =$ the number of hours John took to ride his bicycle to town,
and $3 - x =$ the number of hours John took to walk back from town.

	Rate	×	Time	=	Distance
To town	15 mph		x hours		$15x$
Return trip	3 mph		$3 - x$ hours		$3(3 - x)$

Since John traveled over the same route, the two distances must be equal. Hence,

$$15x = 3(3 - x)$$

$$15x = 9 - 3x$$

$$18x = 9$$

$$x = \frac{9}{18} = \frac{1}{2} \text{ hour}$$

John took $3 - \frac{1}{2} = 2\frac{1}{2}$ hours to walk back.

WORK PROBLEMS

Work problems are solved using the relationship

$$\begin{pmatrix}\text{Rate of}\\ \text{work}\end{pmatrix} \times \begin{pmatrix}\text{Time}\\ \text{worked}\end{pmatrix} = \begin{pmatrix}\text{Part of job}\\ \text{completed}\end{pmatrix}$$

The rate at which work is done is the reciprocal of the amount of time needed to complete the whole job. For example, if a car mechanic takes 3 hours to repair a car, the average rate of work is $\frac{1}{3}\ \frac{\text{job}}{\text{hours}}$. Thus, the mechanic completes $\frac{1}{3}$ of the job in 1 hour, $\frac{2}{3}$ of the job in 2 hours, and $\frac{3}{3}$ of the job in 3 hours. In x hours the mechanic completes $\frac{1}{3}(x)$, or $\frac{x}{3}$, of the job.

EXAMPLE

Jim can wax a car in 3 hours, and Sue can wax the same car in 2 hours. If Jim and Sue work together, how long will it take them to wax the car ?

(A) 60 minutes
(B) 72 minutes
(C) 75 minutes
(D) 105 minutes
(E) 120 minutes

Solution

Since Jim and Sue start and finish at the same time, they work the same number of hours, say x. Make a table as shown below.

	Rate	×	Time	=	Part of job
Jim	$\frac{1}{3}$		x		$\frac{x}{3}$
Sue	$\frac{1}{2}$		x		$\frac{x}{2}$

- For the whole job to be completed, the fractional parts of the job done by Jim and Sue must add up to 1. Thus:

$$\frac{x}{3} + \frac{x}{2} = 1$$

- The fractions can be eliminated by multiplying each member of the equation by 6, the LCD of 3 and 2:

$$6\left(\frac{x}{3}\right) + 6\left(\frac{x}{2}\right) = 6(1)$$

$$2x + 3x = 6$$

$$5x = 6$$

$$x = \frac{6}{5}$$

- Working together, Jim and Sue take $\frac{6}{5}$ hours to wax the car. Since 1 hour is equivalent to 60 minutes, $\frac{6}{5}$ hours is equivalent to

$$\frac{6}{5} \times 60 = 6 \times 12 = 72 \text{ minutes}$$

The correct choice is (B).

DRY MIXTURE PROBLEMS

Sometimes two or more dry ingredients that cost different amounts of money per unit of weight are mixed together. For the mixture,

$$\begin{matrix}\text{Cost of ingredient} \\ \text{per unit of amount}\end{matrix} \times \begin{matrix}\text{Amount of} \\ \text{ingredient}\end{matrix} = \begin{matrix}\text{Value of} \\ \text{ingredient}\end{matrix}$$

EXAMPLE

Regular blend coffee that sells for \$3.00 a pound is to be mixed with a premium blend of coffee that sells for \$4.50 a pound to produce 30 pounds of coffee that sells for \$4.00 a pound. How many pounds of regular coffee should be in the mixture?

Solution

If x represents the number of pounds of regular blend coffee in the mixture, then $30 - x$ is the number of pounds of premium blend in the mixture. Make a table as shown below.

	Price per pound	×	Number of pounds	=	Value
Regular blend	\$3.00		x		$3.00x$
Premium blend	\$4.50		$30 - x$		$4.50(30 - x)$
Mixture	\$4.00		30		$4.00(30)$

- Find the total value of the mixture, which is equal to the sum of the values of both of the ingredients of the mixture. Thus,

$$3.00x + 4.50(30 - x) = 4.00(30)$$

- Multiply each side of the equation by 100 by moving the decimal point in each term two places to the right:

$$300x + 450(30 - x) = 400(30)$$

$$300x + 13{,}500 - 450x = 12{,}000$$

$$-150x + 13{,}500 = 12{,}000$$

$$-150x = -1500$$

$$x = \frac{-1500}{-150} = 10$$

Hence, 10 pounds of regular blend coffee should be in the mixture.

Lesson 5-5 Tune-Up Exercises

Multiple-Choice

1 If four pens cost \$1.96, what is the greatest number of pens that can be purchased for \$7.68?

(A) 12
(B) 14
(C) 15
(D) 16
(E) 17

2 If a car is traveling at a constant rate of 45 miles per hour, how many miles does it travel from 10:40 A.M. to 1:00 P.M. of the same day?

(A) 165
(B) 150
(C) 120
(D) 105
(E) 90

3 If k pencils cost c cents, what is the cost in cents of p pencils?

(A) $\frac{pc}{k}$
(B) $\frac{kc}{p}$
(C) $\frac{c}{kp}$
(D) $c - kp$
(E) $\frac{c}{p-k}$

4 A freight train left a station at 12 noon, going north at a rate of 50 miles per hour. At 1:00 P.M. a passenger train left the same station, going south at a rate of 60 miles per hour. At what time were the trains 380 miles apart?

(A) 3:00 P.M.
(B) 4:00 P.M.
(C) 4:30 P.M.
(D) 5:00 P.M.
(E) 5:30 P.M.

5 Julie can type a manuscript in 4 hours. Pat takes 6 hours to type the same manuscript. If Julie and Pat begin working together at 12 noon, at what time will they complete the typing of the manuscript?

(A) 2:24 P.M.
(B) 2:30 P.M.
(C) 2:40 P.M.
(D) 3:00 P.M.
(E) 3:30 P.M.

6 If x people working together at the same rate can complete a job in h hours, what part of the same job can one person working alone complete in k hours?

(A) $\frac{k}{xh}$
(B) $\frac{h}{xk}$
(C) $\frac{k}{x+h}$
(D) $\frac{kh}{x}$
(E) $\frac{kx}{h}$

7 If 1 cup of milk is added to a 3-cup mixture that is $\frac{2}{5}$ flour and $\frac{3}{5}$ milk, what percent of the 4-cup mixture is milk?

(A) 80%
(B) 75%
(C) 70%
(D) 65%
(E) 60%

8 A freight train and a passenger train start toward each other at the same time from two towns that are 500 miles apart. After 3 hours the trains are still 80 miles apart. If the average rate of speed of the passenger train is 20 miles per hour faster than the average rate of speed of the freight train, what is the average rate of speed, in miles per hour, of the freight train?

(A) 40
(B) 45
(C) 50
(D) 55
(E) 60

9 One machine can seal 360 packages per hour, and an older machine can seal 140 packages per hour. How many MINUTES will the two machines working together take to seal a total of 700 packages?

(A) 48
(B) 72
(C) 84
(D) 90
(E) 108

10 Carrie can inspect a case of watches in 5 hours. James can inspect the same case of watches in 3 hours. After working alone for 1 hour, Carrie stops for lunch. After taking a 40-minute lunch break, Carrie and James work together to inspect the remaining watches. How long do Carrie and James work together to complete the job?

(A) 1 hour and 30 minutes
(B) 1 hour and 45 minutes
(C) 2 hours
(D) 2 hours and 15 minutes
(E) 2 hours and 30 minutes

11 A man who can complete a job in h hours stops working before the job is finished. If the man did not stop, he could have finished the job by working p additional hours. What part of the job did the man complete at the time he stopped working?

(A) $\frac{p}{h}$
(B) $\frac{h-p}{h}$
(C) $\frac{h+p}{h}$
(D) $\frac{h}{h+p}$
(E) $\frac{p}{h+p}$

12 A motor boat traveling at 18 miles per hour traveled the length of a lake in one-quarter of an hour less time than it took when traveling at 12 miles per hour. What was the length in miles of the lake?

(A) 6
(B) 9
(C) 12
(D) 15
(E) 21

13 Carmen went on a trip of 120 miles, traveling at an average of x miles per hour. Several days later she returned over the same route at a rate that was 5 miles per hour faster than her previous rate. If the time for the return trip was one-third of an hour less than the time for the outgoing trip, which equation can be used to find the value of x?

(A) $\frac{120}{x+5} = \frac{1}{3}$
(B) $\frac{x}{120} = \frac{x+5}{120} - \frac{1}{3}$
(C) $\frac{120}{x+(x+5)} = \frac{1}{3}$
(D) $\frac{120}{x} = \frac{120}{x+5} + \frac{1}{3}$
(E) $120(x+5) - 120x = \frac{1}{3}$

Grid-In

1 Fruit for a dessert costs $1.20 a pound. If 5 pounds of fruit are needed to make a dessert that serves 18 people, what is the cost of the fruit needed to make enough of the same dessert to serve 24 people?

2 Candy selling for $2.00 per pound is to be mixed with candy selling for $6.00 per pound. How many pounds of the more expensive candy are needed to produce a 24-pound mixture that sells for $5.00 per pound?

ANSWERS TO CHAPTER 5 TUNE-UP EXERCISES

Lesson 5-1 (*Multiple-Choice*)

1. **(B)** Since Steve has 5 more course credits than Gary and Gary has 8 course credits, Steve has $5 + 8$ or 13 course credits. Carl has 7 fewer than twice the number of course credits that Steve has, so Carl has $(2 \times 13) - 7$ or 19 course credits.
2. **(D)** The phrase "3 less than 2 times x" means $2x$ minus 3 or $2x - 3$.
3. **(D)** When 4 times a number n ($= 4n$) is increased by 9 ($= 4n + 9$), the result is 21. Hence, the equation is $4n + 9 = 21$.
4. **(C)** If $x - 4$ is 2 greater than $y + 1$, then
$$\begin{aligned} x - 4 &= (y + 1) + 2 \\ &= y + 3 \\ x &= y + 7 \end{aligned}$$
Since $x + 6 = (y + 7) + 6$ or $x + 6 = y + 13$, $x + 6$ is greater than y by 13.
5. **(A)** When 3 is subtracted from 5 times a number n ($= 5n - 3$), the result is 27. Hence, the equation is $5n - 3 = 27$.
6. **(C)** *Solution 1*: Let x represent the unknown number. Since $\frac{1}{3}$ of x is 4 less than $\frac{1}{2}$ of x, $\frac{x}{3} = \frac{x}{2} - 4$. Multiplying each member of this equation by 6, the LCD of its denominators, gives $2x = 3x - 24$, so
$$-x = -24 \text{ or, } x = 24$$
Solution 2: Plug each of the answer choices into the statement of the problem until you find one (C) that works: $\frac{1}{3}$ of 24 ($= 8$) is 4 less than $\frac{1}{2}$ of 24 ($= 12$).
7. **(A)** Since Carol weighs 8 pounds less than Judy and Judy weighs x pounds, Carol weighs $x - 8$ pounds. If Susan, who weighs p pounds, gains 17 pounds, she will weigh $p + 17$ pounds, which is as much as Carol weighs. Hence, $p + 17 = x - 8$, so $p = x - 25$.
8. **(C)** The expression "at least" in the given sentence means greater than or equal to. If two times a number n is decreased by 5($= 2n - 5$), the result is at least ($\geq$) 11. Hence, $2n - 5 \geq 11$.
9. **(A)** The expression "at most" in the given sentence means less than or equal to. When 3 times a number n is increased by 7 ($= 3n + 7$), the result is at most ($\leq$) 4 times the number decreased by 1 ($= 4n - 1$). Hence, $3n + 7 \leq 4n - 1$.
10. **(C)** The sum of x and $2y$ is $x + 2y$, and t multiplied by this sum is $t(x + 2y)$. When this expression is divided by $4y$, the result is
$$\begin{aligned} \frac{t(x + 2y)}{4y} &= \frac{xt + 2yt}{4y} \\ &= \frac{xt}{4y} + \frac{\cancel{2y}t}{\underset{2}{\cancel{4y}}} \\ &= \frac{xt}{4y} + \frac{t}{2} \end{aligned}$$
11. **(E)** If 4 subtracted from x is 1 more than y, then $x - 4 = y + 1$, so $x = y + 1 + 4$ or $x = y + 5$.
12. **(B)** If the number $a - 5$ is 3 less than b, then
$$\begin{aligned} a - 5 &= b - 3 \\ a - 5 + 3 &= b \\ a - 2 &= b \end{aligned}$$
Since 1 more than b is $b + 1$, add 1 to each side of the equation $a - 2 = b$, obtaining $a - 2 + 1 = b + 1$ or $a - 1 = b + 1$. Hence, $a - 1$ has the same value as 1 more than b.
13. **(E)** If $x + y$ is 4 more than $x - y$, then $x + y = x - y + 4$ or $y + y = x - x + 4$, so $2y = 4$. Determine whether each Roman numeral statement is true or false.
    - I. The value of x cannot be determined, so statement I is false.
    - II. Statement II is true since $2y = 4$, so $y = 2$.
    - III. Since the value of x is not fixed, xy can have more than one value for different values of x. Hence, statement III is true.

    Only Roman numeral statements II and III are true.
14. **(C)** Jill's weight 1 year ago was $\frac{9}{8}$ of her present weight, so Jill lost $\frac{1}{8}$ of her weight. Since her present weight is 14 pounds less than her weight a year ago, 14 pounds represents $\frac{1}{8}$ of her present weight. Hence, Jill's present weight is 8×14 or 112 pounds.

15. **(A)** If the number x is 3 less than 4 times the number y, then $x = 4y - 3$. If 2 times the sum of x and y is 9, then $2(x + y) = 9$. The two equations in choice (A) can be used to find the values of x and y.

16. **(B)** Since Arthur has 3 times as many marbles as Vladimir, let x represent the number of marbles Vladimir has, and let $3x$ represent the number of marbles Arthur has. If Arthur gives Vladimir 6 marbles, Arthur now has $3x - 6$ marbles and Vladimir has $x + 6$ marbles. Since Arthur is left with 4 more marbles than Vladimir,

$$\begin{aligned} 3x - 6 &= (x + 6) + 4 \\ &= x + 10 \\ 2x &= 16 \\ x &= 8 \end{aligned}$$

Since $x = 8$, $3x = 24$. The total number of marbles that Arthur and Vladimir have is $x + 3x = 8 + 24$ or 32.

17. **(E)** Let n represent the number of rental days. Since the video store charges x dollars for the first day and y dollars for each additional day that the DVD is out, for n days the charge is $1x + (n - 1)y$. Since Sara is charged c dollars, $c = x + (n - 1)y$. Now solve this equation for n by multiplying each term inside the parentheses by y:

$$\begin{aligned} c &= x + ny - y \\ y + c - x &= ny \\ \frac{y}{y} + \frac{c}{y} - \frac{x}{y} &= \frac{ny}{y} \\ 1 + \frac{c - x}{y} &= n \end{aligned}$$

18. **(D)** If x is the number entered, then multiplying x by 3 gives $3x$ and subtracting 4 from that product gives $3x - 4$. Since $3x - 4$ is then entered, the final output is obtained by multiplying $3x - 4$ by 3 and then subtracting 4. Thus, the final output is $3(3x - 4) - 4$, which simplifies to $9x - 12 - 4$ or $9x - 16$.

19. **(A)** Since $\frac{3}{4}$ of $36 = \frac{3}{4} \times 36 = 3 \times 9 = 27$, 27 of the 36 identical cartons must be taped. If every 3 cartons require 2 rolls of tape, then $\frac{27}{3} \times 2 = 9 \times 2 = 18$ rolls of tape will be needed.

20. **(D)** If the original number of people in the room is represented by x, then, after $\frac{2}{3}$ of the x people in the room leave, $\frac{1}{3}x$ people remain. After two more people leave, $\frac{1}{3}x - 2$ people remain. Since this number represents $\frac{1}{4}$ of the x people who originally entered the room, $\frac{1}{3}x - 2 = \frac{1}{4}x$. Multiplying each member of this equation by 12 gives $4x - 24 = 3x$, so $4x - 3x = 24$ and $x = 24$.

21. **(B)** Let x represent the number of sophomores. Then $\frac{2}{3}x$ represents the number of juniors, and $\frac{3}{4}(\frac{2}{3}x)$ or $\frac{1}{2}x$ represents the number of seniors. Since this school has the same number of freshmen as sophomores, the number of freshmen is x. The total number of students is 950 so

$$\begin{aligned} x + x + \frac{2}{3}x + \frac{1}{2}x &= 950 \\ 2x + \frac{2}{3}x + \frac{1}{2}x &= 950 \\ 6(2x) + 6\left(\frac{2}{3}x\right) + 6\left(\frac{1}{2}x\right) &= 6(950) \\ 12x + 4x + 3x &= 5700 \\ 19x &= 5700 \\ x &= \frac{5700}{19} = 300 \end{aligned}$$

Since $\frac{1}{2}x = \frac{1}{2}(300) = 150$, 150 students are seniors.

(Grid-In)

1. **15** If $\frac{5}{4}$ of x is 20, then $\frac{5}{4}x = 20$, so

$$x = \frac{4}{5}(20) = 16$$

Hence, x decreased by 1 is $x - 1$ or 15.

2. **26** If half the difference of two positive numbers is 10, then the difference of the two positive numbers is 20. If the smaller of the two numbers is 3, then the other positive number must be 23 since $23 - 3 = 20$. Hence, the sum of the two numbers is $3 + 23$ or 26.

3. **36** You are given that in a certain college class each student received a grade of A, B, C, D or an "Incomplete." Since $\frac{1}{6} + \frac{1}{4} + \frac{1}{3} + \frac{1}{6}$ or $\frac{11}{12}$ of the number of students in the class received letter grades, $\frac{1}{12}$ of the students in the class received grades of "Incomplete." You are told that 3 students received this grade. Since these 3 students represent $\frac{1}{12}$ of the class, there were 3×12 or 36 students in the class.

Lesson 5-2 (*Multiple-Choice*)

1. **(D)** First find the sum of 25% of 32 and 40% of 15:

$$\begin{array}{rl} 25\% \text{ of } 32 = \frac{1}{4} \times 32 = & 8 \\ + \ 40\% \text{ of } 15 = \frac{2}{5} \times 15 = & 6 \\ \hline & 14 \end{array}$$

Then answer the question "14 is what percent of 35?" by solving the equation $14 = \frac{p}{100} \times 35$, where p is the unknown percent. Since $14 = 0.35p$,

$$p = \frac{14}{0.35} = 40$$

so 14 is 40% of 35.

2. **(D)** 30% of 150 = $0.30 \times 150 = 45$. Answer the question "4.5% of what number is 45?" by solving the equation $0.045 \times n = 45$. Since $0.045n = 45$,

$$n = \frac{45}{0.045} = 1000$$

3. **(C)** If A is 125% of B, then $A = 1.25 \times B$. Since $A = 1.25B$,

$$B = \frac{1}{1.25} \times A = 0.80 \times A$$

so B is 80% of A.

4. **(C)** Since Terry passed 80% of his science tests, he failed 20% or $\frac{1}{5}$ of the tests taken. Terry failed 4 science tests, so 4 represents $\frac{1}{5}$ of the total number of science tests. Hence, Terry took 5×4 or 20 science tests. Since he failed 4 tests, he passed $20 - 4$ or 16 science tests.

5. **(E)** If 30% of x is 21, then $\frac{3}{10}x = 21$, so

$$\frac{1}{10}x = \frac{1}{3}(21) = 7$$

Since $\frac{1}{10}x = 7$, $\frac{8}{10}x$, or 80% of x, is $8 \times 7 = 56$.

6. **(B)** Since the soccer team has won 60% of the 25 games it has played, it has won 0.6×25 or 15 games.

Solution 1: If x represents the minimum number of additional games the team must win in order to finish the season winning 80% of the games it has played, then $15 + x =$ 80% of $(25 + x)$, so $15 + x = 0.80(25 + x)$. Remove the parentheses by multiplying each term inside the parentheses by 0.80:

$$\begin{aligned} 15 + x &= 20 + 0.8x \\ x - 0.8x &= 20 - 15 \\ 0.2x &= 5 \\ x &= \frac{5}{0.2} = 25 \end{aligned}$$

Solution 2: For each answer choice, form and then evaluate the fraction

$$\frac{\text{Wins}}{\text{Total games}} = \frac{\text{Answer Choice} + 15}{\text{Answer Choice} + 25}$$

The correct answer choice is (B), which gives a fraction equal to 0.80:

$$\frac{15}{25} = \frac{25 + 15}{25 + 25} = \frac{40}{50} = 0.80$$

7. **(B)** Answer the question "What percent of 800 is 5?" by solving the equation $\frac{p}{100} \times 800 = 5$, where p is the unknown percent. Since

$$\frac{p}{100} \times 800 = p \times 8 = 5$$

then $p = \frac{5}{8}\%$.

8. **(A)** Answer the question "$\frac{1}{2}$ is what percent of $\frac{1}{5}$?" by solving the equation $\frac{1}{2} = \frac{p}{100} \times \frac{1}{5}$, where p is the unknown percent. Since $\frac{1}{2} = \frac{p}{500}$, then

$$p = \frac{500}{2} = 250\%$$

9. **(C)** In the opinion poll, 70% of the 50 men or $0.70 \times 50 = 35$ men preferred fiction to nonfiction books. In the same poll, 25% of the 40 women or $0.25 \times 40 = 10$ women preferred fiction to nonfiction books. Thus, 45 $(= 35 + 10)$ of the 90 $(= 50 + 40)$ people polled preferred fiction to nonfiction books. Since $\frac{45}{90} = \frac{1}{2} = 50\%$, 50% of the people polled preferred to read fiction.

10. **(B)** 12.5% of $2x = 12.5\% \times 2x$
$$= (12.5\% \times 2) \times x$$
$$= 25\% \times x$$
Hence, 12.5% of $2x = 25\%$ of $x = 12.5$.

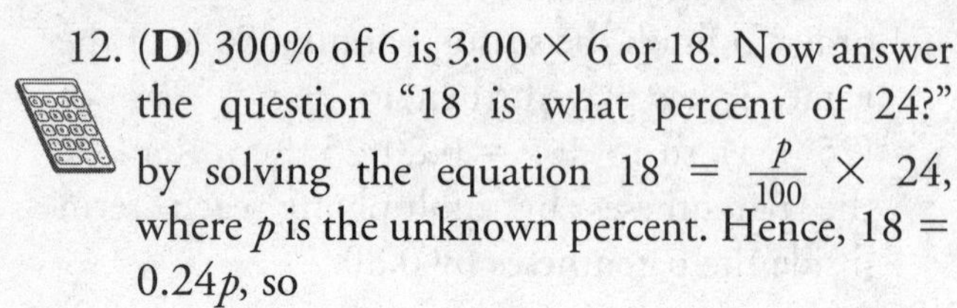

11. **(C)** If $\frac{1}{8}$ of a number is 9, then $\frac{8}{8}$ or the whole number is 8×9 or 72. Hence, 75% of the same number is $0.75 \times 72 = 54$.

12. **(D)** 300% of 6 is 3.00×6 or 18. Now answer the question "18 is what percent of 24?" by solving the equation $18 = \frac{p}{100} \times 24$, where p is the unknown percent. Hence, $18 = 0.24p$, so
$$p = \frac{18}{0.24} = 75\%$$

13. **(E)** Since the regular price of software is 12% off the retail price p, the regular price of the software can be expressed as $0.88p$. During an annual sale, the same software is 25% off the regular price, so the sale price is $0.75 \times 0.88p = 0.66p$.

14. **(C)** Suppose the original price of the stock was \$100. If the price falls 25%, the new price of the stock is $0.75 \times \$100 = \$75$. To reach its original value, the price of the stock must rise \$25, which is $\frac{1}{3}$ or $33\frac{1}{3}\%$ of its \$75 price.

15. **(C)** If a is 30% greater than A, then

$$a = A + 30\% \text{ of } A$$
$$= A + 0.30A$$
$$= 1.3A$$

Similarly, if b is 20% greater than B, then $b = 1.2B$. Thus,
$$ab = (1.3A)(1.2B)$$
$$= 1.56(AB)$$
$$= AB + 0.56AB$$

Since $0.56 = 56\%$, ab is 56% greater than AB.

16. **(C)** If a increased by 20% of a results in b, then
$$a + (20\% \text{ of } a) = b$$
$$a + 0.2a = b$$
$$1.2a = b$$

When b is decreased by $33\frac{1}{3}\%$ of b, the result is c, so
$$b - \frac{1}{3}b = \frac{2}{3}b = c$$

Since
$$c = \frac{2}{3}b = \frac{2}{3}(1.2a) = 0.80a$$
then c is 80% of a.

17. **(B)** To find the number of people who voted for candidate A in the actual election, add the number of people who chose candidate A in the poll to the number of people who voted for candidate A but were undecided in the poll.

- Since 30% of $4000 = 0.30 \times 4000 = 1200$, 1200 people voted for candidate A in the poll as well as in the actual election.
- Since 20% of $4000 = 0.20 \times 4000 = 800$, 800 people were undecided in the poll.
- It is given that in the actual election, the undecided people voted for candidate A in the same proportion as the people who cast votes for candidates in the poll. Hence, of the 800 undecided people, $\frac{30\%}{30\% + 50\%} \times 800 = \frac{3}{8} \times 800 = 300$ voted for candidate A in the actual election.
- Hence, a total of $1200 + 300 = 1500$ people voted for candidate A in the actual election.

(Grid-In)

1. **12** Since the team wins 60% of the first 15 matches, it wins 0.60×15 or 9 of its first 15 matches. Let x represent the number of additional matches it must win to finish the 28-match season winning 75% of its scheduled matches. Since $75\% = \frac{3}{4}$,
$$\frac{\text{Total wins}}{\text{Total games}} = \frac{9 + x}{28} = \frac{3}{4}$$
Solve the equation by cross-multiplying:
$$4(9 + x) = 3(28)$$
$$36 + 4x = 84$$
$$4x = 48$$
$$x = \frac{48}{4}$$
$$= 12$$

2. **55.5** Since
$$80\% \text{ of } 35 = 0.80 \times 35 = 28$$
and
$$25\% \text{ of } 28 = 0.25 \times 28 = 7$$
then 35 (= 28 + 7) of the 63 (35 + 28) boys

and girls have been club members for more than 2 years. Since $\frac{35}{63} = 0.5555\ldots$, 55.5% of the club have been members for more than 2 years.

Lesson 5-3 (*Multiple-Choice*)

1. **(D)** If x is an odd integer, then $x + 2$, $x + 4$, and $x + 6$ represent the next three consecutive odd integers. Since the sum of 3 times the second integer, $x + 2$, and the largest integer, $x + 6$, is 104,

$$3(x + 2) + (x + 6) = 104$$
$$3x + 6 + x + 6 = 104$$
$$4x + 12 = 104$$
$$4x = 92$$
$$x = \frac{92}{4} = 23$$

2. **(B)** If x is Jim's present age, then $x - 4$ represents Jim's age 4 years ago and $x + 7$ represents his age 7 years from now. Since 4 years ago Jim was one-half the age he will be 7 years from now,

$$x - 4 = \frac{1}{2}(x + 7)$$

3. **(E)** If the sum of n consecutive integers starting with -10 is 11, then the consecutive integers must range from -10 to $+11$, inclusive, so that number-pair opposites from -10 to $+10$ cancel each other. Including 0, there are 22 integers in this interval.

4. **(D)** If x is David's present age, then $x + 6$ is his age 6 years from now and $x - 7$ was his age 7 years ago. Since 6 years from now David will be twice his age 7 years ago,

$$x + 6 = 2(x - 7)$$
$$= 2x - 14$$
$$x = 20$$

5. **(E)** If the two consecutive integers are represented by x and $x + 1$, then

$$x + (x + 1) = k$$
$$2x + 1 = k$$
$$2x = k - 1$$
$$x = \frac{k - 1}{2}$$

Since the question asks for the larger of the two integers, you need to find $x + 1$ in terms of k:

$$x + 1 = \overbrace{\frac{k-1}{2}}^{x} + 1$$
$$= \frac{k-1}{2} + \frac{2}{2}$$
$$= \frac{k+1}{2}$$

6. **(E)** If x is the least of the 4 consecutive odd integers, then the next 3 consecutive odd integers are $x + 2$, $x + 4$, and $x + 6$. The greatest of these 4 integers is $x + 6$, and 2 times the sum of the first and the second of these integers is $2[x + (x + 2)]$. Hence, the required equation is

$$x + 6 = 2[x + (x + 2)] + 11$$

7. **(A)** If x is the least of 4 consecutive even integers whose sum is S, then

$$x + (x + 2) + (x + 4) + (x + 6) = S$$
$$4x + 12 = S$$
$$4x = S - 12$$
$$x = \frac{S - 12}{4} = \frac{S}{4} - \frac{\overset{3}{\cancel{12}}}{\cancel{4}} = \frac{S}{4} - 3$$

8. **(B)** Let J represent John's age. If John's age is increased by Mary's age, the result is $J + M$, which equals 2 times John's age 3 years ago or $2(J - 3)$. Thus, $J + M = 2(J - 3)$ or $J + M = 2J - 6$, so

$$M + 6 = 2J - J \quad \text{and} \quad J = M + 6$$

9. **(C)** Suppose R represents Reyna's present age. In $3x$ years, Reyna will be $R + 3x$, which equals $(3y + 4)$ times her present age R. Hence, $R + 3x = (3y + 4)R$. Solve this equation for R:

$$R + 3x = 3yR + 4R$$
$$3x = 3yR + 4R - R$$
$$= 3yR + 3R$$
$$= R(3)(y + 1)$$
$$\frac{3x}{3(y+1)} = R$$
$$\frac{x}{y+1} = R$$

Lesson 5-4 (*Multiple-Choice*)

1. **(C)** When the recipe is adjusted from 4 to 8 servings, the amounts of salt and pepper are *each* doubled, so the ratio of 2 : 3 remains the same.
2. **(E)** *Solution 1*: If x represents the unknown number of miles, then
$$\frac{\text{Inches}}{\text{Miles}} = \frac{\frac{3}{8}}{120} = \frac{1\frac{3}{4}}{x}$$
Cross-multiplying gives
$$\frac{3}{8}x = 1\frac{3}{4}(120) = \frac{7}{4}(120) = 210$$
Then
$$x = \frac{8}{3}(210) = 560 \text{ miles}$$
Solution 2: Since $\frac{3}{8}$ of an inch represents 120 miles, $\frac{1}{8}$ of an inch represents $\frac{1}{3} \times 120 = 40$ miles. Since $1\frac{3}{4} = \frac{14}{8}$, $1\frac{3}{4}$ inches represent 14×40 or 560 miles.
3. **(C)** Let p represent the initial population of bacteria. After 12 minutes the population is $2p$, after 24 minutes it is $4p$, after 36 minutes it is $8p$, after 48 minutes it is $16p$, and after 60 minutes or 1 hour the population is $32p$. Since
$$\frac{32p}{p} = \frac{32}{1}$$
the ratio of the number of bacteria at the end of 1 hour to the number of bacteria at the beginning of that hour is 32 : 1.
4. **(D)** If x represents the number of losses, then
$$\frac{\text{Wins}}{\text{Losses}} = \frac{4}{3} = \frac{12}{x}$$
$$4x = 36$$
$$x = 9$$
The total number of games played was 12 + 9 or 21.
5. **(C)** If $\frac{s}{t} = \frac{4}{7}$, then $s = \frac{4t}{7}$ Since s is an integer, t must be divisible by 7. In the interval $8 < t < 40$, there are four integers that are divisible by 7: 14, 21, 28, and 35.
6. **(D)** If x represents the number of sophomores in the school club, then $3x$ represents the number of juniors, and $2x$ represents the number of seniors. Since the club has 42 members,
$$x + 3x + 2x = 42$$
$$6x = 42$$
$$x = \frac{42}{6} = 7$$
The number of seniors in the club is $2x = 2(7) = 14$.
7. **(A)** If $\frac{c-3d}{4} = \frac{d}{2}$, then
$$2(c - 3d) = 4d$$
$$2c - 6d = 4d$$
$$2c = 10d$$
Since $\frac{c}{d} = 5$, the ratio of c to d is 5:1.
8. **(D)** If x represents the number of pairs of socks that can be purchased for \$22.50, then
$$\frac{\text{Pairs of socks}}{\text{Cost}} = \frac{4}{10} = \frac{x}{22.50}$$
So
$$10x = 4(22.50) = 90$$
$$x = \frac{90}{10} = 9$$
9. **(C)** The number of boys working and the time needed to complete the job are inversely related since, as one of these quantities increases, the other decreases. Let x represent the time three boys take to paint a fence; then
$$3x = 2(5) = 10$$
$$x = \frac{10}{3} = 3\frac{1}{3}$$
10. **(B)** If x represents the number of miles the car travels in 5 hours, then
$$\frac{\text{Distance}}{\text{Time}} = \frac{96}{2} = \frac{x}{5}$$
so
$$2x = 5(96) = 480$$
$$x = \frac{480}{2} = 240$$
11. **(C)** If x represents the number of kilograms of corn needed to feed 2800 chickens, then $2x - 30$ is the number of kilograms needed to feed 5000 chickens. If the amount of feed needed for each chicken is assumed to be constant,
$$\frac{x}{2800} = \frac{2x-30}{5000}$$
so
$$5000x = 2800(2x - 30)$$
$$50x = 28(2x - 30)$$
$$= 56x - 840$$
$$6x = 840$$
$$x = \frac{840}{6} = 140$$

12. **(B)** If x is the least integer in an ordered list of five consecutive positive even integers, then the other integers are $x + 2$, $x + 4$, $x + 6$, and $x + 8$. Since the ratio of the greatest integer to the least integer is 2 : 1,

$$\frac{x+8}{x} = \frac{2}{1}$$
$$2x = x + 8$$
$$x = 8$$

The middle integer in the list is $x + 4 = 8 + 4 = 12$.

13. **(D)** If the ratio of p to q is 3 : 2, then $\frac{p}{q} = \frac{3}{2}$, so

$$2\left(\frac{p}{q}\right) = 2\left(\frac{3}{2}\right) \quad \text{and} \quad \frac{2p}{q} = \frac{3}{1}$$

Hence, the ratio of $2p$ to q is 3 : 1.

14. **(D)** Since $\frac{x}{z} = \frac{1}{3}$, then

$$\frac{x^2}{z} = x\left(\frac{x}{z}\right) = x\left(\frac{1}{3}\right)$$

Determine whether each Roman numeral value is a possible value of $\frac{x^2}{z}$ when x and z are integers.

- I. If $\frac{x^2}{z} = x\left(\frac{1}{3}\right) = \frac{1}{9}$, then $x = \frac{1}{3}$. Since x must be an integer, $\frac{1}{9}$ is not a possible value of $\frac{x^2}{z}$.
- II. If $\frac{x^2}{z} = x\left(\frac{1}{3}\right) = \frac{1}{3}$, then $x = 1$. Hence, $\frac{1}{3}$ is a possible value of $\frac{x^2}{z}$.
- III. If $\frac{x^2}{z} = x\left(\frac{1}{3}\right) = 3$, then $x = 9$. Hence, 3 is a possible value of $\frac{x^2}{z}$.

Only Roman numeral values II and III are possible values of $\frac{x^2}{z}$.

15. **(A)** If $a - 3b = 9b - 7a$, then

$$a + 7a = 9b + 3b$$
$$8a = 12b$$
$$\frac{a}{b} = \frac{12}{8} = \frac{3}{2}$$

Hence, the ratio of a to b is 3 : 2.

16. **(C)** Since the lowest common multiple of 8 and 12 is 24, multiply each term of the ratio $a : 8$ by 3 and multiply each term of the ratio $12 : c$ by 2. The ratio of A to B is then $3a : 24$, and the ratio of B to C is $24 : 2c$, so the ratio of A to C is $3a : 2c$. Since you are given that the ratio of A to C is 2 : 1, then

$$\frac{2}{1} = \frac{3a}{2c}, \text{ so}$$
$$\frac{a}{c} = \frac{2(2)}{3(1)} = \frac{4}{3}$$

Hence, the ratio of a to c is 4 : 3.

17. **(A)** If $8^r = 4^t$, then $2^{3r} = 2^{2t}$ so $3r = 2t$. Dividing both sides of this equation by $3t$ gives

$$\frac{3r}{3t} = \frac{2t}{3t} \quad \text{or} \quad \frac{r}{t} = \frac{2}{3}$$

The ratio of r to t is 2 : 3.

18. **(B)** Rewrite each fraction by dividing each term of the numerator by the denominator:

$$\frac{a+b}{b} = \frac{a}{b} + \frac{b}{b} = \frac{a}{b} + 1 = 4, \frac{a}{b} = 3$$

$$\frac{a+c}{c} = \frac{a}{c} + \frac{c}{c} = \frac{a}{c} + 1 = 3, \frac{a}{c} = 2$$

or $\frac{c}{a} = \frac{1}{2}$

Multiply corresponding sides of the two proportions:

$$\frac{a}{b} \times \frac{c}{a} = 3 \times \frac{1}{2}$$
$$\frac{c}{b} = \frac{3}{2}$$

The ratio of c to b is 3 : 2.

19. **(C)** Since the ratio of mathematics majors to English majors is given as 3 : 8, suppose there are 30 mathematics majors and 80 English majors.

- If the number of mathematics majors increases 20%, then the new number of mathematics majors is $30 + (20\% \times 30) = 30 + 6 = 36$.
- If the number of English majors decreases 15%, then the new number of English majors is $80 - (15\% \times 80) = 80 - 12 = 68$.
- The new ratio of mathematics majors to English majors is $\frac{36}{68} = \frac{9}{17}$ or 9 : 17.

20. **(A)** The ratio of freshmen to juniors who attended the game is 3 : 4, and the ratio of juniors to seniors is 7 : 6. Since the least common multiple of 4 and 7 is 28, multiply the terms of the first ratio by 7 and multiply the terms of the second ratio by 4. Since 21 : 28 is then the ratio of freshmen to juniors and 28 : 24 is the ratio of juniors to seniors, 21 : 24 is the ratio of freshmen to seniors.

Since

$$21 : 24 = \frac{21}{24} = \frac{7}{8}$$

the correct choice is (A).

21. **(B)** It is given that x varies directly as z. If z is doubled, then x is also doubled. Since y varies inversely as x, their product does not change. Hence, since x is doubled or multiplied by 2, then y must be divided by 2.
22. **(B)** The number of men working and the hours needed for the men to complete the job are inversely related. If x represents the number of men needed to complete the job in $5 - 1 = 4$ hours, then $4x = 5 \cdot 12 = 60$. Hence, $x = \frac{60}{4} = 15$. This means that $15 - 12 = 3$ additional men are needed.
23. **(D)** If x represents the number of people renting the limousine when the cost per couple is \$105, then $\frac{\$105}{2}$ represents the cost per person. Hence, $x\left(\frac{\$105}{2}\right) = 9(\$70) =$ \$630, so $\$105x = \$630 \times 2 = \$1{,}260$ and $x = \frac{\$1260}{\$105} = 12$.

(*Grid-In*)

1. **12** Since the lengths of the two pieces of string are in the ratio 2 : 9, let $2x$ and $9x$ represent their lengths. Hence:
$$9x - 2x = 42$$
$$7x = 42$$
$$x = \frac{42}{7} = 6$$
Since $2x = 2(6) = 12$, the length of the shorter piece of string is 12 inches.
2. **6** Since the ratio of a to b is 5 : 9 and the ratio of x to y is 10 : 3, $\frac{a}{b} = \frac{5}{9}$ and $\frac{x}{y} = \frac{10}{3}$, so $\frac{y}{x} = \frac{3}{10}$. Hence,
$$\frac{ay}{bx} = \frac{a}{b} \cdot \frac{y}{x} = \frac{5}{9} \cdot \frac{3}{10} = \frac{15}{90} = \frac{1}{6} \text{ or } 1:6$$
Grid in 6.
3. **9** Since the ratio of dimes to pennies in the purse is 3 : 4, let $3x$ and $4x$ represent the numbers of dimes and pennies, respectively. If 3 pennies are taken out of the purse, then $4x - 3$ pennies remain.
Since the ratio of dimes to pennies is now 1 : 1,
$$\frac{3x}{4x-3} = \frac{1}{1} \quad \text{or} \quad 4x - 3 = 3x$$
so $x = 3$. Hence, the number of dimes in the purse is $3x = 3(3) = 9$.
4. **(15)** Since $2^3 = 2 \times 2 \times 2 = 8$, then $a = 3$ and $b = 2$, so $\frac{a}{b} = \frac{3}{2}$. You are told that $\frac{a}{b} = \frac{c}{d}$. Substituting $\frac{3}{2}$ for $\frac{a}{b}$ and 10 for d gives $\frac{3}{2} = \frac{c}{10}$. Since 10 is 5 times 2, c must be 5 times 3 or 15.
5. **(9)** To find the value of the ratio of $\frac{a}{c}$, multiply $\frac{a}{b}$ by $\frac{b}{c}$. If $6a - 8b = 0$, then $6a = 8b$, so
$$\frac{a}{b} = \frac{8}{6} = \frac{4}{3}$$
Since $c = 12b$, then $\frac{b}{c} = \frac{1}{12}$, so
$$\frac{a}{c} = \frac{a}{b} \cdot \frac{b}{c} = \frac{\overset{1}{\cancel{4}}}{3} \cdot \frac{1}{\underset{3}{\cancel{12}}} = \frac{1}{9}$$
Hence, the ratio of a to c is equivalent to the ratio of 1 to 9.
6. **(4)** Since jars A, B, and C each contain 8 marbles, there are 24 marbles in the three jars. Let x, $2x$, and $3x$ represent the new numbers of marbles in jars A, B, and C, respectively. Hence, $x + 2x + 3x = 24$ or $6x = 24$, so
$$x = \frac{24}{6} = 4$$
To achieve a ratio of 1 : 2 : 3, jars A, B, and C must contain 4, 8, and 12 marbles, respectively. Since jar B already contains 8 marbles, 4 of the 8 marbles originally in jar A must be transferred to jar C.

Lesson 5-5 (*Multiple-Choice*)

1. **(C)** If 4 pens cost \$1.96, the cost of 1 pen is $\frac{\$1.96}{4} = \$0.49$. The greatest number of pens that can be purchased for \$7.68 is the greatest integer that is equal to or less than $\frac{\$7.68}{\$0.49}$. Since $\frac{\$7.68}{\$0.49} = 15.67$, the greatest number of pens that can be purchased is 15.
2. **(D)** From 10:40 A.M. to 1:00 P.M., 2 hours and 20 minutes elapse. Since the car travels at a constant rate of 45 miles per hour, it travels 2×45 or 90 miles in 2 hours. Also, since 20 minutes is $\frac{1}{3}$ of an hour, the car travels $\frac{1}{3} \times 45$ or 15 miles in 20 minutes. The total distance it travels is $90 + 15$ or 105 miles.
3. **(A)** If k pencils cost c cents, then 1 pencil costs $\frac{c}{k}$, so p pencils cost $p \times \frac{c}{k}$ or $\frac{pc}{k}$.

4. **(B)** Since the freight train leaves 1 hour before the passenger train, it travels 1 hour longer. Therefore, when the passenger train has traveled x hours, the freight train has traveled $x + 1$ hours. Make a table.

Rate	×	Time	=	Distance
50 mph		$x + 1$ hours		$50(x + 1)$
60 mph		x hours		$60x$

Since the two trains are traveling in opposite directions, the sum of their distances must equal 380 miles:

$$50(x + 1) + 60x = 380$$
$$50x + 50 + 60x = 380$$
$$110x = 330$$
$$x = \frac{330}{110} = 3$$

Since the passenger train leaves at 1:00 P.M., the two trains are 380 miles apart 3 hours later or at 4:00 P.M.

5. **(A)** Julie can type a manuscript in 4 hours, so her rate of work is $\frac{1}{4}$ of the job per hour. Pat needs 6 hours to type the same manuscript, so his rate of work is $\frac{1}{6}$ of the job per hour. If Julie and Pat work x hours to complete the whole job, then

$$x\left(\frac{1}{4}\right) + x\left(\frac{1}{6}\right) = 1 \quad \text{or} \quad \frac{x}{4} + \frac{x}{6} = 1$$

Multiplying each member of the equation by 12, the LCD of 4 and 6, gives

$$3x + 2x = 12$$
$$5x = 12$$
$$x = \frac{12}{5} = 2.4 \text{ hours}$$

Since Julie and Pat start working at 12:00 P.M. and 0.4 hour equals $0.4 \times 60 = 24$ minutes, the job is completed in 2 hours and 24 minutes past noon or at 2:24 P.M.

6. **(A)** If x people working together at the same rate can complete a job in h hours, then each person working alone can complete the same job in xh hours, so the rate of work is the reciprocal of xh or $\frac{1}{xh}$. Hence, the part of the job one person working alone can complete in k hours is $\frac{k}{xh}$.

7. **(C)** Since $\frac{3}{5} \times 3 = \frac{9}{5}$, the original 3-cup mixture contains $\frac{9}{5}$ cups of milk. After 1 cup of milk is added to the 3-cup mixture, the 4-cup mixture contains

$$\frac{9}{5} + 1 = \frac{9}{5} + \frac{5}{5} = \frac{14}{5} \text{ cups of milk}$$

Since

$$\frac{\frac{14}{5}}{4} = \frac{4}{5} \times \frac{1}{4} = \frac{14}{20} = 0.70$$

70% of the 4-cup mixture is milk.

8. **(E)** Let x be the rate of speed of the freight train. Make a table.

Rate	×	Time	=	Distance
x mph		3 hours		$3x$
$x + 20$ mph		3 hours		$3(x + 20)$

The total distance the two trains travel in 3 hours is $500 - 80$ or 420 miles.

Hence,

$$3x + 3(x + 20) = 420$$
$$3x + 3x + 60 = 420$$
$$6x = 360$$
$$x = \frac{360}{6} = 60$$

9. **(C)** If a machine can seal 360 packages per hour, then it seals $\frac{360}{60}$ or 6 packages per minute. If another machine can seal 140 packages per hour, then it seals $\frac{140}{60}$ or $\frac{7}{3}$ packages per minute. If the two machines working together can seal a total of 700 packages in x minutes, then

$$6x + \frac{7}{3}x = 700$$
$$18x + 7x = 2100$$
$$25x = 2100$$
$$x = \frac{2100}{25} = 84$$

10. **(A)** If x represents the number of hours Carrie and James work together, then $x + 1$ is the total number of hours Carrie works to complete the job. Hence,

$$\underbrace{\left(\frac{1}{5}\right)(x + 1)}_{\text{Part of job Carrie completes}} + \underbrace{\left(\frac{1}{3}\right)x}_{\text{Part of job James completes}} = 1$$

Multiplying each member of the equation by 15, the LCD of 3 and 5, gives $3(x + 1) + 5x = 15$. Then $8x + 3 = 15$, so

$$x = \frac{12}{8} = \frac{3}{2} = 1\frac{1}{2} \text{ hours}$$

or 1 hour and 30 minutes.

11. **(B)** The man's rate of work is the reciprocal of the total number of hours he takes to complete the job working alone, which, in this case, is $\frac{1}{h}$. When the man stopped working, he had worked $h - p$ hours. Hence, the part of the job the man completes is $\frac{h-p}{h}$.

12. **(B)** Make a table.

Rate	× Time	= Distance
18 mph	$x - \frac{1}{4}$ hours	$18\left(x - \frac{1}{4}\right)$
12 mph	x hours	$12x$

Since the length of the lake remains constant,

$$18\left(x - \frac{1}{4}\right) = 12(x)$$

$$18x - \frac{18}{4} = 12x$$

$$6x = \frac{9}{2}$$

$$x = \frac{9}{6(2)} = \frac{9}{12} = \frac{3}{4}$$

Hence,

Rate × Time = Distance

$$12 \times \frac{3}{4} = 9 \text{ miles}$$

13. **(D)** Let x be the time for the outgoing trip. Make a table.

Rate	× Time	= Distance
x mph	$\frac{120}{x}$ hours	120
$x + 5$ mph	$\frac{120}{x+5}$ hours	120

Since the time for the return trip was one-third of an hour less than the time for the outgoing trip,

$$\frac{120}{x} = \frac{120}{x+5} + \frac{1}{3}$$

(Grid-In)

1. **8.0** (or **8**) If 5 pounds of fruit serve 18 people, then $\frac{5}{18}$ pound serves one person, so

$$24 \times \frac{5}{18} = 4 \times \frac{5}{3} = \frac{20}{3} \text{ pounds}$$

serve 24 people. Since the fruit costs $1.20 a pound, the cost of the fruit needed to serve 24 people is

$$\frac{20}{3} \times \$1.20 = 20 \times \$0.40 \text{ or } \$8.00$$

2. **18** Let x represent the number of pounds of the less expensive candy. Since the total mixture is 24 pounds, $24 - x$ represents the number of pounds of the more expensive candy. Make a table.

(Price per pound)	× (Number of pounds)	= Total value
$2.00	x	$2.00x$
$6.00	$24 - x$	$6.00(24 - x)$
$5.00	24	$5.00(24)$

$$2.00x + 6.00(24 - x) = 5.00(24)$$

$$2x + 144 - 6x = 120$$

$$-4x = -24$$

$$x = \frac{-24}{-4} = 6$$

Since $24 - x = 24 - 6 = 18$, 18 pounds of the more expensive candy are needed.

CHAPTER 6

Geometric Concepts and Reasoning

This chapter reviews the geometric relationships and properties of figures that are tested on the SAT. Keep in mind that you will not be required to complete formal geometric proofs or to memorize key formulas.

Unless otherwise stated, figures on the SAT are drawn to scale. If you find a note that tells you the accompanying figure is *not* drawn to scale, try redrawing the figure so that it looks in scale. You may then be able to arrive at the answer by estimating or by eliminating answer choices that appear to contradict the diagram.

LESSONS IN THIS CHAPTER

LESSON

6-1 ANGLE RELATIONSHIPS

OVERVIEW

This lesson reviews the relationships among

- *angles formed when two lines intersect*
- *angles formed when two parallel lines are cut by a third line*
- *the angles of a triangle*

NAMING ANGLES

The **vertex** of an angle is the point at which the two sides of an angle intersect. The symbol ∠ is used as an abbreviation for the word *angle*. An angle may be named by using a number or a letter, as shown below.

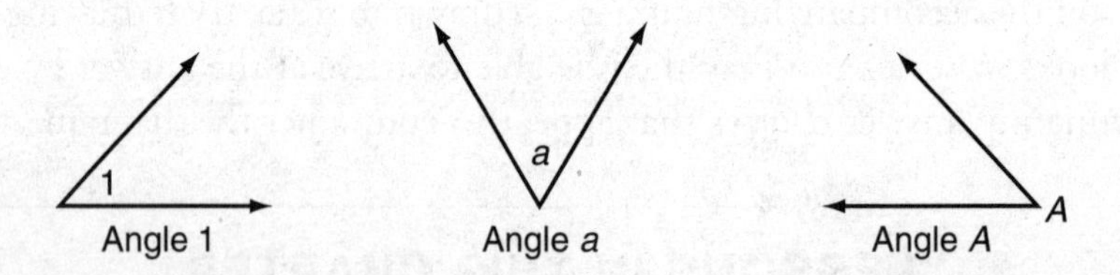

When more than one angle has the same vertex, three letters can be used to name a particular angle, provided that the middle letter is the vertex of the angle. In the accompanying figure, $\angle a$, $\angle BAC$, and $\angle CAB$ all name the same angle.

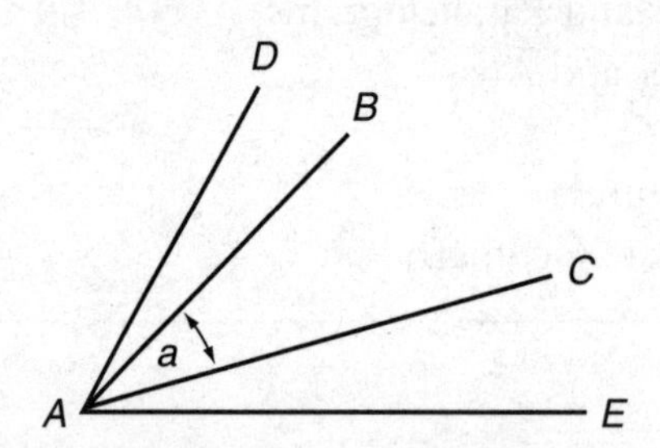

CLASSIFYING ANGLES

Angles can be classified according to their degree measures (see Figure 6.1).

- An **acute angle** is an angle that measures less than 90°.
- A **right angle** is an angle that measures 90°.
- An **obtuse angle** is an angle that measures more than 90° and less than 180°.
- A **straight angle** is an angle that measures 180°, so its sides lie on the same straight line.

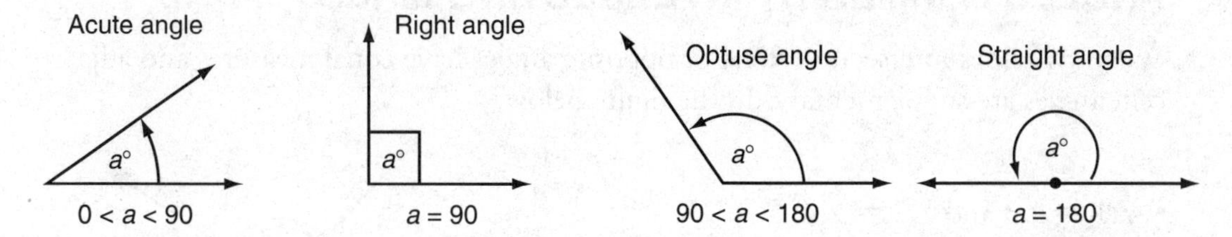

Figure 6.1 Classifying Angles

Pairs of angles whose measures add up to 90° or 180° are given special names.

- Two angles are **complementary** if the sum of the numbers of degrees in the angles is 90°.
- Two angles are **supplementary** if the sum of the numbers of degrees in the angles is 180°.

ADJACENT ANGLES

In Figure 6.2, angles 1 and 2 are **adjacent angles** because they have the same vertex, share a common side, and do not overlap.

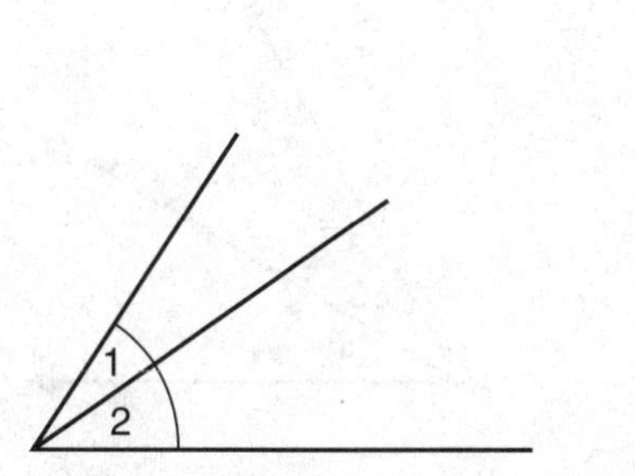

Figure 6.2 Adjacent Angles

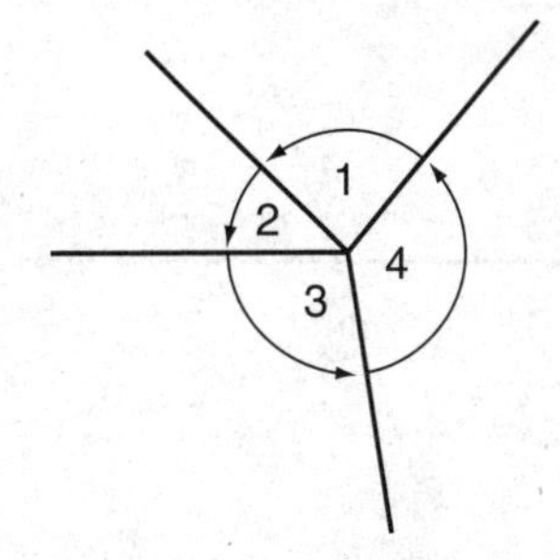

Figure 6.3 Adding Angles About a Point

The sum of the measures of the adjacent angles about a point is 360°. In Figure 6.3,

$$\angle 1 + \angle 2 + \angle 3 + \angle 4 = 360^\circ$$

In the figure at the right, the measure of angle a is $360° - 70°$ or $290°$.

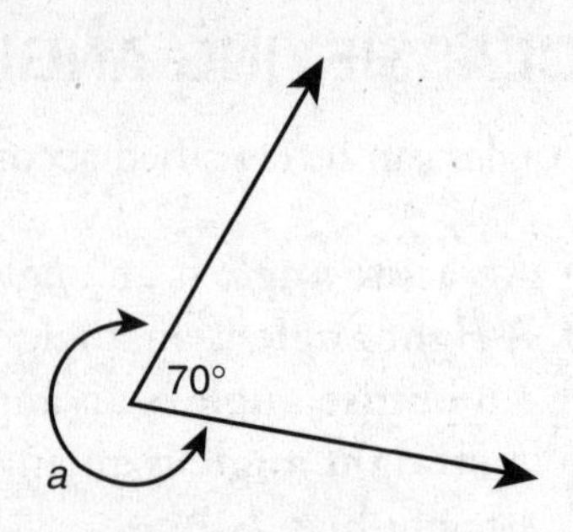

ANGLES FORMED BY INTERSECTING LINES

When two lines intersect, **vertical** or opposite angles have equal measures and adjacent angles are supplementary. In the figure below,

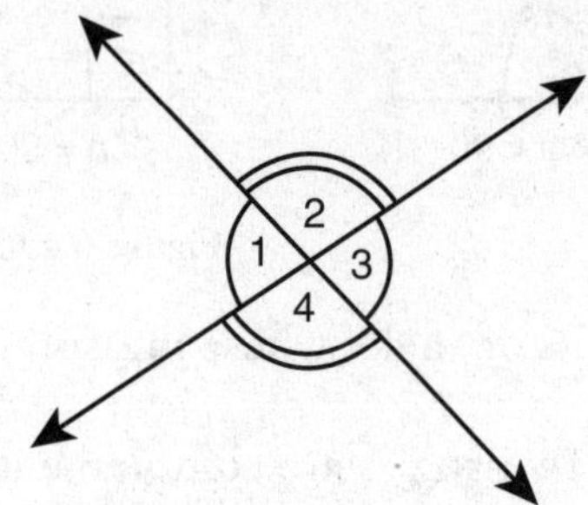

- $\angle 1 = \angle 3$ and $\angle 2 = \angle 4$
- $\angle 1 + \angle 2 = 180°$ and $\angle 1 + \angle 4 = 180°$
- $\angle 3 + \angle 4 = 180°$ and $\angle 2 + \angle 3 = 180°$

BISECTING ANGLES AND LINE SEGMENTS

A line segment or an angle is **bisected** when it is divided into two parts that have equal measures.

- In the figure below left, line segment AB is bisected at point M since line segments AM and MB have the same length.

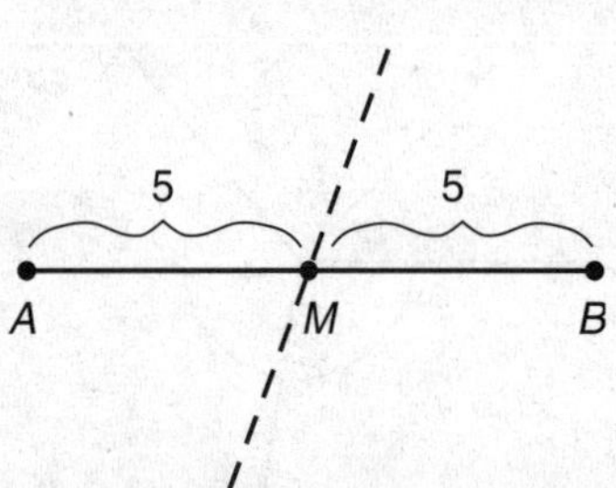

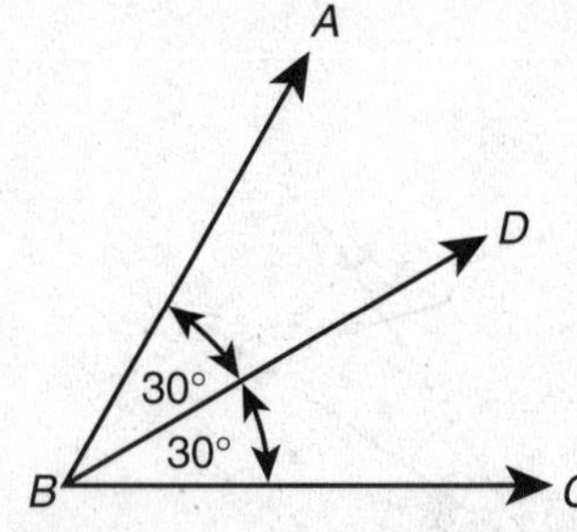

- In the figure above right, $\angle ABC$ is bisected by ray BD since it is divided into two angles that have the same number of degrees.

ANGLES FORMED BY PARALLEL LINES

Parallel lines are always the same distance apart and, as a result, do not intersect. In Figure 6.4, lines ℓ and m are parallel. The notation $\ell \| m$ is read as "ℓ is parallel to m." When two parallel lines are cut by a third line,

- angles that look equal are equal
- angles that do not look equal are supplementary

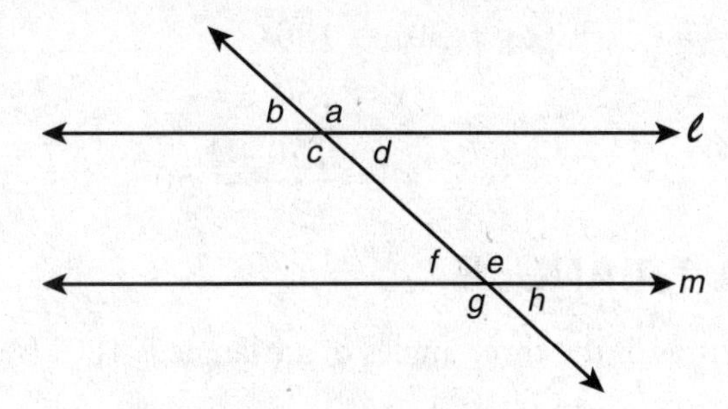

Figure 6.4 Parallel Line Angle Relationships

In Figure 6.4,

- $\angle b = \angle d = \angle f = \angle h$, since all the acute angles formed by parallel lines are equal.
- $\angle a = \angle c = \angle e = \angle g$, since all the obtuse angles formed by parallel lines are equal.
- $\angle a + \angle b = 180$, $\angle a + \angle d = 180$, $\angle a + \angle h = 180$, and so forth, since the measures of any acute angle and any obtuse angle formed by parallel lines always add up to 180.

If a line is perpendicular to one of two parallel lines, it must also be perpendicular to the other parallel line. In the accompanying figure, line ℓ is parallel to line m. If line p is perpendicular to line ℓ, then line p must also be perpendicular to line m.

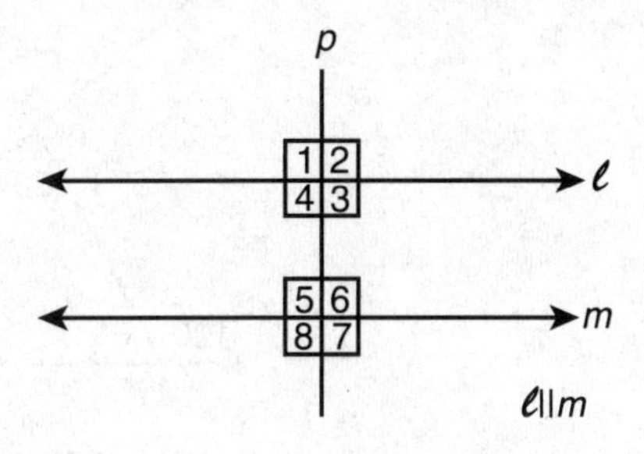

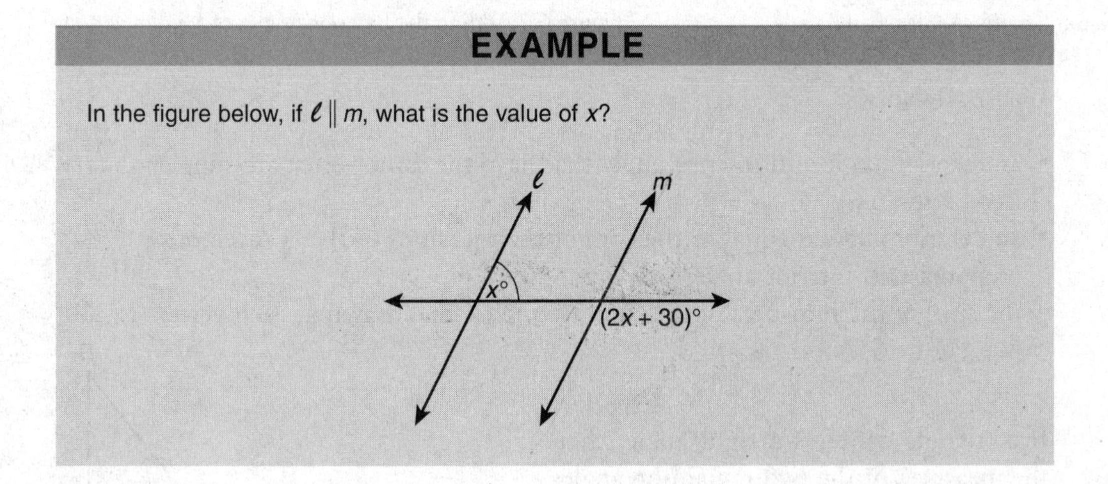

Solution

The angle marked x is acute, and the angle marked $2x + 30$ is obtuse. Since these angles are formed by parallel lines, their degree measures add up to 180. Hence:

$$x + (2x + 30) = 180$$
$$3x = 150$$
$$x = \frac{150}{3} = 50$$

ANGLES OF A TRIANGLE

The sum of the measures of the three angles of a triangle is 180° (see Figure 6.5).

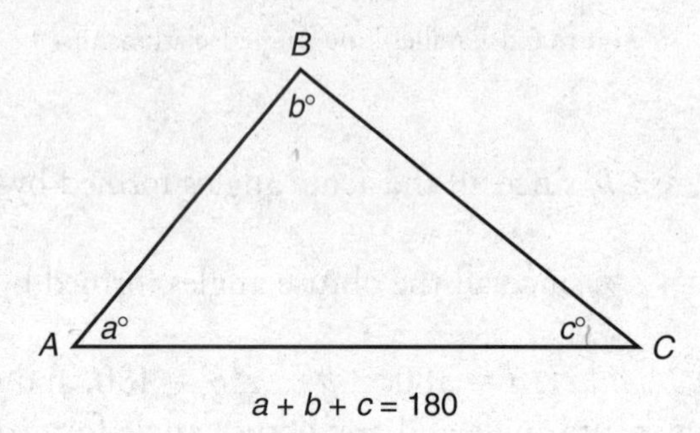

Figure 6.5 Sum of the Angles of a Triangle

An **exterior angle** of a triangle, as shown in Figure 6.6, can be formed at any vertex of the triangle by extending one of its sides.

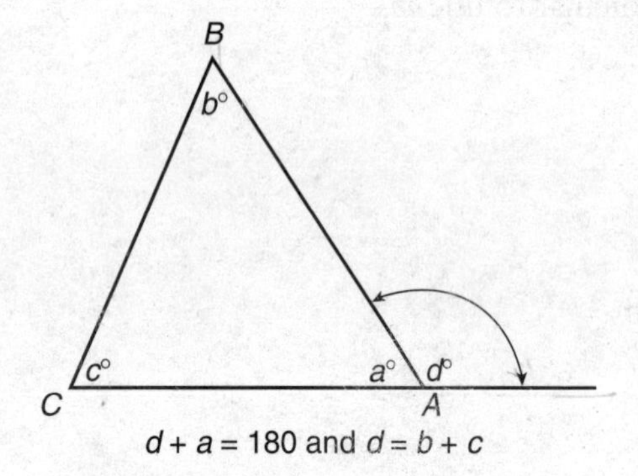

Figure 6.6 Exterior Angle of a Triangle

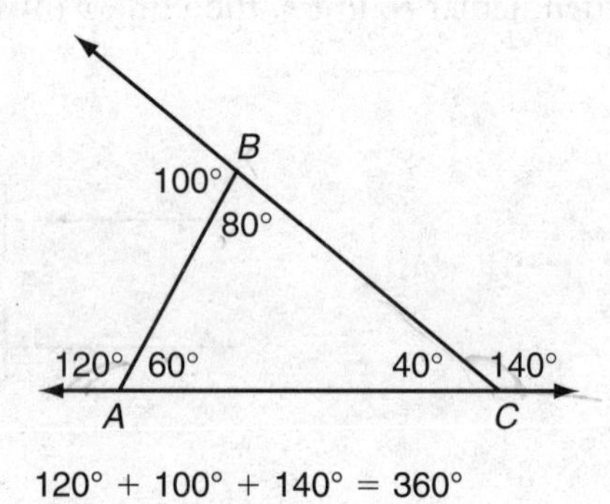

Figure 6.7 Adding the Exterior Angles of a Triangle

In any triangle,

- an exterior angle and interior angle that have the same vertex are supplementary (see Figure 6.6).
- an exterior angle is equal to the sum of the measures of the two remote (nonadjacent) interior angles (see Figure 6.6).
- the sum of the measures of the exterior angles, one drawn at each vertex, is 360° (see Figure 6.7).

If a triangle contains a right angle, then the measures of the two remaining angles of the triangle must add up to 90°. In the figure at the right, since $\angle C$ is a right angle, it measures 90°, so $a + b = 90$.

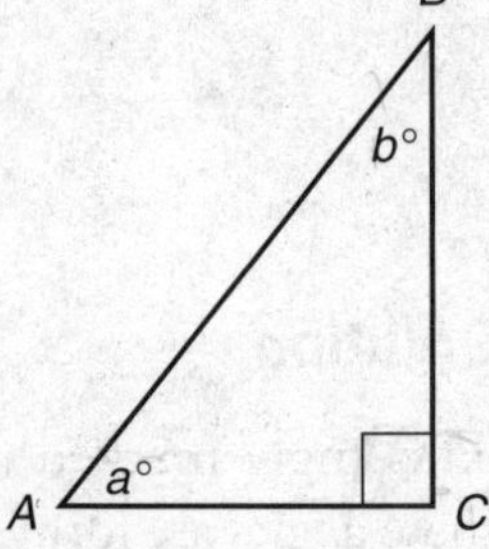

Lesson 6-1 Tune-Up Exercises

Multiple-Choice

1

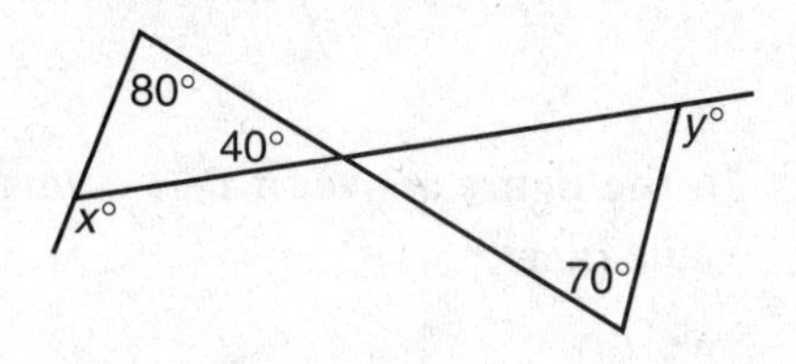

In the figure above, $x + y =$

(A) 270
(B) 230
(C) 210
(D) 190
(E) 180

2

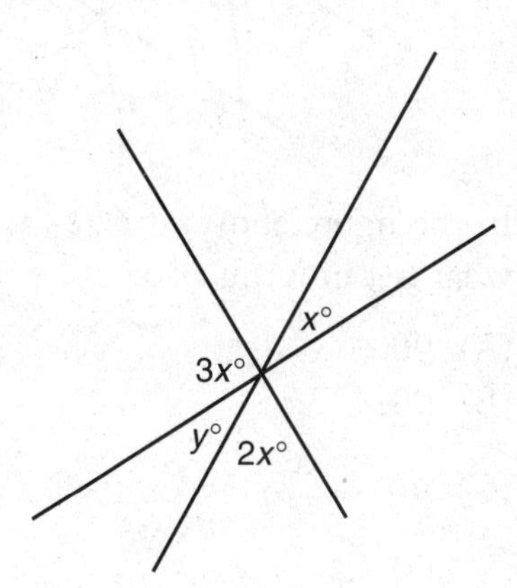

In the figure above, what is the value of y?

(A) 20
(B) 30
(C) 45
(D) 60
(E) It cannot be determined from the information given.

3

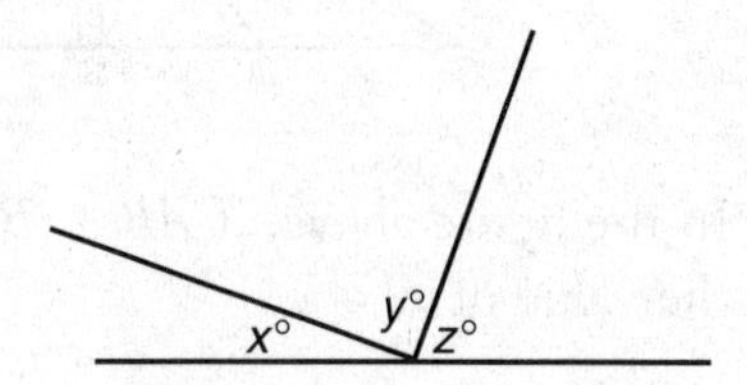

If in the figure above $\frac{y}{x} = 5$ and $\frac{z}{x} = 4$, what is the value of x?

(A) 8
(B) 10
(C) 12
(D) 15
(E) 18

4

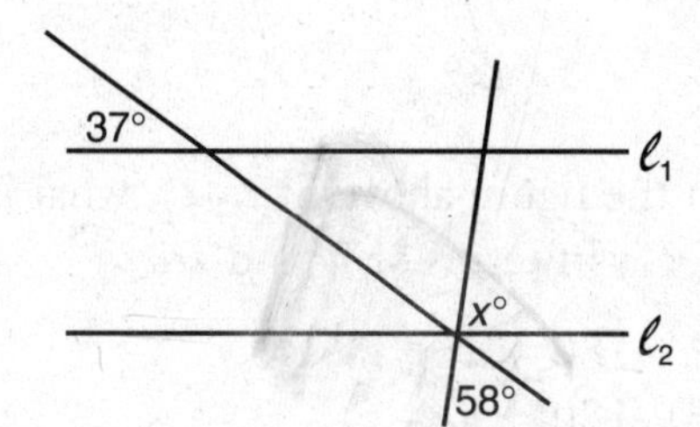

In the figure above, if $\ell_1 \parallel \ell_2$, what is the value of x?

(A) 90
(B) 85
(C) 75
(D) 70
(E) 63

5 If the measures of the angles of a triangle are in the ratio of 3 : 4 : 5, what is the measure of the smallest angle of the triangle?

(A) 25
(B) 30
(C) 45
(D) 60
(E) 75

6

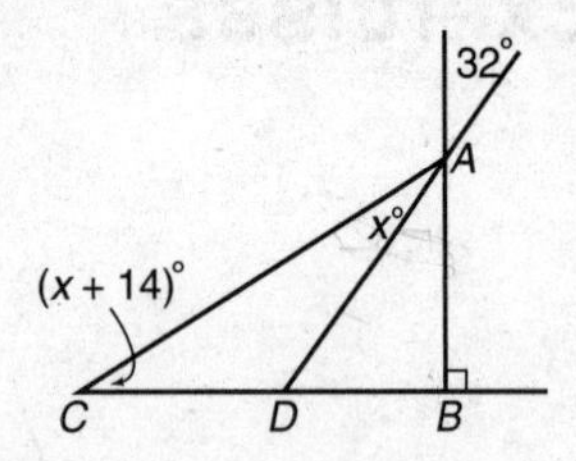

In the figure above, if $AB \perp BC$, what is the value of x?

(A) 18
(B) 22
(C) 25
(D) 27
(E) 32

7

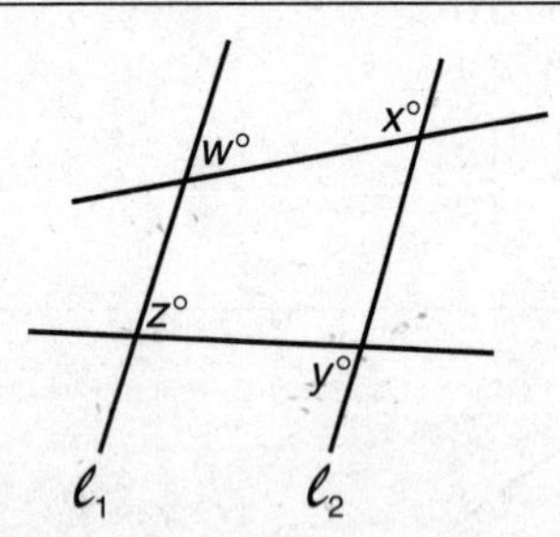

In the figure above, if $\ell_1 \parallel \ell_2$, what is $x + y$ in terms of w and z?

(A) $180 - w + z$
(B) $180 + w - z$
(C) $180 - w - z$
(D) $180 + w + z$
(E) $w + z$

8

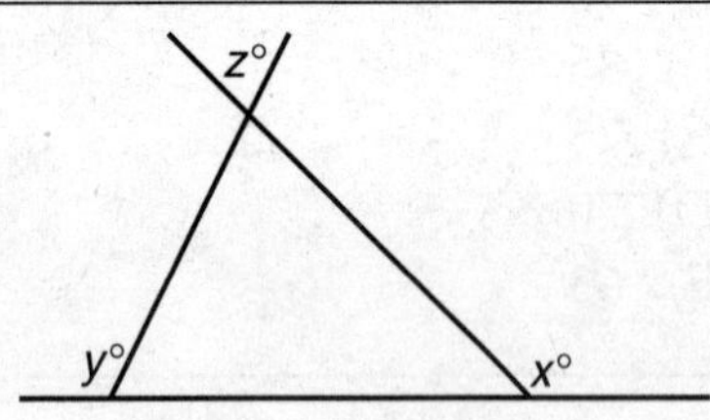

In the figure above, what is z in terms of x and y?

(A) $x + y + 180$
(B) $x + y - 180$
(C) $180 - (x + y)$
(D) $x + y + 360$
(E) $360 - (x + y)$

9

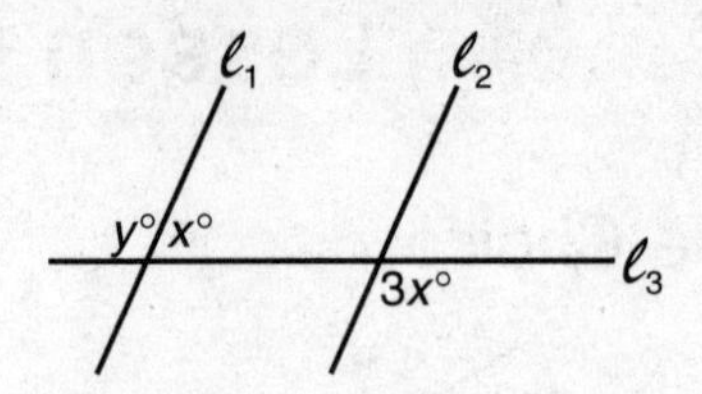

In the figure above, if $\ell_1 \parallel \ell_2$, what is the value of y?

(A) 90
(B) 100
(C) 120
(D) 135
(E) 150

10

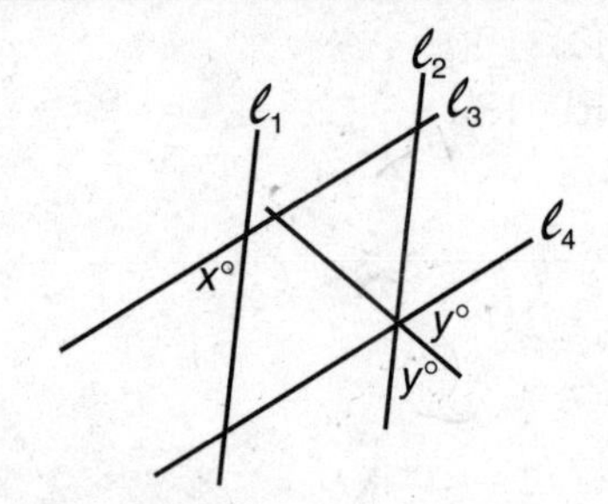

In the figure above, if $\ell_1 \parallel \ell_2$ and $\ell_3 \parallel \ell_4$, what is y in terms of x?

(A) $90 + x$
(B) $90 + 2x$
(C) $90 - \frac{x}{2}$
(D) $90 - 2x$
(E) $180 - 2x$

11

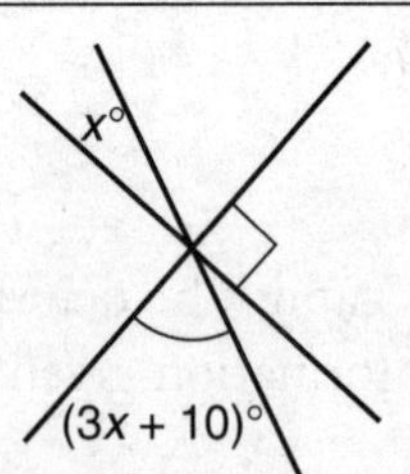

In the figure above, what is the value of x?

(A) 12
(B) 15
(C) 20
(D) 24
(E) 30

12

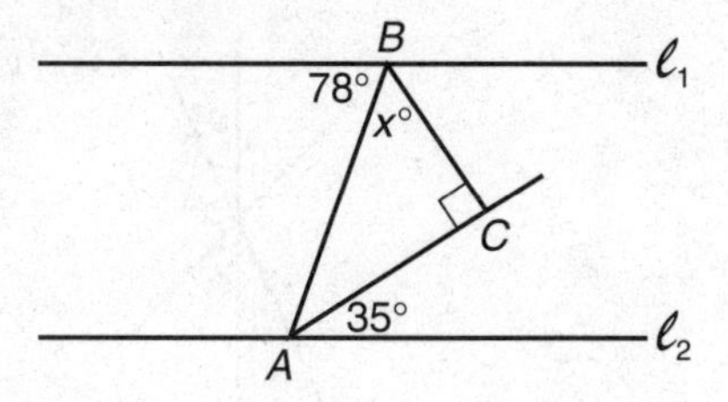

In the figure above, for which value of x is $\ell_1 \parallel \ell_2$?

(A) 37
(B) 43
(C) 45
(D) 47
(E) 55

13

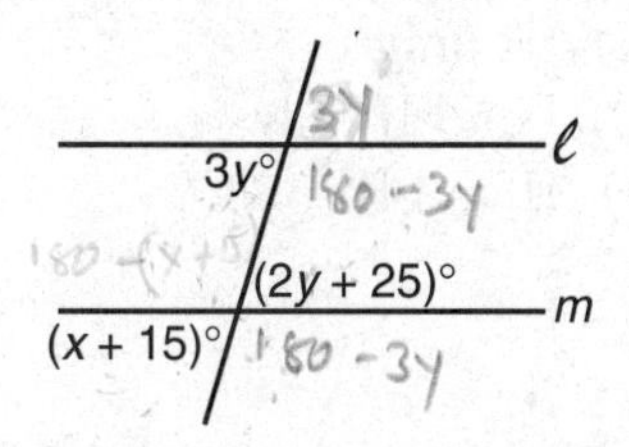

In the figure above, if $\ell \parallel m$, what is the value of x?

(A) 60
(B) 50
(C) 45
(D) 30
(E) 25

14

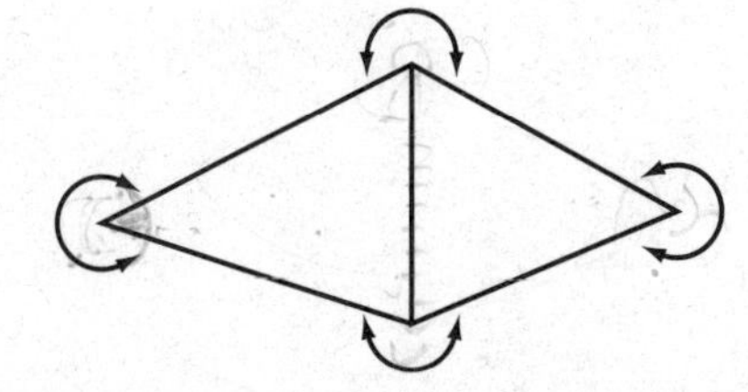

In the figure above, what is the sum of the degree measures of all of the angles marked?

(A) 540
(B) 720
(C) 900
(D) 1080
(E) 1440

15

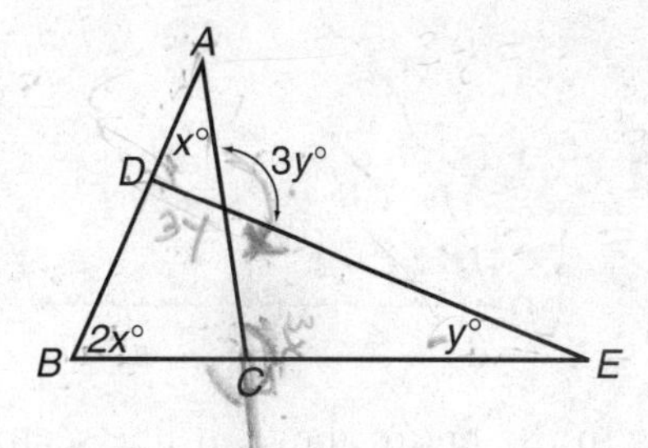

In the figure above, what is y in terms of x?

(A) $\frac{3}{2}x$

(B) $\frac{4}{3}x$

(C) x

(D) $\frac{3}{4}x$

(E) $\frac{2}{3}x$

16

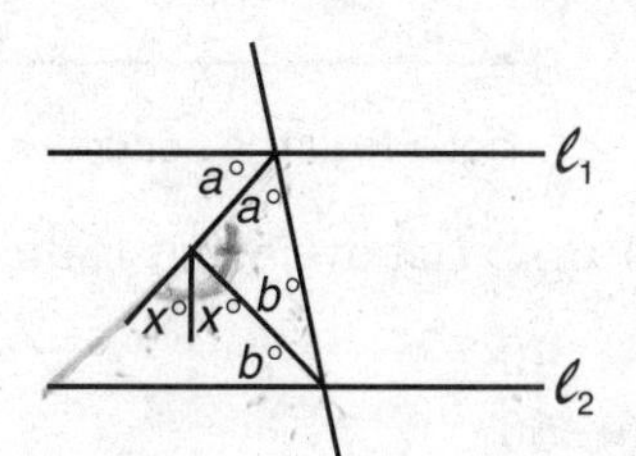

In the figure above, if $\ell_1 \parallel \ell_2$, the value of x is

(A) $22\frac{1}{2}$
(B) 30
(C) 45
(D) 60
(E) It cannot be determined from the information given.

17

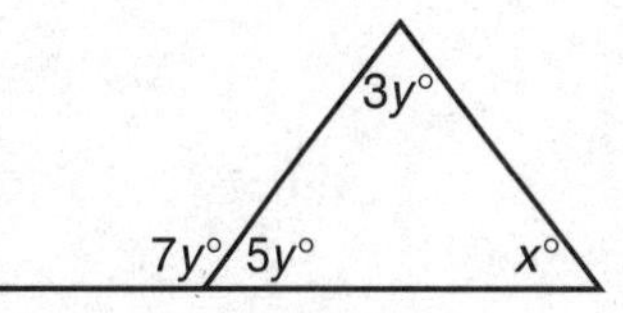

In the figure above, what is the value of x?

(A) 35
(B) 45
(C) 50
(D) 60
(E) 75

18

In the figure above, if line segment AB is parallel to line segment CD, what is the value of y?

(A) 12
(B) 15
(C) 18
(D) 20
(E) 24

19

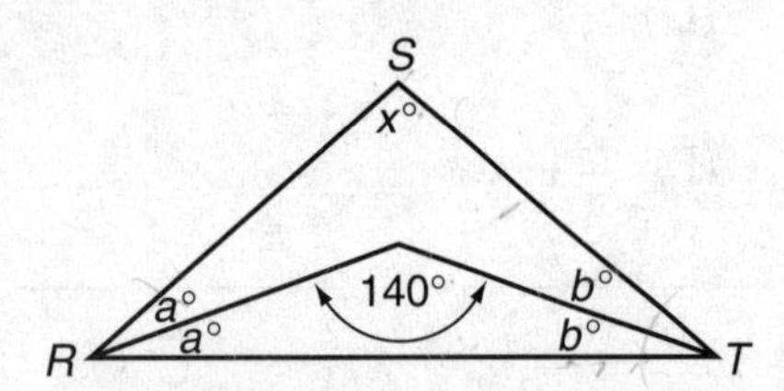

Note: Figure is not drawn to scale.

In $\triangle RST$ above, what is the value of x?

(A) 40
(B) 60
(C) 80
(D) 90
(E) 100

20

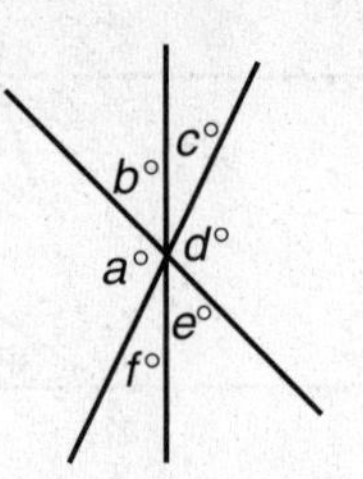

In the figure above, which of the following statements must be true?

I. $a + b = d + c$
II. $a + c + e = 180$
III. $b + f = c + e$

(A) I only
(B) II only
(C) III only
(D) I and II only
(E) II and III only

Grid-In

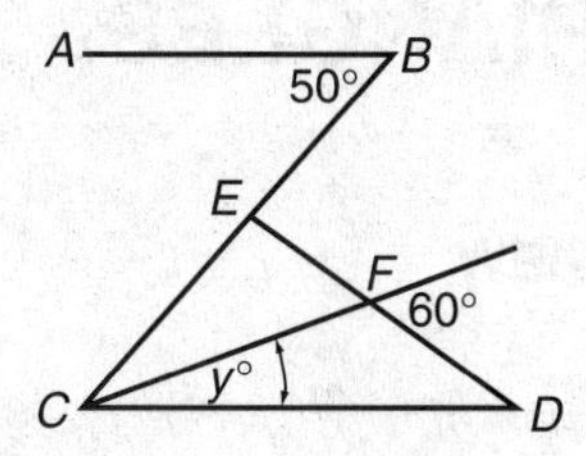

In the figure above, if line segment AB is parallel to line segment CD and $BE \perp ED$, what is the value of y?

2

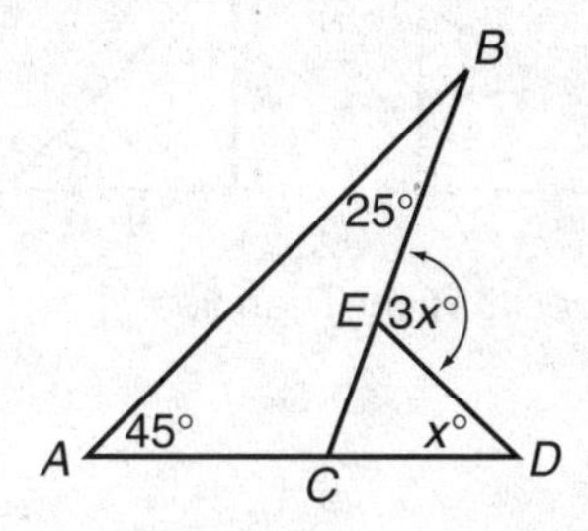

In the figure above, what is the value of x?

Questions 3 and 4.

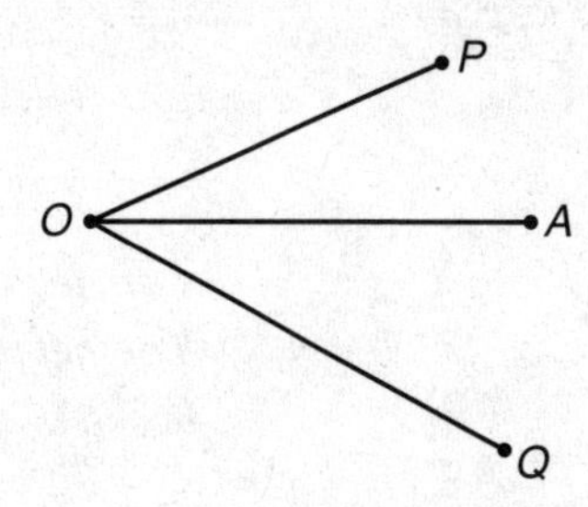

Line segments OP, OA, and OQ, in the figure above, coincide at the time $t = 0$ second. At the same instant of time, OP and OQ rotate in the plane about point O in opposite directions while OA remains fixed. With respect to OA, segment OP rotates at a constant rate of 4° per second and segment OQ rotates at a constant rate of 5° per second.

3 What is the smallest positive value of t, in seconds, for which segments OP and OQ will coincide?

4 When $t = 1$ hour, how many more revolutions does segment OQ complete than segment OP?

LESSON

6-2 SPECIAL TRIANGLES

OVERVIEW

This lesson reviews angle and side relationships in special types of triangles. The following right triangle reference formulas are supplied in the SAT *test booklet:*

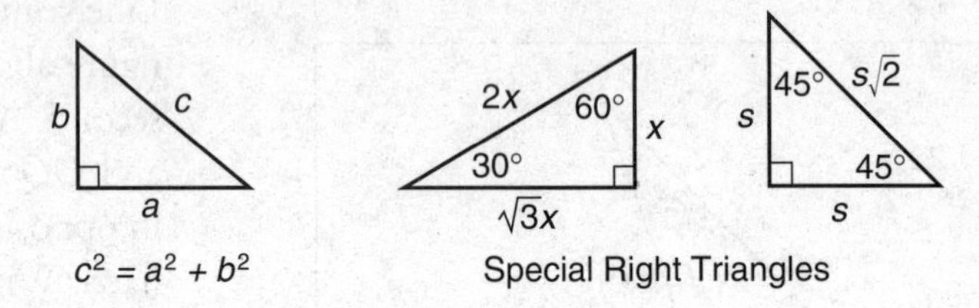

CLASSIFYING TRIANGLES

A triangle can be classified according to the number of equal sides that it contains, as shown in Figure 6.8.

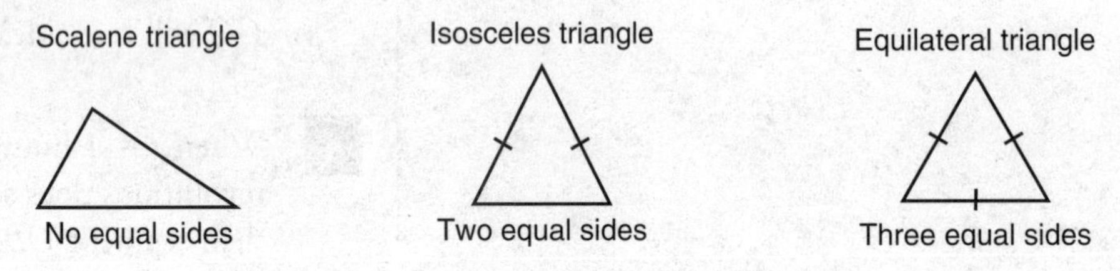

Figure 6.8 Classifying Triangles by Side Lengths

A triangle can also be classified according to the measure of its greatest angle, as shown in Figure 6.9.

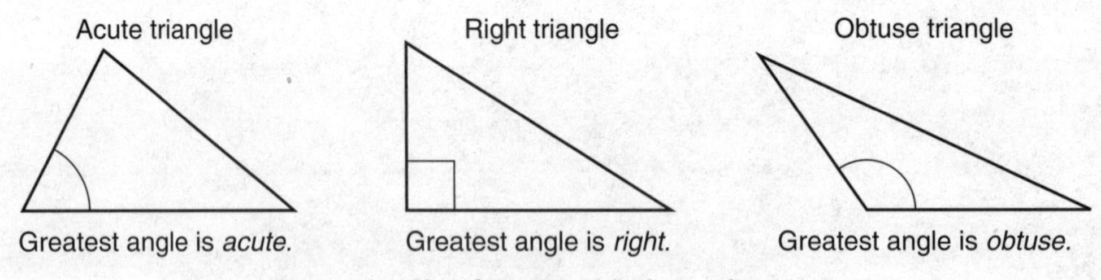

Figure 6.9 Classifying Triangles by Angle Measures

ANGLE AND SIDE RELATIONSHIPS IN AN ISOSCELES TRIANGLE

In an **isosceles triangle,** equal angles are opposite equal sides, as shown in Figure 6.10.

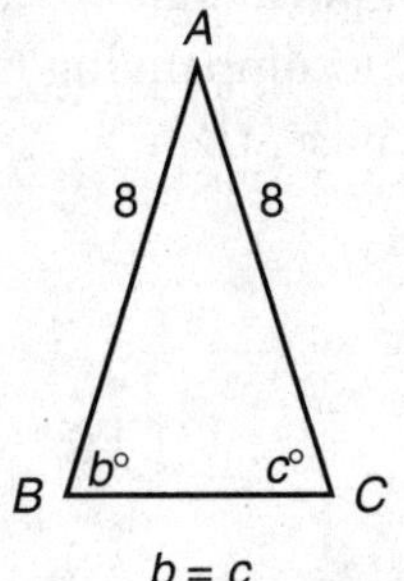

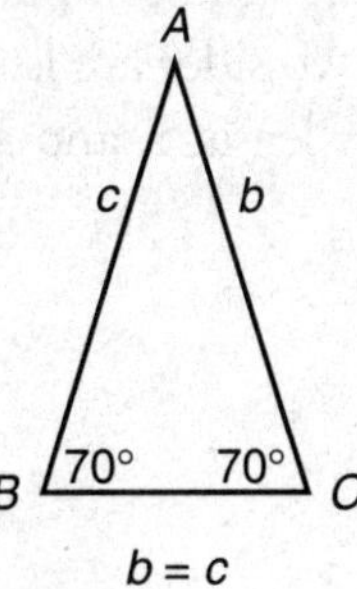

Figure 6.10 Isosceles Triangle Relationships

ANGLE AND SIDE RELATIONSHIPS IN AN EQUILATERAL TRIANGLE

In an **equilateral triangle,** all three sides are equal *and* each of the three angles measures 60°.

If the three sides of a triangle have the same length, you may conclude that the triangle contains three 60° angles (see Figure 6.11).

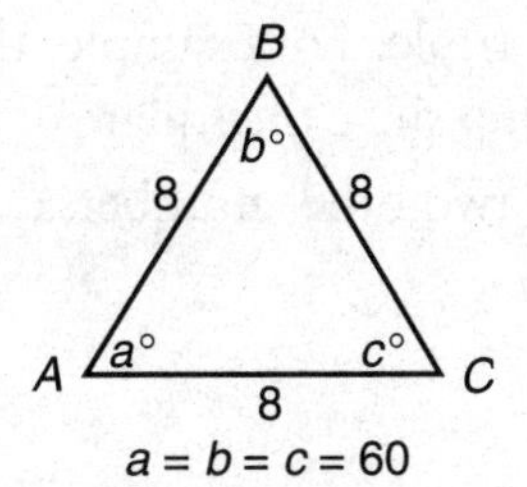

Figure 6.11 Equilateral Triangles Are Equiangular

B
c
a
60°
60°
A
C
b
$\angle B = 60°$ and $a = b = c$

Figure 6.12 Equiangular Triangles Are Equilateral

If two angles of a triangle measure 60°, the third angle also measures 60° and, as a result, the three sides have the same length (see Figure 6.12).

SIDE RELATIONSHIPS IN A RIGHT TRIANGLE

In a **right triangle,** the side opposite the right angle is called the **hypotenuse** and the other two sides are called **legs** (see Figure 6.13).

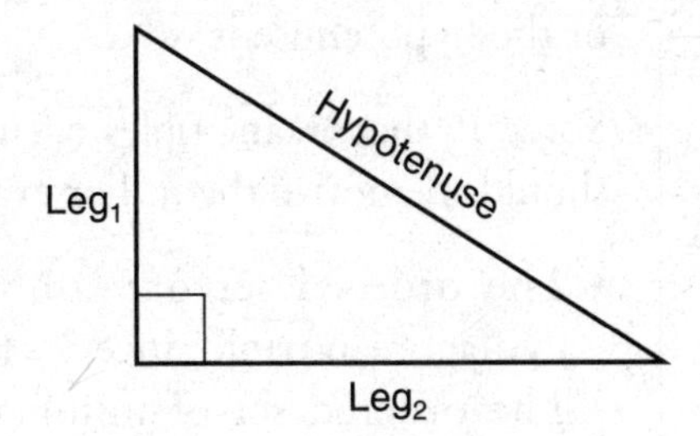

Figure 6.13 Right Triangle

In any right triangle:

$$(\text{leg}_1)^2 + (\text{leg}_2)^2 = (\text{hypotenuse})^2$$

This important relationship is called the **Pythagorean theorem.** The Pythagorean theorem can be used to find the length of any side of a right triangle when the lengths of the other two sides are known. For example, if in the accompanying figure one leg is 3, the other leg is 5, and x is the length of the hypotenuse, then

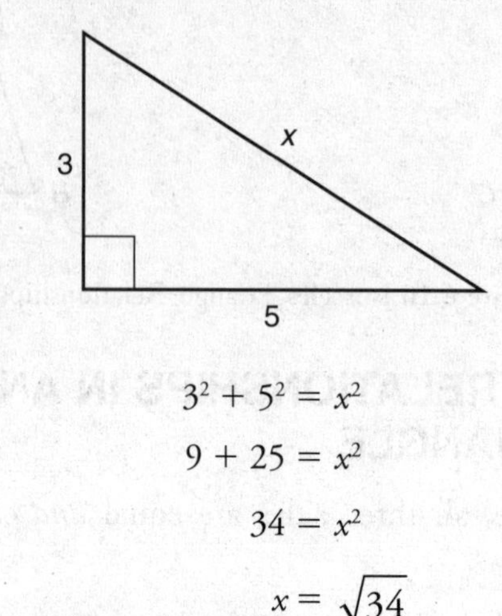

$$3^2 + 5^2 = x^2$$
$$9 + 25 = x^2$$
$$34 = x^2$$
$$x = \sqrt{34}$$

PYTHAGOREAN TRIPLES

If $a^2 + b^2 = c^2$, then a, b, and c form a Pythagorean triple. For example, the set of numbers $(1, 2, \sqrt{5})$ satisfies the Pythagorean relationship since the square of the largest number, $\sqrt{5}$, equals the sum of the squares of the two other numbers:

$$1^2 + 2^2 = 1 + 4 = 5 \quad \text{and} \quad (\sqrt{5})^2 = 5$$

so

$$\underbrace{(1)^2}_{a^2} + \underbrace{(2)^2}_{b^2} = \underbrace{(\sqrt{5})^2}_{c^2}$$

A Pythagorean triple is any set of positive integers that satisfy the relationship $a^2 + b^2 = c^2$.

If the lengths of the sides of a triangle form a Pythagorean triple, then the triangle is a right triangle in which the longest side is the hypotenuse of the triangle. For example, the triangle whose sides measure 1, 2, and $\sqrt{5}$ is a right triangle in which the length of the hypotenuse is $\sqrt{5}$.

TIME SAVER

Recognizing a Pythagorean triple can save you time. If the length and width of a rectangle are 5 and 12, respectively, then you immediately know that since 5–12–*13* is a Pythagorean triple and the diagonal of a rectangle is the hypotenuse of a right triangle, the diagonal must be 13.

Some Pythagorean triples occur so frequently that you should memorize them. For example:

- The ordered set of numbers (3, 4, 5) forms a Pythagorean triple since $3^2 + 4^2 = 5^2$.
- The ordered set of numbers (5, 12, 13) forms a Pythagorean triple since $5^2 + 12^2 = 13^2$.
- The ordered set of numbers (8, 15, 17) forms a Pythagorean triple, since $8^2 + 15^2 = 17^2$.

When each member of a Pythagorean triple is multiplied by the same positive integer, the result is also a Pythagorean triple. Because 3–4–5 is a Pythagorean triple and

$$3 \times 2 = \underline{6},\ 4 \times 2 = \underline{8},\ \text{and } 5 \times 2 = \underline{10},$$

6–8–10 is a Pythagorean triple, as are 9–12–15, 12–16–20, 15–20–25, and so forth. Similarly, multiplying each member of the Pythagorean triples 5–12–13 and 8–15–17 by 2 also produces Pythagorean triples: 10–24–26 and 16–30–34, respectively.

MATH REFERENCE FACT

Other basic Pythagorean triples that you may encounter are 7–24–25 and 9–40–41. Don't be alarmed if you don't remember a particular Pythagorean triple, since you can always find the unknown side of a right triangle by substituting into the Pythagorean theorem and then solving for it.

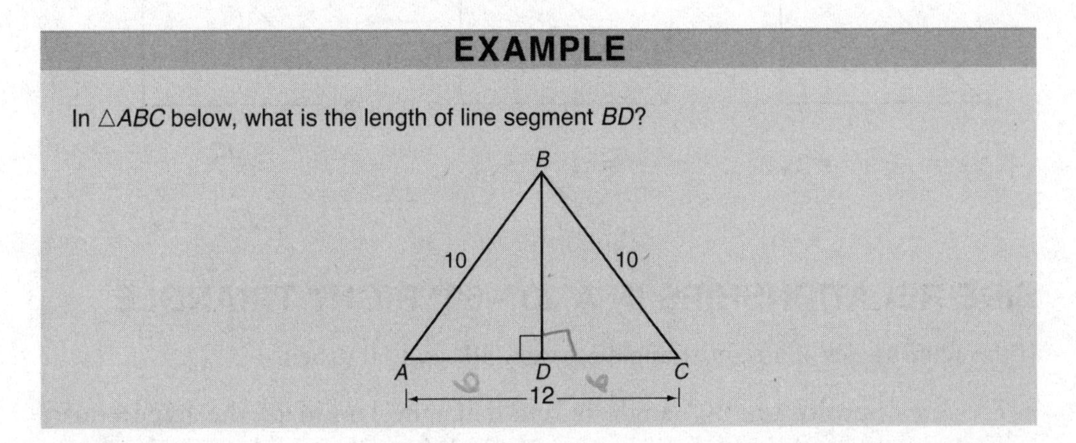

Solution

If two sides of a triangle are equal, a perpendicular drawn to the third side bisects that side. Hence, $AD = DC = 6$. In right triangle *ADB*, the lengths of the sides are members of a (6, *x*, 10) Pythagorean triple where $AD = 6$, hypotenuse $AB = 10$, and $x = BD = 8$.

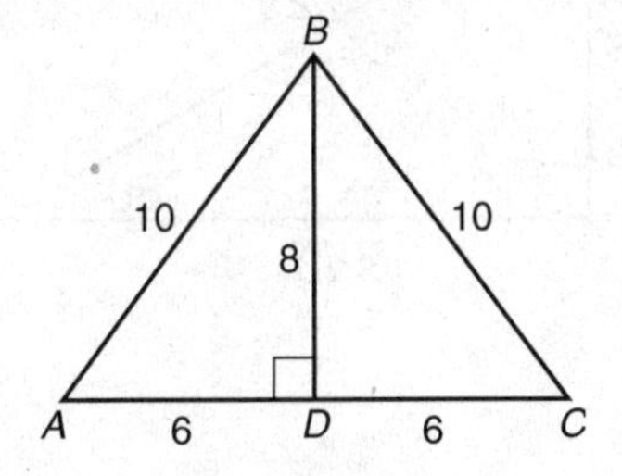

SIDE RELATIONSHIPS IN A 45°– 45° RIGHT TRIANGLE

In an isosceles right triangle the following are true:

- Each acute angle measures 45°.
- The length of the hypotenuse is $\sqrt{2}$ times the length of a leg.
- The length of each leg is $\frac{1}{\sqrt{2}}\left(=\frac{\sqrt{2}}{2}\right)$ times the length of the hypotenuse.

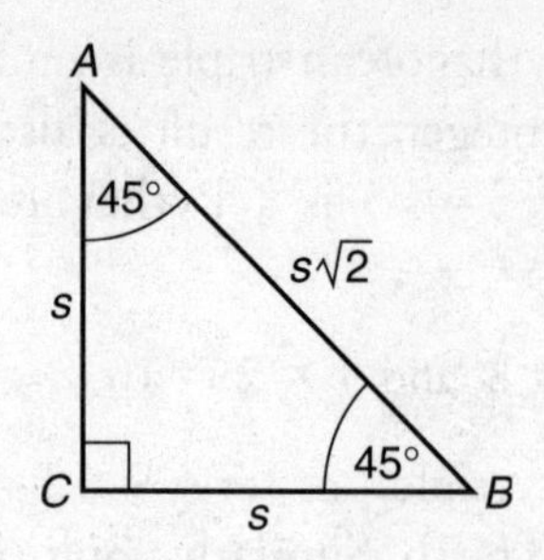

For example,

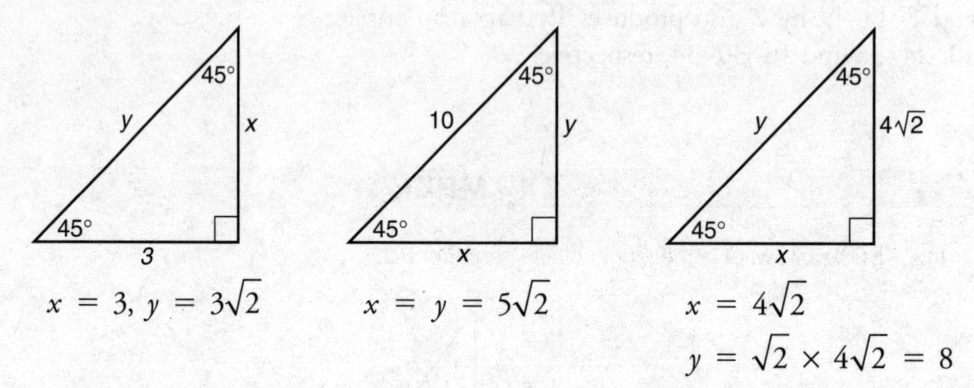

$x = 3,\ y = 3\sqrt{2}$ $\qquad$ $x = y = 5\sqrt{2}$ $\qquad$ $x = 4\sqrt{2}$

$y = \sqrt{2} \times 4\sqrt{2} = 8$

SIDE RELATIONSHIPS IN A 30°–60° RIGHT TRIANGLE

If the acute angles of a right triangle measure 30° and 60°, then

- The leg opposite the 30° angle is one-half the length of the hypotenuse. Equivalently, the length of the hypotenuse is 2 times the length of the leg opposite the 30° angle.
- The leg opposite the 60° angle is $\sqrt{3}$ times the length of the other leg.

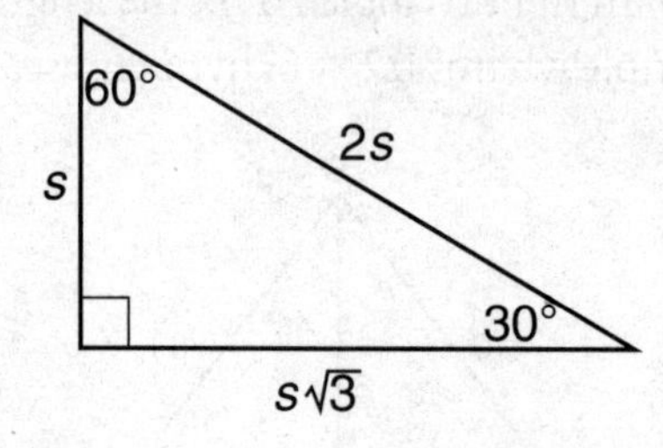

For example:

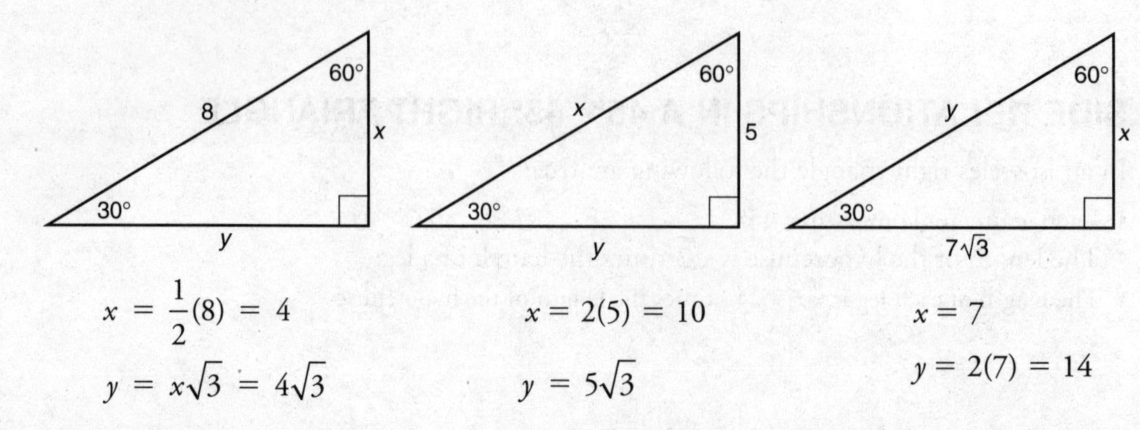

$x = \frac{1}{2}(8) = 4$ $\qquad$ $x = 2(5) = 10$ $\qquad$ $x = 7$

$y = x\sqrt{3} = 4\sqrt{3}$ $\qquad$ $y = 5\sqrt{3}$ $\qquad$ $y = 2(7) = 14$

Lesson 6-2 Tune-Up Exercises

Multiple-Choice

1

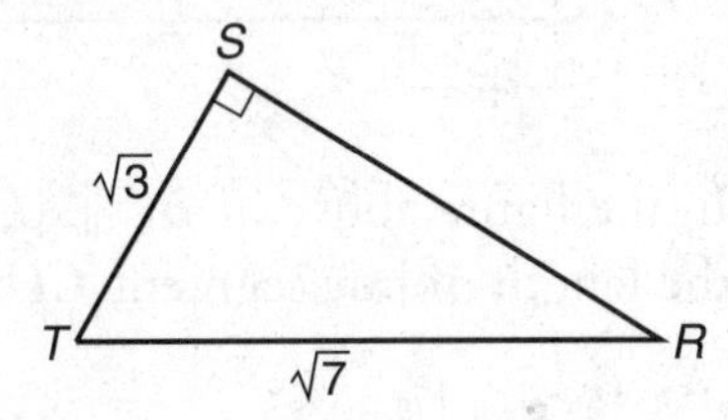

In the figure above, what is the length of RS?

(A) $\sqrt{7} - \sqrt{3}$
(B) 2
(C) 4
(D) $7 + \sqrt{3}$
(E) 10

2

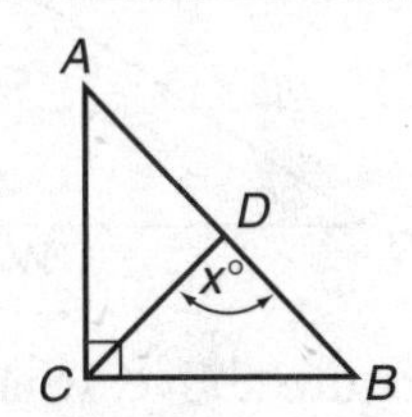

In $\triangle ABC$, $\angle C = 90°$, $CD = BD$, and the ratio of the measure of $\angle A$ to the measure of $\angle B$ is 2 to 3. What is the value of x?

(A) 18
(B) 36
(C) 40
(D) 54
(E) 72

3

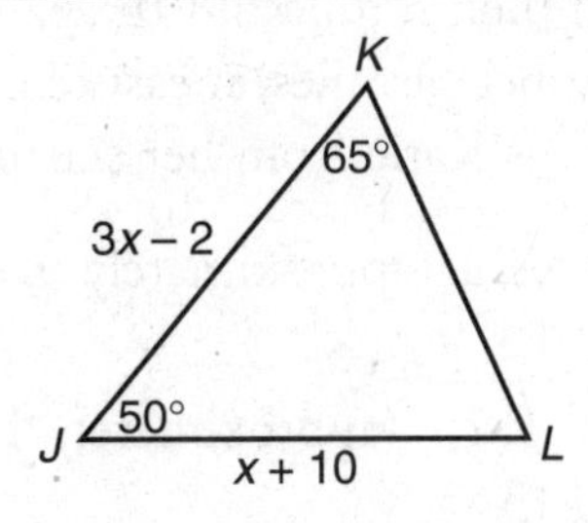

In $\triangle JKL$ above, what is the value of x?

(A) 2
(B) 3
(C) 4
(D) 6
(E) It cannot be determined from the information given.

4

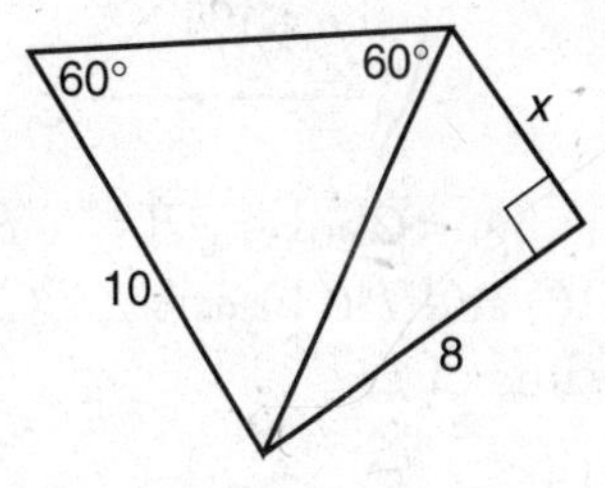

In the figure above, $x =$

(A) 4
(B) 6
(C) $4\sqrt{2}$
(D) $4\sqrt{3}$
(E) $8\sqrt{3}$

5

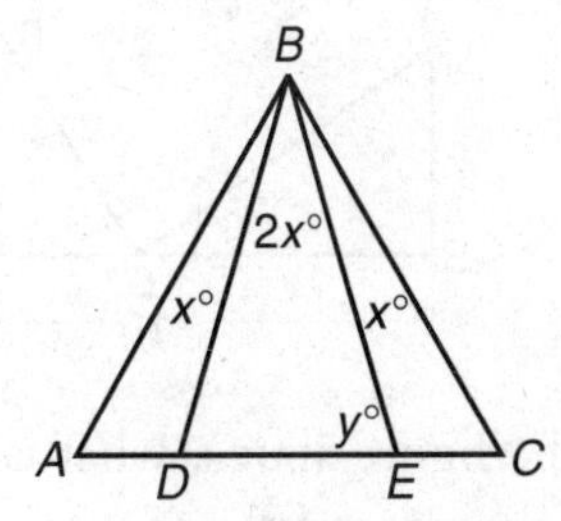

In equilateral triangle ABC above, $y =$

(A) 40
(B) 60
(C) 70
(D) 75
(E) 80

6

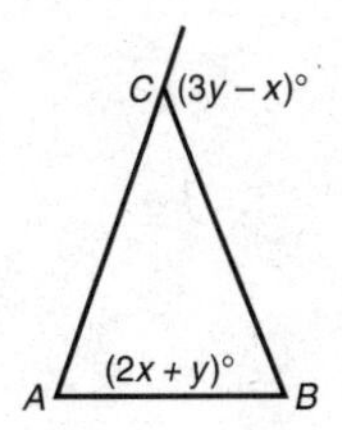

In the figure above, if $AC = BC$ what is y in terms of x?

(A) x
(B) $\frac{3}{2}x$
(C) $2x$
(D) $3x$
(E) $5x$

7

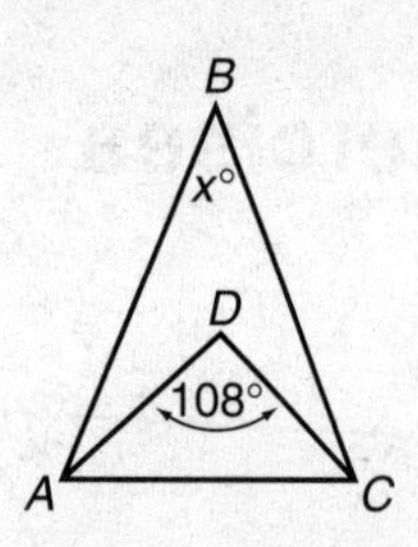

In the figure above, $AB = CB$, DA bisects $\angle BAC$, and DC bisects $\angle BCA$. What is the value of x?

(A) 18
(B) 30
(C) 36
(D) 72
(E) 108

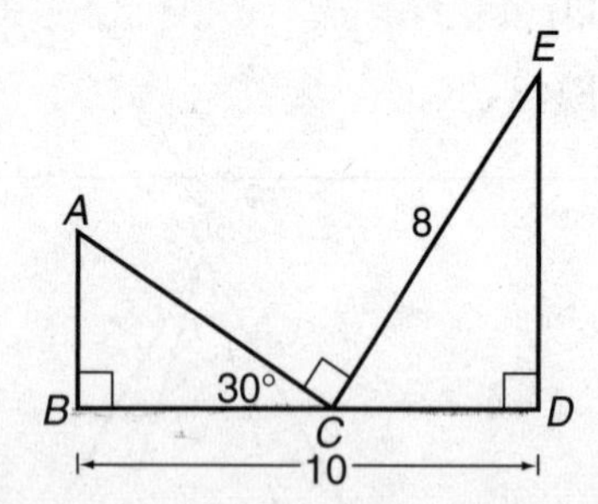

In the figure above, what is the length of line segment AB?

(A) 3
(B) $2\sqrt{3}$
(C) 4
(D) $3\sqrt{3}$
(E) 6

9

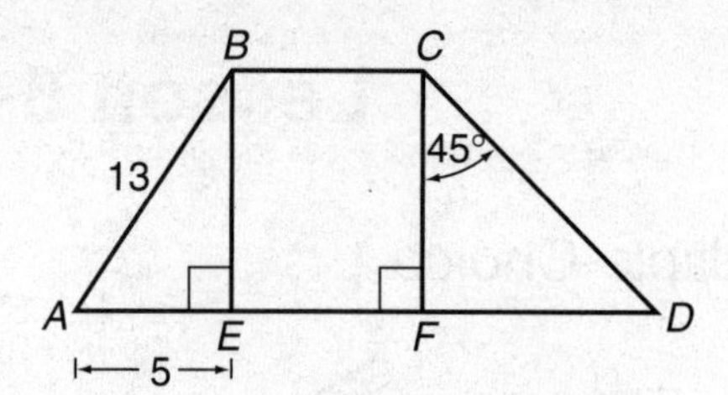

In the figure above, if $BC \parallel AD$, what is the length of line segment CD?

(A) $6\sqrt{2}$
(B) $8\sqrt{2}$
(C) 15
(D) $12\sqrt{2}$
(E) 17

10

Note: Figure is not drawn to scale.

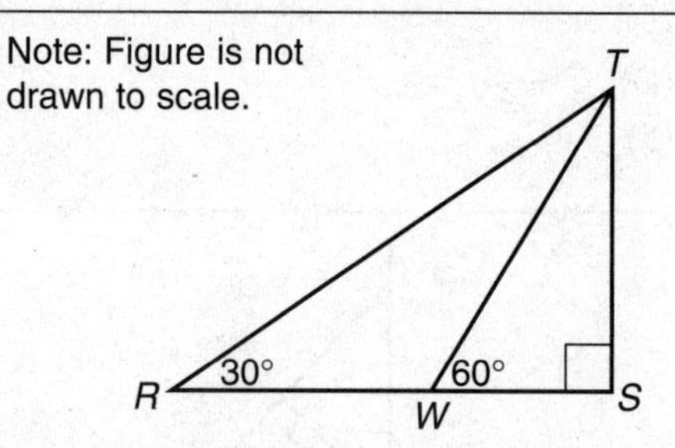

In the figure above, what is the ratio of RW to WS?

(A) $\sqrt{2}$ to 1
(B) $\sqrt{3}$ to 1
(C) 2 to 1
(D) 3 to 1
(E) 3 to 2

11 Katie hikes 5 miles north, 7 miles east, and then 3 miles north again. What number of miles, measured in a straight line, is Katie from her starting point?

(A) $\sqrt{83}$ (approximately 9.11)
(B) 10
(C) $\sqrt{113}$ (approximately 10.63)
(D) 13
(E) 15

Grid-In

1

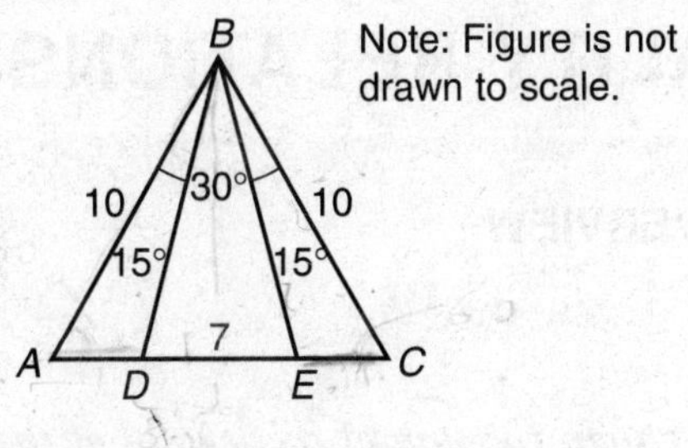

Note: Figure is not drawn to scale.

In the figure above, what is the length of line segment AD?

2

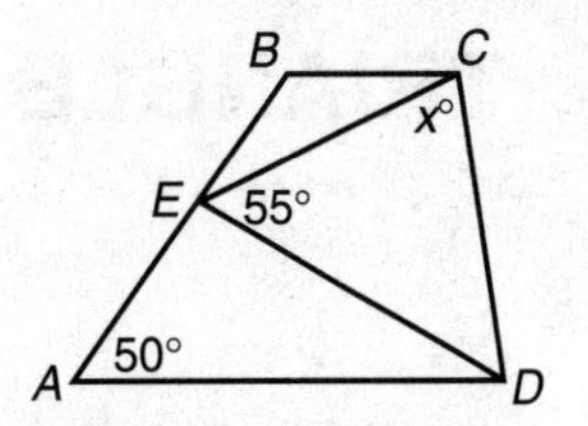

Note: Figure is not drawn to scale.

In the figure above, if $BE = BC$, $BC \parallel AD$, and ED bisects $\angle ADC$, what is the value of x?

3 Two hikers started at the same location. One traveled 2 miles east and then 1 mile north. The other traveled 1 mile west and then 3 miles south. At the end of their hikes, how many miles apart were the two hikers?

LESSON

6-3 TRIANGLE INEQUALITY RELATIONSHIPS

OVERVIEW

In any triangle,

- *the length of each side is less than the sum of the lengths of the other two sides and greater than their difference;*
- *if side lengths or angle measures are unequal, then the longest side is opposite the greatest angle, and the shortest side is opposite the smallest angle.*

TESTING THREE NUMBERS AS POSSIBLE TRIANGLE SIDE LENGTHS

To determine whether a set of three positive numbers can represent the lengths of the sides of a triangle, check that *each* of the three numbers is less than the sum of the other two numbers.

EXAMPLE

Which of the following sets of numbers CANNOT represent the lengths of the sides of a triangle?

(A) 9, 40, 41
(B) 7, 7, 3
(C) 4, 5, 1
(D) 1.6, 1.4, 2.9
(E) 6, 6, 6

Solution

For each choice, check that each of the three given numbers is less than the sum of the other two numbers. In choice (C), $4 < 5 + 1$ and $1 < 4 + 5$, but 5 is *not* less than $1 + 4$.

The correct choice is (C).

SIDE LENGTH RESTRICTIONS

MATH REFERENCE FACT

The length of any side of a triangle is greater than the difference and less than the sum of the lengths of the other two sides.

In any triangle, each side length is limited to a range of values that is determined by the lengths of the other two sides. For example, if the lengths of two sides of a triangle are 4 and 9, then you can conclude that the length of the third side is greater than $9 - 4 = 5$ *and* less than $9 + 4 = 13$.

EXAMPLE

If 3, 8, and x represent the lengths of the sides of a triangle, how many integer values for x are possible?

Solution

- The value of x must be greater than $8 - 3 = 5$ *and* less than $8 + 3 = 11$.
- Since $x > 5$ and $x < 11$, x must be between 5 and 11.
- Because it is given that x is an integer, x can be equal to 6, 7, 8, 9, or 10.
- Hence, there are five possible integer values for x.

EXAMPLE

If the lengths of two sides of an isosceles triangle are 3 and 7, what is the length of the third side?

Solution

Since two sides of an isosceles triangle have the same length, the length of the third side is either 3 or 7. The length of the third side cannot be 3 since 7 is *not* less than $3 + 3$. If the third side is 7, then $3 < 7 + 7$ and $7 < 3 + 7$. Hence, the length of the third side must be 7.

UNEQUAL SIDES ARE OPPOSITE UNEQUAL ANGLES

In a triangle, unequal sides are opposite unequal angles. These relationships are illustrated in the figures below.

MATH REFERENCE FACT

If two sides (or angles) of a triangle are unequal, then the measures of the angles (or sides) opposite them are also unequal, with the larger angle lying opposite the longer side.

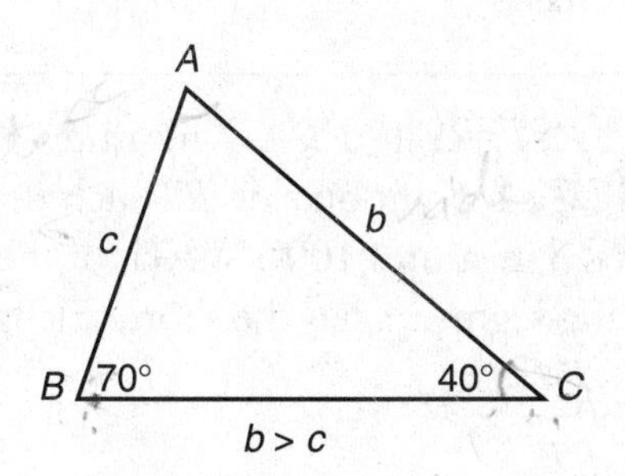

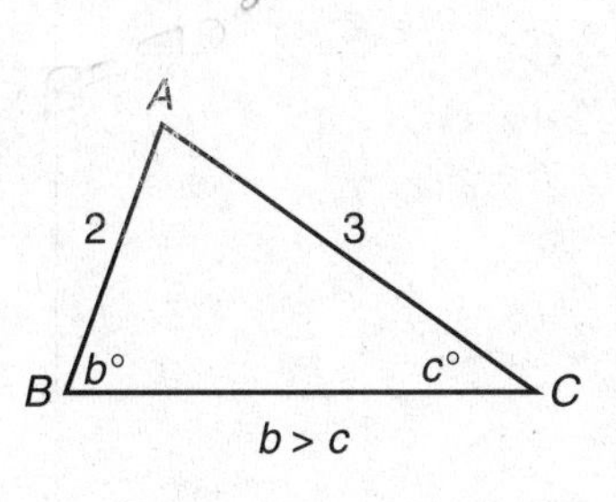

Lesson 6-3 Tune-Up Exercises

Multiple-Choice

1 In obtuse triangle RST, $RT = TS$. Which of the following statements must be true?

I. $\angle R = \angle S$
II. $\angle T$ is obtuse.
III. $RS > TS$

(A) I only
(B) I and II only
(C) I and III only
(D) II and III only
(E) I, II, and III

2 In a triangle in which the lengths of two sides are 5 and 9, the length of the third side is represented by x. Which statement is always true?

(A) $x > 5$
(B) $x < 9$
(C) $5 \leq x \leq 9$
(D) $4 < x < 14$
(E) $5 \leq x < 14$

3 In $\triangle ABC$, $BC > AB$ and $AC < AB$. Which statement is always true?

(A) $\angle A > \angle B > \angle C$
(B) $\angle A > \angle C > \angle B$
(C) $\angle B > \angle A > \angle C$
(D) $\angle C > \angle B > \angle A$
(E) $\angle C > \angle A > \angle B$

4

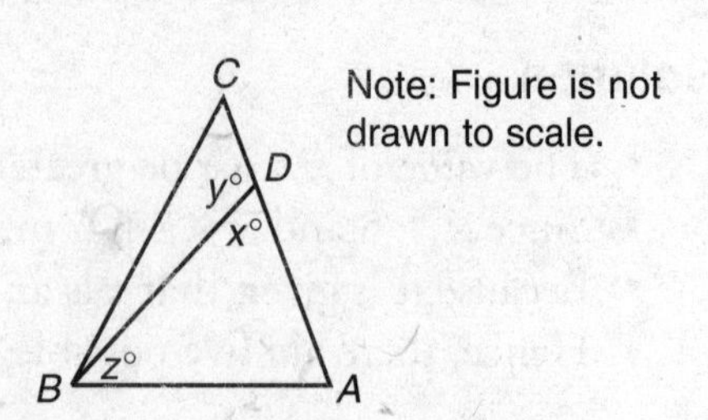

In $\triangle ABC$, if $AB = BD$, which of the following statements must be true?

I. $x > z$
II. $y > x$
III. $AB > BC$

(A) II only
(B) I and II only
(C) I and III only
(D) II and III only
(E) I, II, and III

5 How many different triangles are there for which the lengths of the sides are 3, 8, and n, where n is an integer and $3 < n < 8$?

(A) Two
(B) Three
(C) Four
(D) Five
(E) Six

6 In $\triangle RST$, $RS \perp TS$, $\angle T$ measures 40°, and W is a point on side RT such that $\angle RWS$ measures 100°. Which of the following segments has the shortest length?

(A) RS
(B) ST
(C) RW
(D) TW
(E) SW

Grid-In

1 If the lengths of two sides of an isosceles triangle are 7 and 15, what is the perimeter of the triangle?

2 The perimeter of a triangle in which the lengths of all of the sides are integers is 21. If the length of one side of the triangle is 8, what is the shortest possible length of another side of the triangle?

3 If the integer lengths of the three sides of a triangle are 4, x, and 9, what is the least possible perimeter of the triangle?

4 If the product of the lengths of the three sides of a triangle is 105, what is a possible perimeter of the triangle?

LESSON

6-4 POLYGONS AND PARALLELOGRAMS

OVERVIEW

A triangle is a three-sided polygon. This lesson considers some angle and side relationships in polygons with three or more sides.

TERMS RELATED TO POLYGONS

The accompanying figure shows a polygon with five sides. The corner points A, B, C, D, and E are called **vertices.** Line segments AC and AD are diagonals. A **diagonal** of a polygon is a line segment that connects any two nonconsecutive vertices of the polygon.

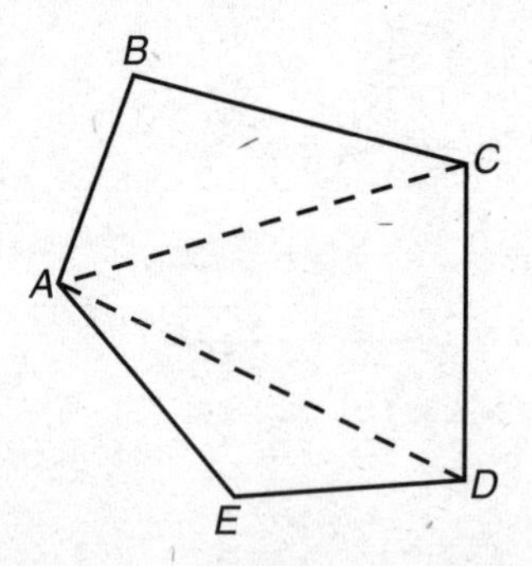

An **equilateral polygon** is a polygon in which all the sides have the same length. An **equiangular polygon** is a polygon in which all the angles are equal.

ANGLES OF A QUADRILATERAL

A **quadrilateral** is a polygon with four sides. Since a quadrilateral can be divided into two triangles, as illustrated in Figure 6.14, the sum of the four interior angles of a quadrilateral is $2 \times 180°$ or $360°$.

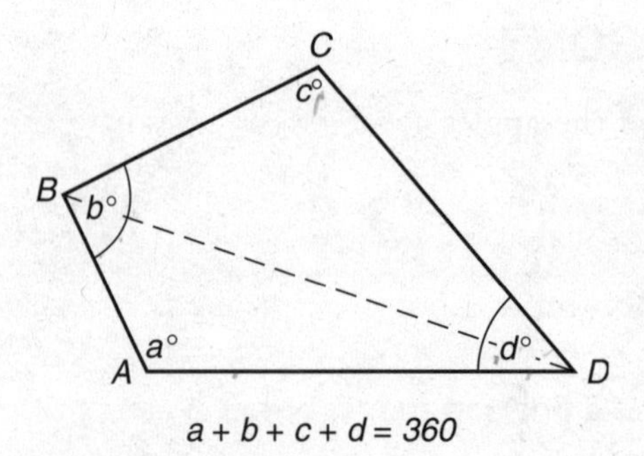

Figure 6.14 Angles of a Quadrilateral

SUM OF THE INTERIOR ANGLES OF A POLYGON

By drawing diagonals from the same vertex, a five-sided polygon can be separated into $5 - 2 = 3$ triangles as illustrated on page 244. Any polygon with n sides can be divided into $n - 2$ nonoverlapping triangles. Since the sum of the angles of a triangle is 180°, the sum S of the interior angles of a polygon with n sides is given by the formula:

$$S = (n - 2) \times 180°$$

For example, to find the sum of the five interior angles of a five-sided polygon, substitute 5 for n in the above formula:

$$\begin{aligned} S &= (n - 2) \times 180° \\ &= (5 - 2) \times 180° \\ &= 3 \times 180° \\ &= 540° \end{aligned}$$

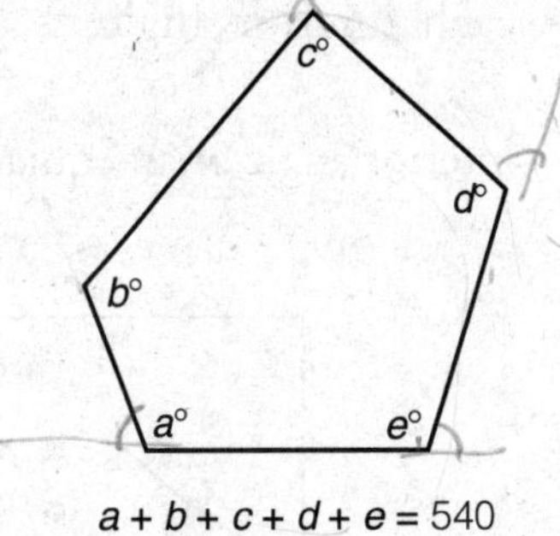

SUM OF THE EXTERIOR ANGLES OF A POLYGON

The sum of the exterior angles of any polygon, with one angle drawn at each vertex of the polygon, is 360° (see Figure 6.15).

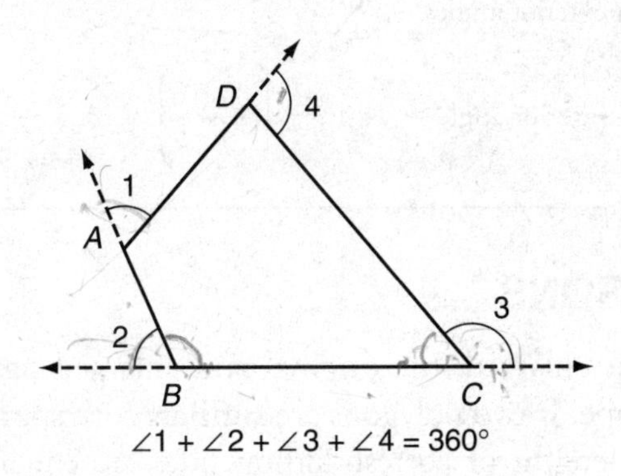

Figure 6.15 Exterior Angles of a Polygon

SPECIAL POLYGONS

All of the sides or all of the angles of a polygon may have the same measure.

- An **equilateral polygon** is a polygon in which all of the sides have the same length.
- An **equiangular polygon** is a polygon in which all of the angles have the same measure.
- A **regular polygon** is a polygon that is both equiangular and equilateral.

INTERIOR AND EXTERIOR ANGLE RELATIONSHIPS

At the vertex of a polygon, the measures of the exterior and interior angles add up to 180°. In an equiangular polygon with n sides:

- each exterior angle $= \dfrac{360^\circ}{n}$. For example, the measure of each exterior angle of an equiangular pentagon is $\dfrac{360^\circ}{5} = 72^\circ$.
- each interior angle $= 180^\circ - \left(\dfrac{360^\circ}{n}\right)$. For example, the measure of each interior angle of an equiangular pentagon is $180^\circ - 72^\circ = 108^\circ$.

Formulas worth remembering for n-sided polygons:

- Sum of interior angles $= (n - 2) \times 180^\circ$
- Sum of exterior angles $= 360^\circ$
- At each vertex,

  interior ∠ + exterior ∠ = 180°

For equiangular n-sided polygons:

- Each exterior angle $= \dfrac{360^\circ}{n}$
- Each interior angle $= 180^\circ - \left(\dfrac{360^\circ}{n}\right)$

SIMILAR POLYGONS

When a photograph is enlarged, the original and enlarged figures are *similar* since they have the same shape. If two polygons are **similar,** corresponding angles are equal and the ratios of the lengths of corresponding sides are equal. Keep in mind that corresponding sides are opposite matching pairs of equal angles. For example, the quadrilaterals below are similar.

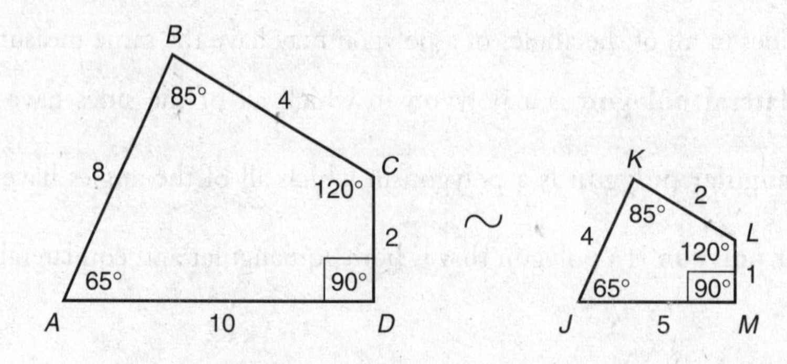

The ratio of the lengths of any pair of corresponding sides of the two similar quadrilaterals is 2 to 1:

$$\frac{\text{side of } ABCD}{\text{corresponding side of } JKLM} = \frac{AB}{JK} = \frac{BC}{KL} = \frac{CD}{LM} = \frac{AD}{JM}$$
$$= \frac{8}{4} = \frac{4}{2} = \frac{2}{1} = \frac{10}{5}$$
$$= 2 : 1$$

SIMILAR TRIANGLES

If *two* angles of one triangle are equal to two angles of the second triangle, the two triangles are similar. For example, the triangles in Figure 6.16 are similar since $\angle B = \angle D$ and the vertical angles at C are equal.

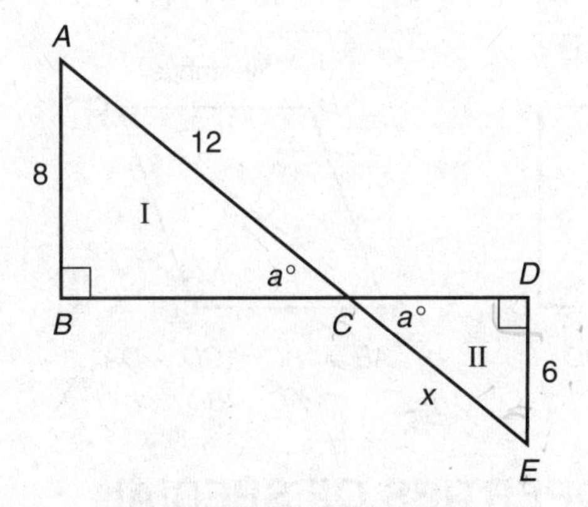

Figure 6.16 Similar Triangles

Since the triangles in Figure 6.16 are similar, the lengths of corresponding sides are in proportion. Length 8 corresponds to length 6 since these sides are opposite equal angles. Also, length 12 corresponds to length x since these sides are opposite equal angles. Hence:

$$\frac{\text{Side of } \triangle \text{ I}}{\text{Side of } \triangle \text{ II}} = \frac{8}{6} = \frac{12}{x}$$
$$8x = 72$$
$$x = \frac{72}{8} = 9$$

PARALLELOGRAMS

A **parallelogram** is a special type of quadrilateral in which both pairs of opposite sides are parallel. A parallelogram has these three additional properties:

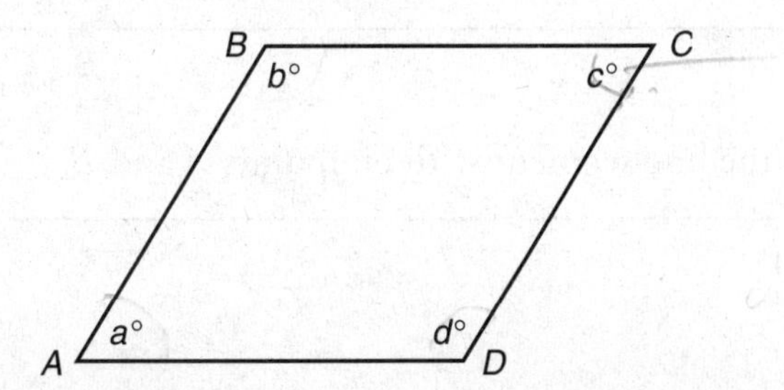

- Opposite sides are equal. Thus:

$$AB = CD \text{ and } AD = BC$$

- Opposite angles are equal. Thus:

$$a = c \text{ and } b = d$$

- Consecutive angles are supplementary. Thus:

$$a + b = b + c = c + d = d + a = 180$$

SPECIAL TYPES OF PARALLELOGRAMS

A **rectangle** is a parallelogram with four right angles. A **rhombus** is a parallelogram with four equal sides. A **square** is a parallelogram with four right angles and four equal sides.

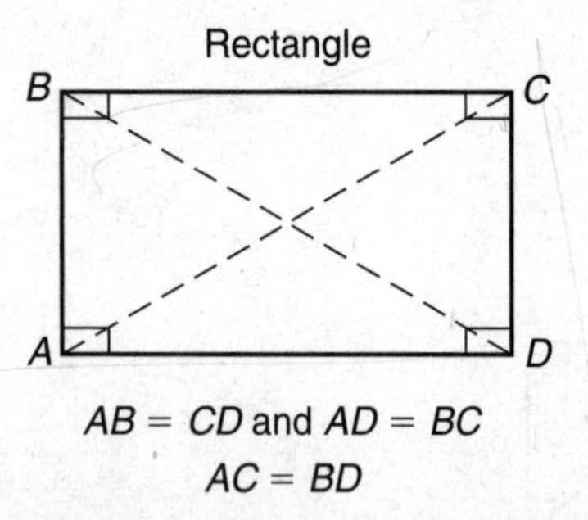

$AB = CD$ and $AD = BC$
$AC = BD$

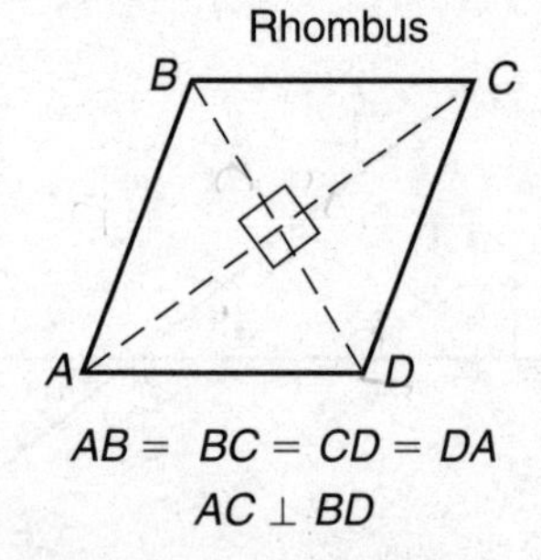

$AB = BC = CD = DA$
$AC \perp BD$

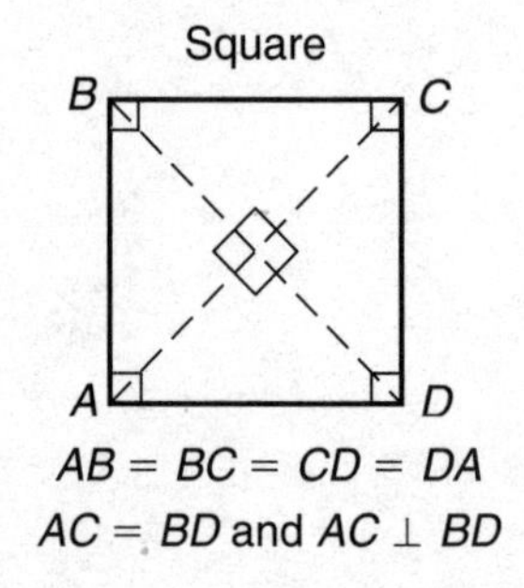

$AB = BC = CD = DA$
$AC = BD$ and $AC \perp BD$

DIAGONAL PROPERTIES OF SPECIAL TYPES OF PARALLELOGRAMS

In any parallelogram, each diagonal cuts the other diagonal in half.

- In a rectangle, the diagonals have the same length.
- In a rhombus, the diagonals intersect at right angles.
- In a square, the diagonals have the same length and intersect at right angles.

GEOMETRIC NOTATION

The SAT uses standard geometric notation for length, segments, lines, rays, and congruence.

- $\overleftrightarrow{AB}$ represents the line that passes through points A and B:

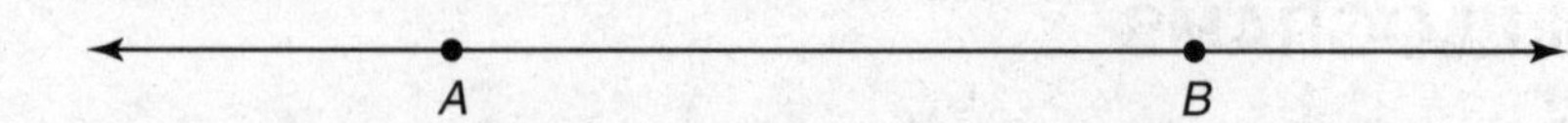

- $\overrightarrow{AB}$ represents the ray beginning at point A and passing through point B:

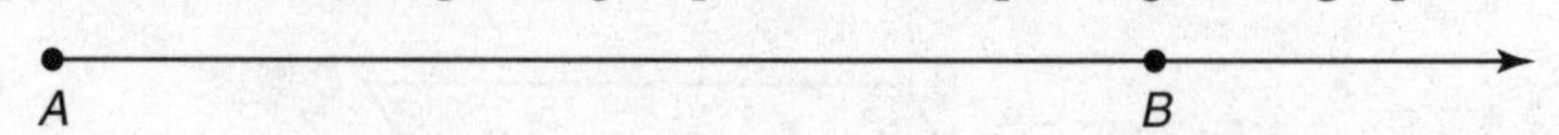

- $\overline{AB}$ represents the line segment with endpoints A and B:

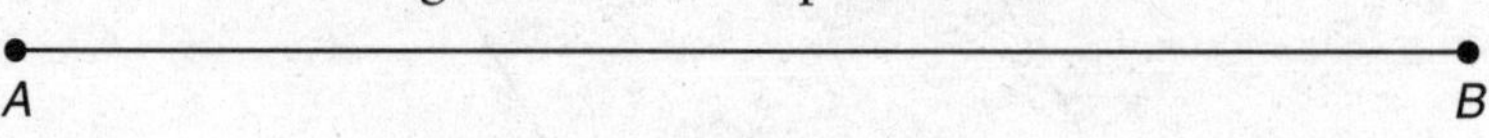

- AB, without an overbar, represents the length of $\overline{AB}$. If the length of $\overline{AB}$ is 6, then $AB = 6$

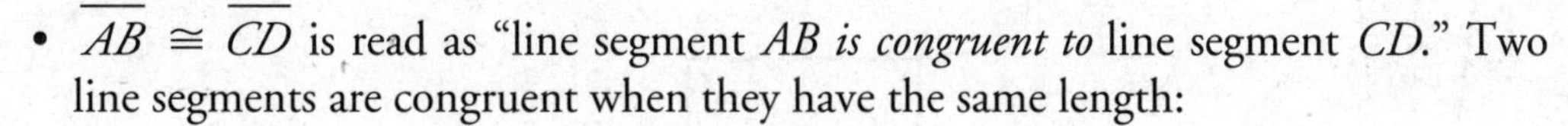

- $\overline{AB} \cong \overline{CD}$ is read as "line segment *AB is congruent to* line segment *CD*." Two line segments are congruent when they have the same length:

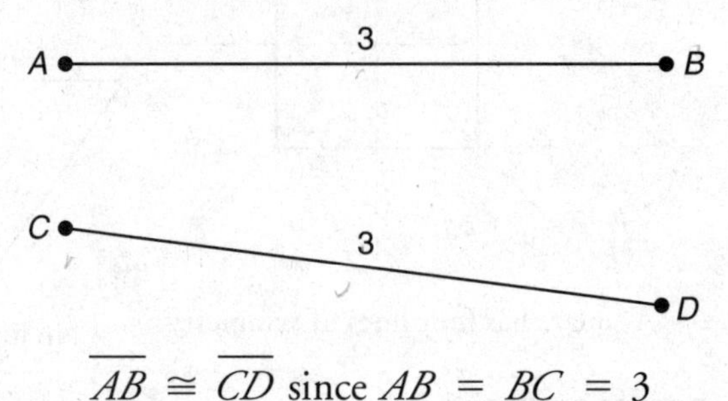

$$\overline{AB} \cong \overline{CD} \text{ since } AB = BC = 3$$

- Two polygons are congruent if all pairs of corresponding sides are congruent and all pairs of corresponding angles are congruent.

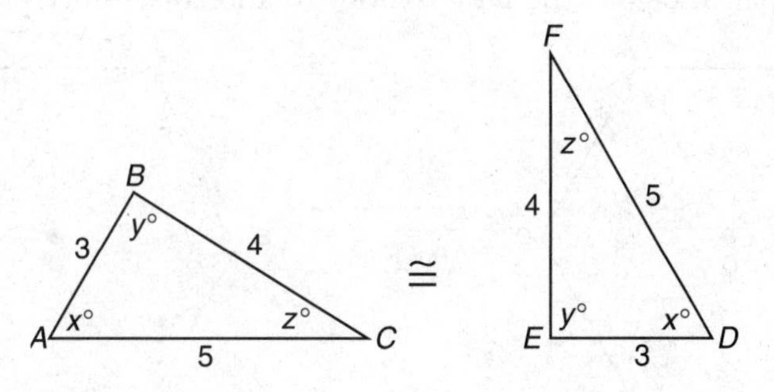

For example, if $\triangle ABC \cong \triangle DEF$, then

$$\begin{aligned} \overline{AB} &\cong \overline{DE} & & \angle A \cong \angle D \\ \overline{BC} &\cong \overline{EF} & \textit{and} \quad & \angle B \cong \angle E \\ \overline{AC} &\cong \overline{DF} & & \angle C \cong \angle F \end{aligned}$$

LINE SYMMETRY

If a line divides a figure into two parts that when folded along the line exactly coincide, then the figure has line symmetry. The line of symmetry may be a horizontal line, a vertical line, or a slanted line, as illustrated in the accompanying figures.

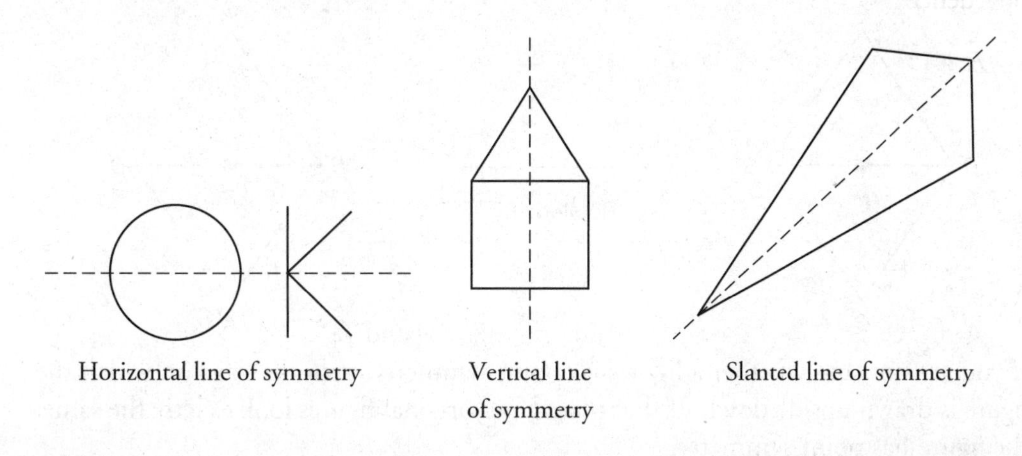

Horizontal line of symmetry — Vertical line of symmetry — Slanted line of symmetry

A figure may have more than one line of symmetry or no lines of symmetry:

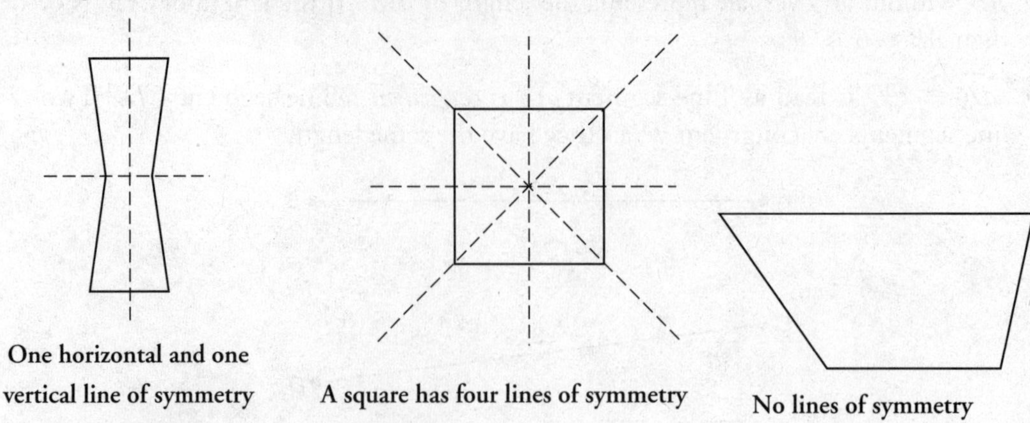

ROTATIONAL SYMMETRY

After a clockwise rotation of 60° about its center *O*, a regular hexagon will coincide with itself. Regular hexagon *ABCDEF* has 60° rotational symmetry:

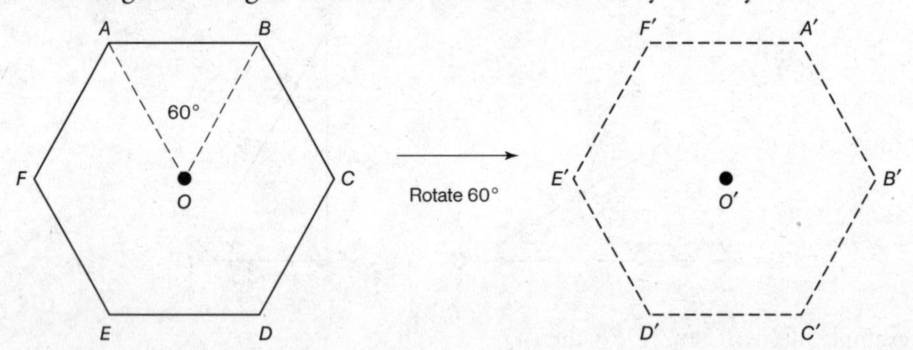

A figure has **rotational symmetry** if it coincides with itself after a rotation through some positive angle less than 360°. Every regular polygon enjoys rotational symmetry about its center for an angle of rotation of $\frac{360^\circ}{n}$, where n is the number of sides of the polygon. For an equilateral triangle, square, regular pentagon, and regular octagon, the angles of rotation are 120°, 90°, 72°, and 45°, respectively.

POINT SYMMETRY

A figure has **point symmetry** with respect to a point when it has 180° rotational symmetry about that point, as illustrated in the accompanying figure:

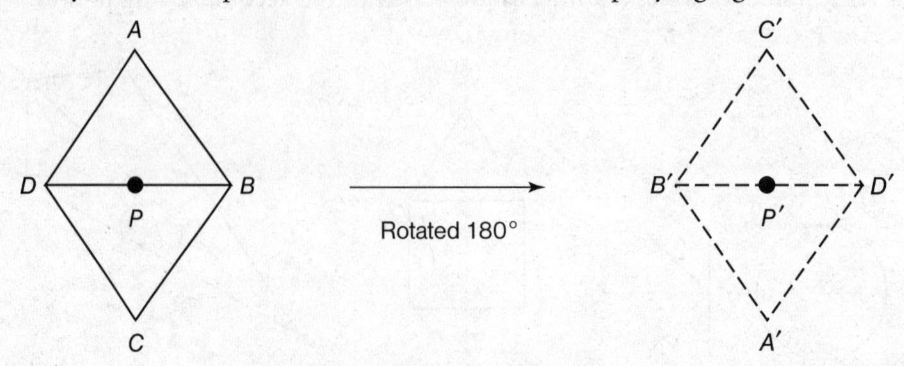

If you are not sure whether a figure has point symmetry, turn the page on which the figure is drawn upside down. If the rotated and original figures look exactly the same, the figure has point symmetry.

Lesson 6-4 Tune-Up Exercises

Multiple-Choice

1 What is the sum of the lengths of the two diagonals in a 9 by 12 rectangle?

(A) 20
(B) 25
(C) 30
(D) 35
(E) 40

2 What is the length of a diagonal of a square with a side length of $\sqrt{2}$?

(A) $\dfrac{\sqrt{2}}{2}$
(B) 1
(C) 2
(D) $2\sqrt{2}$
(E) 4

3 A diagonal of a rectangle forms a 30° angle with each of the longer sides of the rectangle. If the length of the shorter side is 3, what is the length of the diagonal?

(A) $3\sqrt{2}$
(B) $3\sqrt{3}$
(C) 4
(D) 5
(E) 6

4 If the degree measures of the angles of a quadrilateral are $4x$, $7x$, $9x$, and $10x$, what is the sum of the measures of the smallest angle and the largest angle?

(A) 140
(B) 150
(C) 168
(D) 180
(E) 192

5

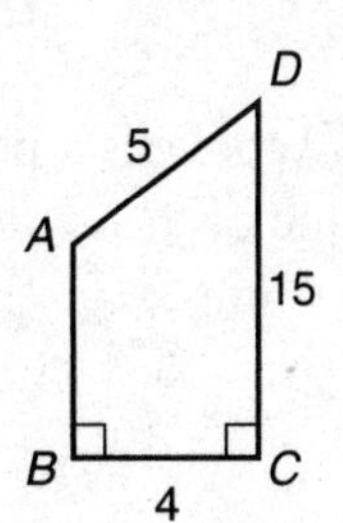

Note: Figure is not drawn to scale.

In the figure above, what is the length of side AB?

(A) 5
(B) 6
(C) 9
(D) 10
(E) 12

6

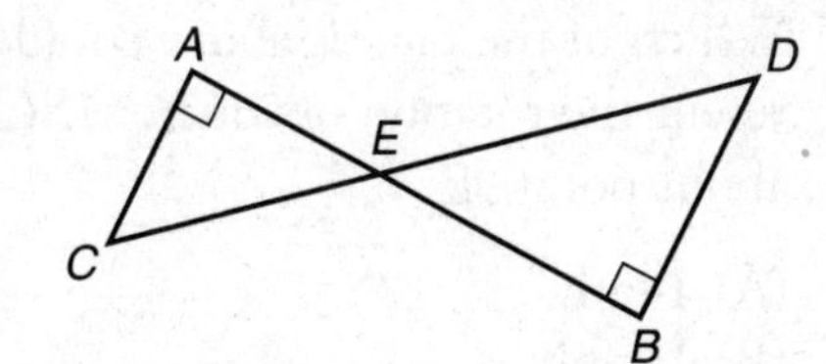

If, in the figure above, $AC = 3$, $DB = 4$, and $AB = 14$, then $AE =$

(A) 19
(B) 12
(C) 10.5
(D) 8
(E) 6

7 Which of the following CANNOT represent the degree measure of an equi-angular polygon?

(A) 165
(B) 162
(C) 140
(D) 125
(E) 90

8

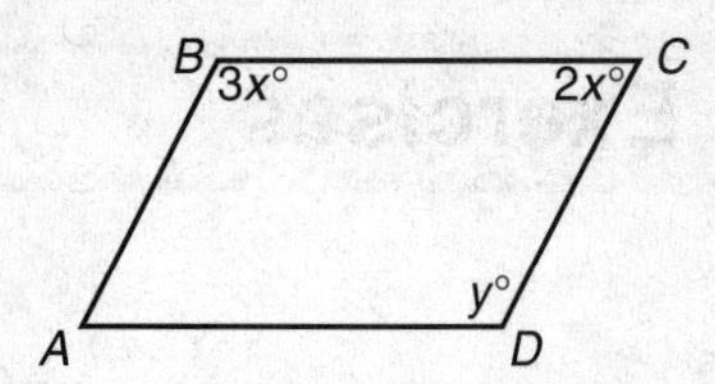

If figure $ABCD$ above is a parallelogram, what is the value of y?

(A) 108
(B) 72
(C) 54
(D) 45
(E) 36

9

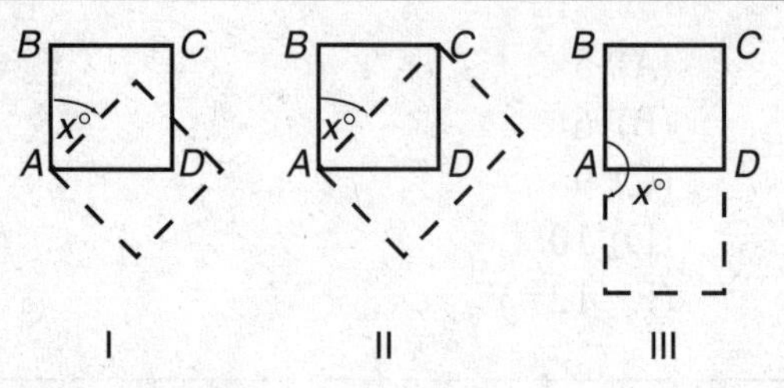

Which of the figures above could represent the rotation of square $ABCD$ $x°$ about point A?

(A) I only
(B) II only
(C) III only
(D) I and III only
(E) I, II, and III

10 Which figures, if any, have BOTH point symmetry and a line of symmetry?

I

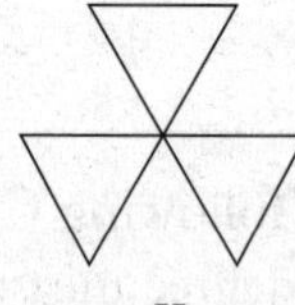
II

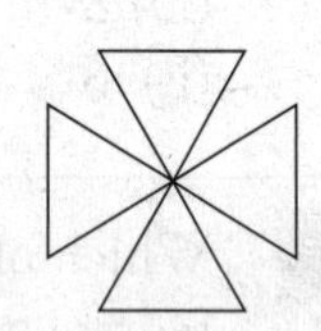
III

(A) I only
(B) III only
(C) I and II only
(D) I and III only
(E) None of the figures

11 What is the number of sides of a polygon in which the sum of the degree measures of the interior angles is 4 times the sum of the degree measures of the exterior angles?

(A) 4
(B) 6
(C) 8
(D) 10
(E) 12

12 In quadrilateral $ABCD$, $\angle A + \angle C$ is 2 times $\angle B + \angle D$. If $\angle A = 40$, then $\angle B =$

(A) 60
(B) 80
(C) 120
(D) 240
(E) It cannot be determined from the information given.

13

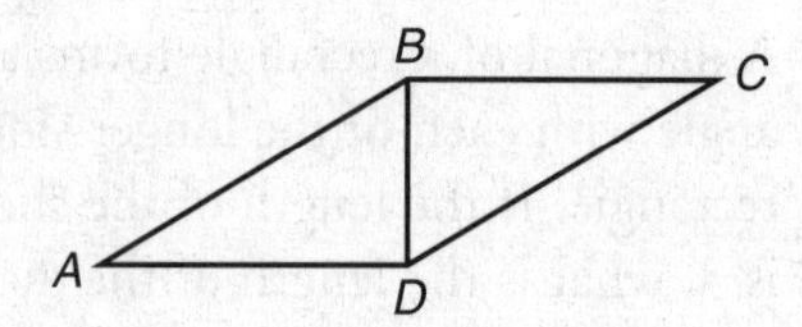

For parallelogram $ABCD$ above, if $AB > BD$, which of the following statements must be true?

I. $CD < BD$
II. $\angle ADB > \angle C$
III. $\angle CBD > \angle A$

(A) None
(B) I only
(C) II and III only
(D) I and III only
(E) I, II, and III

14

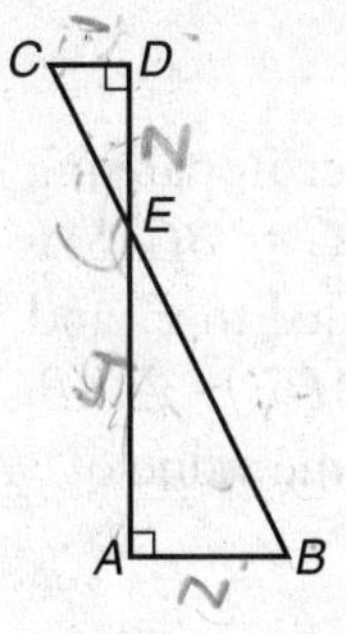

If, in the figure above, $CD = 1$, $AB = 2$, and $AD = 6$, then $BC =$

(A) 5
(B) 9
(C) $2 + \sqrt{5}$
(D) $3\sqrt{5}$
(E) $3\sqrt{2} + 2\sqrt{3}$

15 If the length of each side of a triangle is 4, how many different lines of symmetry can be drawn?

(A) 1
(B) 2
(C) 3
(D) 4
(E) It cannot be determined from the information given.

16 Which letter has point symmetry but no line of symmetry?

(A) E
(B) S
(C) W
(D) I
(E) X

17 Through which set of characters consisting of a letter between two digits can both a horizontal and a vertical line symmetry be drawn?

(A) 3D3
(B) 8S8
(C) 8X8
(D) 8YO
(E) 101

18

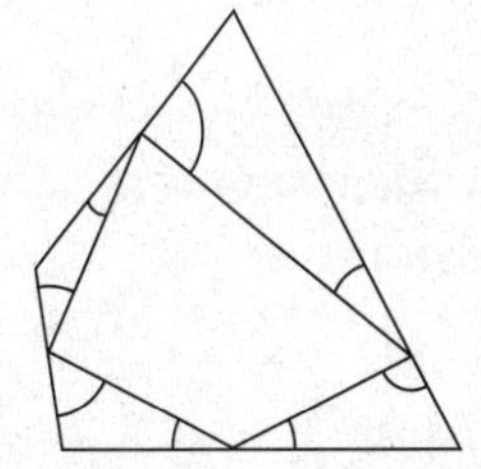

In the figure above, what is the sum of the degree measures of the marked angles?

(A) 120
(B) 180
(C) 360
(D) 540
(E) It cannot be determined from the information given.

Grid-In

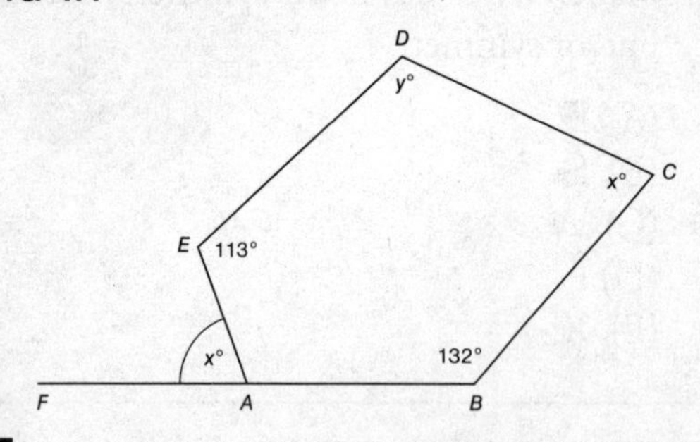

1 In the accompanying figure of pentagon *ABCDE*, points *F*, *A*, and *B* lie on the same line. What is the value of *y*?

2 In the accompanying diagram of triangle *ABC*, $AC = BC$, *D* is a point on $\overline{AC}$, $\overline{AB}$ is extended to *E*, and $\overline{DFE}$ is drawn so that $\triangle ADE \sim \triangle ABC$. If $m\angle C = 30$, what is the value of *x*?

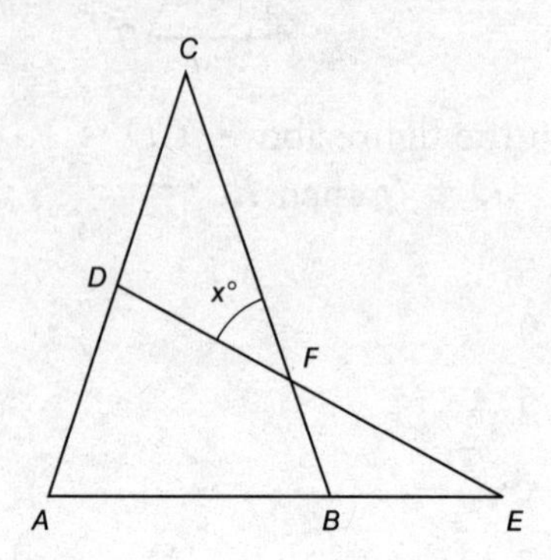

LESSON 6-5

PERIMETER AND AREA

OVERVIEW

*The **perimeter** of a figure is the distance around the figure. To find the perimeter of a polygon, add the lengths of its sides.*

*The **area** of a polygon is the number of square units it encloses. To find the area of a polygon, use one of the formulas given in this lesson.*

At the beginning of each SAT *math section you will find the following reference formulas where A stands for area:*

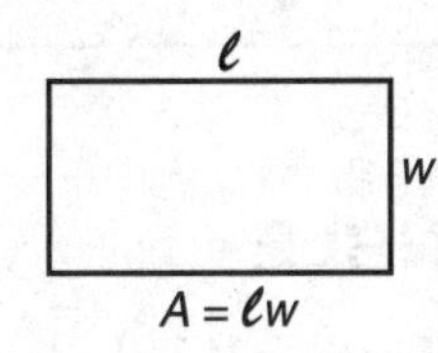

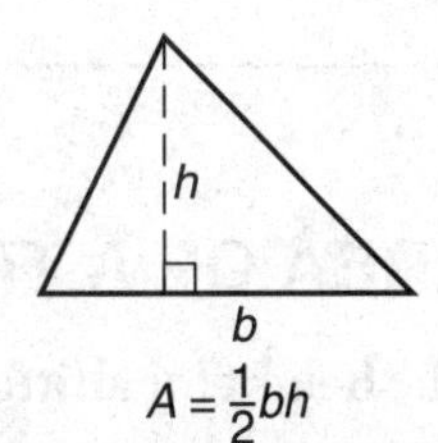

PERIMETER AND AREA OF A RECTANGLE

- To find the perimeter of a rectangle, add the lengths of its four sides.
- To find the area of a rectangle, multiply the lengths of any two adjacent sides.

Formulas for a Rectangle	Examples
• *Perimeter* = 2(ℓ*ength* + *width*)	If $\ell = 7$ and $w = 5$, then
• *Area* = ℓ*ength* × *width*	• $P = 2(\ell + w)$
	$= 2(7 + 5)$
	$= 24$
width	• $A = \ell w$
	$= (7)(5)$
ℓength	$= 35$

A square is a special type of rectangle. To find the area of a square, multiply the length of a side by itself. The diagonals of a square have the same length. If you know the length of diagonal, you can find the area, *A*, of the square using the formula

$$A = \frac{1}{2} \times (\text{diagonal})^2.$$

AREA OF A PARALLELOGRAM

To find the area of a parallelogram, multiply the length of a side (called the **base**) by the length of the perpendicular (called the **height**) drawn to it from the opposite side.

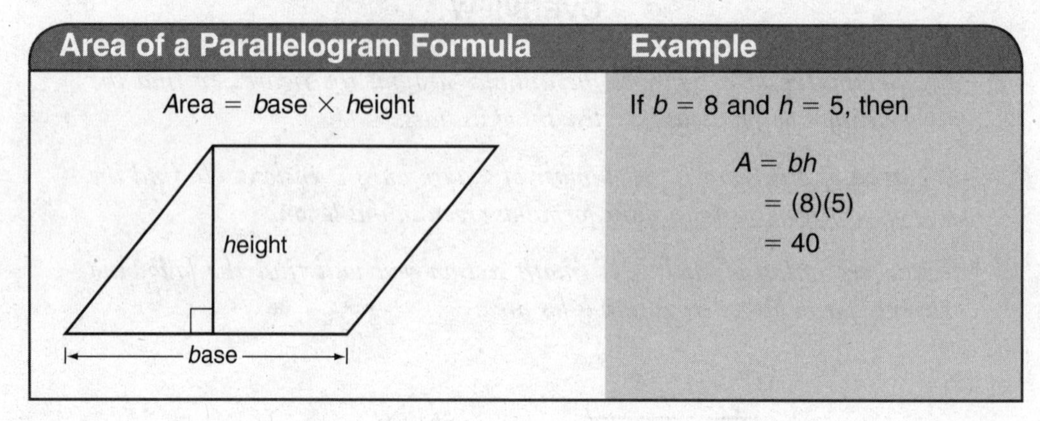

The area, A, of an equilateral triangle is given by the formula

$$A = \frac{x^2}{4}\sqrt{3},$$

where x represents the length of a side.

AREA OF A TRIANGLE

The **height** (or **altitude**) of a triangle is the length of the perpendicular segment drawn from *any* vertex to the opposite side, called the **base.** To find the area of a triangle, multiply the product of its base and height by $\frac{1}{2}$.

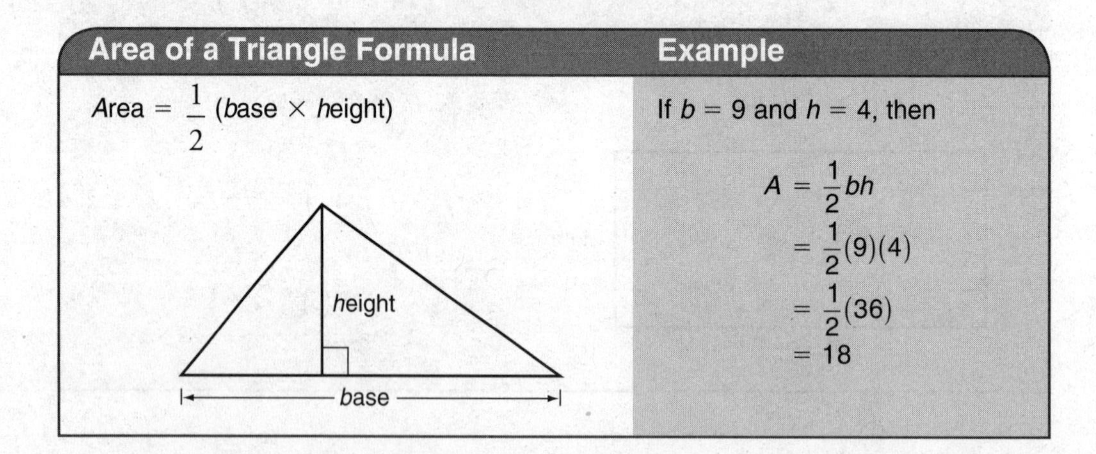

AREA OF A TRAPEZOID

A trapezoid is a quadrilateral in which only one pair of sides, called the **bases,** are parallel. The **height** of a trapezoid is the length of any segment drawn from one base perpendicular to the other base. To find the area of a trapezoid, multiply half the sum of the two bases by the height.

Area of a Trapezoid Formula	Example
$Area = height \times \frac{(\text{sum of } bases)}{2}$	If $h = 5$, $b_1 = 7$ *and* $b_2 = 13$, then $A = 5 \times \frac{(7 + 13)}{2} = 5 \times \frac{20}{2} = 5 \times 10 = 50$

USING AREA FORMULAS

Sometimes it is necessary to figure out the base or height of a figure before an area formula can be used.

EXAMPLE

To find the area of rectangle *ABCD*, note that the diagonal forms a (5, 12, 13) right triangle so the width of the rectangle is 5. Hence:

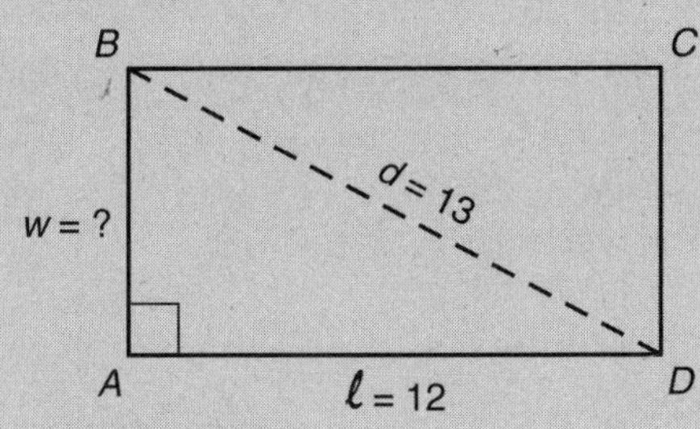

Area of rectangle $= \ell w = 12 \times 5 = 60$

EXAMPLE

To find the area of parallelogram *ABCD*, draw perpendicular segment *BH*, as shown. Since *BH* is the side opposite a 45° angle in a right triangle:

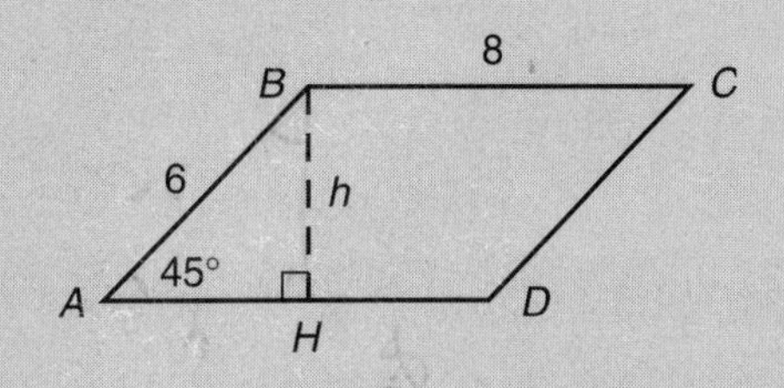

$$h = 6 \times \frac{\sqrt{2}}{2} = 3\sqrt{2}$$

Opposite sides of a parallelogram are equal, so $AD = 8$. Hence,

$$\begin{aligned} \text{Area of parallelogram } ABCD &= bh \\ &= AD \times h \\ &= 8 \times 3\sqrt{2} \\ &= 24\sqrt{2} \end{aligned}$$

EXAMPLE

To find the area of $\triangle ABC$, note that the lengths of the sides of $\triangle ABC$ form a (3, *4*, 5) Pythagorean triple, where $AC = 4$. Hence,

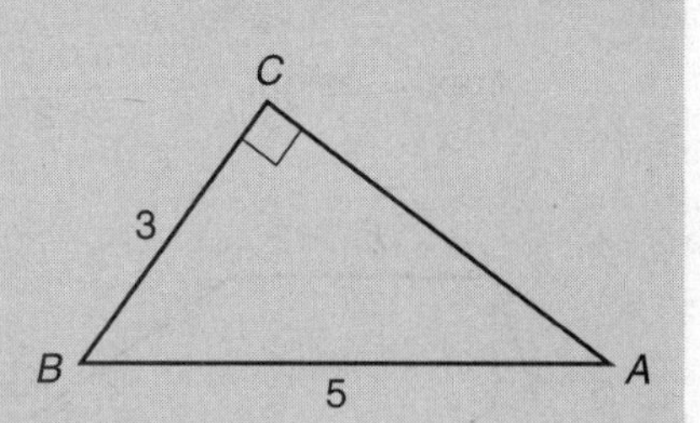

$$\text{Area of } \Delta ABC = \frac{1}{2}bh = \frac{1}{2}(3)(4) = 6$$

Note that in a right triangle either leg is the base and the other leg is the height.

EXAMPLE

To find the area of $\triangle JKL$, drop a perpendicular segment from vertex J to side KL, extending it as necessary. Since $\angle JKH$ measures 30°:

$$h = JH = \frac{1}{2} \times 12 = 6$$

and

$$\text{Area of } \Delta JKL = \frac{1}{2}bh = \frac{1}{2}(8)(6) = 24$$

EXAMPLE

To find the area of trapezoid $ABCD$, use the fact that the lengths of the sides of right triangle AEB form a (5, *12*, 13) Pythagorean triple, where height $BE = 12$. The length of lower base $AD = AE + ED = 5 + 27 = 32$.

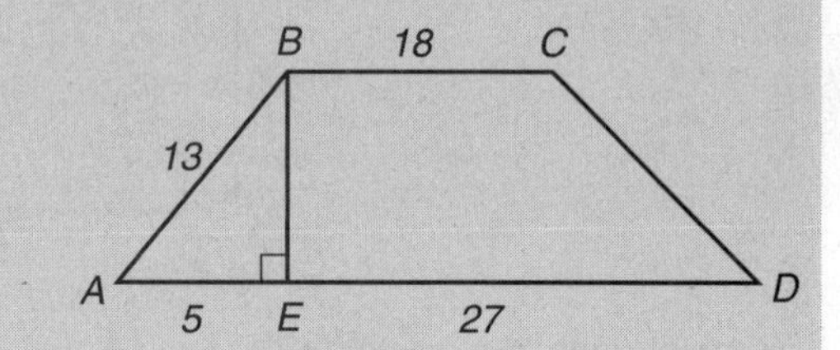

$$\begin{aligned}\text{Area of trapezoid } ABCD &= \text{Height} \times \frac{(\text{Sum of bases})}{2}\\ &= (BE)\frac{(AD + BC)}{2}\\ &= (12)\frac{(32 + 18)}{2}\\ &= 300\end{aligned}$$

EXAMPLE

To find the area of trapezoid $ABCD$, first find the length of base CD by drawing height BH to CD. Since parallel lines are everywhere equidistant, $BH = AD = 15$. The lengths of the sides of right triangle BHC form an (*8*, 15, 17) Pythagorean triple, where $CH = 8$. Thus, $CD = CH + HD = 8 + 10 = 18$.

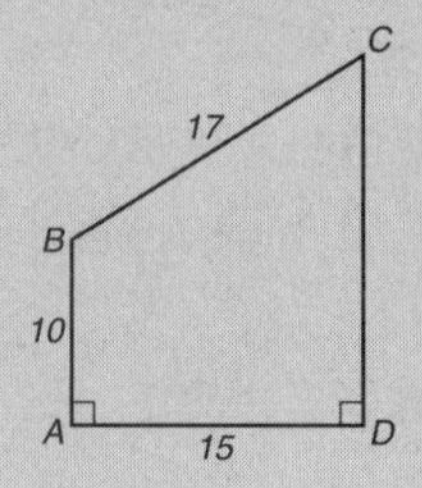

$$\begin{aligned}\text{Area of trapezoid } ABCD &= \text{Height} \times \frac{(\text{Sum of bases})}{2}\\ &= (BH)\frac{(AB + CD)}{2}\\ &= (15)\frac{(10 + 18)}{2}\\ &= 15 \times 14\\ &= 210\end{aligned}$$

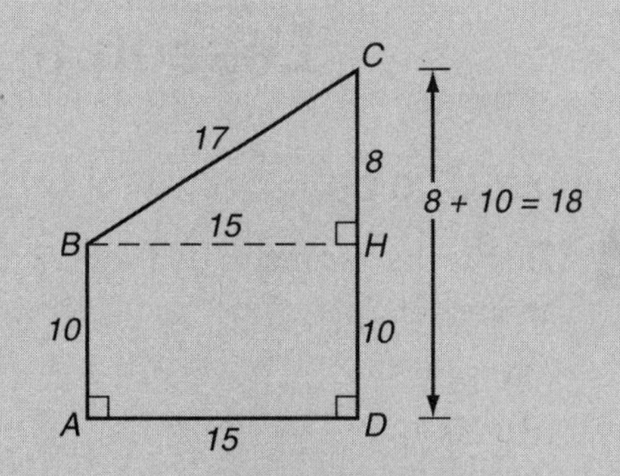

If you didn't remember the formula for the area of a trapezoid, you could also find the area of trapezoid *ABCD* by dividing it into two familiar figures whose area formulas are given in the SAT reference sheet. Drawing $\overline{BH}$ perpendicular to $\overline{CD}$, as performed in the example, separates trapezoid *ABCD* into a rectangle and a right triangle:

$$\begin{aligned}\text{Area of } ABCD &= \text{Area of rectangle } ABCD + \text{area of triangle } BHC\\ &= (AB \times AD) + \left(\frac{1}{2} \times CH \times BH\right)\\ &= (10 \times 15) + \left(\frac{1}{2} \times 8 \times 15\right)\\ &= 150 + 60\\ &= 210\end{aligned}$$

Lesson 6-5 Tune-Up Exercises

Multiple-Choice

1 What is the perimeter of a square that has an area of 25?

(A) 15
(B) 20
(C) 25
(D) $20\sqrt{2}$
(E) $25\sqrt{2}$

2

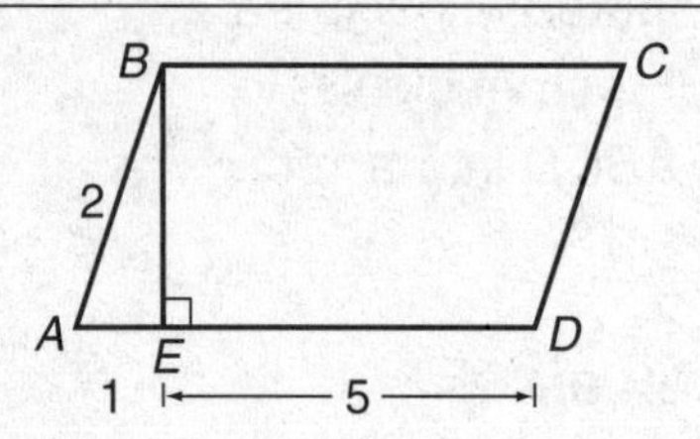

In the figure above, what is the area of parallelogram *ABCD*?

(A) $4\sqrt{2}$
(B) $4\sqrt{3}$
(C) $6\sqrt{2}$
(D) $6\sqrt{3}$
(E) 12

3

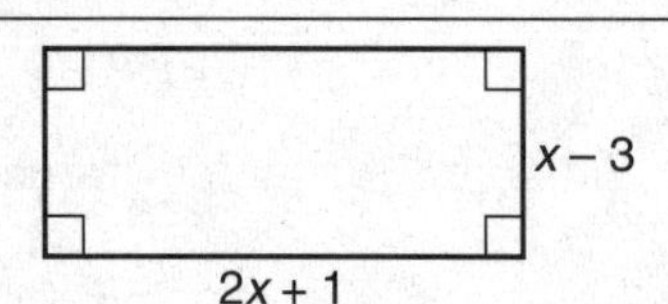

If the perimeter of the rectangle above is 44, then $x =$

(A) 2
(B) 4
(C) 7
(D) 8
(E) 9

4

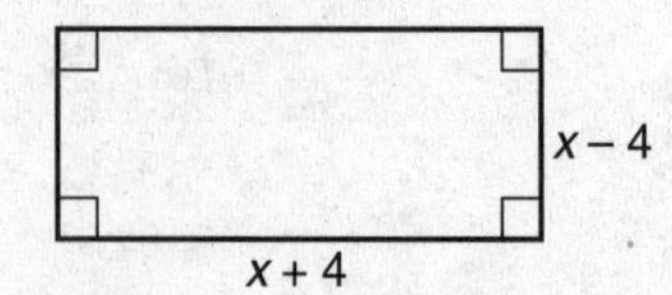

The rectangle above has an area of 65 when $x =$

(A) 9
(B) 8
(C) 6
(D) 5
(E) 3

5 What is the area of a square with a diagonal of $\sqrt{2}$?

(A) $\frac{1}{2}$
(B) 1
(C) $\sqrt{2}$
(D) 2
(E) 4

6

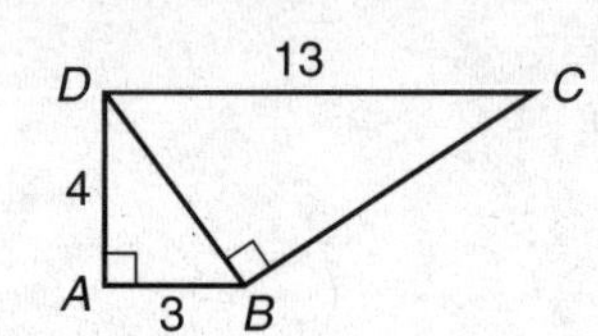

In the figure above, what is the area of quadrilateral *ABCD*?

(A) 28
(B) 32
(C) 36
(D) 42
(E) 60

7

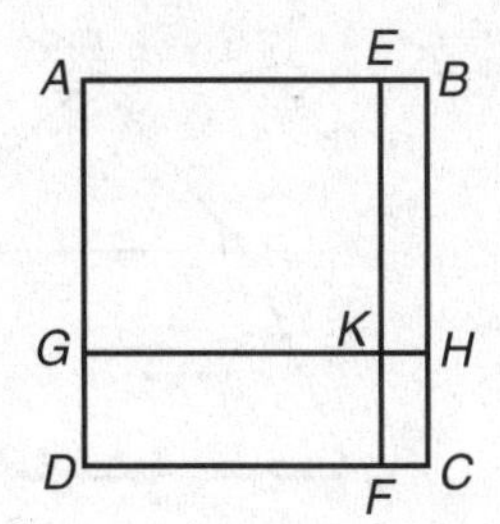

The area of square $ABCD$ in the figure above is 121. Line segments EF and GH are drawn parallel to the sides of the square. If the area of rectangle $FKHC$ is 6 and the area of rectangle $EKHB$ is 16, what is the area of rectangle $GDFK$?

(A) 21
(B) 24
(C) 27
(D) 32
(E) 36

8

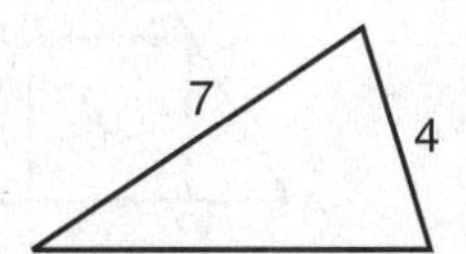

If the perimeter of the triangle above is 18, what is the area of the triangle?

(A) $2\sqrt{33}$
(B) $6\sqrt{5}$
(C) 14
(D) $9\sqrt{5}$
(E) $18\sqrt{5}$

9

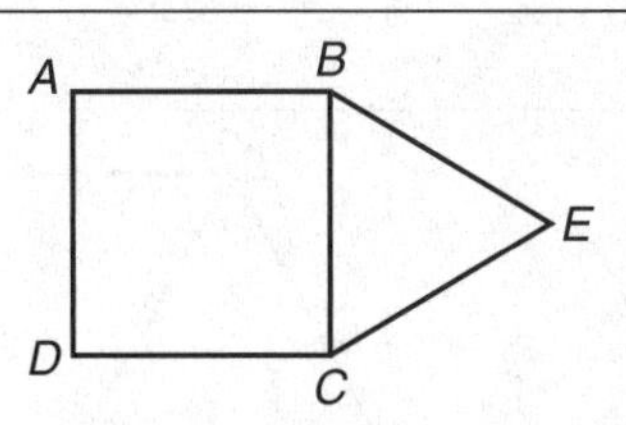

In the figure above, if the area of square $ABCD$ is 64, what is the area of equilateral triangle BEC?

(A) 8
(B) $8\sqrt{3}$
(C) $12\sqrt{3}$
(D) $16\sqrt{3}$
(E) 32

10 If the area of an isosceles right triangle is 8, what is the perimeter of the triangle?

(A) $8 + \sqrt{2}$
(B) $8 + 4\sqrt{2}$
(C) $4 + 8\sqrt{2}$
(D) $12\sqrt{2}$
(E) 10

11 The lengths of the sides of $\triangle ABC$ are consecutive integers. If $\triangle ABC$ has the same perimeter as an equilateral triangle with a side length of 9, what is the length of the shortest side of $\triangle ABC$?

(A) 4
(B) 6
(C) 8
(D) 10
(E) 12

Questions 12 and 13 refer to the figure below.

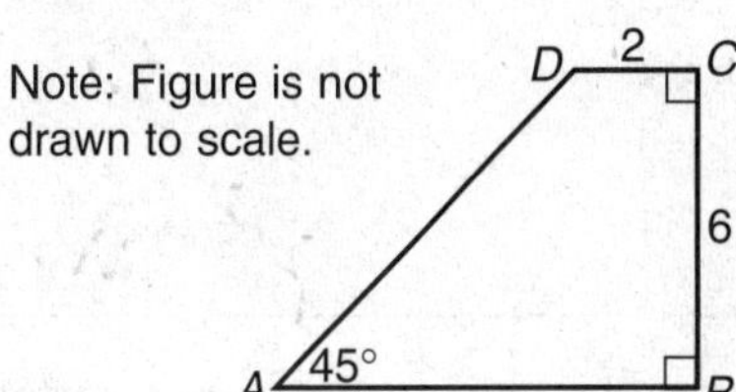

12 What is the perimeter of quadrilateral $ABCD$?

(A) $16 + 3\sqrt{2}$
(B) $16 + 6\sqrt{2}$
(C) 28
(D) $22 + 6\sqrt{2}$
(E) $22\sqrt{2}$

13 What is the area of quadrilateral $ABCD$?

(A) 20
(B) 24
(C) 30
(D) 36
(E) 40

14

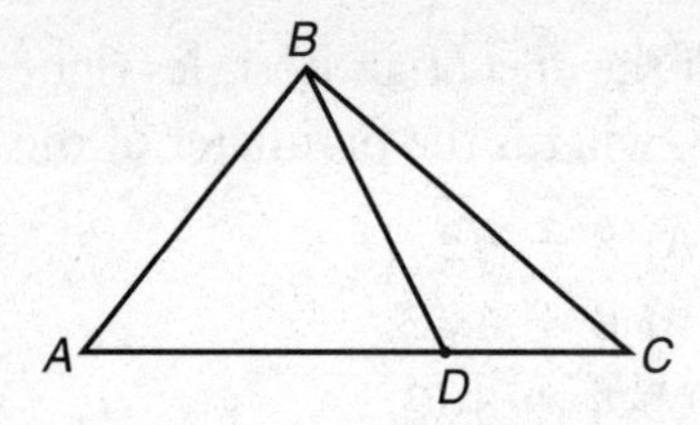

In the figure above, the ratio of AD to DC is 3 to 2. If the area of $\triangle ABC$ is 40, what is the area of $\triangle BDC$?

(A) 16
(B) 24
(C) 30
(D) 36
(E) It cannot be determined from the information given.

15 The perimeter of a rectangle with adjacent side lengths of x and y, where $x > y$, is 8 times as great as the shorter side of the rectangle. What is the ratio of y to x?

(A) 1 : 2
(B) 1 : 3
(C) 1 : 4
(D) 2 : 3
(E) 3 : 4

16 A square has the same area as a rectangle whose longer side is 2 times the length of its shorter side. If the perimeter of the rectangle is 24, what is the perimeter of the square?

(A) $8\sqrt{2}$
(B) $16\sqrt{2}$
(C) 32
(D) $32\sqrt{2}$
(E) 64

17

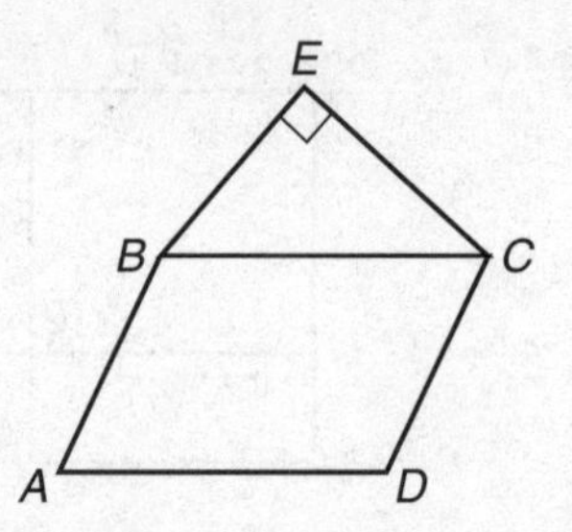

In the figure above, $ABCD$ is a parallelogram with $AB = BE = EC$. If the area of right triangle BEC is 8, what is the perimeter of polygon $ABECD$?

(A) $16 + 8\sqrt{2}$
(B) $12 + 8\sqrt{2}$
(C) $16 + 4\sqrt{2}$
(D) 20
(E) 16

18

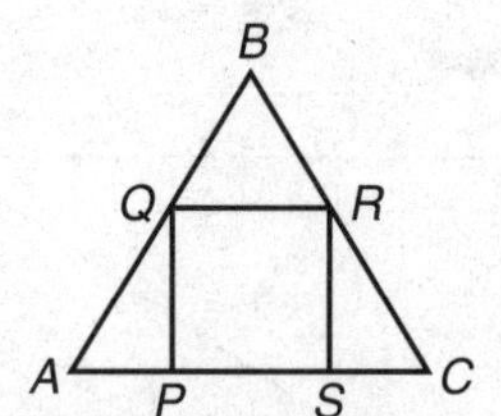

In the figure above, the vertices of square $PQRS$ lie on the sides of equilateral triangle ABC. If the area of the square is 3, what is the perimeter of $\triangle ABC$?

(A) $6\sqrt{3}$
(B) $3 + 6\sqrt{3}$
(C) $6 + 3\sqrt{3}$
(D) 9
(E) 12

19

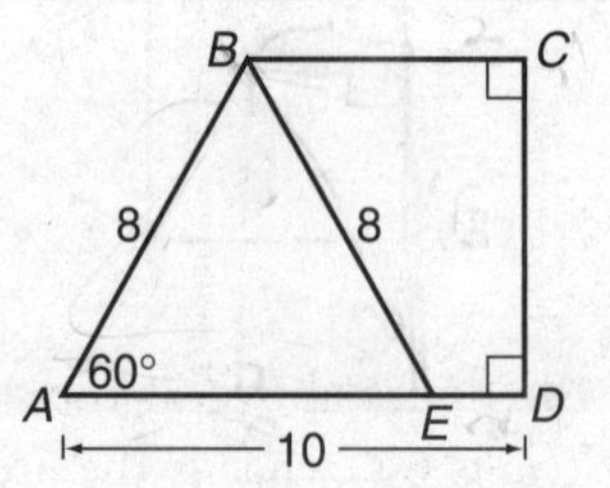

In the figure above, what is the area of quadrilateral $BCDE$?

(A) $8\sqrt{3}$
(B) $16\sqrt{3}$
(C) $8 + 4\sqrt{3}$
(D) $4 + 12\sqrt{3}$
(E) 24

20 If one pair of opposite sides of a square are increased in length by 20% and the other pair of sides are increased in length by 50%, by what percent is the area of the rectangle that results greater than the area of the original square?

(A) 10%
(B) 50%
(C) 70%
(D) 75%
(E) 80%

21

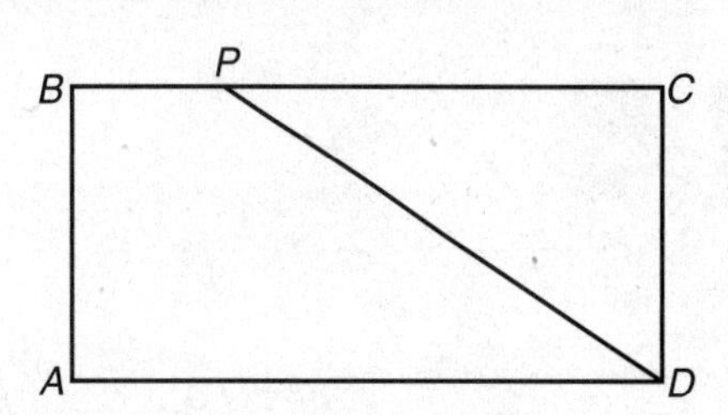

In rectangle $ABCD$, point P divides BC such that BP is 25% of the length of BC. If the area of quadrilateral $ABPD$ is $\frac{3}{4}$, what is the area of rectangle $ABCD$?

(A) $\frac{15}{16}$
(B) $\frac{9}{8}$
(C) $\frac{6}{5}$
(D) $\frac{3}{2}$
(E) $\frac{5}{3}$

Grid-In

1 Brand X paint costs \$14 per gallon, and 1 gallon provides coverage of an area of at most 150 square feet. What is the minimum cost of the amount of brand X paint needed to cover the four walls of a rectangular room that is 12 feet wide, 16 feet long, and 8 feet high?

2

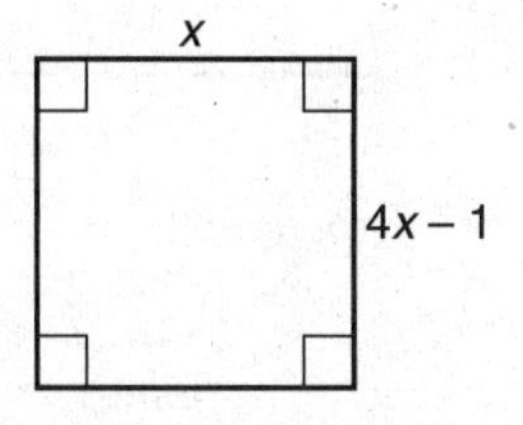

What is the area of the square above?

3

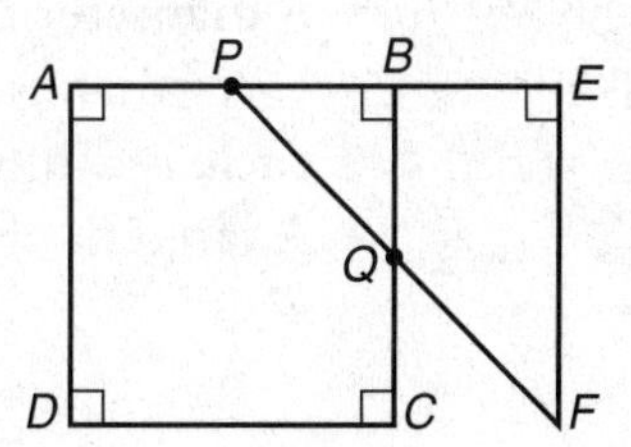

In the figure above, P and Q are the midpoints of sides AB and BC, respectively, of square $ABCD$. Line segment PB is extended by its own length to point E, and line segment PQ is extended to point F so that $FE \perp PE$. If the area of square $ABCD$ is 9, what is the area of quadrilateral $QBEF$?

CIRCLES

OVERVIEW

*A **circle** is a closed figure in which each point on the circle is the same distance from a fixed point called the **center** of the circle. At the beginning of each* SAT *math section you will find the following reference information:*

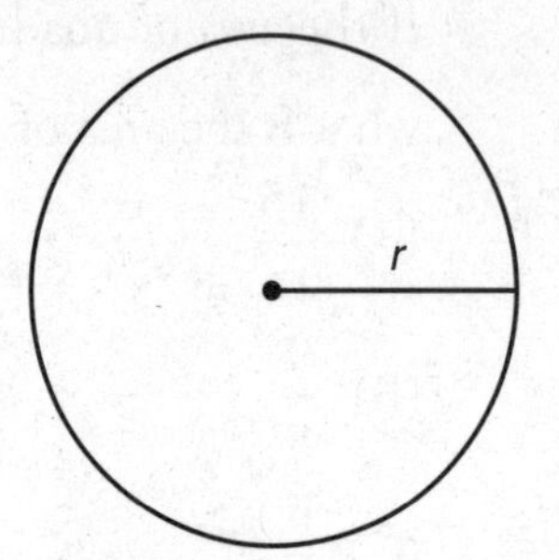

- The number of degrees of arc in a circle is 360.
- Area = πr^2
- Circumference = $2\pi r$

SEGMENTS OF A CIRCLE

A circle is usually denoted by a single capital letter that identifies the center of the circle. Figure 6.17 shows key segments related to circle *O*, where point *O* is the center of this circle.

- A **radius** is any line segment drawn from the center of a circle to any point on the circle. The plural of the word *radius* is *radii.* Hence, all radii of the same circle are equal in length. The distance from the center of a circle to any point on the circle is also called the radius of the circle.
- A **chord** of a circle is a line segment that connects two different points on the circle.
- A **diameter** of a circle is a chord that passes through the center of the circle. Every diameter of a circle is the longest chord of that circle. A diameter of a circle is 2 times the length of a radius of that circle.
- A **tangent** of a circle is a line that intersects a circle in exactly one point.

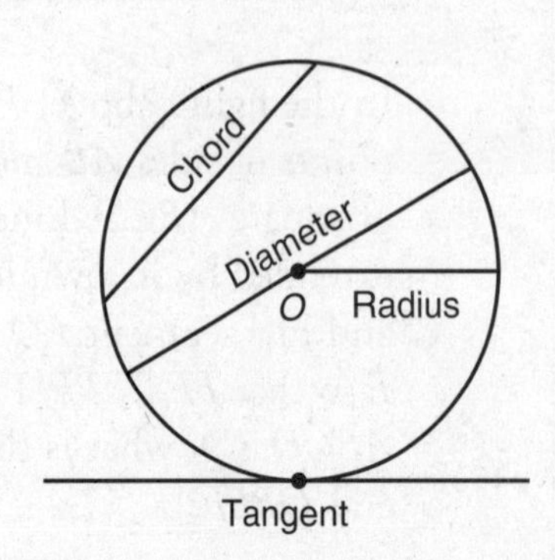

Figure 6.17 Segments Related to a Circle

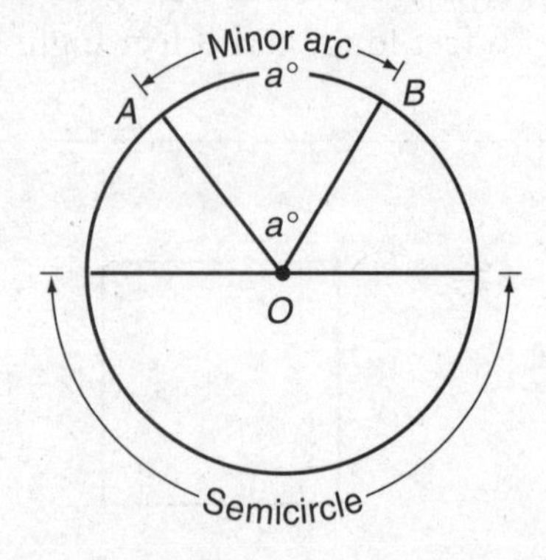

Figure 6.18 Angles and Arcs of a Circle

ANGLES AND ARCS OF A CIRCLE

Figure 6.18 names important angles and arcs of a circle.

- An **arc** of a circle is a curved section of the circle. A diameter of a circle divides a circle into two equal arcs, each of which is called a **semicircle.** An arc smaller than a semicircle is called a **minor arc**. A **major arc** is an arc that is greater than a semicircle.
- A **central angle** of a circle is an angle whose vertex is the center of the circle and whose two sides are radii. In Figure 6.18, $\angle AOB$ is a central angle. A central angle has the same degree measure as the minor arc between its sides.

CIRCLE FORMULAS

The distance around a circle is its **circumference**. The circumference, C, of a circle depends on its diameter (d):

$$C = \pi d = 2\pi r$$

The length, L, of an arc of a circle is a fractional part of the circumference:

$$L = \frac{n}{360} \times \overbrace{2\pi r}^{\text{circumference}}$$

The area of a circle is the number of square units in the region it encloses. The area, A, of a circle depends on its radius:

$$A = \pi r^2$$

A sector of a circle is the region bounded by two radii and their intercepted arc. The area, A_S, of a sector of a circle is a fractional part of the area of the circle:

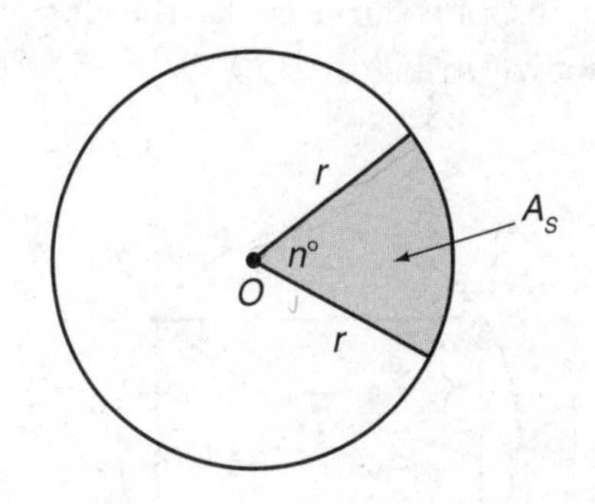

$$A_S = \frac{n}{360} \times \overbrace{\pi r^2}^{\text{area of circle}}$$

EXAMPLE

If in the accompanying figure the length of arc AB is 8π, what is the number of square units in the area of the shaded region?

(A) 16π
(B) 32π
(C) 64π
(D) 96π
(E) 112π

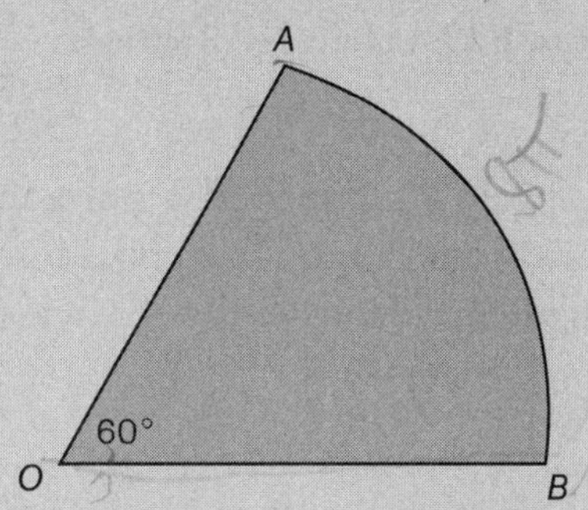

Solution

- Find the radius of circle O.

$$L = \frac{60}{360} \times 2\not{\pi} r = 8\not{\pi}$$
$$\frac{120}{360} \times r = 8$$
$$\frac{1}{3} r = 8$$
$$r = 8 \times 3 = 24$$

- Find the area of the shaded region:

$$A_S = \frac{60}{360} \times \pi(24)^2$$
$$= \frac{\pi}{6} \times 576$$
$$= 96\pi$$

The correct choice is **(D)**.

INSCRIBED POLYGONS AND CIRCLES

A polygon is inscribed in a circle if all its vertices are points on the circle. If a rectangle is inscribed in a circle, the diagonals of the rectangle are diameters of the circle, as shown in Figure 6.19.

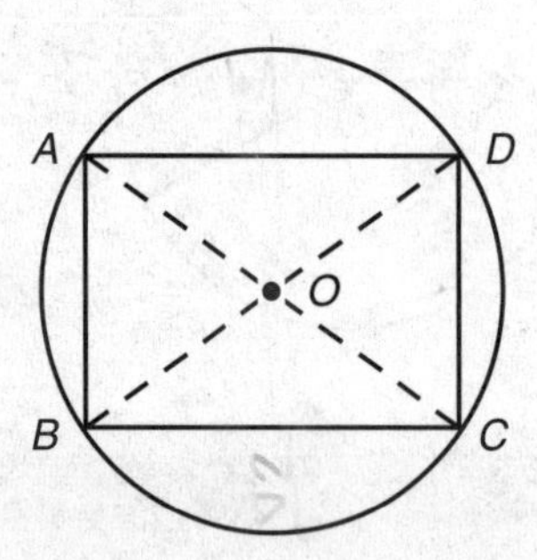

Diagonals AC and BD are diameters of circle O.

Figure 6.19 An Inscribed Rectangle

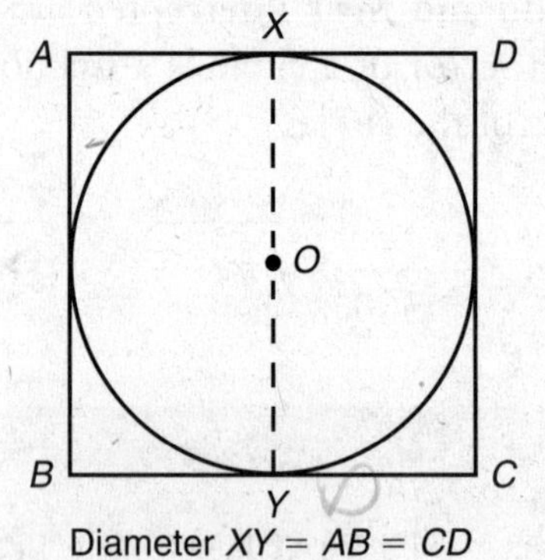

Diameter $XY = AB = CD$

Figure 6.20 A Circle Inscribed in a Square

A circle is inscribed in a polygon if the circle touches each side of the polygon in exactly one point. If a circle is inscribed in a square, a side of the square has the same length as a diameter of the circle (see Figure 6.20).

FINDING AREAS OF SHADED REGIONS INDIRECTLY

To find the area of a shaded region, you may need to subtract the areas of figures that overlap.

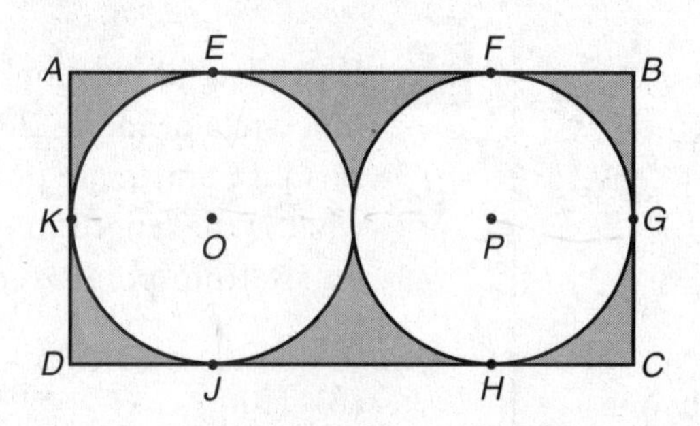

EXAMPLE

In the figure above, the radius of each circle is 1. If circles *O* and *P* touch the sides of rectangle *ABCD* only at the lettered points, what is the area of the shaded region?

Solution

The area of the shaded region is equal to the area of the rectangle minus the sum of the areas of the two circles.

- The width of the rectangle is equal to the diameter of a circle, which is $1 + 1$ or 2.
- The length of the rectangle is equal to the sum of the two diameters, which is $2 + 2$ or 4.
- The area of the rectangle is length $\times$ width $= 4 \times 2 = 8$, and the area of each circle is $\pi r^2 = \pi 1^2 = \pi$.
- Thus,

$$\text{Area of shaded region} = 8 - (\pi + \pi) = 8 - 2\pi$$

RADIUS ⊥ TANGENT

A line tangent to a circle is perpendicular to a radius drawn to the point of contact, as shown in Figure 6.21.

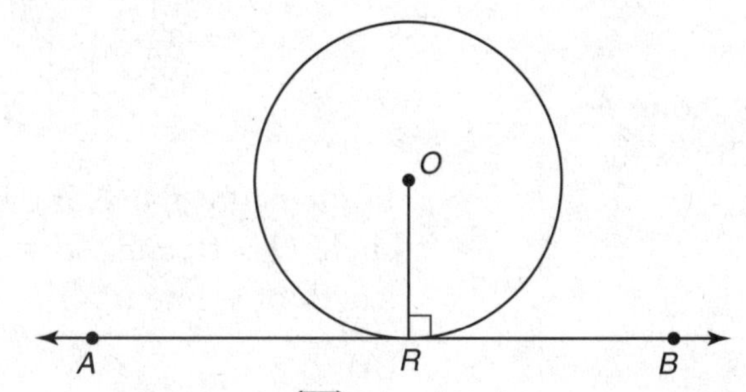

Figure 6.21 Tangent $\overleftrightarrow{AB}$ drawn to circle $O \perp$ radius $\overline{OR}$.

Lesson 6-6 Tune-Up Exercises

Multiple-Choice

1

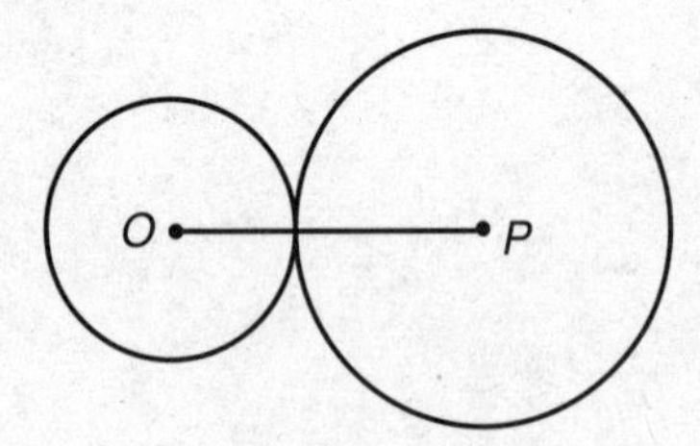

Circles O and P intersect at exactly one point, as shown in the figure above. If the radius of circle O is 2 and the radius of circle P is 6, what is the circumference of any circle that has OP as a diameter?

(A) 4π
(B) 8π
(C) 12π
(D) 16π
(E) 64π

2 What is the area of a circle with a circumference of 10π?

(A) $\sqrt{10\pi}$
(B) 5π
(C) 25π
(D) 100π
(E) $100\pi^2$

3

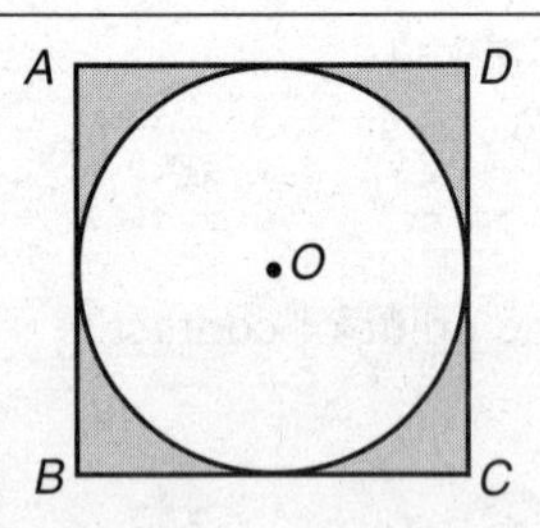

In the figure above, a circle O of radius 4 is inscribed in square $ABCD$. What is the area of the shaded region?

(A) $16 - 4\pi$
(B) $32 - 4\pi$
(C) $32 - 8\pi$
(D) $64 - 8\pi$
(E) $64 - 16\pi$

4 If equilateral polygon $ABCDE$ is inscribed in a circle of radius 20 inches so that A, B, C, D, and E are points on the circle, what is the length in inches of the shortest arc from point A to point C?

(A) 8π
(B) 10π
(C) 12π
(D) 16π
(E) 24π

5 What is the circumference of a circle in which a 5 by 12 rectangle is inscribed?

(A) 13π
(B) 17π
(C) 26π
(D) 34π
(E) 60π

6 What is the perimeter of a square that has the same area as a circle with a circumference of π?

(A) $2\sqrt{\pi}$
(B) 4
(C) 2π
(D) $4\sqrt{\pi}$
(E) 4π

7

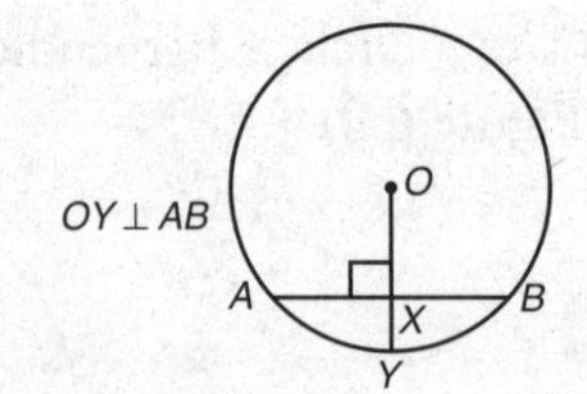

In the figure above, if the radius length of circle O is 10 and $AB = 16$, what is the length of segment XY?

(A) 2
(B) 3
(C) 4
(D) 6
(E) It cannot be determined from the information given.

8 If a bicycle wheel has traveled $\frac{f}{\pi}$ feet after n complete revolutions, what is the length in feet of the diameter of the bicycle wheel?

(A) $\frac{f}{n\pi^2}$

(B) $\frac{\pi^2}{fn}$

(C) $\frac{nf}{\pi^2}$

(D) nf

(E) $\frac{f}{n}$

9

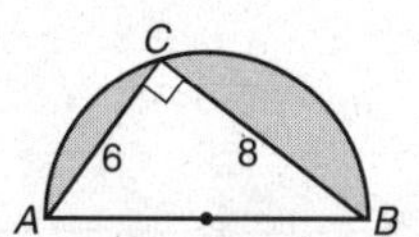

In the figure above, arc ACB is a semicircle of which AB is a diameter. If $AC = 6$ and $BC = 8$, what is the area of the shaded region?

(A) $100\pi - 48$
(B) $50\pi - 24$
(C) $25\pi - 48$
(D) $12.5\pi - 24$
(E) $12.5\pi - 48$

10

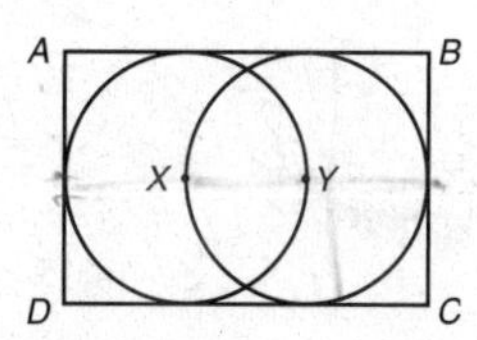

In the figure above, X and Y are the centers of two overlapping circles. If the area of each circle is 7, what is the area of rectangle $ABCD$?

(A) $14 - \frac{7}{\pi}$

(B) $7 + \frac{14}{\pi}$

(C) $\frac{28}{\pi}$

(D) $\frac{42}{\pi}$

(E) It cannot be determined from the information given.

11

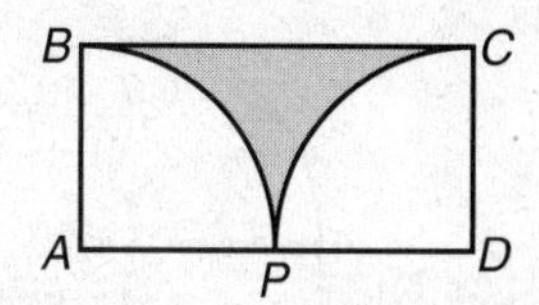

In rectangle $ABCD$ above, arcs BP and CP are quarter circles with centers at points A and D, respectively. If the area of each quarter circle is π, what is the area of the shaded region?

(A) $4 - \frac{\pi}{2}$
(B) $4 - \pi$
(C) $8 - \pi$
(D) $8 - 2\pi$
(E) 8

12

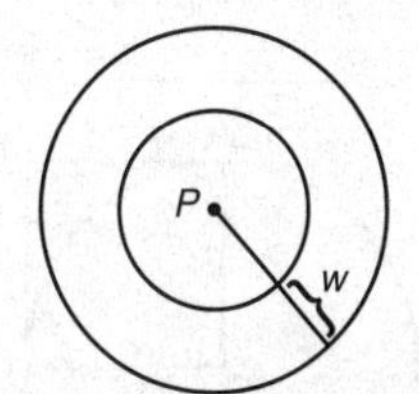

In the figure above, point P is the center of each circle. The circumference of the larger circle exceeds the circumference of the smaller circle by 12π. What is the width, w, of the region between the two circles?

(A) 4
(B) 6
(C) 8
(D) 9
(E) 12

13

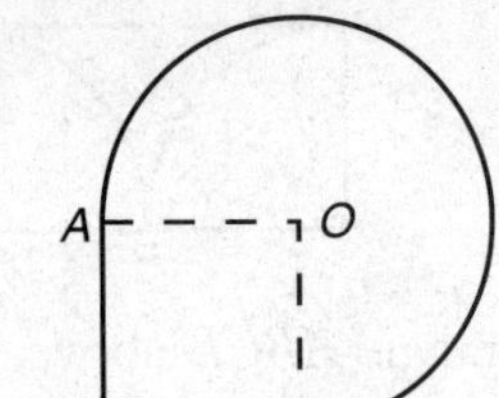

In the figure above, $OACB$ is a square with area $4x^2$. If OA and OB are radii of a sector of a circle O, what is the perimeter, in terms of x, of the unbroken figure?

(A) $x(4 + 3\pi)$
(B) $x(3 + 4\pi)$
(C) $x(6 + 4\pi)$
(D) $4(x + 2\pi)$
(E) $7\pi x$

14

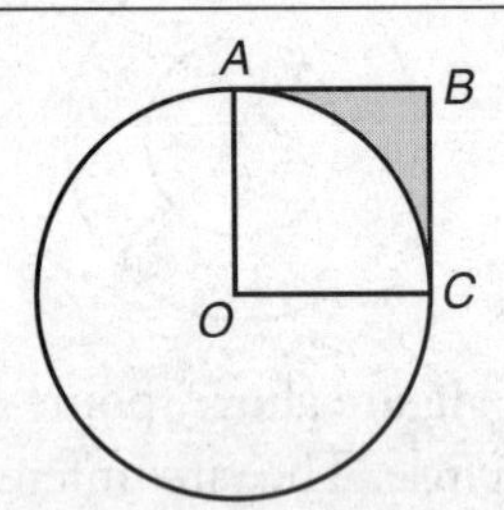

In the figure above, $OABC$ is a square. If the area of circle O is 2π, what is the area of the shaded region?

(A) $\frac{\pi}{2} - 1$
(B) $2 - \frac{\pi}{2}$
(C) $\pi - 2$
(D) $\frac{\pi - 1}{2}$
(E) $\frac{2\pi - 1}{4}$

15 If the circumference of a circle of radius r inches is equal to the perimeter of a square with a side length of s inches, $\frac{r}{s} =$

(A) $\frac{4}{\pi}$
(B) $\frac{2}{\pi}$
(C) $\sqrt{\frac{2}{\pi}}$
(D) $\frac{\sqrt{2}}{\pi}$
(E) $\frac{1}{\pi}$

16

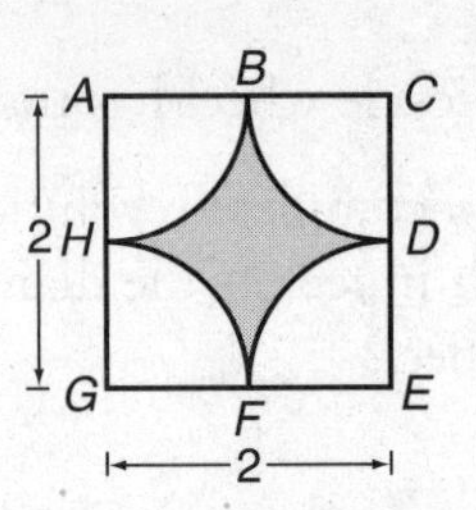

In the figure above, the vertices of square $ACEG$ are the centers of four quarter circles of equal area. What is the best approximation for the area of the shaded region? (Use $\pi = 3.14$.)

(A) 0.64
(B) 0.79
(C) 0.86
(D) 1.57
(E) 2.36

17

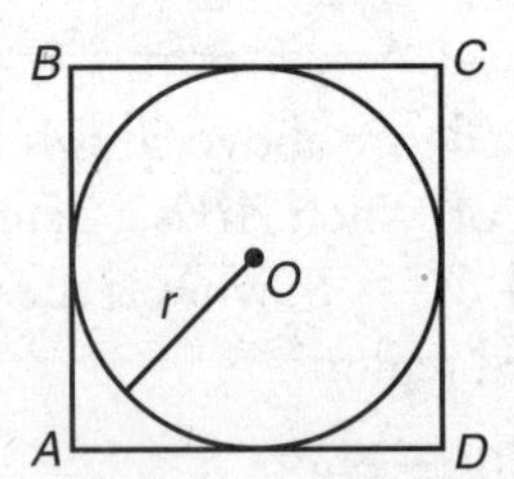

In the figure above, if circle O is inscribed in square $ABCD$ in such a way that each side of the square is tangent to the circle, which of the following statements must be true?

I. $AB \times CD < \pi \times r \times r$
II. Area $ABCD = 4r^2$
III. $r < \frac{2(CD)}{\pi}$

(A) I and II
(B) I and III
(C) II and III
(D) II only
(E) III only

18

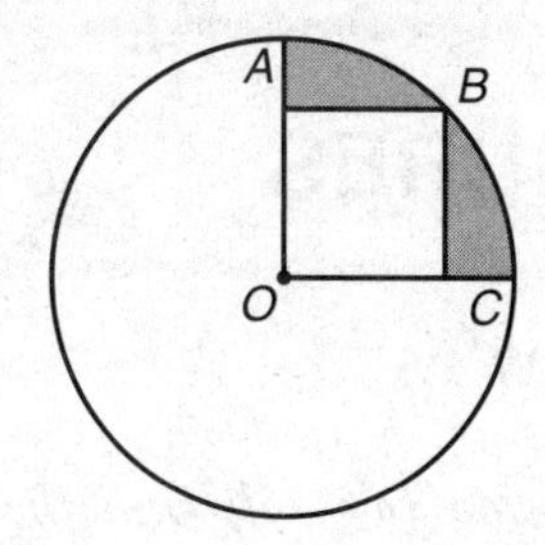

In the figure above, $OABC$ is a square and B is a point on the circle with center O. If $AB = 6$, what is the area of the shaded region?

(A) $9(\pi - 2)$
(B) $9(\pi - 1)$
(C) $12(\pi - 2)$
(D) $18(\pi - 2)$
(E) $15(2\pi - 1)$

19

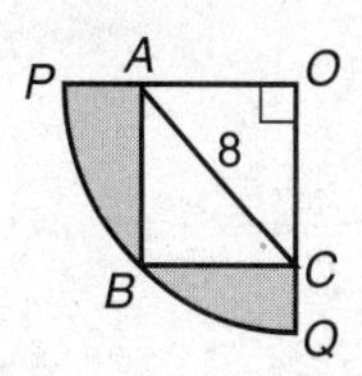

In the figure above, arc PBQ is one-quarter of a circle with center at O, and $OABC$ is a rectangle. If AOC is an isosceles right triangle with $AC = 8$, what is the perimeter of the figure that encloses the shaded region?

(A) $24 - 4\pi$
(B) $24 - 4\sqrt{2} + 4\pi$
(C) $16 - 4\sqrt{2} + 4\pi$
(D) $16 + 4\pi$
(E) $16\sqrt{2} + 8\pi$

Grid-In

1

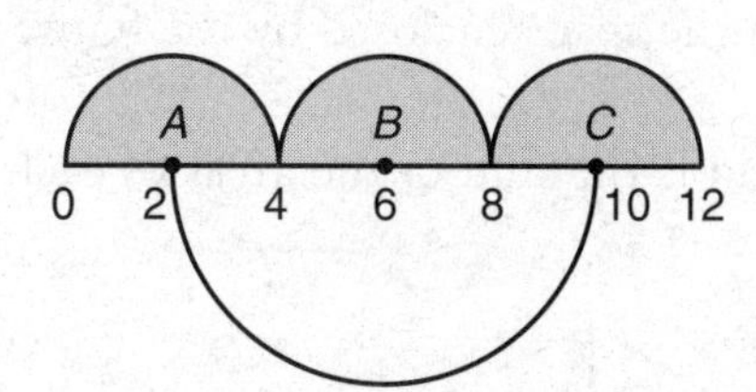

In the figure above, the sum of the areas of the three shaded semicircles with centers at A, B, and C is X, and the area of the larger semicircle below the line is Y. If $Y - X = k\pi$, what is the value of k?

2

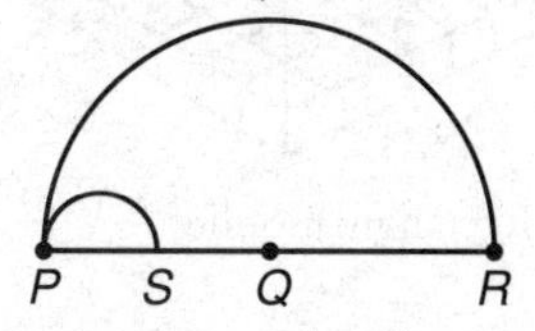

In the figure above, each arc is a semicircle. If S is the midpoint of PQ and Q is the midpoint of PR, what is the ratio of the area of semicircle PS to the area of semicircle PR?

3 Through how many degrees does the minute hand of a clock move from 1:25 P.M. to 1:37 P.M. of the same day?

4

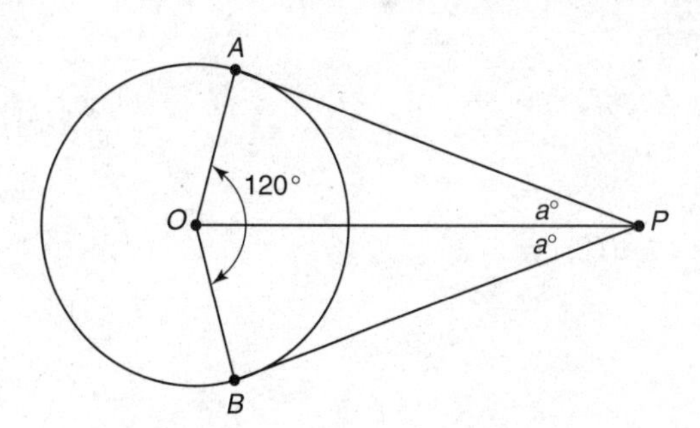

In the figure above, $\overrightarrow{PA}$ is tangent to circle O at point A, $\overrightarrow{PB}$ is tangent to circle O at point B. Angle AOB measures 120° and $OP = \frac{24}{\pi}$. What is the length of minor arc AB?

LESSON

6-7 SOLID FIGURES

OVERVIEW

At the beginning of each SAT *math section you will find the following reference formulas, where V stands for volume:*

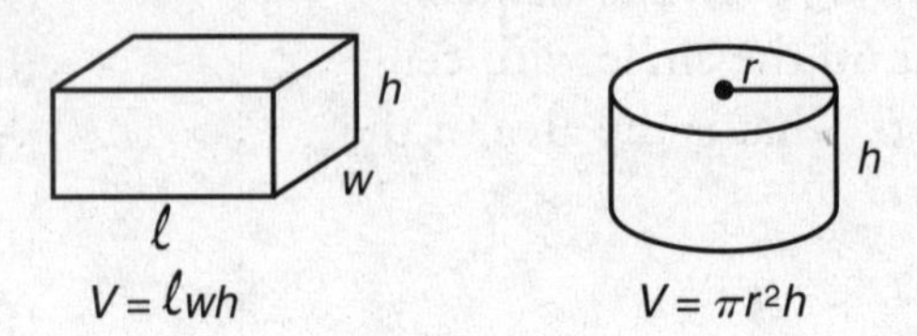

VOLUME AND SURFACE AREA OF A RECTANGULAR SOLID

- The **volume** of a solid figure represents the number of cubes each with an edge length of 1 unit that can be placed in the space enclosed by the figure (see Figure 6.22).
- The **surface area** of a rectangular solid is the sum of the areas of each of its six surfaces.

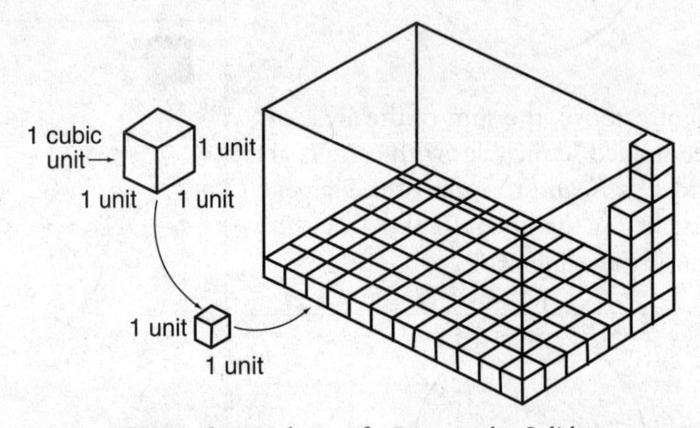

Figure 6.22 Volume of a Rectangular Solid

Formulas for a Rectangular Solid

- Volume

$\text{Volume}_{box} = \ell\text{ength} \times \text{width} \times \text{height}$

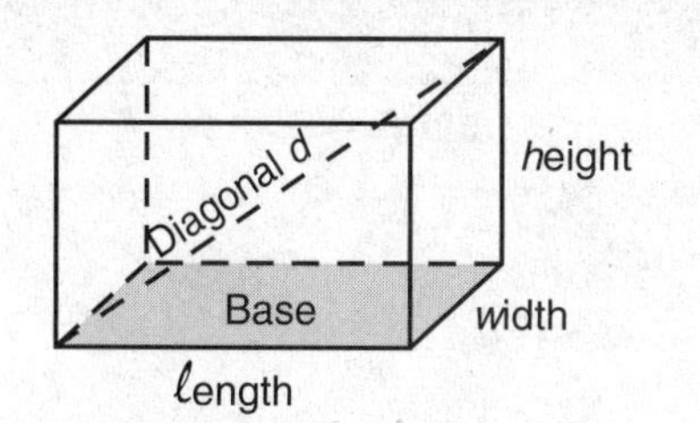

- Surface area

$\text{Area}_{surface} = 2[(\ell \times w) + (\ell \times h) + (w \times h)]$

- Length of diagonal d

$\text{Diagonal } d = \sqrt{\ell^2 + w^2 + h^2}$

Examples

If $\ell = 6$, $w = 3$, and $h = 2$, then

- $V = \ell wh$
  $= 6 \times 3 \times 2$
  $= 36$
- $A = 2[(\ell \times w) + (\ell \times h) + (w \times h)]$
  $= 2[(6 \times 3) + (6 \times 2) + (3 \times 2)]$
  $= 2(18 + 12 + 6)$
  $= 2(36)$
- $d = \sqrt{\ell^2 + w^2 + h^2}$
  $= \sqrt{6^2 + 3^2 + 2^2}$
  $= \sqrt{36 + 9 + 4}$
  $= \sqrt{49}$
  $= 7$

VOLUME AND SURFACE AREA OF A CUBE

- A **cube** is a rectangular solid in which each of its six square faces has the same edge length. Its volume is the cube of the edge length of any face.
- The surface area of a cube is 6 times the area of any square face of the cube.

Formulas for a Cube

- Volume

$\text{Volume}_{cube} = \text{edge} \times \text{edge} \times \text{edge} = e^3$

- Surface area

$\text{Area}_{surface} = 6(\text{edge})^2$

Examples

If $e = 4$, then

- $V = e^3$
  $= 4 \times 4 \times 4$
  $= 64$
- $A = 6e^2$
  $= 6(4)^2$
  $= 6(16)$
  $= 96$

VOLUME OF A CYLINDER

The volume of a **cylinder** is the area of its circular base times its height.

Volume of a Cylinder Formula	Example
$\text{Volume}_{\text{cylinder}} = \pi \times (\text{radius})^2 \times \text{height}$	If $r = 3$, and $h = 5$, then $V = \pi r^2 h = \pi(3)^2(5) = 45\pi$

VOLUME OF A CONE

The volume of a **cone** is one-third times the area of its circular base times its height.

The point of a cone is called the **vertex** and the length of a segment connecting the vertex to a point on its circular base is called the **slant height.**

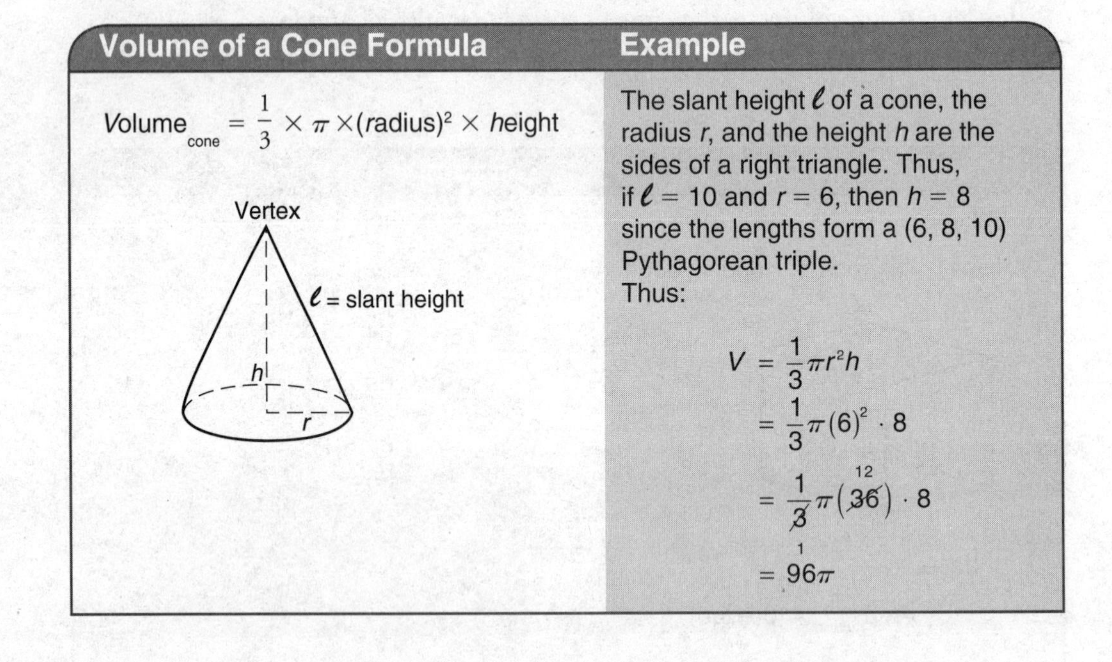

Volume of a Cone Formula	Example
$\text{Volume}_{\text{cone}} = \frac{1}{3} \times \pi \times (\text{radius})^2 \times \text{height}$	The slant height ℓ of a cone, the radius r, and the height h are the sides of a right triangle. Thus, if $\ell = 10$ and $r = 6$, then $h = 8$ since the lengths form a (6, 8, 10) Pythagorean triple. Thus: $V = \frac{1}{3}\pi r^2 h = \frac{1}{3}\pi(6)^2 \cdot 8 = \frac{1}{3}\pi(36) \cdot 8 = 96\pi$

Lesson 6-7 Tune-Up Exercises

Multiple-Choice

1 What is the volume of a cube whose surface area is 96?

(A) $16\sqrt{2}$
(B) 32
(C) 64
(D) 125
(E) 216

2 The length, width, and height of a rectangular solid are in the ratio of 3 : 2 : 1. If the volume of the box is 48, what is the total surface area of the box?

(A) 27
(B) 32
(C) 44
(D) 64
(E) 88

3 If X is the center point of a face of a cube with a volume of 8, and Y is the center point of the opposite face of this cube, what is the distance from X to Y?

(A) $\sqrt{2}$
(B) 2
(C) $2\sqrt{2}$
(D) 4
(E) 6

4 A cube whose volume is $\frac{1}{8}$ cubic foot is placed on top of a cube whose volume is 1 cubic foot. The two cubes are then placed on top of a third cube, whose volume is 8 cubic feet. What is the height, in *inches*, of the stacked cubes?

(A) 30
(B) 40
(C) 42
(D) 44
(E) 64

5 Note: Figure is not drawn to scale.

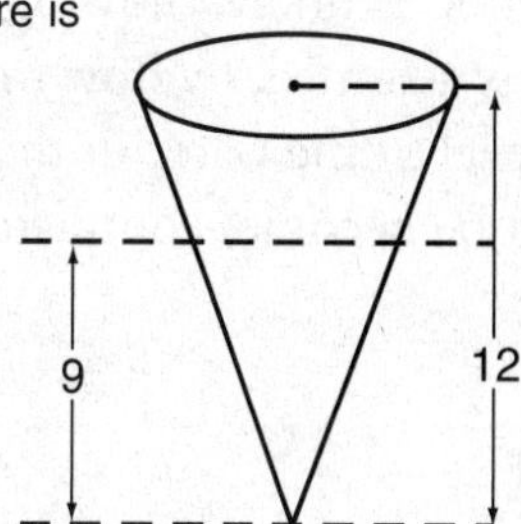

The height of the solid cone above is 12 inches, and the area of the circular base is 64π square inches. What is the area, in *square inches*, of the base of the cone formed when a plane parallel to the base cuts through the original cone 9 inches above the vertex of the cone?

(A) 9π
(B) 16π
(C) 25π
(D) 36π
(E) 49π

6 If the height of a cylinder is doubled, by what number must the radius of the base be multiplied so that the resulting cylinder has the same volume as the original cylinder?

(A) 4
(B) 2
(C) $\frac{1}{\sqrt{2}}$
(D) $\frac{1}{2}$
(E) $\frac{1}{4}$

7 A rectangular fish tank has a base 2 feet wide and 3 feet long. When the tank is partially filled with water, a solid cube with an edge length of 1 foot is placed in the tank. If no overflow of water from the tank is assumed, by how many *inches* will the level of the water in the tank rise when the cube becomes completely submerged?

(A) $\frac{1}{6}$
(B) $\frac{1}{2}$
(C) 2
(D) 3
(E) 4

8 The volume of a cylinder of radius r is $\frac{1}{4}$ of the volume of a rectangular box with a square base of side length x. If the cylinder and the box have equal heights, what is r in terms of x?

(A) $\frac{x^2}{2\pi}$
(B) $\frac{x}{2\sqrt{\pi}}$
(C) $\frac{\sqrt{2x}}{\pi}$
(D) $\frac{\pi}{2\sqrt{x}}$
(E) $\sqrt{2\pi x}$

9 The height of sand in a cylinder-shaped can drops 3 inches when 1 cubic foot of sand is poured out. What is the diameter, in *inches*, of the cylinder?

(A) $\frac{2}{\sqrt{\pi}}$
(B) $\frac{4}{\sqrt{\pi}}$
(C) $\frac{16}{\pi}$
(D) $\frac{32}{\sqrt{\pi}}$
(E) $\frac{48}{\sqrt{\pi}}$

10 The height h of a cylinder equals the circumference of the cylinder. In terms of h, what is the volume of the cylinder?

(A) $\frac{h^2}{4\pi}$
(B) $\frac{h^2}{2\pi}$
(C) $\frac{h^3}{2}$
(D) $h^2 + 4\pi$
(E) πh^3

11

X: 4 by 3; Y: 3 by 3; Z: 5 by 4

For which of the following combinations of rectangular faces X, Y, and Z in the figures above can a rectangular solid be formed?

I. Two of face X, two of face Y, and two of face Z
II. Four of face X and two of face Y
III. Two of face Y and four of face Z

(A) I only
(B) II only
(C) I and III only
(D) II and III only
(E) None

12 A cylinder with radius r and height h is closed on the top and bottom. Which of the following expressions represents the total surface area of this cylinder?

(A) $2\pi r(r + h)$
(B) $\pi r(r + 2h)$
(C) $\pi r(2r + h)$
(D) $\pi r^2 + 2h$
(E) $2\pi r^2 + h$

13

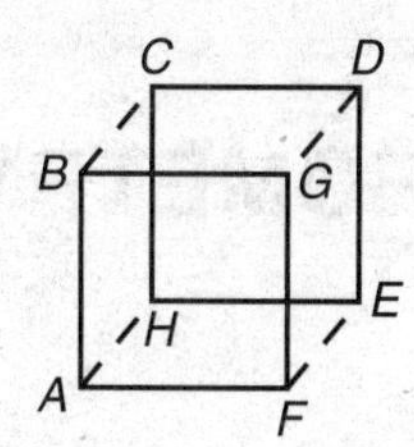

In the figure above, if the edge length of the cube is 4, what is the shortest distance from A to D?

(A) $4\sqrt{2}$
(B) $4\sqrt{3}$
(C) 8
(D) $4\sqrt{2} + 4$
(E) $8\sqrt{3}$

14

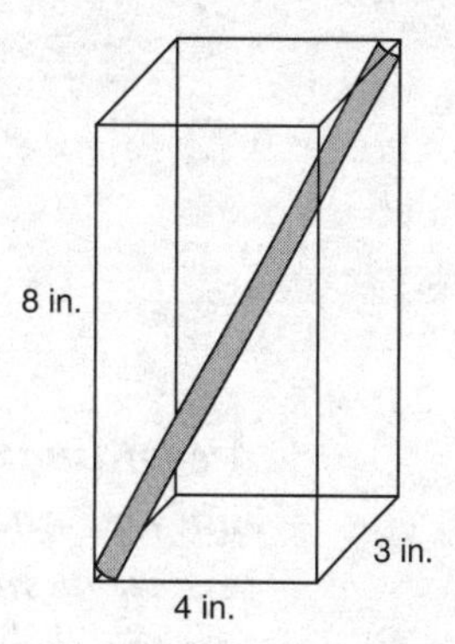

A cylindrical tube with negligible thickness is placed into a rectangular box that is 3 inches by 4 inches by 8 inches, as shown in the accompanying diagram. If the tube fits exactly into the box diagonally from the bottom left corner to the top right back corner, what is the best approximation of the number of inches in the length of the tube?

(A) 3.9
(B) 5.5
(C) 7.8
(D) 9.4
(E) 15.0

Grid-In

1 The dimensions of a rectangular box are integers greater than 1. If the area of one side of this box is 12 and the area of another side is 15, what is the volume of the box?

2 What is the number of inches in the minimum length of $\frac{1}{4}$-inch-wide tape needed to cover completely a cube whose volume is 8 cubic inches?

3 By what percent does the volume of a cube increase when the length of each of its sides is doubled?

4

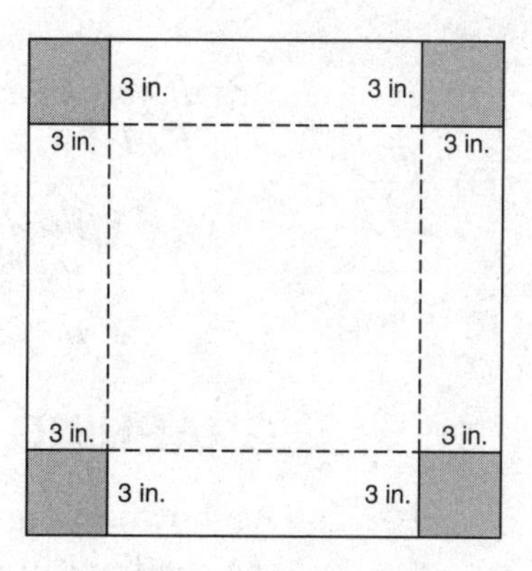

Note: Figure is not drawn to scale.

A box is constructed by cutting 3-inch squares from the corners of a square sheet of cardboard, as shown in the accompanying diagram, and then folding the sides up. If the volume of the box is 75 cubic inches, find the number of square inches in the area of the *original* sheet of cardboard.

LESSON

6-8 COORDINATE GEOMETRY

OVERVIEW

*A **coordinate plane** is represented by a grid of square boxes that is divided into four **quadrants** by a horizontal number line, called the x-axis, and a vertical number line, called the y-axis, that intersect at their* 0 *points. Each point in the coordinate plane is located by using an ordered pair of numbers, called **coordinates**, of the form (x,y). For example, point* P(3,5) *is located* 3 *units to the right of the origin and* 5 *units above the origin.*

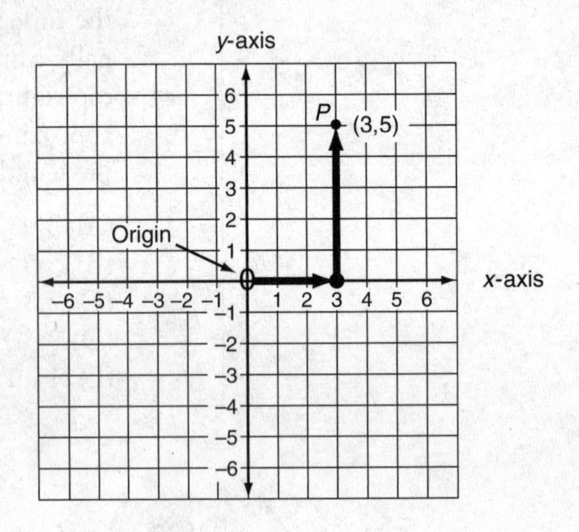

If you know the coordinates of points A and B, you can use a formula to find:

- *the distance from A to B; or*
- *the "slant" or **slope** of the line that passes through A and B; or*
- *the midpoint of line segment AB.*

GRAPHING ORDERED PAIRS

As shown in Figure 6.23, the signs of the *x*- and *y*-coordinates of a point determine the quadrant in which the point lies.

- A point lies to the right of the *y*-axis when *x* is *positive*, and lies to the left of the *y*-axis when *x* is *negative*.
- A point lies *above* the *x*-axis when *y* is *positive*, and lies *below* the *x*-axis when *y* is *negative*.

Point	Sign of Coordinates	Quadrant
$A(2,3)$	(+, +)	I
$B(-4,5)$	(–, +)	II
$C(-3,-6)$	(–, –)	III
$D(3,-3)$	(+, –)	IV

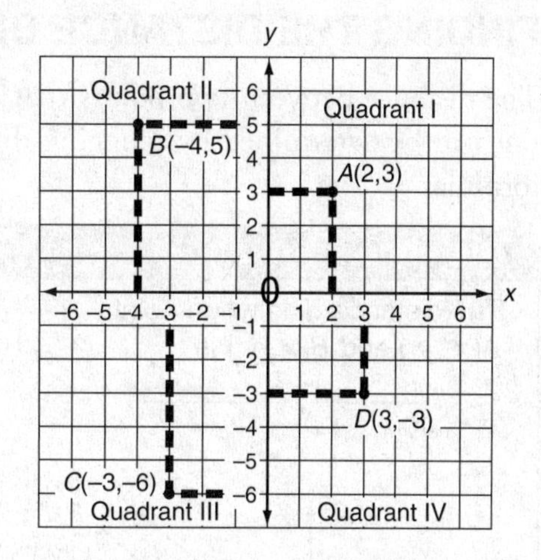

Figure 6.23 Locating Points

FINDING LENGTHS OF HORIZONTAL AND VERTICAL SEGMENTS

Points that have the same x-coordinate lie on the same vertical line, and points that have the same y-coordinate lie on the same horizontal line. To find the length of a horizontal or vertical segment, take the positive difference between the two unequal coordinates.

EXAMPLE

In the figure below, what is the perimeter of $\triangle ABC$?

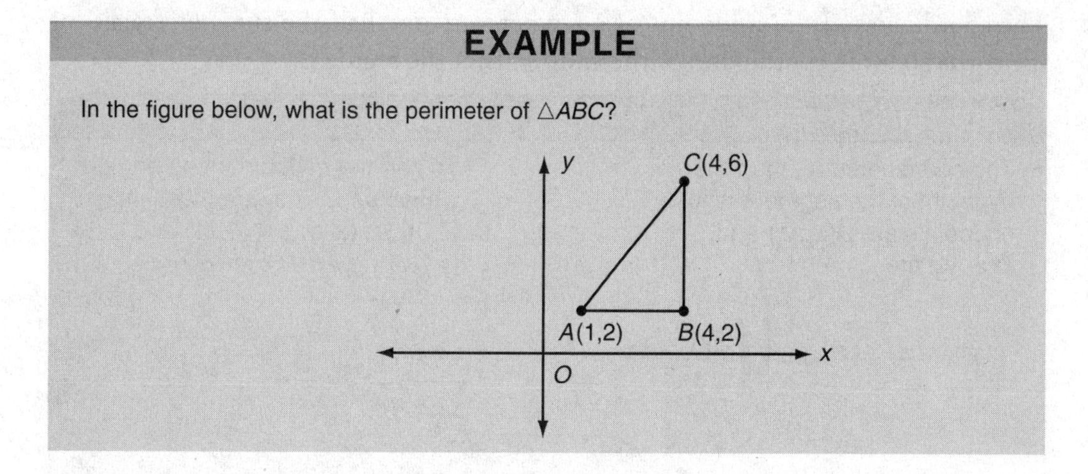

Solution

- The length of horizontal segment AB is $4 - 1 = 3$.
- The length of vertical segment BC is $6 - 2 = 4$.
- Since $\angle ABC$ is formed by horizontal and vertical line segments, it is a right angle. Use the Pythagorean relationship to find the length of AC:

$$\begin{aligned} AC &= \sqrt{(AB)^2 + (BC)^2} \\ &= \sqrt{(3)^2 + (4)^2} \\ &= \sqrt{25} \\ &= 5 \end{aligned}$$

Hence, the perimeter of $\triangle ABC$ is $3 + 4 + 5 = 12$.

FINDING THE DISTANCE BETWEEN TWO POINTS

The distance between two points that do not lie on the same vertical or horizontal line can be determined by plugging the coordinates of the points into the distance formula.

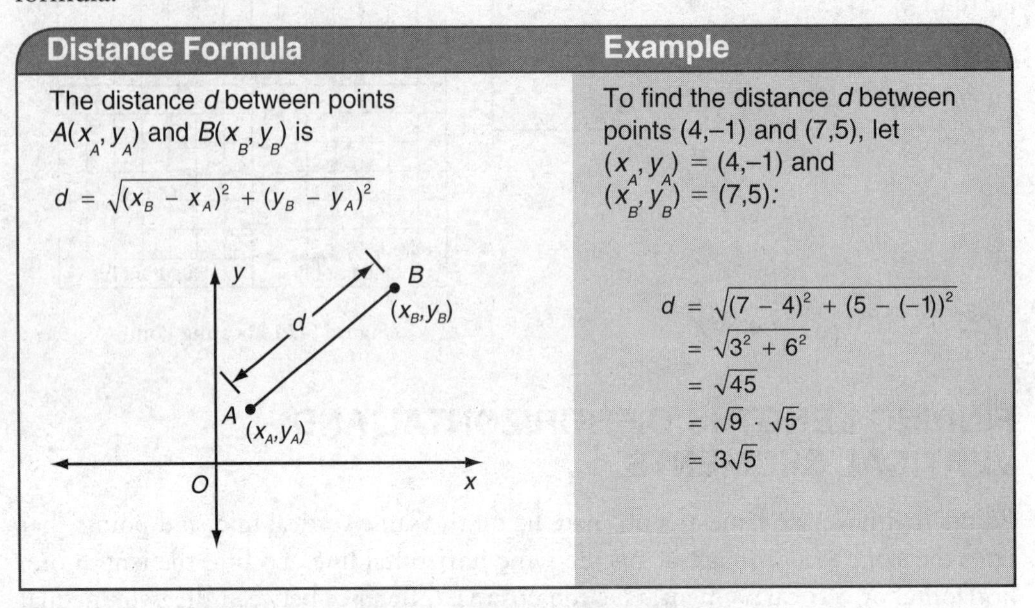

FINDING THE MIDPOINT OF A SEGMENT

The coordinates of the midpoint of a segment are one-half the sums (averages) of the corresponding coordinates of the endpoints of the segment.

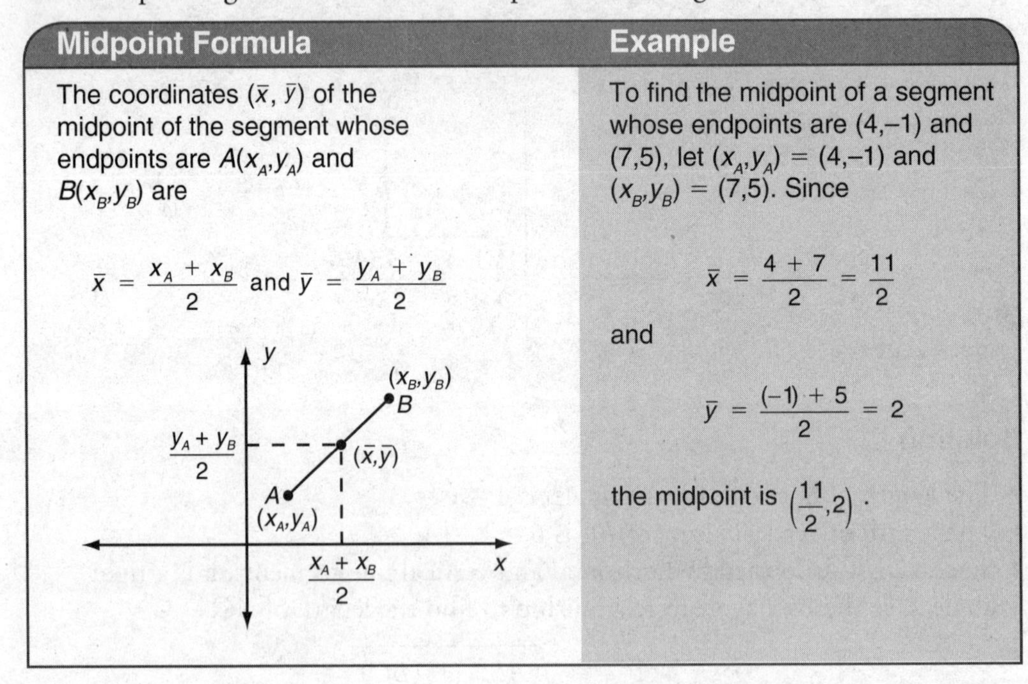

FINDING AREAS USING COORDINATES

If a side of a triangle or parallelogram lies on an axis or is parallel to an axis, you can calculate the area of the figure by finding the lengths of the base and height and then using the appropriate area formula.

EXAMPLE

What is the area of the shaded region in the figure below?

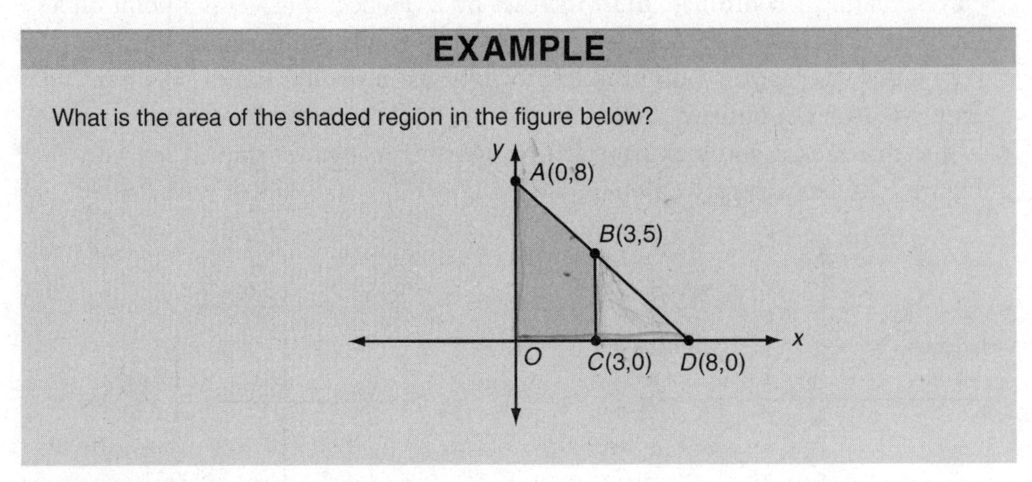

Solution

$$\begin{aligned}\text{Area of } OABC &= \text{Area of } \Delta AOD - \text{Area of } \Delta BCD\\ &= \frac{1}{2}(OD)(OA) - \frac{1}{2}(CD)(BC)\\ &= \frac{1}{2}(8)(8) - \frac{1}{2}(5)(5)\\ &= 32 - 12.5\\ &= 19.5\end{aligned}$$

FINDING THE SLOPE OF A LINE

Slope is a number that represents the steepness of a line. To find the slope of a line, write the change in the *y*-coordinates of any two points on the line over the corresponding change in the *x*-coordinates of the same two points. Then simplify the fraction that results.

Slope Formula	Example
The slope of a nonvertical line that contains points $A(x_A,y_A)$ and $B(x_B,y_B)$ is given by the formula $\text{Slope} = \dfrac{\text{Vertical change}}{\text{Horizontal change}} = \dfrac{y_B - y_A}{x_B - x_A}$	To find the slope of the line that contains points (4,–1) and (7,5), let $(x_A,y_A) = (4,-1)$ and $(x_B,y_B) = (7,5)$. Then: $\text{Slope} = \dfrac{y_B - y_A}{x_B - x_A} = \dfrac{5-(-1)}{7-4} = \dfrac{5+1}{3} = 2$

SOME FACTS ABOUT SLOPE

- If you know the slope and the coordinates of one point of a line, you can find the coordinates of other points on the same line. For example, if the slope of a line is 2, then for each 1-unit increase in the x-coordinate of a point on this line, the corresponding y-coordinate must increase by 2. Hence, if (4,–1) is a point on a line with slope 2, then point $(4 + 1, -1 + 2) = (5,1)$ is on the same line.
- A line that *rises* as you look from left to right has a *positive* slope. Line p in Figure 6.24 has a positive slope.
- A line that *falls* as you look from left to right has a *negative* slope. Line n in Figure 6.24 has a negative slope.

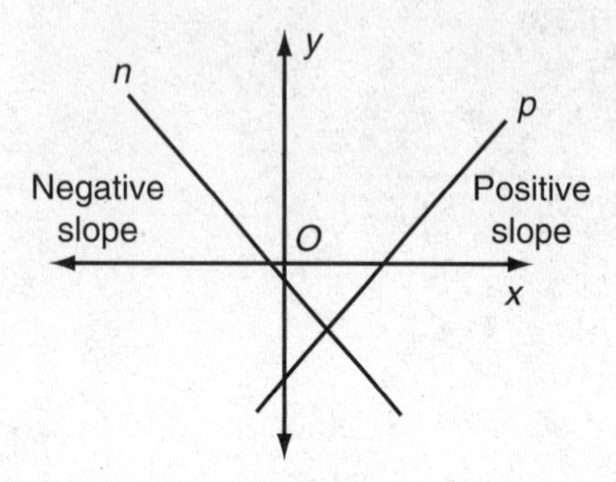

Figure 6.24 Lines with Positive and Negative Slopes

Figure 6.25 Lines with 0 and Undefined Slope

- A horizontal line has a slope of 0, and the slope of a vertical line is not defined. (See Figure 6.25.)
- The slope of a line does not change when different points on the same line are used to calculate the slope. If points A, B, and C lie on the same nonvertical line, then

$$\text{Slope of } AB = \text{Slope of } BC = \text{Slope of } AC$$

- Two nonvertical lines will intersect only if the lines have different slopes. **Lines that have the same slope are parallel.**
- Pairs of numbers such as $\frac{3}{4}$ and $\frac{-4}{3}$ are negative reciprocals because their product is –1. **Lines that have slopes that are negative reciprocals are perpendicular.**

EQUATION OF A LINE: $y = mx + b$

- If a line that passes through the origin contains the point (a,b), then the slope of the line is $\frac{b}{a}$.
- If the slope of line ℓ is m, then the slope of a line parallel to ℓ is also m, and the slope of a line perpendicular to ℓ is $-\frac{1}{m}$, provided $m \neq 0$.

If a nonvertical line has a slope of m and intersects the y-axis at $(0,b)$, then the equation $y = mx + b$ describes the set of all points (x,y) that the line contains. For example:

- If an equation of a line is $y = -5x + 3$, then $m = -5$ and $b = 3$, so the slope of the line is –5 and the line crosses the y-axis at (0,3).
- If an equation of a line is $y = 2x - 4$, then $m = 2$. The slope of a line parallel to the given line is also 2.

- If an equation of a line is $3x + y = 7$, then solving for y gives $y = -3x + 7$, so $m = -3$ and $b = 7$. The slope of a line perpendicular to the given line is the negative reciprocal of -3, which is $\frac{1}{3}$.

WRITING AN EQUATION OF A LINE: $y = mx + b$

The y-coordinate of the point at which a line crosses the y-axis is called the ***y*-intercept** of the line. If you know the slope (m) and the y-intercept (b) of a line, you can form its equation by writing $y = mx + b$.

- Suppose you know that the slope of a line is 2 and that the line contains the point $(-1,1)$. Since it is given that $m = 2$, the equation of the line must look like $y = 2x + b$. Because the line contains the point $(-1,1)$, when $x = -1$, $y = 1$:

$$\begin{aligned} y &= 2x + b \\ 1 &= 2(-1) + b \\ 1 &= -2 + b \\ 3 &= b \end{aligned}$$

  Hence, an equation of this line is $y = 2x + 3$.
- If you know that the line contains the points $(4,0)$ and $(-1,5)$, you can find its equation by first determining the slope of the line. If Δx represents the difference in the x-coordinates of the two points and Δy stands for the difference in the y-coordinates of these points, then:

$$m = \frac{\Delta y}{\Delta x} = \frac{5 - 0}{-1 - 4} = \frac{5}{-5} = -1$$

  Since $m = -1$, the equation of the line must look like $y = -x + b$. To find the y-intercept, b, substitute the coordinates of either given point into the equation. Since $y = 0$ when $x = 4$:

$$\begin{aligned} y &= -x + b \\ 0 &= -4 + b \\ 4 &= b \end{aligned}$$

  Hence, an equation of the line is $y = -x + 4$.

WRITING AN EQUATION OF A LINE FROM ITS GRAPH

You can determine the y-intercept and the slope of a line from its graph.

- To figure out the slope of the line shown in Figure 6.26, form a right triangle by moving 1 unit to the right from any point on the line, say $(0,3)$, and then moving up or down until the line is reached. The horizontal side of the right triangle is formed by moving 1 unit to the *right*, so

Memorize the formulas for midpoint, distance, and slope, as these formulas are *not* provided in the math reference section of the actual test. You are expected to be able to recall and apply these formulas, including in problems that involve forming or analyzing the equation of a line.

$\Delta x = +1$. The vertical side of the right triangle is formed by moving 4 units *up*, so $\Delta y = +4$. Hence, the slope of the line is

$$m = \frac{\Delta y}{\Delta x} = \frac{4}{1} = 4$$

Since $m = 4$ and $b = 3$, an equation of the line is $y = 4x + 3$.

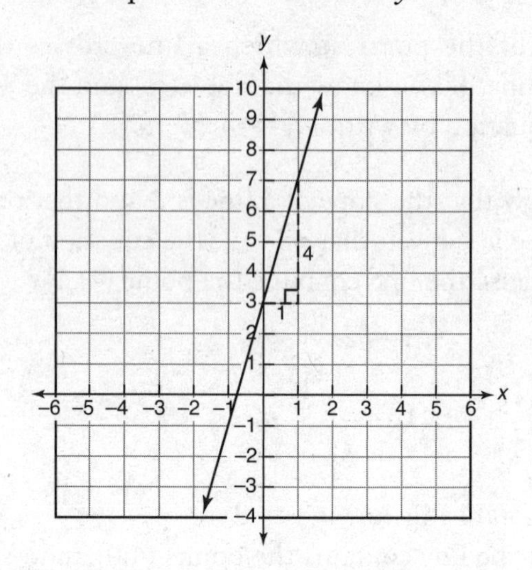

Figure 6.26 Determining the equation $y = 4x + 3$ from its graph

- To figure out the slope of the line shown in Figure 6.27, form a right triangle by moving 1 unit to the right from any point on the line, say (0,6), and then moving up or down until the line is reached. The horizontal side of the right triangle is formed by moving 1 unit to the *right*, so $\Delta x = +1$. The vertical side of the right triangle is formed by moving 3 units *down*, so $\Delta y = -3$. Hence, the slope of the line is

$$m = \frac{\Delta y}{\Delta x} = \frac{-3}{1} = -3$$

Since $m = -3$ and $b = 6$, an equation of the line is $y = -3x + 6$.

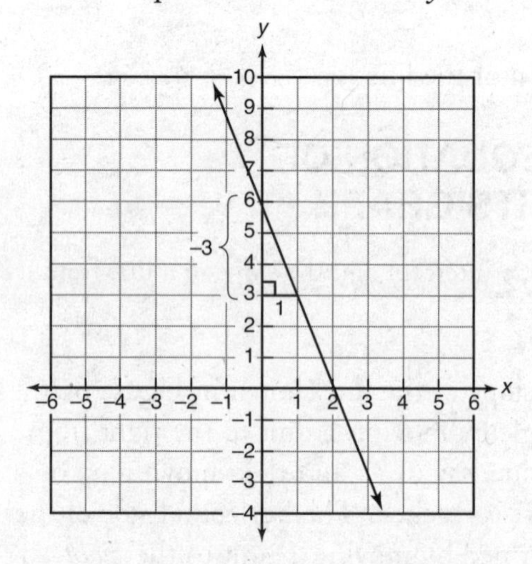

Figure 6.27 Determining the equation $y = -3x + 6$ from its graph

EXAMPLE

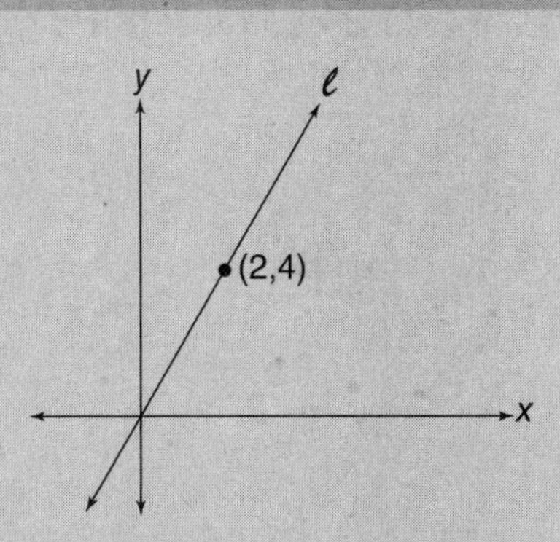

In the accompanying figure, line ℓ passes through the origin and the point (2,4). Line *m* (not shown) is perpendicular to line ℓ at (2,4). Line *m* intersects the *x*-axis at which point?

(A) (5,0)
(B) (6,0)
(C) (8,0)
(D) (10,0)
(E) (12,0)

Solution

The slope of line ℓ is $\dfrac{4-0}{2-0} = 2$

- Since the slopes of perpendicular lines are negative reciprocals, the slope of line *m* is $-\dfrac{1}{2}$.
  The equation of line *m* has the form $y = -\dfrac{1}{2}x + b$.
- Because line *m* contains the point (2,4), the coordinates of this point must satisfy its equation:

$$\begin{aligned} y &= -\frac{1}{2}x + b \\ 4 &= -\frac{1}{2}(2) + b \\ 4 &= -1 + b \\ 5 &= b \end{aligned}$$

  An equation of line *m* is $y = -\dfrac{1}{2}x + 5$.
- Line *m* intersects the *x*-axis where $y = 0$:

$$\begin{aligned} 0 &= -\frac{1}{2}x + 5 \\ -5 &= -\frac{1}{2}x \\ (-2)(-5) &= (-2)\left(-\frac{1}{2}x\right) \\ 10 &= x \end{aligned}$$

Since line *m* intersects the *x*-axis at (10,0), the correct choice is **(D)**.

Lesson 6-8 Tune-Up Exercises

Multiple-Choice

1 The length of the line segment whose endpoints are (3,–1) and (6,5) is

(A) 3
(B) 5
(C) $3\sqrt{5}$
(D) $5\sqrt{3}$
(E) $\sqrt{97}$

2 What is the area of a rectangle whose vertices are (–2,5), (8,5), (8,–2), and (–2,–2)?

(A) 45
(B) 50
(C) 55
(D) 60
(E) 70

3 What is the area of a parallelogram whose vertices are (–4,–2), (–2,6), (10,6), and (8,–2)?

(A) 32
(B) 48
(C) 72
(D) 96
(E) 104

4 What is the area of a triangle whose vertices are (–4,0), (2,4), and (4,0)?

(A) 8
(B) 12
(C) 16
(D) 32
(E) 64

5 If $A(-3,0)$ and $C(5,2)$ are the endpoints of diagonal AC of rectangle $ABCD$, with B on the x-axis, what is the perimeter of rectangle $ABCD$?

(A) 24
(B) 20
(C) 16
(D) 14
(E) 10

6

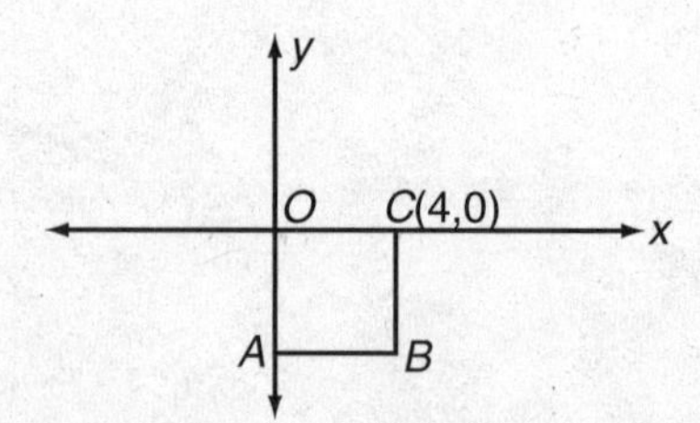

In the figure above, if A, B, C, and O are the vertices of a square and the coordinates of A are (k, p), what are the values of k and p?

(A) $k = -4$ and $p = 0$
(B) $k = 0$ and $p = -4$
(C) $k = -2$ and $p = 0$
(D) $k = 0$ and $p = -2$
(E) $k = 2$ and $p = -2$

7

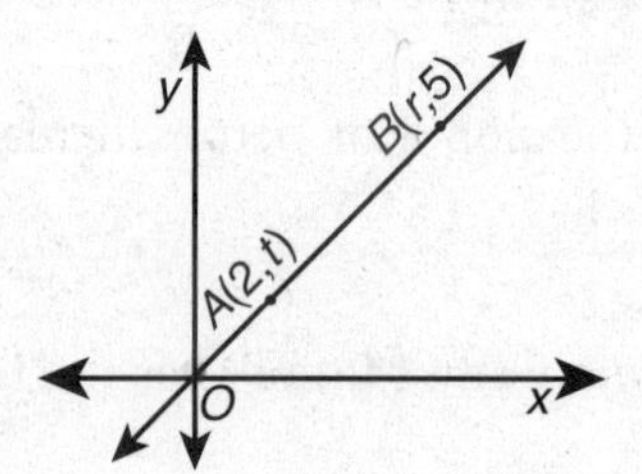

In the graph above, what is r in terms of t?

(A) $\frac{5}{2}t$

(B) $\frac{2}{5}t$

(C) $\frac{t}{10}$

(D) $10t$

(E) $\frac{10}{t}$

8

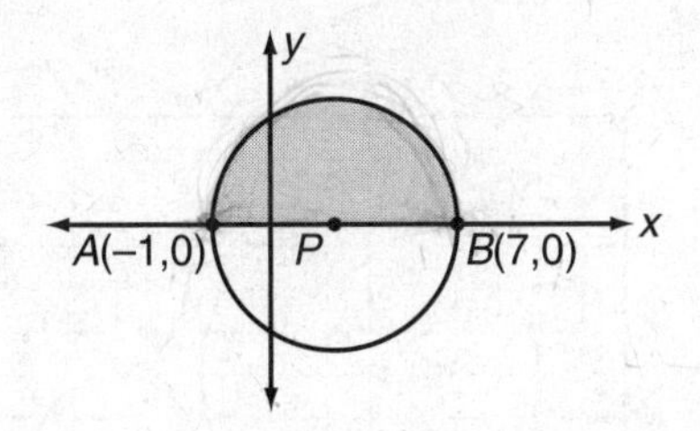

In the figure above, if AB is a diameter of circle P, what is the perimeter of the shaded region?

(A) $4\pi + 8$
(B) $8\pi + 4$
(C) $8\pi + 8$
(D) $16\pi + 4$
(E) $16\pi + 8$

9 If points $A(2,0)$ and $B(8,-4)$ are the endpoints of diameter AB of circle O, what is the area of circle O?

(A) 10π
(B) 13π
(C) 24π
(D) 26π
(E) 52π

10 The point whose coordinates are (4,–2) lies on a line whose slope is $\frac{3}{2}$. Which of the following are the coordinates of another point on this line?

(A) (1,0)
(B) (2,1)
(C) (6,1)
(D) (7,0)
(E) (1,4)

11 Line segment AB is a diameter of circle O. If the coordinates of O are (–2,1) and the coordinates of point A are (1,2), what are the coordinates of point B?

(A) (0,5)
(B) (–3,4)
(C) (–5,0)
(D) (1,–3)
(E) (–3,–1)

12 If point $E(5,h)$ is on the line that contains $A(0,1)$ and $B(-2,-1)$, what is the value of h?

(A) –1
(B) 0
(C) 1
(D) 3
(E) 6

13 If the line whose equation is $y = x + 2k$ passes through point (1,–3), then $k =$

(A) –2
(B) –1
(C) 1
(D) 2
(E) 4

14 A circle that has its center at the origin and passes through (–8,–6) will also pass through which point?

(A) (1,10)
(B) (4,9)
(C) (7,7)
(D) $(9,\sqrt{19})$
(E) $(\sqrt{37},8)$

15

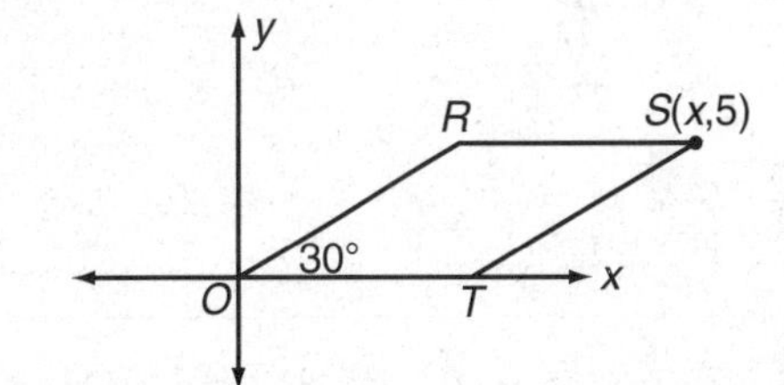

In the figure above, $ORST$ is a parallelogram with $OR = OT$. What is the perimeter of parallelogram $ORST$?

(A) 20
(B) 32
(C) $20\sqrt{3}$
(D) 40
(E) 64

16

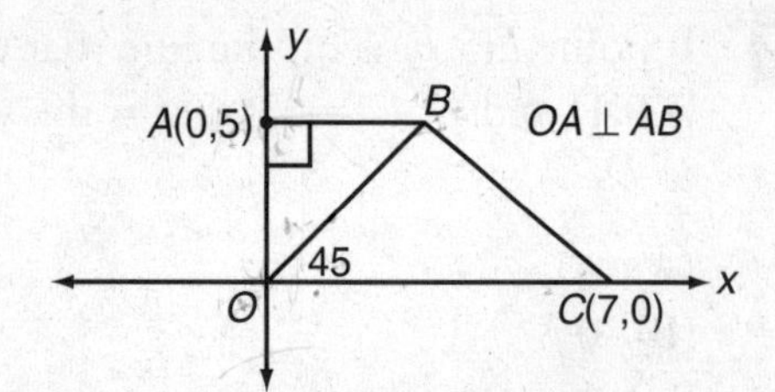

In the figure above, what is the area of quadrilateral $OABC$?

(A) 15
(B) 20
(C) 25
(D) 30
(E) 40

17

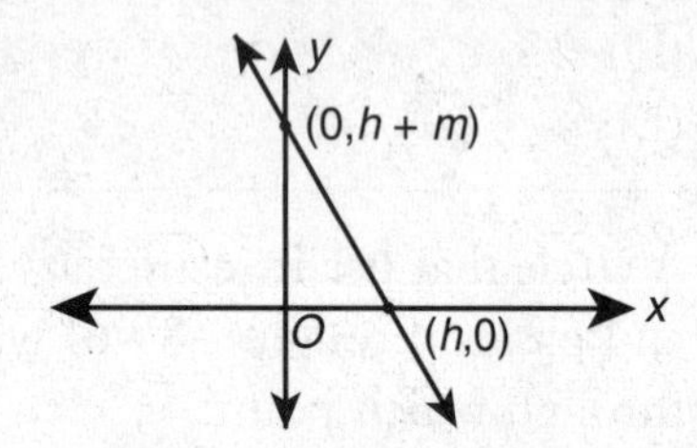

In the figure above, if the slope of line ℓ is m, what is m in terms of h?

(A) $\dfrac{h}{1+h}$

(B) $\dfrac{-h}{1+h}$

(C) $\dfrac{h}{1-h}$

(D) $1 + h$

(E) $1 - h$

18 Points $A(0,0)$, $B(5,8)$, and $C(10,4)$ lie in the coordinate plane. What is the ratio of the slope of $\overleftrightarrow{AB}$ to the slope of $\overleftrightarrow{AC}$?

(A) $-\dfrac{1}{2}$

(B) $-\dfrac{1}{4}$

(C) $\dfrac{1}{2}$

(D) 2

(E) 4

19

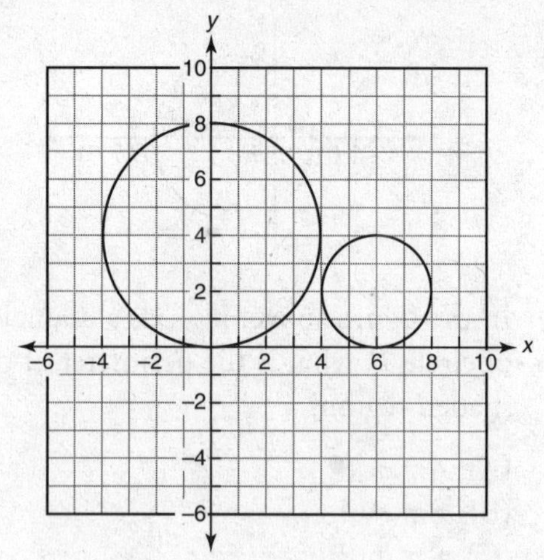

In the accompanying diagram, the diameter of the larger circle is 8 and the diameter of the smaller circle is 4. The circles are tangent to the x-axis at (0,0) and (6,0). What is the x-coordinate of the point at which the line (not shown) that contains the centers of the two circles intersects the x-axis?

(A) 10
(B) 11
(C) 12
(D) 13
(E) 14

20 The line $2y + x = b$ is perpendicular to a line that passes through the origin. If the two lines intersect at the point $(k + 2, 2k)$, what is the value of k?

(A) $-\dfrac{3}{2}$

(B) $-\dfrac{2}{3}$

(C) $\dfrac{2}{5}$

(D) $\dfrac{2}{3}$

(E) $\dfrac{3}{2}$

21 Which of the following is an equation of the line that is parallel to the line $y - 4x = 0$ and has the same y-intercept as the line $y + 3 = x + 1$?

(A) $y = 4x - 2$
(B) $y = 4x + 1$
(C) $y = -\frac{1}{4}x + 1$
(D) $y = -\frac{1}{4}x - 2$
(E) $y = -4x + 2$

22

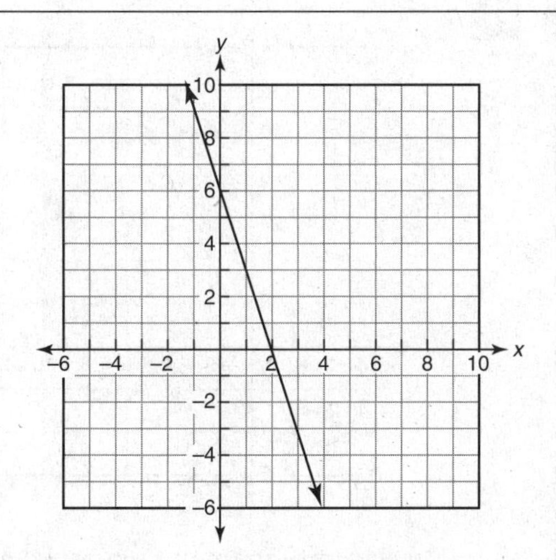

Which of the following is an equation of the line shown in the accompanying figure?

(A) $y = 3x + 6$
(B) $y = -3x - 6$
(C) $y = -3x + 6$
(D) $y = -6x + 3$
(E) $y = -3x + 2$

23

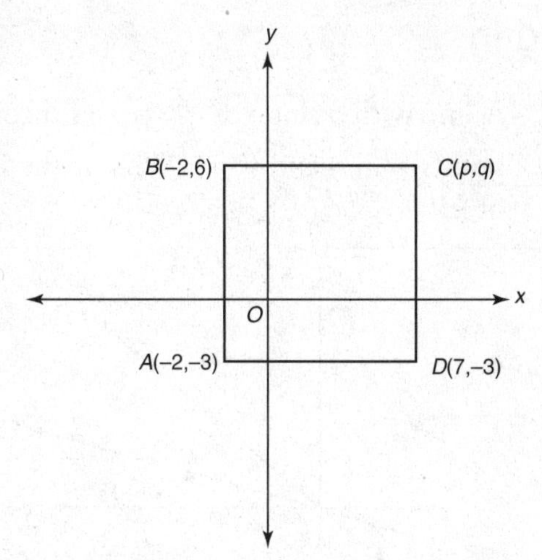

Which of the following is an equation of the line that contains diagonal $\overline{AC}$ of square $ABCD$ shown in the accompanying figure?

(A) $y = 2x + 1$
(B) $y = -x + 1$
(C) $y = \frac{1}{2}x - 2$
(D) $y = 2x - 8$
(E) $y = x - 1$

24

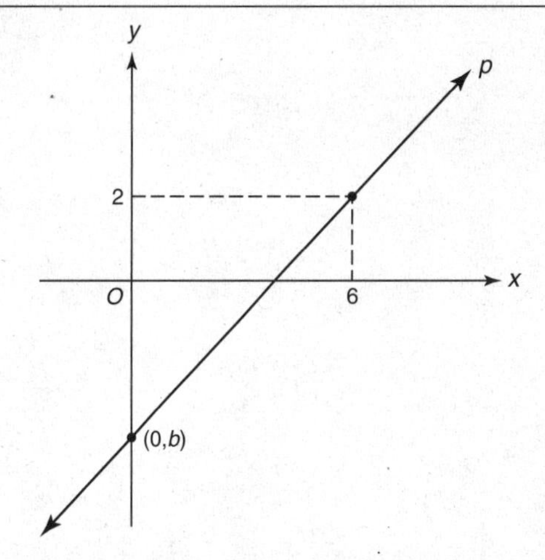

Note: Figure is not drawn to scale.

If the slope of line p shown in the figure above is $\frac{3}{2}$, what is the value of b?

(A) –8
(B) –7
(C) –5
(D) –3
(E) –2

Grid-In

1 A line with a slope of $\frac{3}{14}$ passes through points $(7,3k)$ and $(0,k)$. What is the value of k?

2

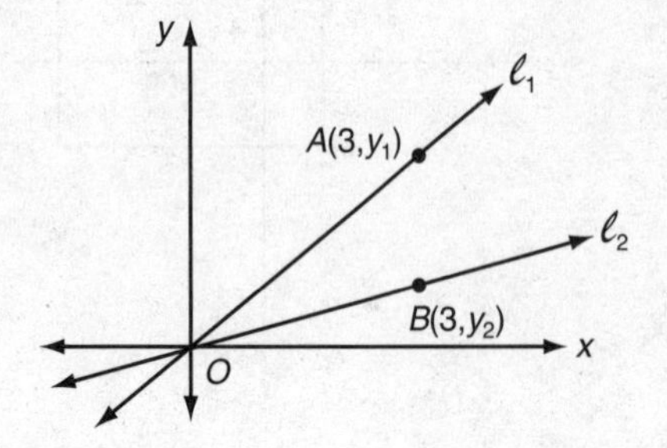

In the figure above, the slope of line ℓ_1 is $\frac{5}{6}$ and the slope of line ℓ_2 is $\frac{1}{3}$. What is the distance from point A to point B?

3

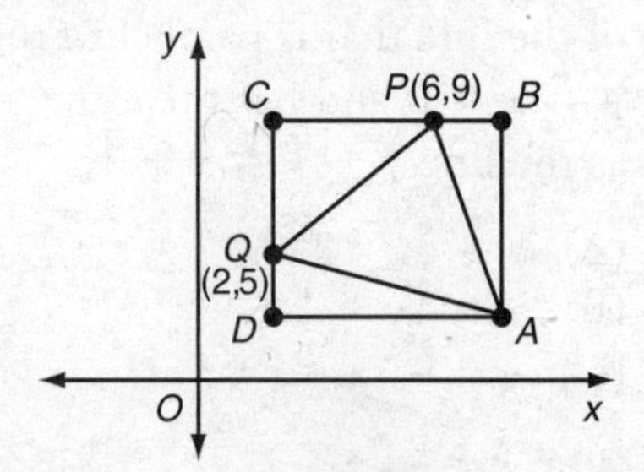

In the figure above, if the area of square $ABCD$ is 36, what is the area of $\triangle APQ$?

4

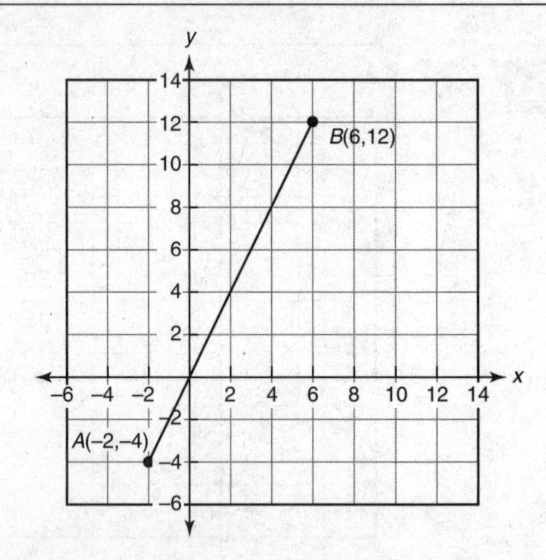

In the accompanying figure, what is the y-coordinate of the point at which the line that is perpendicular to $\overline{AB}$ (not shown) at its midpoint crosses the y-axis?

ANSWERS TO CHAPTER 6 TUNE-UP EXERCISES

Lesson 6-1 (*Multiple-Choice*)

1. **(B)** Since vertical angles are equal, the angle opposite the 40° angle also measures 40°. The measure of an exterior angle of a triangle is equal to the sum of the measures of the two nonadjacent interior angles of the triangle. Hence:
$$x° = 80° + 40° = 120°$$
and
$$y° = 40° + 70° = 110°$$
so
$$x + y = 120 + 110 = 230$$

2. **(B)** Since vertical angles are equal, the angle opposite the angle marked $3x°$ also measures $3x°$. Further, the angle opposite the angle marked $2x°$ also measures $2x°$. The sum of the measures of the angles about a point is 360°. Hence:
$$3x + 2x + x + 3x + 2x + y = 360$$
or $11x + y = 360$. Since vertical angles are equal, $y = x$. Thus:
$$11x + x = 360$$
$$12x = 360$$
$$x = \frac{360}{12} = 30$$
Hence, $y = x = 30$.

3. **(E)** Since $x + y + z$ forms a straight angle, $x + y + z = 180$. You are given that $\frac{y}{x} = 5$ and $\frac{z}{x} = 4$, so $y = 5x$ and $z = 4x$. Hence:
$$x + 5x + 4x = 180$$
$$10x = 180$$
$$x = \frac{180}{10} = 18$$

4. **(B)** First find the measures of the two angles that lie above ℓ_2 and that have the same vertex as angle x. The vertical angle opposite the 58° angle also measures 58°. The acute angle above line ℓ_2 that is adjacent to the 58° angle measures 37° since acute angles formed by parallel lines have equal measures. Since the sum of the measures of the angles that form a straight line is 180°,
$$37° + 58° + x° = 180°$$
so $x = 180 - 95 = 85$.

5. **(C)** If the angles of a triangle are in the ratio of $3 : 4 : 5$, let $3x$, $4x$, and $5x$ represent the measures of the three angles. Since the sum of the measures of the angles of a triangle is 180,
$$3x + 4x + 5x = 180$$
$$12x = 180$$
$$x = \frac{180}{12} = 15$$
Hence, the measure of the smallest angle of the triangle is $3x = 3(15) = 45$.

6. **(B)** Since $AB \perp BC$, $\angle ABD = 90$. Vertical angles have equal measures, so $\angle DAB = 32$. Since the measures of the acute angles of a right triangle must add up to 90,
$$(x + 14) + (x + 32) = 90$$
$$2x + 46 = 90$$
$$x = \frac{44}{2} = 22$$

7. **(A)** Since $\ell_1 \parallel \ell_2$, $w + x = 180$ and $y = z$. Adding corresponding sides of these two equations gives
$$w + x + y = 180 + z$$
or $x + y = 180 - w + z$.

8. **(B)** The measures of the three interior angles of the triangle are $(180 - x)$, $(180 - y)$, and z. Since the sum of the measures of the angles of a triangle is 180,
$$(180 - x) + (180 - y) + z = 180$$
$$360 - x - y + z = 180$$
$$360 + z = 180 + x + y - 360$$
$$= x + y - 180$$

9. **(D)** Angles x and y form a straight line, so $x + y = 180$. Since $\ell_1 \parallel \ell_2$, obtuse angle y equals obtuse angle $3x$. Substituting $3x$ for y in $x + y = 180$ gives $4x = 180$, so $\frac{180}{4} = 45$. Hence:
$$y = 3x = 3(45) = 135$$

10. **(C)** Since $\ell_1 \parallel \ell_2$, the acute angle formed by lines ℓ_2 and ℓ_3 equals acute angle x. Also, since $\ell_3 \parallel \ell_4$, the obtuse angle whose measure is $y + y$ is supplementary to acute angle x, so $x + 2y = 180$. Solving this equation for y gives $2y = 180 - x$ or
$$y = \frac{180}{2} - \frac{x}{2} = 90 - \frac{x}{2}$$

11. **(C)** Since the measures of vertical angles are equal, the angle opposite the $(3x + 10)°$ angle also measures $(3x + 10)°$. Since the angles marked $x°$, $(3x +10)°$, and $90°$ (by the right-angle mark) form a straight line:

$$x^{\circ} + (3x + 10)^{\circ} + 90^{\circ} = 180^{\circ}$$
$$4x + 100 = 180$$
$$x = \frac{80}{4} = 20$$

12. **(D)**

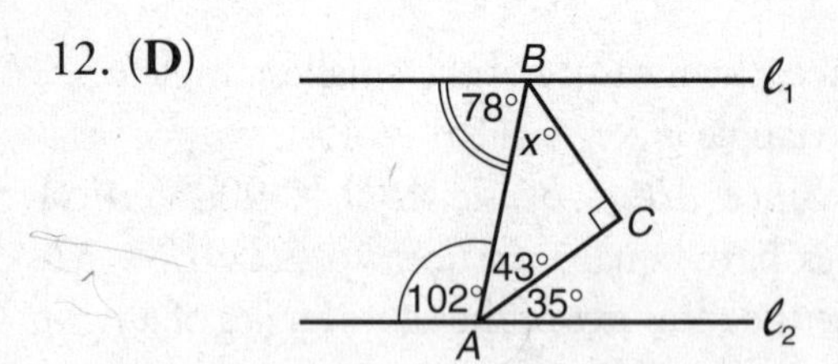

Suppose $\ell_1 \parallel \ell_2$. Then the obtuse angle formed at vertex A is supplementary to the acute 78° angle formed at vertex B, so it measures $180° - 78°$ or $102°$. Hence, the measure of $\angle BAC$ is $180° - 102° - 35°$ or $43°$. Since the acute angles of a right triangle are supplementary,
$$x = 90 - 43 = 47$$

13. **(A)** Since $\ell \parallel m$, the acute angles marked $3y$ and $2y + 25$ must have the same measure, so $3y = 2y + 25$ and $y = 25$. Then
$$2y + 25 = 2(25) + 25 = 75$$
Since vertical angles have the same measures, $x + 15 = 75$, so $x = 60$.

14. **(D)** Since the sum of the measures of the angles about a point is 360, the sum of the measures of the marked angles and the unmarked angles at the four vertices is 4×360 or 1440. The unmarked angles are the interior angles of the two triangles, so their sum is 2×180 or 360. Hence, the sum of the measures of the marked angles is $1440 - 360$ or 1080.

15. **(A)** The measure of $\angle ACE$ is $x + 2x$ or $3x$ since it is an exterior angle of $\triangle ABC$. Also, $3y = 3x + y$ since the angle marked $3y°$ is an exterior angle of the triangle in which $3x°$ and $y°$ are the measures of the two nonadjacent interior angles. Hence, $2y = 3x$, so $y = \frac{3}{2}x$.

16. **(C)** Since $\ell_1 \parallel \ell_2$, the obtuse angle that measures $a + a$ or $2a°$ is supplementary to the acute angle that measures $b + b$ or $2b°$. Hence, $2a + 2b = 180$, so $a + b = 90$. The angle that measures $x + x$ or $2x°$ is an exterior angle of the triangle in which the two nonadjacent angles are marked $a°$ and $b°$. Hence, $2x = a + b = 90$, so $x = \frac{90}{2} = 45$.

17. **(D)** Since the angles marked $7y°$ and $5y°$ form a straight line:
$$7y + 5y = 180$$
$$12y = 180$$
$$y = \frac{180}{12} = 15$$
An exterior angle of a triangle is equal to the sum of the measures of the two nonadjacent angles. Hence, $7y = 3y + x$ or $x = 4y$. Since $y = 15$, then $x = 4(15) = 60$.

18. **(D)** Since acute angle ABD is supplementary to obtuse angle CDB:
$$(y + 2y + y) + 5y = 180$$
$$9y = 180$$
$$y = \frac{180}{9} = 20$$

19. **(E)** In the smaller triangle, $a + b + 140 = 180$, so $a + b = 40$. In $\triangle RST$, $x + 2a + 2b = 180$. Dividing each member of this equation by 2 gives $\frac{x}{2} + a + b = 90$. Since $a + b = 40$, then
$$\frac{x}{2} + 40 = 90$$
$$\frac{x}{2} = 50$$
$$x = 100$$

20. **(E)** Determine whether each Roman numeral statement is always true.
   - I. Since the measures of vertical angles are equal, $a = d$. Since we cannot tell whether $b = c$, we do not know whether statement I, $a + b = d + c$, is always true.
   - II. Since angles a, b, and c form a straight line, $a + b + c = 180$. Angles b and e are vertical angles, so $b = e$. Substituting e for b in $a + b + c = 180$ gives $a + c + e = 180$, so statement II is always true.
   - III. Angles b and e are vertical angles, as are f and c. Since $b = e$ and $f = c$, then $b + f = c + e$, so statement III is always true.

Only Roman numeral statements II and III are always true.

(Grid-In)

1. **20** The measures of vertical angles are equal, so $\angle EFC = 60$. In right triangle *CEF*, the measures of the acute angles add up to 90, so $\angle ECF + 60 = 90$ or
$$\angle ECF = 90 - 60 = 30$$
Since the measures of acute angles formed by parallel lines are equal, $y + \angle ECF = 50$. Hence, $y + 30 = 50$, so $y = 20$.
2. **35** In $\triangle ABC$,
$$\angle ACB = 180 - 25 - 45 = 110$$
Since angles *ACB* and *DCE* form a straight line,
$$\angle DCE = 180 - 110 = 70$$
Angle *BED* is an exterior angle of $\triangle ECD$. Hence:
$$3x = 70 + x$$
$$2x = 70$$
$$x = \frac{70}{2} = 35$$
3. **40** Since *OP* and *OQ* are rotating in opposite directions, for the least value of t for which these segments coincide, $4t + 5t$ represents one complete revolution or 360°. Thus:
$$4t + 5t = 360$$
$$9t = 360$$
$$t = \frac{360}{9} = 40 \text{ seconds}$$
4. **10** Change 1 hour to 3600 seconds. Segment *OP* turns $\frac{4°}{\text{sec}} \times 3600\,\text{sec} = 14{,}400°$, which

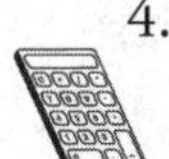

is equivalent to $\frac{14{,}400°}{360°} = 40$ revolutions.
Since *OQ* turns $\frac{5°}{\text{sec}} \times 3600\,\text{sec} = 18{,}000°$ and $\frac{18{,}000°}{360°} = 50$ revolutions, *OQ* completes $50 - 40 = 10$ more revolutions than *OP*.

Lesson 6-2 (*Multiple-Choice*)

1. **(B)** Since $\triangle RST$ is a right triangle,
$$(RS)^2 + (\sqrt{3})^2 = (\sqrt{7})^2$$
$$(RS)^2 + 3 = 7$$
$$RS = \sqrt{4} = 2$$
2. **(E)** You are given that the measures of $\angle A$ and $\angle B$ are in the ratio of 2 to 3. Let $2y$ represent the measure of $\angle A$, and $3y$ represent the measure of $\angle B$. Since the measures of the acute angles of a right triangle add up to 90,
$$2y + 3y = 90$$
$$5y = 90$$
$$y = \frac{90}{5} = 18$$
Hence, $\angle B = 3y = 3(18) = 54$. In $\triangle CDB$, $CD = BD$ so $\angle DCB = \angle B = 54$. Since the sum of the measures of the three angles of a triangle is 180,
$$x = 180 - 54 - 54 = 72$$
3. **(D)** In $\triangle JKL$:
$$\angle L = 180 - 50 - 65 = 65$$
Since $\angle K = 65$ and $\angle L = 65$, then $\angle K = \angle L$, so the sides opposites these angles, *JK* and *JL*, must be equal in length. Hence:
$$JK = JL$$
$$3x - 2 = x + 10$$
$$2x = 12$$
$$x = \frac{12}{2} = 6$$
4. **(B)** If two angles of a triangle each measure 60°, then the third angle also measures 60° because $180 - 60 - 60 = 60$. Since an equiangular triangle is also equilateral, the length of each side of the acute triangle in the figure is 10. Since the hypotenuse of the right triangle in the figure is 10, the lengths of the sides of the right triangle form a (6, 8, 10) Pythagorean triple, where $x = 6$.
5. **(D)** Since you are told that $\triangle ABC$ is equilateral, each angle of the triangle measures 60°. Hence:
$$\angle B = x + 2x + x = 4x = 60$$
so $x = \frac{60}{4} = 15$. The angle that measures $y°$ is an exterior angle of $\triangle BEC$, so
$$y = x + \angle C = 15 + 60 = 75$$
6. **(E)** If $AC = BC$, then $\angle A = \angle B = 2x + y$. Since the angle that measures $3y - x$ degrees is an exterior angle of $\triangle ABC$,
$$3y - x = (2x + y) + (2x + y)$$
$$= 4x + 2y$$
$$3y - 2y = 4x + x$$
$$y = 5x$$
7. **(C)** Since $\angle D = 108$, the measures of angles *DAC* and *DCA* must add up to 72 since

$180 - 108 = 72$. Since DA bisects $\angle BAC$ and DC bisects $\angle BCA$, the sum of the measures of $\angle BAC$ and $\angle BCA$ is 2×72 or 144. Hence, $x = 180 - 144 = 36$.

8. **(B)** Since the sum of the measures of the angles that have C as a vertex is 180, $\angle ECD = 60$, so $\angle E = 30$. Hence, CD, the side opposite the 30° angle in a 30°-60° right triangle, is $\frac{1}{2} \times 8$ or 4 so $BC = 10 - 4 = 6$. In right triangle ABC, $\angle A = 60$. Since BC is the side opposite the 60° angle, it must be $\sqrt{3}$ times as long as AB. Hence, $AB = \frac{6}{\sqrt{3}}$ because $\frac{6}{\sqrt{3}} \times \sqrt{3} = 6$.

   Since $\frac{6}{\sqrt{3}}$ is not one of the choices, simplify:

$$\frac{6}{\sqrt{3}} = \frac{6}{\sqrt{3}} \times \frac{\sqrt{3}}{\sqrt{3}} = \frac{6\sqrt{3}}{3} = 2\sqrt{3}$$

9. **(D)** The lengths of the sides of right triangle AEB form a (5, *12*, 13) Pythagorean triple, where $BE = 12$. Since $BC \parallel AD$, the distance between these segments must always be the same, so $CF = BE = 12$. In a 45°-45° right triangle, the hypotenuse is $\sqrt{2}$ times the length of either leg. Hence:

$$CD = CF \times \sqrt{2} = 12\sqrt{2}$$

10. **(C)** Let the length of WS be any convenient number. If $WS = 1$, then $TS = 1 \times \sqrt{3} = \sqrt{3}$. In right triangle RST, RS is the side opposite the 60° angle, so

$$RS = \sqrt{3} \times TS = \sqrt{3} \times \sqrt{3} = 3$$

   Hence, $RW = RS - WS = 3 - 1 = 2$. Since $RW = 2$ and $WS = 1$, the ratio of RW to WS is 2 to 1.

11 **(C)** Draw a diagram in which the four key points on Katie's trip are labeled A through D.

- To determine how far, in a straight line, Katie is from her starting point at A, find the length of $\overline{AD}$.

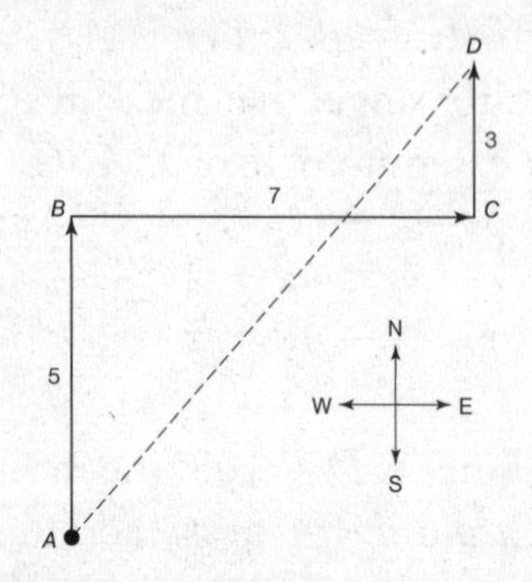

- Form a right triangle in which $\overline{AD}$ is the hypotenuse by completing rectangle $BCDE$, as shown in the accompanying diagram.

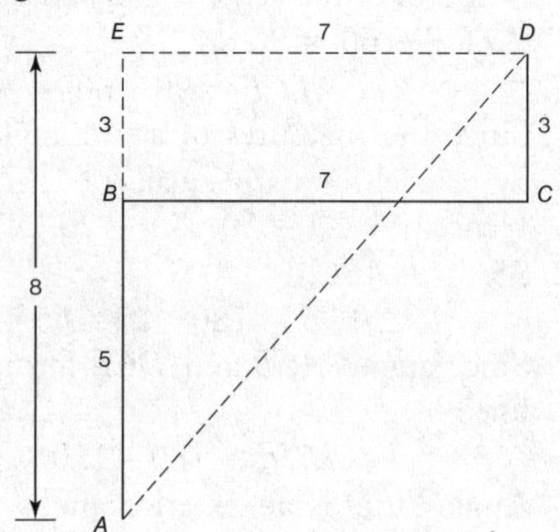

- Because opposite sides of a rectangle have the same length, $ED = BC = 7$, and $BE = CD = 3$. Thus, $AE = 5 + 3 = 8$.
- Since AD is the hypotenuse in right triangle AED,

$$\begin{aligned}(AD)^2 &= (AF)^2 + (FD)^2 \\ &= 8^2 + 7^2 \\ &= 64 + 49 \\ &= 113 \\ AD &= \sqrt{113}\end{aligned}$$

(Grid-In)

1. **1.5** Angle B measures $15 + 30 + 15$ or 60°, so the sum of the measures of angles A and C is 120. Since $AB = BC = 10$, then $\angle A = \angle C = 60$, so $\triangle ABC$ is equiangular. A triangle that is equiangular is also equilateral, so $AC = 10$. Angles BDE and BED each measure $60 + 15$ or 75° since they are exterior angles of triangles ADB and CEB. Hence triangles ADB and CEB have the same shape and size, so $AD = CE$. Since you are given that $DE = 7$, then $AD + CE = 3$, so $AD = 1.5$.

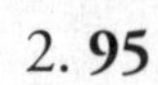

2. **95**

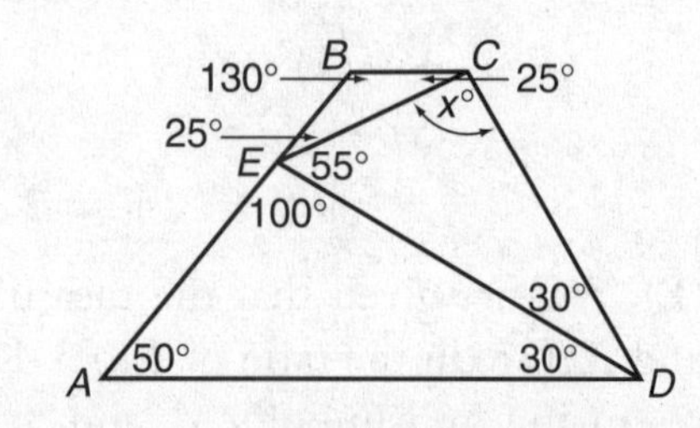

The measures of acute angle A and obtuse angle B must add up to 180 since the angles are formed by parallel lines. Since $\angle A = 50$, then $\angle B = 130$. You are told that $BE = BC$, so $\angle BEC = \angle BCE$. In $\triangle BEC$, the measures of angles BEC and BCE must add up to 50, so $\angle BEC = \angle BCE = 25$. Since the three adjacent angles about point E form a straight angle, $25 + 55 + \angle AED = 180$, so $\angle AED = 100$. In $\triangle AED$,

$$\angle ADE = 180 - 100 - 50 = 30$$

Since you are also told that ED bisects $\angle ADC$, $\angle CDE = \angle ADE = 30$. In $\triangle CED$,

$$x = 180 - 55 - 30 = 95$$

3. **5** Draw a diagram in which the starting point is labeled point X.

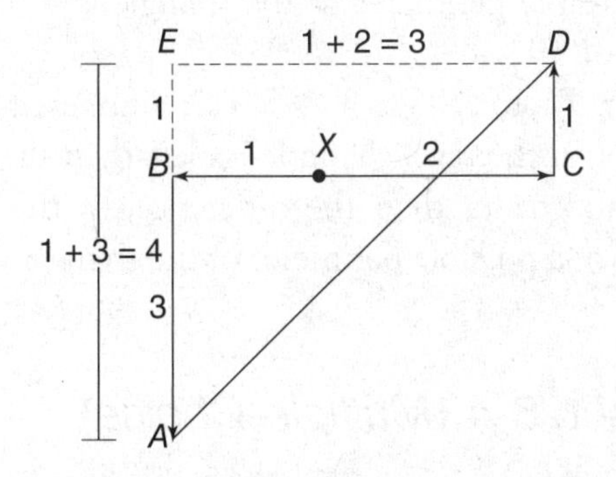

- Label the diagram with the given information. Complete rectangle $BCDE$.
- Because opposite sides of a rectangle have the same length, $ED = BC = 3$, and $BE = CD = 1$. Thus, $AE = 1 + 3 = 4$.
- Since AD is the hypotenuse of a right triangle in which the legs measure 3 and 4, $AD = 5$.

Lesson 6-3 (*Multiple-Choice*)

1. **(E)** In obtuse triangle RST, $RT = TS$. If the obtuse angle were opposite one of the equal sides, there would be another obtuse angle opposite the other equal side, an impossible case. Hence, the obtuse angle must be opposite RS.

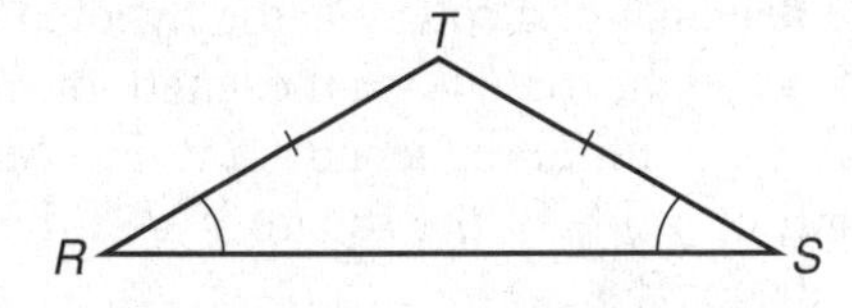

Determine whether each Roman numeral statement is always true.

- I. Since $RT = TS$, the angles opposite these sides are equal, so $\angle R = \angle S$. Hence, statement I is always true.
- II. The obtuse angle must be the angle opposite the unequal side. Since $\angle T$ is opposite side RS, $\angle T$ is obtuse. Hence, statement II is always true.
- III. Since RS is opposite obtuse angle T, it is the longest side of the triangle. Hence, $RS > TS$ and statement III is always true.

Roman numeral statements I, II, and III are always true.

2. **(D)** In a triangle the length of any side is less than the sum of the lengths of the other two sides. If the lengths of two sides are 5 and 9, and the length of the third side is x, then
   - $x < 5 + 9$ or $x < 14$
   - $5 < x + 9$
   - $9 < x + 5$ or $4 < x$

   Since $x < 14$ and $4 < x$, $4 < x < 14$.
3. **(B)** You are given that, in $\triangle ABC$, $BC > AB$ and $AC < AB$. Hence, $BC > AB > AC$. The measures of the angles opposite these sides are ordered in the same way. Hence, $\angle A$ (opposite BC) $> \angle C$ (opposite AB) $> \angle B$ (opposite AC).
4. **(A)** In $\triangle ABC$, if $AB = BD$, then $\angle A = x$. Determine whether each Roman numeral statement is always true.
   - I. You are not given any information that tells you whether $x > z$, $x < z$, or $x = z$. Statement I may or may not be true.
   - II. Since an exterior angle of a triangle is greater than either nonadjacent interior angle of the triangle, $y > \angle A$. Since $\angle A = x$, then $y > x$, so statement II is always true.
   - III. Angle x is an exterior angle of $\triangle CBD$, so $x > \angle C$. Since $\angle A = x$, then $\angle A > \angle C$, so BC (the side opposite $\angle A$) $> AB$ (the side opposite $\angle C$). Since it is not true that $AB > BC$, statement III is false. Hence, only Roman numeral statement II is always true.
5. **(A)** Since n is an integer and $3 < n < 8$, n may be equal to 4, 5, 6, or 7. Test whether a triangle can be formed for each possible value of n.

- If $n = 4$, then $4 < 3 + 8$ and $3 < 4 + 8$, but 8 is not less than $3 + 4$, so $n \neq 4$.
- If $n = 5$, then $5 < 3 + 8$ and $3 < 5 + 8$, but 8 is not less than $3 + 5$, so $n \neq 5$.
- If $n = 6$, then $6 < 3 + 8$, $3 < 6 + 8$, *and* $8 < 3 + 6$, so n may be equal to 6.
- If $n = 7$, then $7 < 3 + 8$, $3 < 7 + 8$, *and* $8 < 3 + 7$, so n may be equal to 7.

Hence, two different triangles are possible.

6. **(C)** Draw $\triangle RST$ so that $RS \perp TS$, $\angle T$ measures 40°, and W is a point on side RT such that $\angle RWS$ measures 100°.

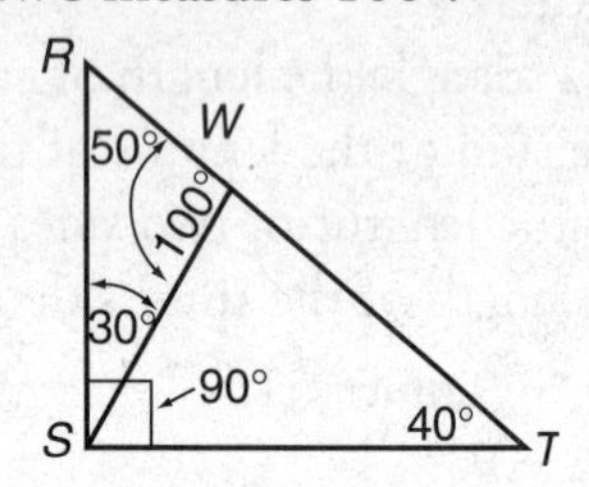

- In $\triangle RWS$, $\angle RSW = 180 - 50 - 100 = 30$. Since $\angle RSW$ is the smallest angle of $\triangle RWS$, then $RW < RS$ and $RW < SW$. Hence, eliminate choices (A) and (E).
- Since $\angle T < \angle R$, $RS < ST$. Since $RW < RS$ and $RS < ST$, then $RW < ST$, so you can also eliminate choice (B).
- Since $\angle TSW = 90 - 30 = 60$, then, $\angle T < \angle TSW$. Hence, in $\triangle TSW$, $SW < TW$. Since $RW < SW$ and $SW < TW$, then $RW < TW$, so you can also eliminate choice (D). Hence, RW is the shortest segment.

(*Grid-In*)

1. **37** If the lengths of two sides of an isosceles triangle are 7 and 15, then the third side must be 7 or 15. Since 15 is not less than $7 + 7$, the third side cannot be 7. Hence, the lengths of the three sides of the triangle must be 7, 15, and 15. The perimeter of this triangle is $7 + 15 + 15$ or 37.
2. **3** You are told that the perimeter of a triangle with integer side lengths is 21 and that the length of one side is 8. Hence, the sum of the lengths of the other two sides must be 13. Test possible combinations of integer side lengths to find one in which the lengths of two of the sides add up to 13 and the third side is 8.
   - Suppose the lengths of the three sides are 1, 12, and 8. This combination is not possible since 12 is not less than $1 + 8$.
   - Suppose the lengths of the three sides are 2, 11, and 8. This combination is not possible since 11 is not less than $2 + 8$.
   - Suppose the lengths of the three sides are 3, 10, and 8. This combination is possible since $3 < 10 + 8$, $10 < 3 + 8$, and $8 < 3 + 10$.

   Hence, the shortest possible length of a side is 3.
3. **19** In the given triangle, $9 - 4 < x < 9 + 4$ or, equivalently, $5 < x < 13$. Since the smallest possible integer value of x is 6, the least possible perimeter of the triangle is $4 + 6 + 9$ or 19.
4. **15** Factor 105 as $3 \times 5 \times 7$. Since each number in the set 3, 5, and 7 is less than the sum, and greater than the difference, of the other two, a possible perimeter of the triangle is $3 + 5 + 7 = 15$.

Lesson 6-4 (*Multiple-Choice*)

1. **(C)**

12
$3 \cdot 3 = 9$
$3 \cdot 5 = 15$
9
$3 \cdot 4 = 12$

The diagonal of a 9 by 12 rectangle is the hypotenuse of a (9, 12, *15*) right triangle. Since the two diagonals of a rectangle have the same length, the sum of the lengths of the diagonals of a 9 by 12 rectangle is $15 + 15$ or 30.

2. **(C)**

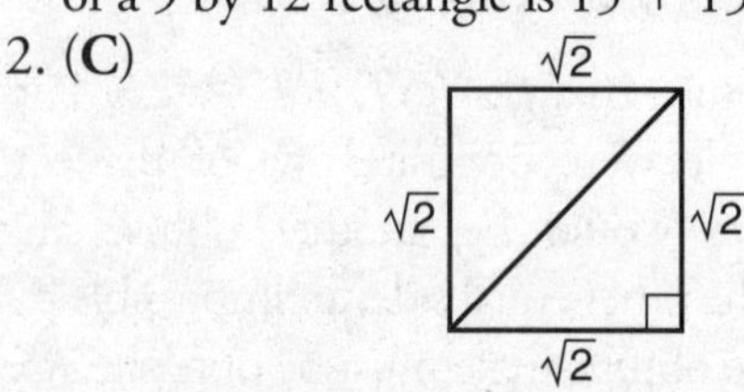

The diagonal of a square is the hypotenuse of a 45°–45° right triangle, so the length of the diagonal is $\sqrt{2}$ times the length of a side. Since the length of a side of the square is $\sqrt{2}$, the length of a diagonal is $\sqrt{2} \times \sqrt{2}$ or 2.

3. **(E)**

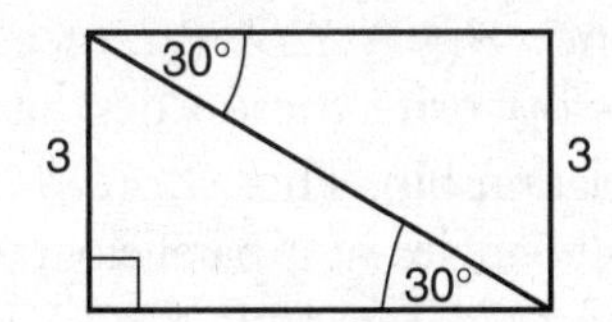

If a diagonal of a rectangle forms a 30° angle with each of the longer sides of the rectangle, the shorter side is the leg of a right triangle that is opposite the 30° angle. Since the length of the shorter side is 3, the length of the diagonal, or hypotenuse of the right triangle, is 2×3 or 6.

4. **(C)** The sum of the measures of the four angles of a quadrilateral is 360. If the degree measures of the angles of a quadrilateral are $4x$, $7x$, $9x$, and $10x$, then

$$\begin{aligned} 4x + 7x + 9x + 10x &= 360 \\ 30x &= 360 \\ x &= \frac{360}{30} = 12 \end{aligned}$$

The degree measure of the smallest angle is $4x = 4(12) = 48$, and the degree measure of the largest angle is $10x = 10(2) = 120$. Hence, the sum of the degree measures of the smallest and largest angles of the quadrilateral is $48 + 120$ or 168.

5. **(E)** From point A, draw a perpendicular to side CD. Call the point of intersection E.

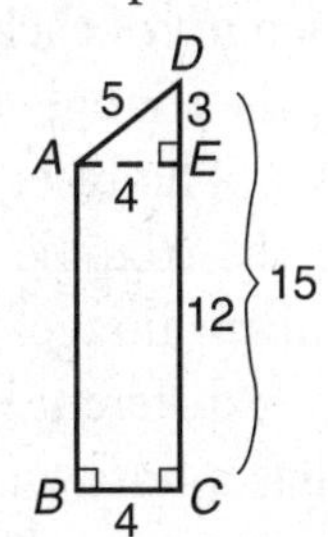

Since $ABCE$ is a rectangle, $AE = BC = 4$. The lengths of the sides of right triangle AED form a (3, 4, 5) Pythagorean triple in which $DE = 3$, so

$$AB = EC = 15 - 3 = 12$$

6. **(E)** Since the right angles and the vertical angles at E are equal, triangles AEC and BED are similar. The lengths of corresponding sides of similar triangles are in proportion. Let x represent the length of AE. Then

$$\begin{aligned} \frac{AC}{DB} &= \frac{AE}{BE} \\ \frac{3}{4} &= \frac{x}{14 - x} \\ 4x &= 3(14 - x) \\ &= 42 - 3x \\ 7x &= 42 \\ x &= \frac{42}{7} = 6 \end{aligned}$$

Hence, $AE = 6$.

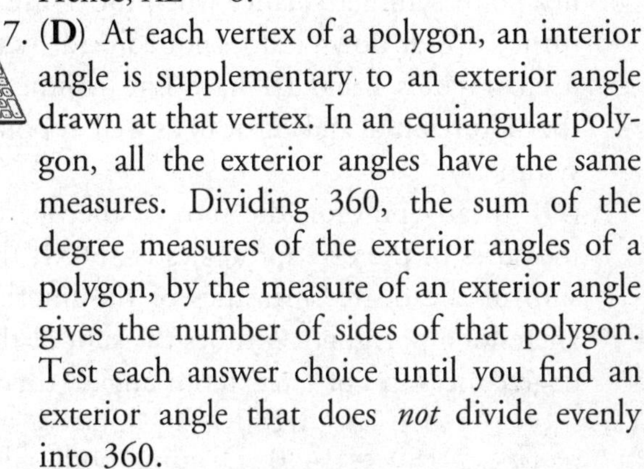

7. **(D)** At each vertex of a polygon, an interior angle is supplementary to an exterior angle drawn at that vertex. In an equiangular polygon, all the exterior angles have the same measures. Dividing 360, the sum of the degree measures of the exterior angles of a polygon, by the measure of an exterior angle gives the number of sides of that polygon. Test each answer choice until you find an exterior angle that does *not* divide evenly into 360.

- (A): The measure of an exterior angle is $180 - 165$ or 15 and $\frac{360}{15} = 24$, so the polygon has 24 sides.
- (B): The measure of an exterior angle is $180 - 162$ or 18 and $\frac{360}{18} = 20$, so the polygon has 20 sides.
- (C): The measure of an exterior angle is $180 - 140$ or 40 and $\frac{360}{40} = 9$, so the polygon has 9 sides.
- (D): The measure of an exterior angle is $180 - 125$ or 55, but $\frac{360}{55}$ does not give an integer answer.

Hence, a polygon in which each interior angle measures 125 is not possible.

8. **(A)** Since consecutive angles of a parallelogram are supplementary,

$$\begin{aligned} 3x + 2x &= 180 \\ 5x &= 180 \\ x &= \frac{180}{5} = 36 \end{aligned}$$

Since opposite angles of a parallelogram are equal,

$$y = 3x = 3(36) = 108$$

9. **(A)** Since side AC of the rotated square cannot have the same length as a diagonal of the

original square that contains AC as a side, Figure II *cannot* represent a rotation of square $ABCD$ about point A. Figure III is not correct since if $x = 180$, then

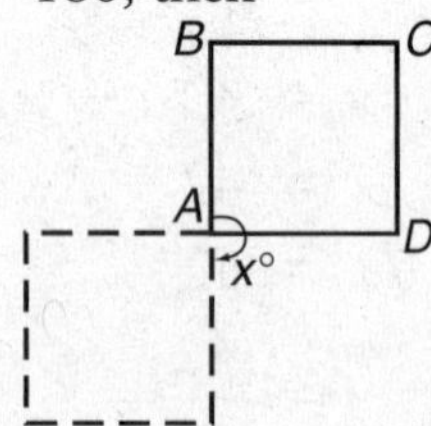

10. **(D)** Figure II has vertical line symmetry but not point symmetry, since when the figure is turned upside down it does not coincide with itself. Figures I and III have line symmetry (both horizontal and vertical) as well as point symmetry.
11. **(D)** In any polygon the sum of the degree measures of the exterior angles is 360. If the sum of the degree measures of the interior angles of a polygon is 4 times the sum of the degree measures of the exterior angles, then
$$S = (n - 2)180 = 4 \times 360$$
or $(n - 2)180 = 1440$. Dividing both sides of the equation by 180 gives
$$(n - 2) = \frac{1440}{180} = 8$$
Since $n - 2 = 8$, then $n = 8 + 2$ or 10.
12. **(E)** You are given that, in quadrilateral $ABCD$, $\angle A + \angle C$ is 2 times $\angle B + \angle D$, and $\angle A = 40$, so
$$40 + \angle C = 2(\angle B + \angle D)$$
Since the sum of the degree measures of the four angles of a quadrilateral is 360,
$$40 + \angle C + (\angle B + \angle D) = 360$$
Substituting $2(\angle B + \angle D)$ for $40 + \angle C$ gives
$$2(\angle B + \angle D) + (\angle B + \angle D) = 360$$
Since the last equation contains two unknowns, the measures of angles B and D, it is not possible to find the measure of $\angle B$ (or $\angle D$).
13. **(C)** Determine whether each Roman numeral statement is true, given that $ABCD$ is a parallelogram and $AB > BD$.
   - I. Since opposite sides of a parallelogram have the same length, $CD = AB$. Substituting CD for AB in $AB > BD$ gives $CD > BD$. Hence, statement I is false.
   - II. Since $AB > BD$, the measures of the angles opposite these sides have the same size relationship. Thus, $\angle ADB > \angle A$. Since opposite angles of a parallelogram are equal, $\angle A = \angle C$, so $\angle ADB > \angle C$. Hence, statement II is true.
   - III. Acute angles formed by parallel lines are equal, so $\angle CBD = \angle ADB$. Since $\angle ADB > \angle A$, then $\angle CBD > \angle A$, so statement III is true.

   Only Roman numeral statements II and III must be true.
14. **(D)** Right triangles CDE and BAE are similar, so the lengths of corresponding sides are in proportion. Since $CD = 1$ and $AB = 2$, each side of $\triangle BAE$ is 2 times the length of the corresponding side of $\triangle CDE$. Since $AD = 6$, it must be the case that $AE = 4$ and $DE = 2$. Since $BC = CE + BE$, use the Pythagorean theorem to find the lengths of CE and BE.
   - In right triangle CDE,
   $$(CE)^2 = 1^2 + 2^2 = 1 + 4 = 5$$
   so $CE = \sqrt{5}$.
   - In right triangle BAE,
   $$(BE)^2 = 4^2 + 2^2 = 16 + 4 = 20$$
   so $BE = \sqrt{20} = \sqrt{4} \cdot \sqrt{5} = 2\sqrt{5}$

   Hence,
   $$BC = CE + BE = \sqrt{5} + 2\sqrt{5} = 3\sqrt{5}$$
15. **(C)** Since the length of each side of the triangle is the same, the triangle is both equilateral and equiangular. The bisector of each of the three angles of the triangle divides the triangle into two "mirror image" parts. Hence, the triangle has three different lines of symmetry.
16. **(B)** Letter E has a horizontal line symmetry but no point symmetry. Letter S has point symmetry but no line of symmetry. Letter W has a vertical line of symmetry but no point symmetry. Letter I has both a line of symmetry and point symmetry, as does Letter X.
17. **(C)** 3D3 has only a horizontal line of symmetry, 8S8 does not have a line of symmetry, 8X8 has both a horizontal and a vertical line of symmetry, 8Y0 has no line of symmetry, and 101 has only a vertical line of symmetry.

18. **(C)** The sum of the degree measures of the three adjacent angles at each of the marked vertices is equivalent to the sum of the measures of four straight angles, which equals 4×180 or 720°. Since the sum of 720° includes the sum of the measures of the four angles of the inscribed quadrilateral, which is 360, the sum of the degree measures of only the marked angles is $720 - 360$ or 360.

(*Grid-In*)

1. **115** The measure of interior angle A is $(180 - x)°$. The sum of the measures of the interior angles of a 5-sides polygon is $(5-2) \times 180° = 540°$. Hence,

$$(180 - x) + 132 + x + y + 113 = 540$$
$$425 + y = 540$$
$$y = 540 - 425 = 115$$

2. **45** Since $AB = BC$, $m\angle A = m\angle B = x$. It is given that $m\angle C = 30$, so $x + x + 30 = 180$ and $x = 75$.
   - Because $\Delta ADE \sim \Delta ABC$, corresponding angles E and C have the same measure, so $m\angle E = m\angle C = 30$.
   - Angle ABC is an exterior angle of ΔEBF. Hence,

$$m\angle ABC = 75 = m\angle E + \angle EFB$$
$$75 = 30 + m\angle EFB$$
$$45 = m\angle EFB$$

   - Since vertical angles are equal in measure, $x = m\angle EFB = 45$.

Lesson 6-5 (*Multiple-Choice*)

1. **(B)** If the area of a square is 25, then the length of a side of that square is $\sqrt{25}$ or 5. The perimeter of the square is 4×5 or 20.
2. **(D)** The height of parallelogram $ABCD$ is the length of perpendicular segment BE. The length of base AD is $1 + 5$ or 6. In right triangle AEB:

$$(BE)^2 + 1^2 = 2^2$$

or $(BE)^2 + 1 = 4$, so $BE = \sqrt{3}$. Hence:

$$\text{Area of parallelogram } ABCD = BE \times AD$$
$$= \sqrt{3} \times 6 \text{ or } 6\sqrt{3}$$

3. **(D)** If the perimeter of the rectangle is 44, then 2 times the length plus 2 times the width is 44. Hence:

$$2(2x + 1) + 2(x - 3) = 44$$
$$4x + 2 + 2x - 6 = 44$$
$$6x - 4 = 44$$
$$6x = 48$$
$$x = \frac{48}{6} = 8$$

4. **(A)** Since the area of the rectangle is 65,

$$(x + 4)(x - 4) = 65$$
$$x^2 - 16 = 65$$
$$x^2 = 81$$
$$x = \sqrt{81} = 9$$

5. **(B)** The area of a square is one-half the product of the lengths of its equal diagonals. If the length of a diagonal of a square is $\sqrt{2}$, the area of the square is $\frac{1}{2}(\sqrt{2})(\sqrt{2})$ or $\frac{1}{2}(2)$, which equals 1.

6. **(C)** To figure out the area of quadrilateral $ABCD$, add the areas of right triangles DAB and DBC.
   - The area of right triangle DAB is $\frac{1}{2} \times 3 \times 4$ or 6.
   - The lengths of the sides of right triangle DAB form a (3, 4, 5) Pythagorean triple in which hypotenuse $BD = 5$. The lengths of the sides of right triangle DBC form a (5, *12*, 13) Pythagorean triple in which $BC = 12$. Hence, the area of right triangle DBC is $\frac{1}{2} \times 5 \times 12$ or 30.

   The area of quadrilateral $ABCD$ is $6 + 30$ or 36.
7. **(C)** Since the area of rectangle $GDFK = KF \times DF$, you need to find the lengths of line segments DF and KF.
   - The area of square $ABCD$ is 121, so

$$AB = BC = \sqrt{121} = 11$$

Since the area of $FKHC$ is 6 and the area of $EKHB$ is 16, the area of rectangle $FEBC$ is $6 + 16$ or 22. Also, the area of rectangle $FEBC = BC \times FC = 22$, where $BC = 11$, so $FC = 2$. Hence:

$$DF = DC - FC = 11 - 2 = 9$$

   - Since the area of rectangle $FKHC = FC \times KF = 6$ and $FC = 2$, then $KF = 3$.

Area of rectangle $GDFK = KF \times DF = 3 \times 9$ or 27.

8. **(B)** Since the perimeter of the triangle is 18, the length of the third side is 18 – 7 – 4 or 7. Draw a segment perpendicular to the shortest side from the opposite vertex. In an isosceles triangle, the perpendicular drawn to the base bisects the base.

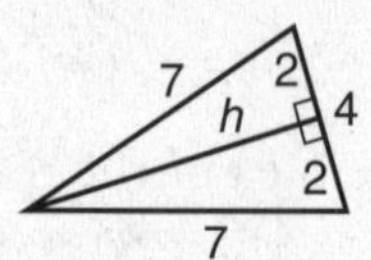

Hence, if h represents the height, then

$$h^2 + 2^2 = 7^2$$
$$h^2 = 49 - 4$$
$$= 45$$
$$h = \sqrt{45} = \sqrt{9} \bullet \sqrt{5} = 3\sqrt{5}$$

Since the base is 4, the area of the triangle is

$$\frac{1}{2} \times 4 \times 3\sqrt{5} = 6\sqrt{5}.$$

9. **(D)** If the area of square $ABCD$ is 64, then the length of each side is $\sqrt{64}$ or 8. Since $BC = 8$, the length of each side of equilateral triangle BEC is 8. Hence:

$$\text{Area equilateral } \triangle BEC = \frac{(\text{side})^2}{4} \times \sqrt{3}$$
$$= \frac{(8)^2}{4} \times \sqrt{3}$$
$$= \frac{64}{4} \times \sqrt{3}$$
$$= 16\sqrt{3}$$

10. **(B)** If x represents the lengths of the equal legs of an isosceles triangle whose area is 8, then

$$\frac{1}{2}(x)(x) = 8$$
$$x^2 = 16$$
$$x = \sqrt{16} = 4$$

In an isosceles (45°–45°) right triangle, the length of the hypotenuse is $\sqrt{2}$ times the length of a leg. Since the length of each leg is 4, the length of the hypotenuse is $4\sqrt{2}$ The perimeter of the triangle is $4 + 4 + 4\sqrt{2}$ or $8 + 4\sqrt{2}$.

11. **(C)** Since the lengths of the sides of $\triangle ABC$ are consecutive integers, let x, $x + 1$, and $x + 2$ represent the lengths of the three sides. Since the perimeter of $\triangle ABC$ has the same perimeter as an equilateral triangle with a side length of 9,

$$x + (x + 1) + (x + 2) = 9 + 9 + 9$$
$$3x + 3 = 27$$
$$3x = 24$$

Then x, the length of the shortest side of $\triangle ABC$, is $\frac{24}{3}$ or 8.

12. **(B)** Draw $\overline{DE} \perp \overline{AB}$.

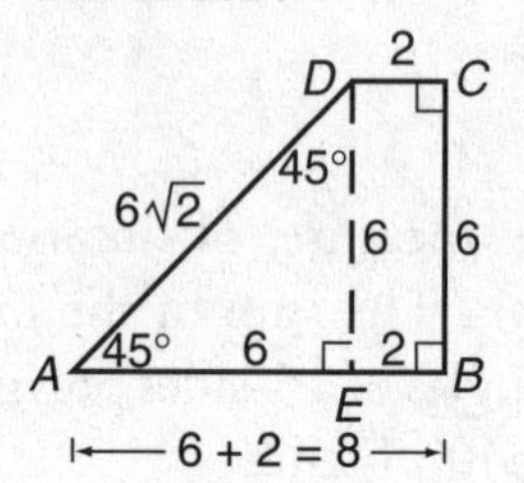

Since quadrilateral $DEBC$ is a rectangle, $EB = DC = 2$ and $DE = CD = 6$. In isosceles right triangle $\triangle AED$, $AE = DE = 6$, so

$$AB = AE + EB = 6 + 2 = 8$$

Also, hypotenuse $AD = 6\sqrt{2}$. Hence:

$$\text{Perimeter } ABCD = AD + DC + BC + AB$$
$$= 6\sqrt{2} + 2 + 6 + 8$$
$$= 16 + 6\sqrt{2}$$

13. **(C)**

$$\text{Area quadrilateral } ABCD = \text{Area rectangle } DEBC + \text{Area } \triangle AED$$
$$= (2 \times 6) + \left(\frac{1}{2} \times 6 \times 6\right)$$
$$= 12 + 18$$
$$= 30$$

14. **A** Since the ratio of AD to DC is 3 to 2, DC is $\frac{2}{3+2}$ or $\frac{2}{5}$ of AC. The base of $\triangle BDC$ is $\frac{2}{5}$ of the base of $\triangle ABC$, and the heights of triangles ABC and DBC are the same. Hence, the area of $\triangle BDC$ is $\frac{2}{5}$ of the area of $\triangle ABC$, so

$$\text{Area } \triangle BDC = \frac{2}{5} \times 40 = 16$$

15. **(B)** If the lengths of the adjacent sides of a rectangle are x and y, where $x > y$, then the perimeter of the rectangle is $2x + 2y$. The perimeter is 8 times as great as the shorter side of the rectangle, so

$$2x + 2y = 8y$$
$$2x = 6y$$
$$\frac{2}{6} = \frac{y}{x}$$

Since $\frac{2}{6} = \frac{1}{3}$, the ratio of y to x is 1 : 3.

16. **(B)** Since the longer side of a rectangle whose perimeter is 24 is 2 times the length of the shorter side,

$$2x + x + 2x + x = 24$$

where x is the length of the shorter side. Hence, $6x = 24$ or $x = \frac{24}{6} = 4$. Since the shorter side is 4, the longer side is 2×4 or 8, so the area of the rectangle is 4×8 or 32. If a square has the same area as this rectangle, then the length of a side of this square is

$$\sqrt{32} = \sqrt{16} \bullet \sqrt{2} = 4\sqrt{2}$$

The perimeter of the square is $4 \times 4\sqrt{2}$ or $16\sqrt{2}$.

17. **(C)** The area of right triangle BEC is 8, so $8 = \frac{1}{2} \times BE \times EC$ or $16 = BE \times EC$. You are given that $AB = BE = EC$, so

$$AB = BE = EC = 4$$

In isosceles right triangle BEC, hypotenuse $BC = 4\sqrt{2}$. Since opposite sides of a parallelogram have the same length,

$$CD = AB = 4$$

and

$$AD = BC = 4\sqrt{2}$$

Then

$$\begin{aligned} \text{perimeter } ABECD &= AB + BE + EC + CD + AD \\ &= 4 + 4 + 4 + 4 + 4\sqrt{2} \\ &= 16 + 4\sqrt{2} \end{aligned}$$

18. **(C)** You are given that the area of square $PQRS$ is 3, so the length of each side of the square is $\sqrt{3}$. You also know that $\triangle ABC$ is equilateral, so $\angle A = \angle B = \angle C = 60$. Triangle APQ is a 30°–60° right triangle in which the length of the longer leg is $\sqrt{3}$. Since the length of the longer leg is always $\sqrt{3}$ times the length of the shorter leg, $AP = 1$ since $PQ = 1 \times \sqrt{3} = \sqrt{3}$. Similarly, in right triangle RSC, $SC = AP = 1$. Hence:

$$\begin{aligned} AC &= AP + PS + SC \\ &= 1 + \sqrt{3} + 1 \\ &= 2 + \sqrt{3} \end{aligned}$$

Since $\triangle ABC$ is equilateral, its perimeter is 3 times the length of any side. Hence, the perimeter of $\triangle ABC$ is $3(2 + \sqrt{3})$ or $6 + 3\sqrt{3}$.

19. **(B)** You are given that $AB = BE = 8$ and $\angle A = 60$. Since $\triangle ABE$ is isosceles, $\angle E = 60$, so

$$\angle B = 180 - 60 - 60 = 60$$

Since $\triangle ABE$ is equiangular, it is also equilateral, $AE = 8$ and $ED = 10 - 8 = 2$.

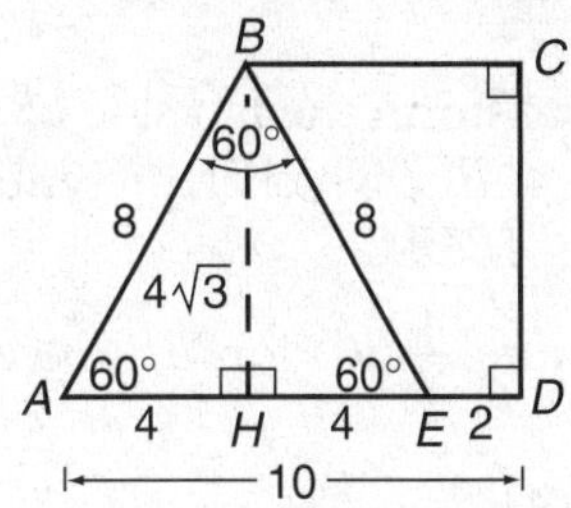

From B draw a perpendicular to AD, intersecting it at point H. Since BH is the side opposite the 60° angle in right triangle BHA,

$$BH = \frac{1}{2} \times 8 \times \sqrt{3} = 4\sqrt{3}$$

Also, since HE is the side opposite the 30° angle in right triangle BHE,

$$HE = \frac{1}{2} \times 8 = 4$$

so $HD = 4 + 2 = 6$.

$$\begin{aligned} \underset{\text{quadrilateral } BCDE}{\text{Area}} &= \underset{\text{rectangle } BHDC}{\text{Area}} - \underset{\text{right } \triangle BHE}{\text{Area}} \\ &= BH \times HD - \frac{1}{2} \times BH \times HE \\ &= 4\sqrt{3} \times 6 - \frac{1}{2} \times 4\sqrt{3} \times 4 \\ &= 24\sqrt{3} - 8\sqrt{3} \\ &= 16\sqrt{3} \end{aligned}$$

20. **(E)** Suppose the length of a side of the original square is 10; then the area of that square is 10×10 or 100. Since one pair of opposite sides of the square are increased in length by 20%, and 20% of 10 is 2, the length of that pair of opposite sides of the new rectangle is 10 + 2 or 12. Since the other pair of sides are increased in length by 50%, and 50% of 10 is 5, the length of the other pair of opposite sides of the new rectangle is 10 + 5 or 15. The area of the new rectangle formed is 12×15 or 180, which is 80% greater than 100, the area of the original square.

21. **(C)**

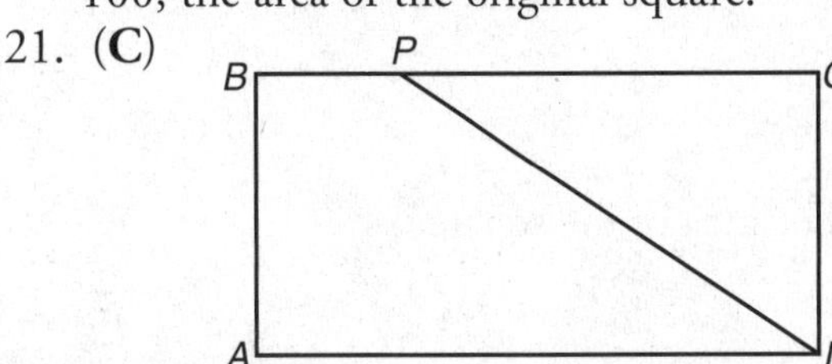

Quadrilateral $ABPD$ is a trapezoid with bases BP and AD, and height AB. Hence,

$$\text{Area trapezoid } ABPD = AB \times \frac{(BP + AD)}{2} = \frac{3}{4}$$

Try to transform the left side of the equation into $AB \times BC$, which represents the area of rectangle $ABCD$.

- Substitute $\frac{1}{4}BC$ for BP since BP is 25% of BC:

$$\frac{AB}{2} \times \left(\frac{1}{4}BC + AD\right) = \frac{3}{4}$$

- Since $AD = BC$, replace AD with BC:

$$\frac{AB}{2} \times \left(\frac{1}{4}BC + BC\right) = \frac{3}{4}$$

- Simplify:

$$\frac{AB}{2} \times \left(\frac{5}{4}BC\right) = \frac{3}{4}$$

$$\frac{5}{8} \times \left(AB \times BC\right) = \frac{3}{4}$$

$$\underbrace{AB \times BC}_{\text{Area rect. } ABCD} = \frac{8}{5} \times \frac{3}{4} = \frac{6}{5}$$

(Grid-In)

1. **42** First find the sum of the areas of the four walls:

$$\begin{aligned} 12 \times 8 &= 96 \\ 12 \times 8 &= 96 \\ 16 \times 8 &= 128 \\ 16 \times 8 &= 128 \\ \hline \text{Sum of areas} &= 448 \end{aligned}$$

Since 1 gallon of paint provides coverage of an area of at most 150 square feet and $\frac{448}{150} = 2.9...$, a minimum of 3 gallons of paint is needed. The paint costs \$14 per gallon, so the minimum cost of the paint needed is $3 \times \$14$ or \$42. Grid in as 42.

2. **1/9** Since the figure is a square,

$$\begin{aligned} x &= 4x - 1 \\ 3x &= 1 \\ x &= \frac{1}{3} \end{aligned}$$

The area of the square is

$$x^2 = \left(\frac{1}{3}\right)^2 = \frac{1}{9}$$

Grid in as 1/9.

3. **27/8** The area of quadrilateral $QBEF$ equals the area of right triangle PEF minus the area of right triangle PBQ.

- The area of square $ABCD$ is 9, so the length of each side of the square is $\sqrt{9}$ or 3. Since P and Q are midpoints, $PB = QB = \frac{3}{2}$, so

$$\text{Area right } \triangle PBQ = \frac{1}{2} \times \frac{3}{2} \times \frac{3}{2} \text{ or } \frac{9}{8}$$

- Since $PB = QB$, right triangle PBQ is isosceles, so $\angle EPF = 45°$. Hence, right triangle PEF is also isosceles, so $EF = PE = 3$, and

$$\text{Area right } \triangle PEF = \frac{1}{2} \times 3 \times 3 = \frac{9}{2}$$

$$\begin{aligned} \text{Area } QBEF &= \text{Area right } \triangle PEF - \text{Area right } \triangle PBQ \\ &= \frac{9}{2} - \frac{9}{8} \\ &= \frac{36}{8} - \frac{9}{8} = \frac{27}{8} \end{aligned}$$

Grid in as 27/8.

LESSON 6-6 (Multiple-Choice)

1. (**B**) Since the radius of circle O is 2 and the radius of circle P is 6, $OP = 2 + 6$ or 8. Hence, the circumference of any circle that has OP as a diameter is 8π.
2. (**C**) If the circumference of a circle is 10π, its diameter is 10 and its radius is 5. Hence, its area is $\pi(5^2) = 25\pi$.
3. (**E**) The area of the shaded region equals the area of the square minus the area of the circle.
   - Since a circle of radius 4 is inscribed in square $ABCD$, AB has the same length as a diameter of the circle, so $AB = 4 + 4 = 8$. Hence the area of square $ABCD$ is 8×8 or 64.
   - The area of the inscribed circle is $\pi 4^2$ or 16π.
   - The area of the shaded region $= 64 - 16\pi$.
4. (**D**) Equilateral polygon $ABCDE$ has five sides of equal length. If the polygon is inscribed in a circle, it divides the circle into five arcs of equal

length. The circumference of the circle is $2\pi \times 20$ or 40π, so the length of each of the five intercepted arcs is $\frac{1}{5}$ of 40π or 8π. The length of the shortest arc from point *A* to point *C* is the sum of the lengths of arcs *AB* and *BC*, which is $8\pi + 8\pi$ or 16π.

5. **(A)** The length of a diagonal of a 5 by 12 rectangle is 13 since, when a diagonal is drawn, the lengths of the sides of the right triangle that results form a (5, 12, *13*) Pythagorean triple with 13 as the hypotenuse. If this rectangle is inscribed in a circle, the diagonals of the rectangle are diameters of the circle, so the length of a diameter of the circle is 13. The circumference of a circle with diameter 13 is 13π.

6. **(A)** Since

$$\text{Circumference} = \pi \times \text{Diameter}$$

then, if the circumference of a circle is π,

$$\text{Diameter} = \frac{\text{Circumference}}{\pi} = \frac{\pi}{\pi} = 1$$

Since the diameter of the circle is 1, the radius of the circle is $\frac{1}{2}$. The area of the circle is $\pi\left(\frac{1}{2}\right)^2$ or $\frac{1}{4}\pi$. If a square has the same area as this circle, then $(\text{side})^2 = \frac{1}{4}\pi$ or

$$\text{Side} = \sqrt{\frac{1}{4}\pi} = \frac{1}{2}\sqrt{\pi}$$

The perimeter of the square is 4 times the length of a side or

$$4\left(\frac{1}{2}\sqrt{\pi}\right) = 2\sqrt{\pi}$$

7. **(C)** To find the length of segment *XY*, first find the length of segment *OX*. Draw radii *OA* and *OB*.

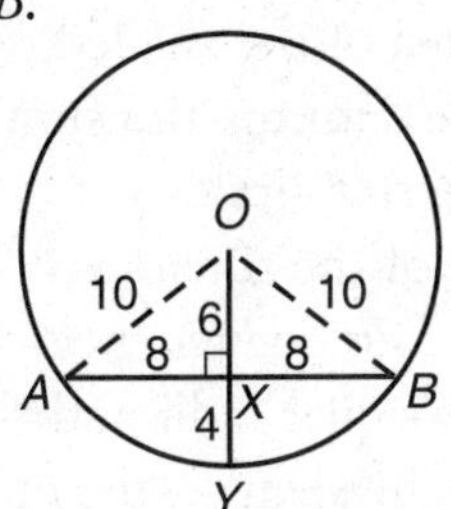

Since all radii of a circle have the same length, ΔAOB is isosceles, so perpendicular segment *OX* bisects chord *AB*. In right triangle *AXO*, radius $OA = 10$ and

$$AX = \frac{1}{2}(16) = 8$$

The lengths of the sides of right triangle *AXO* form a (*6*, 8, 10) Pythagorean triple where $OX = 6$. Since radius $OY = 10$,

$$XY = OY - OX = 10 - 6 = 4$$

8. **(A)** Let *d* represent the length in feet of the diameter of the bicycle wheel; then πd is the circumference of the wheel. After completing *n* revolutions, the wheel has traveled $n \times \pi d$ feet. Since it is given that the bicycle wheel has traveled $\frac{f}{\pi}$ feet after *n* complete revolutions:

$$n\pi d = \frac{f}{\pi}$$

$$\frac{\overset{1}{\cancel{n\pi}} d}{\cancel{n\pi}} = \frac{f}{n\pi \cdot \pi}$$

$$d = \frac{f}{n\pi^2}$$

9. **(D)** The area of the shaded region is the area of the semicircle minus the area of the inscribed right triangle. Since $AC = 6$ and $BC = 8$, the lengths of the sides of right triangle *ABC* form a (6, 8, *10*) Pythagorean triple in which diameter $AB = 10$. Hence, the radius of the circle is $\frac{1}{2} \times 10$ or 5.
   - The area of right triangle $ABC = \frac{1}{2} \times 6 \times 8 = 24$.
   - The area of semicircle *ACB* is one-half the area of the circle that contains *AB* as a diameter. Hence:

$$\text{Area semicirlce} = \frac{1}{2}\left(\pi 5^2\right) = \frac{25}{2}\pi = 12.5\pi$$

   - The area of the shaded region $= 12.5\pi - 24$.

10. **(D)** Since the area of each circle is 7, $7 = \pi r^2$, so the radius of each circle is $\sqrt{\frac{7}{\pi}}$.
   - $AD =$ Vertical diameter of circle $X = 2\sqrt{\frac{7}{\pi}}$
   - $AB =$ Horizontal diameter of circle $X +$ Horizontal radius of circle

$$Y = 2\sqrt{\frac{7}{\pi}} + \sqrt{\frac{7}{\pi}} = 3\sqrt{\frac{7}{\pi}}$$

- Area $ABCD = AD \times AB$

$$= 2\sqrt{\frac{7}{\pi}} \times 3\sqrt{\frac{7}{\pi}}$$
$$= 6 \times \frac{7}{\pi}$$
$$= \frac{42}{\pi}$$

11. (**D**) The area of the shaded region is the area of the rectangle minus the sum of the areas of the two quarter circles, *BP* and *CP*.
   - Since the area of quarter circle *BP* is π,

$$\frac{1}{4} \times \pi (AB)^2 = \pi$$

   Hence, $(AB)^2 = 4$ so $AB = AP = 2$.
   - The area of quarter circle *CP* is also π, so $DP = AP = 2$.
   - Since $AD = AP + PD = 2 + 2 = 4$, the area of rectangle $ABCD = AB \times AD = 2 \times 4 = 8$.
   - The sum of the areas of the two quarter circles is $\pi + \pi$ or 2π.
   - The area of the shaded region is $8 - 2\pi$.
12. (**B**) If *R* stands for the radius of the smaller circle, then $R + w$ represents the radius of the larger circle. Since the circumference of the larger circle exceeds the circumference of the smaller circle by 12π,

$$2\pi(R + w) - 2\pi R = 12\pi$$
$$(2\pi R + 2\pi w) - 2\pi R = 12\pi$$
$$2\pi w = 12\pi$$
$$w = \frac{12\pi}{2\pi} = 6$$

13. (**A**) The perimeter of the unbroken figure is the sum of the lengths of *AC*, *BC*, and major arc *AB*.
   - The area of square *OACB* is $4x^2$, so

$$AC = BC = \sqrt{4x^2} = 2x$$

   - Since *OACB* is a square, it is equilateral, so radius $OA = AC = 2x$. The circumference of circle *O* is $2 \times \pi \times 2x$ or $4\pi x$. Angle *AOB* measures 90°, so the central angle that intercepts major arc *AB* is 360 – 90 or 270°. Hence:

$$\text{Length major arc } AB = \frac{270^\circ}{360^\circ} \times 4\pi x$$
$$= \frac{3}{4} \times 4\pi x$$
$$= 3\pi x$$

   - The perimeter of the unbroken figure is $2x + 2x + 3\pi x$ or $4x + 3\pi x$ or, factoring out *x*,

$$\text{Perimeter} = x(4 + 3\pi)$$

14. (**B**) The area of the shaded region is the area of square *OABC* minus the area of sector *AOC*.
   - Since the area of circle *O* is given as 2π,

$$\text{Area sector } AOC = \frac{90^\circ}{360^\circ} \times 2\pi$$
$$= \frac{1}{4} \times 2\pi$$
$$= \frac{\pi}{2}$$

   - The area of circle $O = 2\pi = \pi(OA)^2$, so $2 = (OA)^2$. Then

$$\text{Area square } OABC = (OA)^2 = 2$$

   - The area of the shaded region $= 2 - \frac{\pi}{2}$.
15. (**B**) Since the circumference of a circle with radius *r* inches is equal to the perimeter of a square with a side length of *s* inches:

$$2\pi r = 4s$$
$$\pi r = 2s$$
$$\frac{r}{s} = \frac{2}{\pi}$$

16. (**C**) The area of the shaded region is the area of the square minus the sum of the areas of the four quarter circles.
   - The area of the square is 2×2 or 4.
   - Since the four quarter circles have equal areas, they have equal radii and the sum of their areas is equivalent to the area of one whole circle with the same radius length. The length of side *GE* is given as 2. Then the length of the radius of each quarter circle is 1 since *B*, *D*, *F*, and *H* are midpoints of the sides of the square. Hence, the sum of the areas of the four quarter circles is $\pi(1^2)$ or π.

- The area of the shaded region $= 4 - \pi$. Using $\pi = 3.14$, you find that the best approximation for the area of the shaded region is $4 - 3.14$ or 0.86.

17. **(C)** Determine whether each of the Roman numeral statements is always true.
- I. Since all four sides of a square have the same length, the area of the square is the product of the lengths of any two of its sides. Hence, the left side of the inequality $AB \times CD < \pi \times r \times r$ represents the area of square *ABCD*, and the right side of the inequality represents the area of inscribed circle *O*. Since the area of the square is greater, not less, than the area of the inscribed circle, statement I is not true.
- II. Since a diameter of the circle can be drawn whose endpoints are points at which circle *O* intersects two sides of the square, the length of a side of the square is $2r$, so the area of the square can be represented as $2r \times 2r$ or $4r^2$. Hence, statement II is true.
- III. Since the circumference of the circle is less than the perimeter of the square:

$$2\pi r < 4(CD)$$
$$r < \frac{4(CD)}{2\pi}$$
$$r < \frac{2(CD)}{\pi}$$

Hence, statement III is true.

Only Roman numeral statements II and III must be true.

18. **(D)** Draw radius *OB*. Since $\triangle OAB$ is an isosceles right triangle and side $AB = 6$, hypotenuse $OB = 6\sqrt{2}$ = radius of circle *O*. Thus:

$$\begin{aligned}\text{Area of shaded region} &= \text{Area of quarter circle} - \text{Area of square } OABC \\ &= \frac{1}{4} \times \pi \times \left(6\sqrt{2}\right)^2 - (6 \times 6) \\ &= \frac{1}{4} \times \pi \times 72 - 36 \\ &= 18\pi - 36 \\ &= 18(\pi - 2)\end{aligned}$$

19. **(D)** The perimeter of the figure that encloses the shaded region is the length of arc *PBQ* plus the sum of the lengths of segments *AP*, *CQ*, *AB*, and *BC*.
- Draw radius *OB*. You are told that *OABC* is a rectangle. Since the diagonals of a rectangle have the same length, $OB = AC = 8$. The circumference of the circle that has its center at *O* and that contains arc *PBQ* is $2 \times \pi \times 8$ or 16π. Since a 90° central angle intercepts arc *PBQ*:

$$\begin{aligned}\text{Length arc } PBQ &= \frac{90^\circ}{360^\circ} \times 16\pi \\ &= \frac{1}{4} \times 16\pi \\ &= 4\pi\end{aligned}$$

- In isosceles right triangle *AOC*,

$$OA = OC = \frac{1}{2} \times 8 \times \sqrt{2} = 4\sqrt{2}$$

Hence:

$$AP = OP - OA = 8 - 4\sqrt{2}$$

and

$$CQ = OQ - OC = 8 - 4\sqrt{2}$$

- In isosceles right triangle *ABC*,

$$AB = BC = \frac{1}{2} \times 8 \times \sqrt{2} = 4\sqrt{2}$$

- The perimeter of the figure that encloses the shaded region $=$ arc $PBQ + AP + CQ + AB + BC = 4\pi + (8 - 4\sqrt{2}) + (8 - 4\sqrt{2}) + 4\sqrt{2} + 4\sqrt{2} = 4\pi + 16$ or $16 + 4\pi$.

(Grid-In)

1. **2** The length of the radius of each semicircle above the line is 2, so

$$\begin{aligned}\text{Area each semicircle} &= \frac{1}{2} \times \pi \times 2^2 \\ &= \frac{1}{2} \times 4\pi \\ &= 2\pi\end{aligned}$$

Hence,

$$X = 2\pi + 2\pi + 2\pi = 6\pi$$

The diameter of the larger semicircle below the line is 8, so its radius is 4. Then

$$\text{Area larger semicircle} = \frac{1}{2} \times \pi \times 4^2$$
$$= \frac{1}{2} \times 16\pi$$

so $Y = 8\pi$.

Hence,

$$Y - X - 8\pi - 6\pi = 2\pi = k\pi$$

so $k = 2$.

2. **2/32** (or **1/16**) Pick an easy number for the length of the diameter of the smaller semicircle. Then find the areas of semicircles *PS* and *PR*.
   - If diameter $PS = 4$, the length of the radius of the smaller semicircle is 2, so:

$$\text{Area smaller semicircle} = \frac{1}{2} \times \pi \times 2^2$$
$$= \frac{1}{2} \times 4\pi$$
$$= 2\pi$$

   - Since $PS = 4$ and S is the midpoint of PQ, the length of radius PQ of the larger semicircle is 8. Then:

$$\text{Area larger semicircle} = \frac{1}{2} \times \pi \times 8^2$$
$$= \frac{1}{2} \times 64\pi$$
$$= 32\pi$$

   - The ratio of the area of semicircle *PS* to the area of semicircle *PR* is $\frac{2\pi}{32\pi}$ or $\frac{2}{32}\left(\text{or } \frac{1}{16}\right)$. Grid in as 2/32 (or 1/16).

3. **72** From 1:25 P.M. to 1:37 P.M. of the same day, the minute hand of the clock moves 12 minutes since $37 - 25 = 12$. There are 60 minutes in 1 hour, so 12 minutes represents $\frac{12}{60}$ of a complete rotation. Since there are 360° in a complete rotation, the minute hand moves

$$\frac{12}{\cancel{60}} \times \overset{6}{\cancel{360}}$$

or 12×6 or 72°.

4. **8** First find the radius of the circle.
   - Since a radius of a circle is perpendicular to a tangent at the point of contact, angles *OAP* and *OBP* measure 90°. The sum of the measures of the four angles of a quadrilateral is 360°. Thus, $120° + 90° + 90° + \angle P = 360°$, so $\angle P = 2a = 60°$ and $a = 30°$.
   - In right triangle *OAP*, $a = 30°$, so $\angle AOP = 60°$. In a 30°–60° right triangle, the length of the side opposite the 30° angle is one-half the length of the hypotenuse. Hence, radius $OA = \frac{1}{2}OP = \frac{1}{2} \times \frac{24}{\pi} = \frac{12}{\pi}$.
   - The cirumference of the circle O is $2\pi r = 2\pi$ $2\pi \times \frac{12}{\pi} = 24$.
   - Since $\frac{120°}{360°} = \frac{1}{3}$, the length of minor arc *AB* is $\frac{1}{3}$ of the circumference of the circle. Hence, the length of minor arc $AB = \frac{1}{3} \times 24 = 8$.

LESSON 6-7 (*Multiple-Choice*)

1. **(C)** A cube has six surfaces. If the surface area of the cube is 96, then the area of each square surface is $\frac{96}{6}$ or 16, so the edge length is $\sqrt{16}$ or 4. Hence, the volume of the cube is $4 \times 4 \times 4$ or 64.
2. **(E)** If the length, width, and height of a rectangular solid are in the ratio of 3 : 2 : 1, then let $3x$, $2x$, and x represent the dimensions of the solid. You are given that the volume of the box is 48. Hence:

$$(3x)(2x)(x) = 48$$
$$6x^3 = 48$$
$$x^3 = \frac{48}{6} = 8$$

The dimensions of the box are 2, 4, and 6 since $x = 2$, $2x = 2(2) = 4$, and $3x = 3(2) = 6$. The surface area of a rectangular box is 2 times the sum of the products of the three pairs of dimensions:

Surface area of box

$$= 2[(\ell \times w) + (\ell \times h) + (h \times w)]$$
$$= 2(2 \times 4 + 2 \times 6 + 4 \times 6)$$
$$= 2(8 + 12 + 24)$$
$$= 2(44)$$
$$= 88$$

3. **(B)** If the volume of a cube is 8, then the edge length of the cube is 2 since $2 \times 2 \times 2 = 8$. The distance from point X at the center of a face of a cube to point Y at the center of the opposite face of the cube equals the edge length, which is 2.
4. **(C)** The height of the stacked cubes will be the sum of the edge lengths of the three cubes.
   - If the volume of the first cube is $\frac{1}{8}$ cubic feet, its edge length is $\frac{1}{2}$ of a foot or 6 inches.
   - If the volume of the second cube is 1 cubic foot, its edge length is 1 foot or 12 inches.
   - If the volume of the third cube is 8 cubic feet, its edge length is 2 feet or 2×12 or 24 inches.

   The height of the stacked cubes, in *inches*, is $6 + 12 + 24$ or 42.
5. **(D)** First find the radius of the circular base of the smaller cone formed when a plane parallel to the base cuts through the original cone.
   - Since the height of the original cone is 12 inches and the height of the smaller cone is 9 inches, the radius of the circular base of the smaller cone is $\frac{9}{12}$ or $\frac{3}{4}$ of the radius of the original cone.
   - You are given that the area of the original cone is 64π, so the radius of its base is $\sqrt{64}$ or 8. Hence, the radius of the circular base of the smaller cone is $\frac{3}{4} \times 8$ or 6.

   Since the radius of the base of the smaller cone is 6 inches, the area, in *square inches*, of the base is $\pi 6^2$ or 36π.
6. **(C)** The volume V of a cylinder is given by the formula $V = \pi r^2 h$, where r is the radius of the circular base and h is the height. If h is replaced by $2h$ and the radius is multiplied by a constant k, so that the volume V does not change, then

$$V = \pi r^2 h = \pi (kr)^2 (2h)$$

so

$$\pi r^2 h = \pi (k^2 r^2)(2h)$$
$$1 = 2k^2$$
$$\frac{1}{2} = k^2$$
$$\frac{1}{\sqrt{2}} = k$$

7. **(C)** When a solid cube with an edge length of 1 foot is placed in the fish tank, it displaces a volume of water that is equal to the volume of the cube, which is $1 \times 1 \times 1$ or 1 cubic foot. If h represents the change in the number of feet in the height of the water in the fish tank, then

$$\text{Volume of displaced water} = 2 \times 3 \times h = 1$$
$$h = \frac{1}{6} \text{ foot}$$

   Hence, the number of *inches* the level of the water in the tank will rise is $\frac{1}{6} \times 12$ or 2.
8. **(B)** Let h represent the equal heights of the cylinder and the rectangular box. Since you are given that the volume of the cylinder of radius r is $\frac{1}{4}$ of the volume of the rectangular box with a square base of side length x,

$$\overbrace{\pi r^2 h}^{\text{Volume of cylinder}} = \frac{1}{4}\overbrace{(x \cdot x \cdot h)}^{\text{Volume of box}}$$
$$\pi r^2 h = \frac{x^2}{4} h$$
$$r^2 = \frac{x^2}{4\pi}$$
$$r = \frac{\sqrt{x^2}}{\sqrt{4\pi}} = \frac{x}{2\sqrt{\pi}}$$

9. **(E)** You are given that the height of the sand in the cylinder-shaped can drops 3 inches or $\frac{1}{4}$ foot when 1 cubic foot of sand is poured out. If r represents the length in feet of the

radius of the circular base, then the volume of the sand poured out must be equal to 1. Hence:

$$\begin{aligned}
\text{Volume of sand} &= \pi r^2\left(\frac{1}{4}\right) = 1 \\
\pi r^2 &= 1 \times 4 \\
r^2 &= \frac{4}{\pi} \\
r &= \frac{\sqrt{4}}{\sqrt{\pi}} = \frac{2}{\sqrt{\pi}} \text{ feet} \\
&= \frac{2}{\sqrt{\pi}} \text{ feet} \times 12 \text{ inches/feet} \\
&= \frac{24}{\sqrt{\pi}} \text{ inches}
\end{aligned}$$

Hence the radius of the cylinder is $\frac{2}{\sqrt{\pi}}$ feet $\times$ 12 or $\frac{24}{\sqrt{\pi}}$ inches, and the diameter in *inches* is $2 \times \frac{24}{\sqrt{\pi}}$ or $\frac{48}{\sqrt{\pi}}$.

10. **(A)** If the height h of a cylinder equals the circumference of the cylinder with radius r, then $h = 2\pi r$, so $r = \frac{h}{2\pi}$. Hence:

$$\begin{aligned}
\text{Volume of cylinder} &= \pi r^2 h \\
&= \pi\left(\frac{h}{2\pi}\right)^2 h \\
&= \pi\left(\frac{h^2}{4\pi^2}\right) h \\
&= \frac{\pi h^3}{4\pi^2} \\
&= \frac{h^3}{4\pi}
\end{aligned}$$

11. **(B)** For each Roman numeral combination determine whether a rectangular solid can be formed.
    - I. If the set of six faces of a rectangular solid includes two of face X and two of face Y, then the rectangular solid must be a 3 by 3 by 4 solid. Since face Z is 4 by 5, it cannot be used to complete the rectangular solid. Hence, the six faces in combination I cannot be used to form a rectangular solid.
    - II. If the set of six faces of a rectangular solid include four of face X, then these four faces can be arranged so that they form an open box with square 3 by 3 bases at opposite ends of the box. Since face Y is a 3 by 3 square, it can be used to complete the rectangular solid. Hence, the six faces in combination II can be used to form a rectangular solid.
    - III. If the set of six faces of a rectangular solid includes four of face Z, then these four faces can be arranged so that they form an open box with either 4 by 4 squares or 5 by 5 squares at opposite ends of the box. Since face Y is a 3 by 3 square, it cannot be used to complete the rectangular solid. Hence, the six faces in combination III cannot be used to form a rectangular solid.

    Only Roman numeral combination II gives faces from which a rectangular solid can be formed.

12. **(A)** If a cylinder with radius r and height h is closed on the top and bottom, then the sum of the areas of the circular top and bottom is $\pi r^2 + \pi r^2$ or $2\pi r^2$. The area of the curved surface is the height h times the distance around the curved surface, $2\pi r$. Hence, the total surface area is $2\pi r^2 + 2\pi rh$ or $2\pi r(r + h)$.

13. **(B)** *Solution 1.* The shortest distance from A to D is the length of AD. Draw AD, which is the hypotenuse of a right triangle whose legs are AE and ED. Since the edge length of the cube is given as 4, $ED = AF = 4$. Segment AE is the diagonal of square $AFEH$, so $AE = 4\sqrt{2}$.

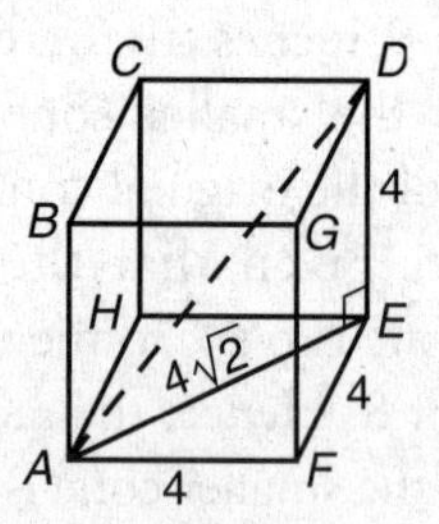

In right triangle AED,

$$\begin{aligned}
(AD)^2 &= (AE)^2 + (ED)^2 \\
&= \left(4\sqrt{2}\right)^2 + (4)^2 \\
&= (16 \cdot 2) + 16 \\
&= 32 + 16 \\
AD &= \sqrt{48} = \sqrt{16} \cdot \sqrt{3} = 4\sqrt{3}
\end{aligned}$$

Solution 2.

$$\begin{aligned} \text{Diagonal } d &= \sqrt{\ell^2 + w^2 + h^2} \\ \text{Let } \ell = w = h = 4{:} &= \sqrt{4^2 + 4^2 + 4^2} \\ &= \sqrt{16 + 16 + 16} \\ &= \sqrt{48} \\ &= \sqrt{16} \cdot \sqrt{3} \\ &= 4\sqrt{3} \end{aligned}$$

14. **(D)** Let $\ell = 4$, $w = 3$, $h = 8$:

$$\begin{aligned} \text{Diagonal } d &= \sqrt{\ell^2 + w^2 + h^2} \\ &= \sqrt{4^2 + 3^2 + 8^2} \\ &= \sqrt{16 + 9 + 64} \\ &= \sqrt{89} \\ &\approx 9.4 \end{aligned}$$

(Grid-In)

1. **60** You are given that all the dimensions of a rectangular box are integers greater than 1. Since the area of one side of this box is 12, the dimensions of this side must be either 2 by 6 or 3 by 4. The area of another side of the box is given as 15, so the dimensions of this side must be 3 by 5. Since the two sides must have at least one dimension in common, the dimensions of the box are 3 by 4 by 5, so its volume is $3 \times 4 \times 5$ or 60.
2. **96** A cube whose volume is 8 cubic inches has an edge length of 2 inches since $2 \times 2 \times 2 = 8$. Since a cube has six square faces of equal area, the surface area of this cube is 6×2^2 or 6×4 or 24. The minimum length L of $\frac{1}{4}$-inch-wide tape needed to completely cover the cube must have the same surface area as the cube. Hence, $L \times \frac{1}{4} = 24$ and $L = 24 \times 4 = 96$ inches.
3. **700** Pick an easy number for the edge length of the cube. If the edge length is 1, the volume of the cube is $1 \times 1 \times 1$ or 1. If the length of each side of this cube is doubled, a cube with an edge length of 2 results. The volume of the new cube is $2 \times 2 \times 2$ or 8. Hence:

$$\begin{aligned} \%\text{ increase in volume} &= \frac{\text{Increase in volume}}{\text{Original volume}} \times 100\% \\ &= \frac{8 - 1}{1} \times 100\% \\ &= 700\% \end{aligned}$$

Grid in as 700.

4.

- Since 3-inch squares from the corners of the square sheet of cardboard are cut and folded up to form a box, the height of the box thus formed is 3 inches.
- If x represents the length of a side of the square sheet of cardboard, then the length and width of the box is $x - 6$.
- Since the volume of the box is 75 cubic inches:

$$\begin{aligned} \text{length} \times \text{width} \times \text{height} &= \text{Volume of box} \\ (x - 6)(x - 6)3 &= 75 \\ (x - 6)^2 &= \frac{75}{3} = 25 \\ x - 6 &= \sqrt{25} = 5 \\ x &= 5 + 6 = 11 \end{aligned}$$

Because the length of each side of the original square sheet of cardboard is 11 inches, the area of the square sheet of cardboard is $11 \times 11 = 121$ square inches.

LESSON 6-8 (*Multiple-Choice*)

1. **(C)** Let $(x_A, y_A) = (3,-1)$ and $(x_B, y_B) = (6,5)$. Use the distance formula to find the distance d between these points:

$$\begin{aligned} d &= \sqrt{(x_B - x_A)^2 + (y_B - y_A)^2} \\ &= \sqrt{(6 - 3)^2 + (5 - (-1))^2} \\ &= \sqrt{3^2 + (5 + 1)^2} \\ &= \sqrt{9 + 36} \\ &= \sqrt{45} = \sqrt{9}\sqrt{5} = 3\sqrt{5} \end{aligned}$$

The length of the line segment whose endpoints are (3,–1) and (6,5) is $3\sqrt{5}$.

2. (**E**) Sketch the rectangle whose vertices are (–2,5), (8,5), (8,–2), and (–2,–2).

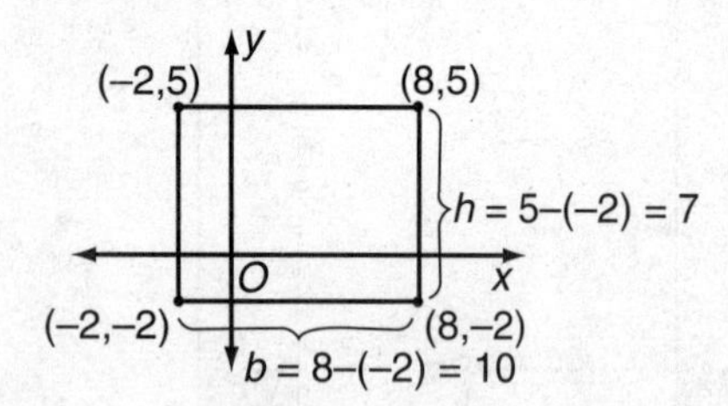

Since

$$b = 8 - (-2) = 8 + 2 = 10$$

and

$$h = 5 - (-2) = 5 + 2 = 7$$

then

$$\text{Area of rectangle} = b \times h = 10 \times 7 = 70$$

3. (**D**) Sketch the parallelogram whose vertices are (–4,–2), (–2,6), (10,6), and (8,–2).

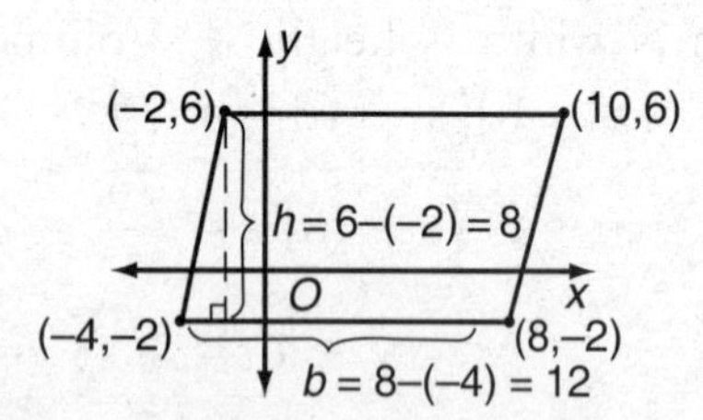

Since

$$b = 8 - (-4) = 8 + 4 = 12$$

and

$$h = 6 - (-2) = 6 + 2 = 8$$

then

$$\text{Area of parallelogram} = b \times h = 12 \times 8 = 96$$

4. (**C**) Sketch the triangle whose vertices are (–4,0), (2,4), and (4,0).

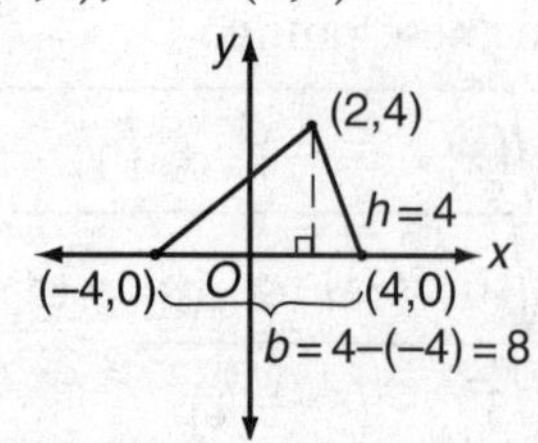

Since

$$b = 4 - (-4) = 4 + 4 = 8$$

and $h = 4$, then

$$\text{Area of triangle} = \frac{1}{2} \times b \times h = \frac{1}{2} \times 8 \times 4 = 16$$

5. (**B**) If $A(-3,0)$ and $C(5,2)$ are the endpoints of diagonal AC of rectangle $ABCD$, then the other two vertices are (–3,2) and (5,0).

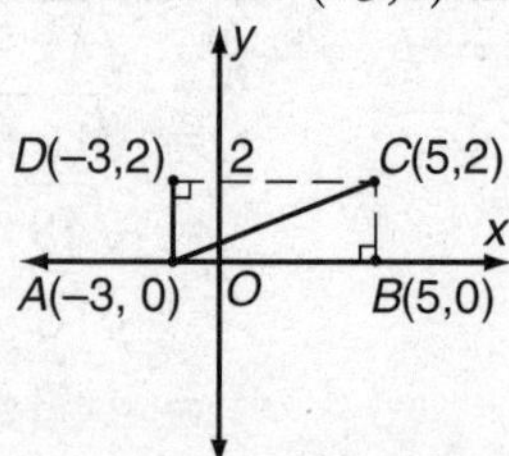

Since the length of one side of the rectangle is 8 and the length of an adjacent side is 2, the perimeter of rectangle $ABCD$ is $2(8 + 2)$ or 20.

6. (**B**) Since point A is on the y-axis and below the x-axis, its x-coordinate is 0 and its y-coordinate is negative. You are told that $OABC$ is a square, so $OA = OC = 4$. Hence, the y-coordinate of A is –4, making $(0,-4) = (k,p)$, so $k = 0$ and $p = -4$.

7. (**E**) Since points O, A, and B lie on the same line,

$$\text{Slope } OA = \text{Slope } OB$$

$$\frac{\text{Change in } y}{\text{Change in } x} = \frac{t-0}{2-0} = \frac{5-0}{r-0}$$

$$\frac{t}{2} = \frac{5}{r}$$

$$r \times t = 2 \times 5$$

$$r = \frac{10}{t}$$

8. (**A**) The perimeter of the shaded region is the sum of the circumference of semicircle P and the length of diameter AB. Since

$$AB = 7 - (-1) = 8$$

the diameter of circle P is 8, so

$$\text{Circumference semicircle } P = \frac{1}{2}(\pi D) = \frac{1}{2}(8\pi) = 4\pi$$

Hence, the perimeter of the shaded region is $4\pi + 8$.

9. (**B**) If points $A(2,0)$ and $B(8,-4)$ are the endpoints of diameter AB of circle O, then let $(x_A, y_A) = (2,0)$ and $(x_B, y_B) = (8,-4)$. Use the distance formula to find the length of diameter AB:

$$AB = \sqrt{(x_B - x_A)^2 + (y_B - y_A)^2}$$
$$= \sqrt{(8 - 2)^2 + (-4 - 0)^2}$$
$$= \sqrt{6^2 + (-4)^2}$$
$$= \sqrt{52}$$

Since the length of a diameter of circle O is $\sqrt{52}$, the length of a radius of circle O is $\frac{\sqrt{52}}{2}$. Hence:

$$\text{Area circle } O = \pi r^2 = \pi\left(\frac{\sqrt{52}}{2}\right)^2$$
$$= \pi\left(\frac{52}{4}\right)$$
$$= 13\pi$$

10. **(C)** Test each point in the set of answer choices until you find the point that makes the slope m of the line containing that point and (4,–2) equal to $\frac{3}{2}$. Let $(x_A, y_A) = (4,-2)$. Choice (C) works since, if $(x_B, y_B) = (6,1)$, then

$$m = \frac{y_B - y_A}{x_B - x_A} = \frac{1 - (-2)}{6 - 4} = \frac{1 + 2}{2} = \frac{3}{2}$$

The coordinates of another point on the line are (6,1).

11. **(C)** The center of a circle is the midpoint of any diameter of the circle. Let $(x_A, y_A) =$ (1,2) and (x_B, y_B) represent the coordinates of point B.
Solution 1: Since the coordinates of O are (–2,1),

- $-2 = \frac{1 + x_B}{2}$, so $1 + x_B = -4$ or $x_B = -5$.
- $1 = \frac{2 + y_B}{2}$, so $2 + y_B = 2$ or $y_B = 0$.
- The coordinates of point $B = (x_B, y_B) =$ (–5,0).

Solution 2: Find the midpoint of each segment whose endpoints are point A, and test each point in the set of answer choices until you obtain O(–2,1) as the midpoint of AB. Choice (C) works since

$$-2 = \frac{1 - 5}{2}$$

and

$$1 = \frac{2 - 0}{2}$$

12. **(E)** Since it is given that point $E(5,h)$ is on the line that contains A(0,1) and B(–2,–1), Slope of $\overline{EA}$ = Slope of $\overline{AB}$

$$= \frac{-1 - 1}{-2 - 0} = \frac{-2}{-2} = 1$$

Hence,

$$\text{Slope of } \overline{EA} = \frac{h - 1}{5 - 0} = 1$$
$$h - 1 = 5$$
$$h = 6$$

13. **(A)** If the line whose equation is $y = x + 2k$ passes through point (1,–3), then substituting –3 for y and 1 for x will make the resulting equation a true statement. Hence:

$$-3 = 1 + 2k$$
$$-4 = 2k$$
$$k = \frac{-4}{2} = -2$$

14. **(D)** If a circle with center at the origin passes through (–8,–6), the length of a radius of the circle is the length of the segment whose endpoints are A(0,0) and B(–8,–6). Use the distance formula to find the length of $\overline{AB}$:

$$AB = \sqrt{(-8 - 0)^2 + (-6 - 0)^2}$$
$$= \sqrt{64 + 36}$$
$$= \sqrt{100}$$
$$= 10$$

Any point that lies on the circle will be the same distance from the origin as point B. Find the distance from the origin to each point in the set of answer choices until you obtain a distance of 10. Choice (D) works since

$$\sqrt{(9 - 0)^2 + (\sqrt{19} - 0)^2} = \sqrt{81 + 19}$$
$$= \sqrt{100} = 10$$

The circle passes through $(9, \sqrt{19})$ since the distance of this point from the origin is 10.

15. (**D**) Drop a perpendicular to the x-axis from point S. Call the point where it intersects the x-axis point H. Since point S is 5 units above the x-axis, $SH = 5$. Opposite sides of a parallelogram are parallel, so $\angle STH = \angle O = 30°$. In 30°–60° right triangle SHT, hypotenuse ST is 2 times the length of SH (the side opposite the 30° angle). Hence, $ST = 2 \times 5 = 10$. Since opposite sides of a parallelogram have the same length, $OR = ST = 10$. Also, since $OR = OT$, then $OT = RS = 10$. Hence, the perimeter of parallelogram $ORST$ is $10 + 10 + 10 + 10$ or 40.

16. (**D**) The area of quadrilateral $OABC$ is the sum of the areas of triangles OAB and OBC.
    - Since $\angle BOC = 45$, then

    $$AOB = 90 - 45 = 45$$

    Also, since point A is 5 units above the x-axis, $OA = OB = 5$, so

    $$\begin{aligned}\text{Area right } \triangle OAB &= \frac{1}{2} \times 5 \times 5 \\ &= 12.5\end{aligned}$$

    - Since $\overline{AB} \perp y$-axis, point B is 5 units above the x-axis, so the height of $\triangle OBC$ is 5 and the length of base OC is 7. Hence:

    $$\begin{aligned}\text{Area } \triangle OBC &= \frac{1}{2} \times 7 \times 5 \\ &= 17.5\end{aligned}$$

    - The area of quadrilateral $OABC$ is $12.5 + 17.5$ or 30.

17. (**B**) If the slope of line ℓ is m, then

    $$m = \frac{(h + m) - 0}{0 - h} = \frac{h + m}{-h}$$

    Hence, $-hm = h + m$ or $-h = m + mh$. Factoring out m from the right side of the equation gives $-h = m(1 + h)$ so

    $$m = \frac{-h}{1 + h}$$

18. (**E**) Given the points $A(0,0)$, $B(5,8)$, and $C(10,4)$:

    $$\text{slope of } \overleftrightarrow{AB} = \frac{\Delta y}{\Delta x} = \frac{8 - 0}{5 - 0} = \frac{8}{5}, \text{and}$$

    $$\text{slope of } \overleftrightarrow{AC} = \frac{\Delta y}{\Delta x} = \frac{4 - 0}{10 - 0} = \frac{4}{10} = \frac{2}{5}$$

    Therefore,

    $$\frac{\text{slope of } \overleftrightarrow{AB}}{\text{slope of } \overleftrightarrow{AC}} = \frac{8}{5} \div \frac{2}{5} = \frac{\cancel{8}^{\,4}}{\cancel{5}} \times \frac{\cancel{5}^{\,1}}{\cancel{2}} = 4$$

19. (**C**) First find an equation $y = mx + b$ of the line that contains the centers of the two circles:
    - The center of the larger circle is at (0,4) and the center of the smaller circle is at (6,2). Hence, the slope of the line that contains these points is:

    $$m = \frac{\Delta y}{\Delta x} = \frac{2 - 4}{6 - 0} = -\frac{2}{6} = -\frac{1}{3}$$

    - Since the y-intercept of the line is (0,4), $b = 4$, so an equation of the line is $y = -\frac{1}{3}x + 4$.
    - The line will intersect the x-axis when $y = 0$. To find the corresponding value of x, replace y with 0 in the equation of the line. Thus, $0 = -\frac{1}{3}x + 4$, so $\frac{1}{3}x = 4$ and $x = 3 \cdot 4 = 12$.

20. (**D**) If $y + 2x = b$, then $y = -2x + b$, so the slope of this line is –2. The slope of a line perpendicular to this line is $\frac{1}{2}$, the negative reciprocal of –2. If this line passes through the origin, its y-intercept is 0, so its equation is $y = \frac{1}{2}x$. Because the point of intersection of the two lines is $(k + 2, 2k)$, the coordinates of this point must satisfy both equations. Find the value of k by substituting the coordinates of this point into the equation $y = \frac{1}{2}x$:

    $$\begin{aligned}2k &= \frac{1}{2}(k + 2) \\ 4k &= k + 2 \\ 3k &= 2 \\ \frac{3k}{3} &= \frac{2}{3} \\ k &= \frac{2}{3}\end{aligned}$$

21. (**A**) Let $y = mx + b$ represent the equation of the desired line.
    - If $y - 4x = 0$, then $y = 4x$, so the slope of this line is 4. Since parallel lines have equal slopes, $m = 4$.

- If $y + 3 = -x + 1$, then $y = -x - 2$, so its y-intercept is -2. Since the desired line has the same y-intercept, $b = -2$.
- The equation of the desired line is $y = 4x - 2$.

22. **(C)** From the graph, the y-intercept is at (0,6). Hence, the equation of the line has the form $y = mx + 6$. To find the slope of the line, apply the slope formula to the x- and y-intercepts: $\frac{\Delta y}{\Delta x} = \frac{6-0}{0-2} = -3$.
Since $m = -3$ and $b = 6$, an equation of the line is $y = -3x + 6$.

23. **(E)** Since $ABCD$ is a square, C has the same x-coordinate as D and the same y-coordinate as B. Hence, $p = 7$ and $q = 6$.
- To find an equation of the line that contains diagonal $\overline{AC}$, first find the slope of $\overline{AC}$:

$$m = \frac{\Delta y}{\Delta x} = \frac{6-(-3)}{7-(-2)} = \frac{9}{9} = 1$$

Hence, the equation of $\overline{AC}$ has the form $y = x + b$.
<u>Note</u>: Instead of using the slope formula, you could reason that since the diagonal of a square forms two right triangles in which the vertical and horizontal sides always have the same length, their ratio is always 1. Since $\overline{AC}$ rises from left to right, its slope is positive. Therefore, the slope of $\overline{AC}$ is 1 (and the slope of diagonal $\overline{BD}$ is -1).
- To find b, substitute the coordinates of $C(7,6)$ into $y = x + b$, which makes $6 = 7 + b$, so $b = -1$.
- An equation of $\overline{AC}$ is $y = x - 1$.

24. **(B)** The general form of an equation of line p is $y = mx + b$. It is given that the slope of the line is $\frac{3}{2}$ and the line contains the point (6,2). Substitute $m = \frac{3}{2}$, $x = 6$, and $y = 2$ in the equation of the line and solve for b:

$$\begin{aligned} y &= mx + b \\ 2 &= \frac{3}{2}(6) + b \\ 2 &= 9 + b \\ -7 &= b \end{aligned}$$

(Grid-In)

1. **3/4** Since the line that passes through points $(7,3k)$ and $(0,k)$ has a slope of $\frac{3}{14}$,

$$\begin{aligned} \frac{3k-k}{7-0} &= \frac{3}{14} \\ \frac{2k}{7} &= \frac{3}{14} \\ 28k &= 21 \\ k &= \frac{21}{28} = \frac{3}{4} \end{aligned}$$

Grid in as 3/4.

2. **3/2** Since the slope of line ℓ_1 is $\frac{5}{6}$, then

$$\frac{y_1 - 0}{3 - 0} = \frac{5}{6} \quad \text{or} \quad y_1 = \frac{15}{6} = \frac{5}{2}$$

The slope of line ℓ_2 is $\frac{1}{3}$, so

$$\frac{y_2 - 0}{3 - 0} = \frac{1}{3} \quad \text{or} \quad y_2 = \frac{3}{3} = 1$$

Since points A and B have the same x-coordinates, they lie on the same vertical line, so

$$\begin{aligned} \text{Distance from } A \text{ to } B &= y_1 - y_2 \\ &= \frac{5}{2} - 1 \text{ or } \frac{3}{2} \end{aligned}$$

Grid in as 3/2.

3. **16** The area of $\triangle APQ$ is equal to the area of square $ABCD$ minus the sum of the areas of the three corner right triangles.

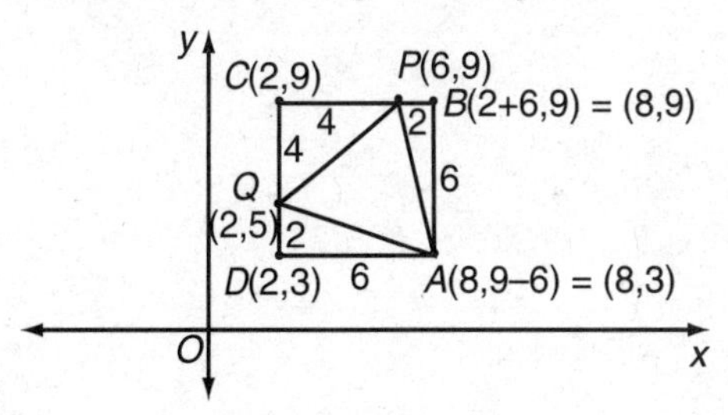

- Find the coordinates of the vertices of square $ABCD$, as shown in the accompanying diagram.
- Find the area of each of the right triangles.

$$\begin{aligned} \text{Area right } \triangle QDA &= \text{Area right } \triangle PBA \\ &= \frac{1}{2} \times 2 \times 6 = 6 \end{aligned}$$

$$\text{Area right } \triangle QCP = \frac{1}{2} \times 4 \times 4 = 8$$

- Then

$$\text{Area } \triangle APQ = 36 - (6 + 6 + 8) = 36 - 20 = 16$$

4. **5** Find the value of the y-intercept "b" in the equation of the line that is perpendicular to $\overline{AB}$ at its midpoint.
   - The midpoint of $\overline{AB}$ is

$$M\left(\frac{-2+6}{2}, \frac{-4+12}{2}\right) = M(2,4)$$

The desired line contains the point $M(2,4)$.

- The slope of $\overline{AB}$ =

$$\frac{\Delta y}{\Delta x} = \frac{12-(-4)}{6-(-2)} = \frac{12+4}{6+2} = \frac{16}{8} = 2$$

Hence, the slope of a line perpendicular to $\overline{AB}$ is the negative reciprocal of 2, which is $-\frac{1}{2}$.

- The equation of the desired line has the form $y = -\frac{1}{2}x + b$. To find b, substitute the coordinates of $M(2,4)$ into the equation. Since $4 = -\frac{1}{2}(2) + b$, $4 = -1 + b$, so $b = 5$.
- Since $b = 5$, 5 is the y-coordinate of the point at which the line that is perpendicular to $\overline{AB}$ at its midpoint crosses the y-axis.

Special Problem Types

CHAPTER 7

This chapter reviews some special types of problems that frequently appear on SAT exams.

LESSONS IN THIS CHAPTER

LESSON

7-1 AVERAGE PROBLEMS

OVERVIEW

There are three types of statistics that can be used to describe a set of N numbers:

- *The **average (arithmetic mean)**, which is calculated by dividing the sum of the numbers in the set by N. The term* average *always refers to the arithmetic mean of a set of numbers.*
- *The **median**, which is the middle value when the numbers in the set are listed in size order.*
- *The **mode**, which is the number in the set that occurs most frequently.*

FINDING THE AVERAGE (ARITHMETIC MEAN)

To find the average of a set of n numbers, add all the numbers together and then divide the sum by n. Thus:

$$\text{Average} = \frac{\text{Sum of the } n \text{ values}}{n}$$

For example, the average of three exam grades of 70, 80, and 72 is 74 since

$$\text{Average} = \frac{70 + 80 + 72}{3} = \frac{222}{3} = 74$$

FINDING AN UNKNOWN NUMBER WHEN AN AVERAGE IS GIVEN

If you know the average of a set of n numbers, you can find the sum of the n values by using this relationship:

$$\text{Sum of the } n \text{ values} = \text{Average} \times n$$

EXAMPLE

The average of a set of four numbers is 78. If three of the numbers in the set are 71, 74, and 83, what is the fourth number?

Solution

If the average of four numbers is 78, the sum of the four numbers is $78 \times 4 = 312$. The sum of the three given numbers is $71 + 74 + 83 = 228$. Since $312 - 228 = 84$, the fourth number is 84.

EXAMPLE

The average of *w*, *x*, *y*, and *z* is 31. If the average of *w* and *y* is 24, what is the average of *x* and *z*?

Solution 1

- Since $\frac{w + x + y + z}{4} = 3$, $w + x + y + z = 4 \times 31 = 124$.
- Since $\frac{w + y}{2} = 24$, $w + y = 2 \times 24 = 48$.
- Thus, $x + z + 48 = 124$, so $x + z = 76$.
- Since $x + z = 76$, $\frac{x + z}{2} = \frac{76}{2} = 38$.

Hence, the average of x and z is 38.

Solution 2

Since the average of two of the four numbers is 7 (= 31 − 24) less than the average of the four numbers, the average of the other two numbers must be 7 more than the average of the four numbers. Hence, the average of x and z is $31 + 7 = 38$.

FINDING THE WEIGHTED AVERAGE

A **weighted average** is the average of two or more sets of numbers that do not contain the same number of values. To find the weighted average of two or more sets of numbers, proceed as follows:

- Multiply the average of each set by the number of values in that set, and then add the products.
- Divide the sum of the products by the total number of values in all of the sets.

EXAMPLE

In a class, 18 students had an average midterm exam grade of 85 and the 12 remaining students had an average midterm exam grade of 90. What is the average midterm exam grade of the entire class?

Solution

$$\text{Weighted average} = \frac{\overbrace{(\text{Average} \times \text{number})}^{\text{Group 1}} + \overbrace{(\text{Average} \times \text{number})}^{\text{Group 2}}}{\text{Total number of students}}$$

$$= \frac{(85 \times 18) + (90 \times 12)}{30}$$

$$= \frac{1530 + 1080}{30}$$

$$= \frac{2610}{30}$$

$$= 87$$

FINDING THE MEDIAN

To find the median of a set of numbers, first arrange the numbers in size order.

- If a set contains an odd number of values, the median is the middle value. For example, the median of the set of numbers

$$8, 12, 15, \underbrace{17}_{\text{Median}}, 19, 20, 25$$

is 17 since the number of values in the set that are less than 17 is the same as the number of values in the set that are greater than 17.

- If a set contains an even number of values, the median is the average (arithmetic mean) of the two middle values. For example, the set of numbers

$$10, 20, 24, 30, 40, 50$$

contains six values. Since the two middle values in the set are 24 and 30, the median is their average:

$$10, 20, \underbrace{24, 30}, 40, 50$$

$$\text{Median} = \frac{24 + 30}{2} = 27$$

FINDING THE MODE

The mode of a set of numbers is the number that appears the greatest number of times in the set. For example, the mode of the set

$$7, 2, 6, 3, 2, 6, 7, 3, 9, 6$$

is 6, since 6 appears more times than any other number in the set.

Lesson 7-1 Tune-Up Exercises

Multiple-Choice

1 The average (arithmetic mean) of a set of seven numbers is 81. If one of the numbers is discarded, the average of the remaining numbers is 78. What is the value of the number that was discarded?

(A) 98
(B) 99
(C) 100
(D) 101
(E) 102

2

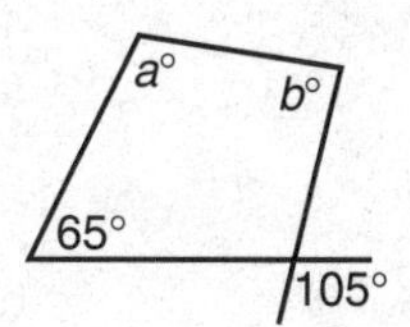

In the figure above, what is the average of a and b?

(A) 75
(B) 80
(C) 85
(D) 90
(E) 95

3 If the average of the two-digit numbers $1N$, $N7$, and NN is 35, then $N =$

(A) 2
(B) 3
(C) 4
(D) 5
(E) 6

4 What is the area of the circle whose radius is the average of the radii of two circles with areas of 16π and 100π?

(A) 25π
(B) 36π
(C) 49π
(D) 64π
(E) 81π

5 If k is a positive integer, which of the following represents the average of 3^k and 3^{k+2}?

(A) $\frac{1}{2} \bullet 3^{k+1}$
(B) $5 \bullet 3^k$
(C) $6^{\frac{3}{2}k}$
(D) $\frac{1}{2} \bullet 3^{3k}$
(E) $\left(\frac{3}{2}\right)^{k+1}$

6 When x is subtracted from $2y$, the difference is equal to the average of x and y. What is the value of $\frac{x}{y}$?

(A) $\frac{1}{2}$
(B) $\frac{2}{3}$
(C) 1
(D) $\frac{3}{2}$
(E) 2

7 If the average of x, y, and z is 32 and the average of y and z is 27, what is the average of x and $2x$?

(A) 42
(B) 45
(C) 48
(D) 50
(E) 63

8 For which set of numbers do the average, median, and mode all have the same value?

(A) 2, 2, 2, 2, 4
(B) 1, 3, 3, 3, 5
(C) 1, 1, 2, 5, 6
(D) 1, 1, 1, 2, 5
(E) 1, 1, 3, 5, 10

9 A man drove a car at an average rate of speed of 45 miles per hour for the first 3 hours of a 7-hour car trip. If the average rate of speed for the entire trip was 53 miles per hour, what was the average rate of speed in miles per hour for the remaining part of the trip?

(A) 50
(B) 55
(C) 57
(D) 59
(E) 62

10 For the set of numbers 2, 2, 4, 5, and 12, which statement is true?

(A) Mean = Median
(B) Mean > Mode
(C) Mean < Mode
(D) Mode = Median
(E) Mean = Mode

11 Susan received grades of 78, 93, 82, and 76 on four math exams. What is the lowest score she can receive on her next math exam and have an average of at least 85 on the five exams?

(A) 96
(B) 94
(C) 92
(D) 90
(E) 88

12 What is the average of $(x + y)^2$ and $(x - y)^2$?

(A) $\frac{x + y}{2}$
(B) xy
(C) $x^2 - y^2$
(D) $\frac{xy}{2}$
(E) $x^2 + y^2$

13 If the average of x and y is $\frac{11}{2}$ and the average of $\frac{1}{x}$ and $\frac{1}{y}$ is $\frac{11}{24}$, then $xy =$

(A) 4
(B) 6
(C) 11
(D) 12
(E) 14

14 If the ratio of a to b is $\frac{1}{2}$ and the ratio of c to b is $\frac{1}{3}$, what is the average of a and c in terms of b?

(A) $\frac{b}{6}$
(B) $\frac{b}{4}$
(C) $\frac{5b}{12}$
(D) $\frac{b}{2}$
(E) $\frac{5b}{6}$

15 The average of a, b, c, d, and e is 28. If the average of a, c, and e is 24, what is the average of b and d?

(A) 31
(B) 32
(C) 33
(D) 34
(E) 36

16 If $2a + b = 7$ and $b + 2c = 23$, what is the average of a, b, and c?

(A) 5
(B) 7.5
(C) 10
(D) 12.25
(E) 15

17 The average final exam grade of 10 students is x, and the average final exam grade of 15 students is y. If the average final exam grade of the 25 students is t, what is x in terms of y and t?

(A) $\frac{5t - 3y}{2}$
(B) $\frac{3t - 5y}{2}$
(C) $\frac{3t - 2y}{5}$
(D) $\frac{t - 15y}{10}$
(E) $\frac{t - 10y}{15}$

18 The average of a, b, c, and d is p. If the average of a and c is q, what is the average of b and d in terms of p and q?

(A) $2p + q$
(B) $2p - q$
(C) $2q + p$
(D) $2q - p$
(E) $\frac{2p + q}{3}$

Grid-In

1 What is the greatest of four consecutive even integers whose average is 19?

2 If the average of x, y, and z is 12, what is the average of $3x$, $3y$, and $3z$?

3 If the average of three consecutive multiples of 4 is 32, what is the LEAST of these multiples?

4 The median of seven test scores is 82, the mode is 87, the lowest score is 70, and the average is 80. If the scores are integers, what is the greatest possible test score?

LESSON

7-2 COUNTING PROBLEMS

OVERVIEW

Some counting problems can be solved simply by compiling a list and then adding up the number of items on the list. Rather than listing the number of ways in which two or more events can happen, it may be faster to multiply together the numbers of ways in which all these events can occur. Overlapping circles can be used to help solve counting problems that involve sets with members in common. Special formulas make it easy to count the number of possible arrangements and selections of objects.

COUNTING BY MAKING A LIST

Some counting problems can be solved by systematically listing and then counting all possible items that fit the conditions of the problem.

EXAMPLE

If $1 \leq x \leq 5$, $0 \leq y \leq 3$, and x and y are integers, how many ordered pairs (x, y) can be formed in which the sum of x and y is less than 4?

Solution

Make a list by picking 1, 2, and 3 as values for x and matching possible values of y that make their sums equal to 3 or less:

(1, 0) (2, 0) (3, 0)

(1, 1) (2, 1)

(1, 2)

Hence, there are six such ordered pairs.

USING THE MULTIPLICATION PRINCIPLE OF COUNTING

When it is not practical to count by making a list, you may be able to use the multiplication principle of counting. This principle states that, if one event can happen in p ways and another event can happen in q ways, then both events can happen in $p \times q$ ways. For example:

- If there are three roads that can be traveled by car to get from town A to town B, and two roads that can be traveled by car to get from town B to town C, then there is a total of $3 \times 2 = 6$ ways in which a car can travel from town A to town B, and then to town C.
- A man with four different ties, five different shirts, and three different pairs of slacks can put together $4 \times 5 \times 3 = 60$ different outfits consisting of one tie, one shirt, and one pair of slacks.
- The total number of different three-digit combinations possible for a combination lock is $10 \times 10 \times 10 = 1000$ since each of the three lock settings can accept any one of the ten digits from 0 to 9.

EXAMPLE

Using the numbers 1, 2, 3, 4, 5, 6, and 7, how many three-digit whole numbers greater than 500 can be formed if the same digit cannot be used more than once?

Solution

To find the number of three-digit whole numbers that satisfy the conditions of the problem, multiply together the numbers of possible digits that can be used to fill all of the three positions of the number that is being formed.

- Since the three-digit number must be greater than 500, there are only three possible choices for the first digit: 5, 6, and 7.
- After a digit is selected for the first position, any one of the remaining six digits can be used to fill the middle position of the three-digit number. After the middle digit is selected, any one of the five remaining digits can be used to fill the last position.
- Using the multiplication principle of counting gives

  $$3 \times 6 \times 5 = 90$$

  Thus, 90 three-digit whole numbers greater than 500 can be formed without reusing any of the given digits.

If digits could be repeated, then $3 \times 7 \times 7 = 147$ represents the number of three-digit whole numbers greater than 500 that could be formed from the digits 1, 2, 3, 4, 5, 6, and 7.

COUNTING USING VENN DIAGRAMS

For any sets A and B, let

$n(A)$ = the number of items in set A;

$n(B)$ = the number of items in set B; and

$n(A \text{ or } B)$ = the number of items that belong to both sets.

These relationships of the elements in sets A and B are shown in the **Venn diagram** below.

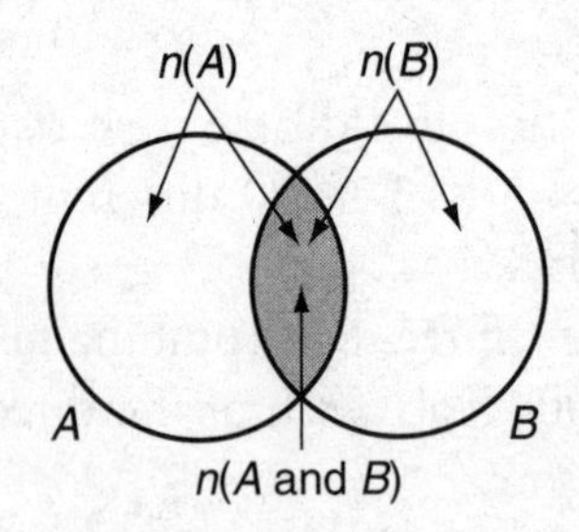

The number of items in set A *or* in set B or in *both* sets A and B is denoted by $n(A \text{ or } B)$, where

$$n(A \text{ or } B) = n(A) + n(B) - n(A \text{ and } B)$$

We subtract $n(A \text{ and } B)$ from the sum of $n(A)$ and $n(B)$ in order not to count the number of items common to both sets two times.

EXAMPLE

In a class of 30 students, 20 students are studying French, 12 students are studying Spanish, and 7 students are studying both French and Spanish. How many students in this class are studying neither French nor Spanish?

Solution

If set A represents students who are studying French and set B represents students who are studying Spanish, then $n(A) = 20$, $n(B) = 12$, and $n(A \text{ and } B) = 7$. Hence,

$$\begin{aligned} n(A \text{ or } B) &= n(A) + n(B) - n(A \text{ and } B) \\ &= 20 + 12 - 7 \\ &= 25 \end{aligned}$$

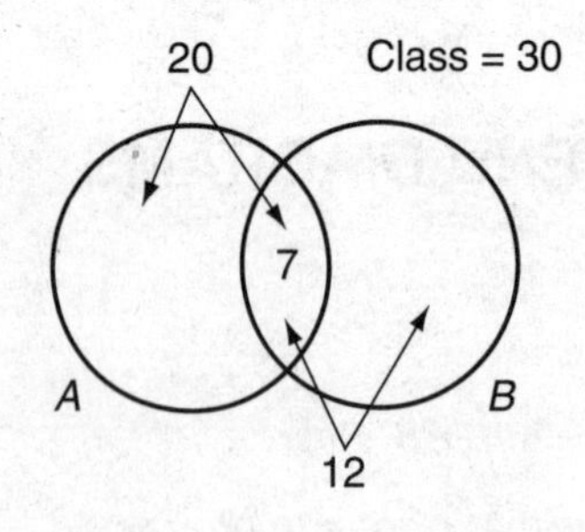

Since 25 of the 30 students in the class are studying either French or Spanish or both French and Spanish, 5 (= 30 − 25) students in the class are studying neither French nor Spanish.

EXAMPLE

If in the preceding example Spanish and French are the only languages studied, how many students study exactly one language?

Solution

- $20 - 7 = 13$ students study French but not Spanish.
- $12 - 7 = 5$ students study Spanish but not French.
- $13 + 5 = 18$ students study French or Spanish but not both.

Hence, 18 students study exactly one language.

COUNTING WHEN THREE SETS OVERLAP

When working with sets, the symbol $\cup$ means *or* and $\cap$ means *and*. In the accompanying Venn diagram of three intersecting sets, region 1 represents $A \cap B$, region 2 corresponds to $B \cap C$, region 3 is where $A \cap C$, and the shaded region represents $A \cap B \cap C$.

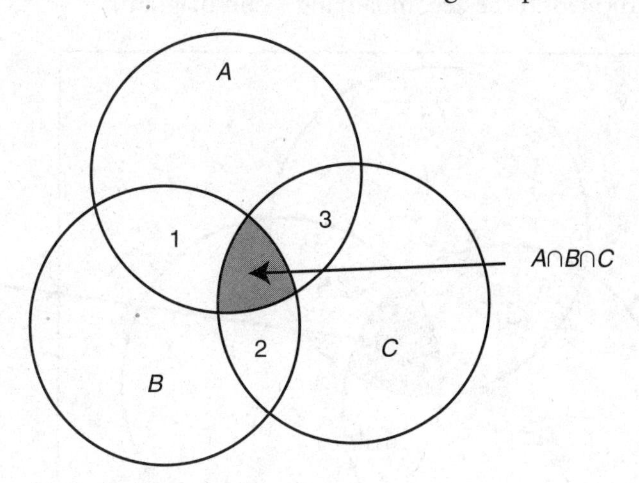

The number of elements in sets *A*, *B*, or *C* is given by this formula:

$$n(A \cup B \cup C) = \underbrace{\left[n(A) + n(B) + n(C)\right]}_{\text{The shaded region is included in each set, so it is being added 3 times.}} - \underbrace{\left[n(A \cap B) + n(A \cap C) + n(B \cap C)\right]}_{\text{The shaded region is included in each set intersection, so it is being subtracted 3 times.}}$$

$$+\underbrace{\left[n(A \cap B \cap C)\right]}_{\text{The shaded region is added back, so it is counted exactly once.}}$$

The formula avoids counting members of overlapping regions more than once by subtracting the intersections of sets taken two at a time (second bracketed term) from the sum of the number of elements in the three individual sets (first bracketed term). Because the number of elements in the intersection of all three sets is counted three times in the first bracketed term and then subtracted three times in the second bracketed term, it ultimately needs to be added back, which accounts for the last bracketed term.

EXAMPLE

In a survey of 100 people, it was found that 74 people subscribed to a news magazine, 65 people subscribed to an entertainment magazine, and 51 people subscribed to a sports magazine. Of the 100 people, 37 subscribed to both a news and an entertainment magazine, 25 subscribed to both a news and a sports magazine, and 45 subscribed to both an entertainment magazine and a sports magazine. If 8 people did not subscribe to any of the three types of magazines, how many people subscribed to the following:

a. All three types of magazines?

b. Only a news magazine?

Solution

Let N, E, and S represent the sets of people who subscribed to a news, entertainment, or sports magazine, respectively. It is given that $n(N) = 74$, $n(E) = 65$, $n(S) = 51$, $n(N) \cap n(E) = 37$, $n(N) \cap n(S) = 25$, and $n(E) \cap n(S) = 45$.

a. Let x represent the number of people who subscribed to all three types of magazines, as indicated in the accompanying Venn diagram.

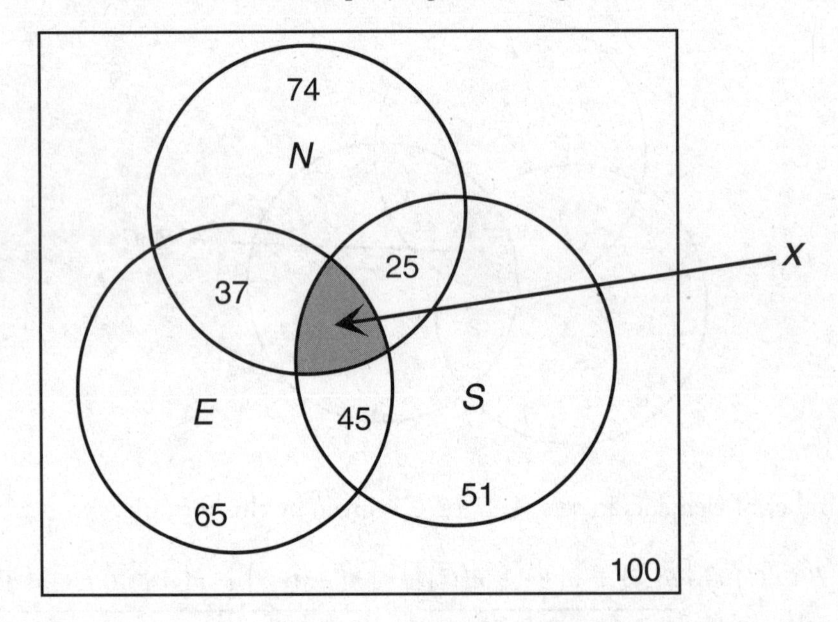

Since you are told that 8 people did not subscribe to any of the three types of magazines, the union of the three sets is $100 - 8 = 92$. Thus:

$$92 = (74 + 65 + 51) - (37 + 25 + 45) + x$$
$$92 = 190 - 107 + x$$
$$92 = 83 + x$$
$$x = 9$$

Thus, **9** people subscribed to all three types of magazines.

b. Let y represent the number of people who subscribed only to a news magazine. To find y, subtract from the number of people who subscribed to a news magazine the number of people who also subscribed to a sports magazine or to an entertainment magazine:

$$n(N) - [n(N) \cap n(E)] - [n(N) \cap n(S)] = 74 - 37 - 25 = 12$$

Since $n(N) \cap n(E) \cap n(S) = 9$ is included in each of the two bracketed terms, it is being subtracted one extra time. Therefore, it must be added back once to get y:

$$y = 12 + 9 = 21$$

Thus, **21** people subscribed only to a news magazine.

n-FACTORIAL

The product of consecutive positive integers from n down to 1, inclusive, is called ***n*-factorial** and is written as $n!$. For example, $4! = 4 \times 3 \times 2 \times 1 = 24$. By definition, $0! = 1$.

Arranging *n* Objects in *n* Slots

The lock combination 5–19–34 is different than the lock combination 34–19–5. Each of these lock combinations represents a *permutation* of the same three numbers: 5, 19, and 34. A **permutation** is an arrangement of objects in which the order of the objects matters. To figure out the total number of different lock combinations that consists of the three numbers 5, 19, and 34, use the counting principle and reason as follows:

- Any one of the three numbers can fill the first position of the lock combination:

$$\underbrace{\boxed{3} \times \boxed{?} \times \boxed{?}}_{\text{total number of choices for each position of the lock combination}}$$

- After one of the three numbers is used, any one of the remaining two numbers can fill the second position of the lock combination:

$$\boxed{3} \times \boxed{2} \times \boxed{?}$$

- Since the remaining number must be used to fill the third position, there is only one possible choice for the third position of the lock combination:

$$\boxed{3} \times \boxed{2} \times \boxed{1} = 6$$

Hence, there are six possible lock combinations, assuming that the digits 5, 19, and 34 are used exactly one time in each possible combination.

In general, the number of different ways in which n different objects can be arranged in a line is $n!$. For example, five people can be arranged in a line in $5! = 5 \times 4 \times 3 \times 2 \times 1 = 120$ different ways.

Arranging *n* Objects in Fewer than *n* Slots

The number of objects or people to be arranged may be greater than the available number of slots. For example, if seven students run in a race in which there are no ties, then the number of different arrangements of first, second, and third place that are possible is

$$\boxed{7} \times \boxed{6} \times \boxed{5} = 210$$

EXAMPLE

Shari remembers the last four digits of a seven-digit telephone number but only knows that each of the first three digits of the telephone number is an odd number. Find the maximum number of telephone calls she must make before she dials the correct number when the first three digits are different.

Solution

There are five odd digits: 1, 3, 5, 7, and 9. Hence, we must find the number of arrangements of the five odd digits in the three available slots that represent the first three digits of the telephone number:

$$\boxed{5} \times \boxed{4} \times \boxed{3} = 60 \text{ arrangements}$$

Shari must dial a maximum of 60 telephone numbers.

PERMUTATIONS SUBJECT TO CONDITIONS

When counting the arrangements of a set of objects, there may be conditions that require certain objects in the set to fill particular positions in each arrangement.

EXAMPLE

Seven students with unequal heights are arranged in a line. In how many ways can the students be arranged in a line so that the shortest student is first and the tallest student is last?

Solution

The shortest student must fill the first of the seven positions and the tallest student must fill the last position. The remaining five positions can be filled in 5! ways:

$$\underbrace{\boxed{1}}_{\text{shortest}} \times \underbrace{\boxed{5} \times \boxed{4} \times \boxed{3} \times \boxed{2} \times \boxed{1}}_{\text{remaining 5 students}} \times \underbrace{\boxed{1}}_{\text{tallest}} = 120$$

Thus, the students can be arranged in 120 different ways.

COMBINATIONS

If from a group of five people a committee consisting of Joe, Susan, and Elizabeth is selected, this *combination* of three people is the *same* as a committee consisting of Elizabeth, Joe, and Susan. A **combination** is a selection of people or objects in which the *identity,* rather than the order, of the people or objects is important.

EXAMPLE

How many committees of two people can be selected from a group of four people?

Solution

Make an organized list in which the four people are referred to as *A, B, C,* and *D.* Keep in mind that order does not matter:

A and *B*	*B* and *C*	*C* and *D*
A and *C*	*B* and *D*	
A and *D*		

Thus, a total of six committees can be formed.

You can count the number of combinations using the combinations formula

$$_{n}C_{r} = \frac{n!}{(n-r)! \times r!}$$

where a set of r things are selected from a set of n things and the notation $_{n}C_{r}$ is read, "the number of combinations of n things taken r at a time." In the previous example, $n = 4$ and $r = 2$, so

$$_{4}C_{2} = \frac{4!}{(4-2)! \times 2!} = \frac{24}{2 \times 2} = 6$$

TIME SAVER

Rather than memorize formulas, use your graphing calculator to perform the combinations calculations for given values of r and n. Usually this involves entering the value of n, pressing the MATH key, moving the cursor to highlight the PRoBability menu, choosing $_{n}C_{r}$ from it, and then entering the value of r. From the probability menu, you can also select $_{n}P_{r}$ for permutations or ! for factorials, as needed for a problem. Performing these calculations with a calculator is quick and easy.

Lesson 7-2 Tune-Up Exercises

Multiple-Choice

1 An ice cream parlor makes a sundae using one of six different flavors of ice cream, one of three different flavors of syrup, and one of four different toppings. What is the total number of different sundaes that this ice cream parlor can make?

(A) 72
(B) 36
(C) 30
(D) 26
(E) 13

2 Five blocks are painted red, blue, green, yellow, and white. In how many different ways can the blocks be arranged in a row if the red block is always first?

(A) 120
(B) 96
(C) 48
(D) 24
(E) 12

3

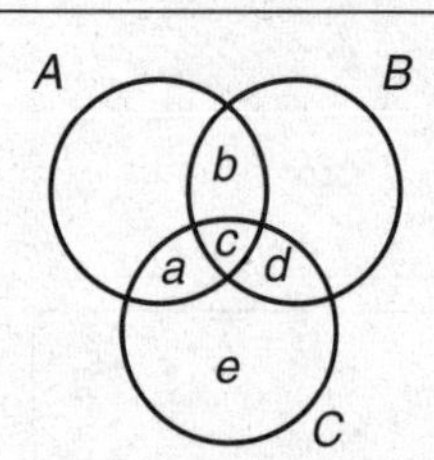

In the figure above, five nonoverlapping regions, labeled *a* through *e*, are formed by sets *A*, *B*, and *C*. Which region represents the set of elements that belong to sets *B* and *C* but not to set *A*?

(A) *a*
(B) *b*
(C) *c*
(D) *d*
(E) *e*

4 In how many different ways can five students be seated in three chairs?

(A) 15
(B) 18
(C) 30
(D) 60
(E) 120

5 A quarter, a dime, a nickel, and a penny are placed in a box. One coin is drawn from the box and then put back before a second coin is drawn. In how many different ways can two coins be drawn so that the sum of the values of the two coins is at least 25 cents?

(A) 9
(B) 7
(C) 6
(D) 5
(E) 4

6 The students in a certain physical education class are on either the basketball team or the tennis team, are on both these teams, or are not on either team. If 15 students are on the basketball team, 18 students are on the tennis team, 11 students are on both teams, and 14 students are not on either of these teams, how many students are in the class?

(A) 48
(B) 40
(C) 36
(D) 30
(E) 28

7 How many four-digit numbers greater than 1000 can be formed from the digits 0, 1, 2, 3, 4, and 5 if the same digit cannot be used more than once?

(A) 1296
(B) 625
(C) 360
(D) 300
(E) 120

8 From the letters of the word "TRIANGLE," how many three-letter arrangements can be formed if the first letter must be *T*, one of the other letters must be *A*, and no letter can be used more than once in any arrangement?

(A) 6
(B) 10
(C) 12
(D) 42
(E) 84

9 In a French class, each student belongs to the French Club or to the Spanish Club. In this class, *F* students are in the French Club, *S* students are in the Spanish Club, and *N* of these students are in both the French and the Spanish clubs. What fraction of the class belongs to the French Club but not to the Spanish Club?

(A) $\frac{F-N}{F+S}$
(B) $\frac{F-N}{F+S+N}$
(C) $\frac{F-N}{F+S-N}$
(D) $\frac{F}{F+S-2N}$
(E) $\frac{F-N}{F+S-2N}$

10 How many numbers greater than 400 but less than 9999 can be formed using the digits 0, 1, 3, 4, and 5 if no digit may be used more than once in any number?

(A) 24
(B) 48
(C) 96
(D) 120
(E) 180

11

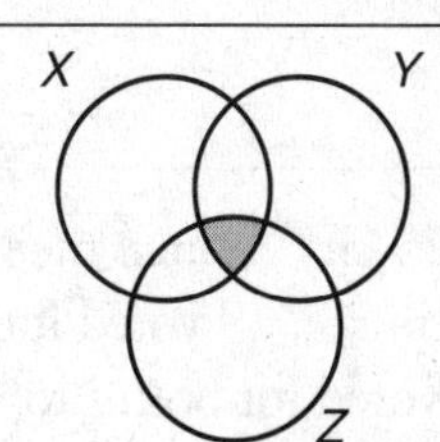

In the figure above, circle *X* represents the set of all positive odd integers, circle *Y* represents the set of all numbers whose square roots are integers, and circle *Z* represents the set of all positive multiples of 5. Which of the following numbers is a member of the set represented by the shaded region?

(A) 25
(B) 49
(C) 75
(D) 81
(E) 100

12

$$X = \{1, 2, 4\}$$
$$Y = \{1, 3, 4\}$$

If, in the sets above, *x* is any number in set *X* and *y* is any number in set *Y*, how many different values of $x + y$ are possible?

(A) Five
(B) Six
(C) Seven
(D) Eight
(E) Nine

13 In a music class of 30 students, 50% study woodwinds, 40% study strings, and 30% study both woodwinds and strings. What percent of the students in the class do NOT study either woodwind or string instruments?

(A) 25%
(B) 40%
(C) 50%
(D) 60%
(E) 75%

14 Of the 25 city council members, 12 voted for proposition *A*, 19 voted for proposition *B*, and 10 voted for both propositions. Which of the following statements must be true?

I. Seven members voted for proposition *B* but not for proposition *A*.
II. Four members voted against both propositions.
III. Eleven members voted for one proposition and against the other proposition.

(A) I only
(B) II only
(C) I and II only
(D) I and III only
(E) II and III only

15 In a survey of 63 people, 33 people subscribed to magazine *A*, 30 people subscribed to magazine *B*, and 17 subscribed to magazine *C*. For any two of the magazines, 9 people subscribed to both magazines. If 5 people in the survey did not subscribe to any of the three magazines, how many people subscribed to all three magazines?

(A) 9
(B) 7
(C) 5
(D) 2
(E) It cannot be determined from the information given.

16 How many different five-member teams can be made from a group of eight students, if each student has an equal chance of being selected?

(A) 26
(B) 40
(C) 56
(D) 336
(E) 6,720

17 Six people are running in an election to fill two vacancies on a local school board. In the election booth, a voter may cast a ballot for any two of the six candidates, or cast a ballot for exactly one of the six candidates, or cast a ballot for none of the six candidates. What is the total number of different choices a voter has when casting a ballot?

(A) 9
(B) 20
(C) 22
(D) 36
(E) 37

Grid-In

1

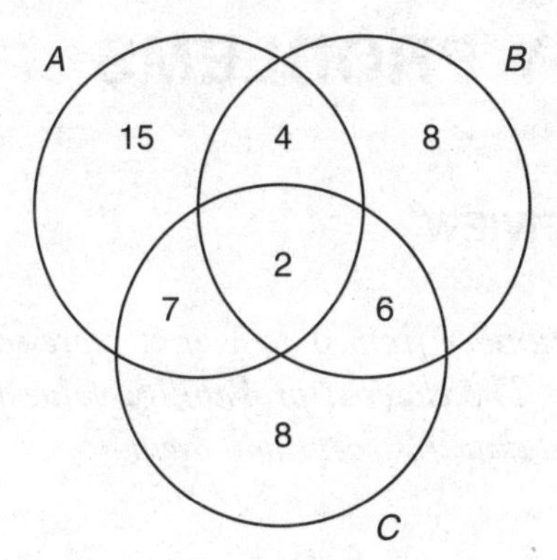

The accompanying diagram shows the number of students who are enrolled in three different courses. All students in circle *A* are enrolled in mathematics, all students in circle *B* are enrolled in biology, and all students in circle *C* are enrolled in computer science. What percentage of the students are enrolled in mathematics or computer science?

2 When Kim bought her new car, she found that there were 72 different ways in which her car could be equipped. Her selections included four choices of engine and three choices of transmission. If her only other selection was color, how many choices of color did she have?

3 All seven-digit telephone numbers in a certain town begin with 245. How many different telephone numbers may be assigned if the last four digits of each telephone number do not begin or end in a zero?

4

In the accompanying diagram, five points lie in the same plane with no three points on the same line. A circle can be drawn through any three of the five points. What is the total number of different circles that can be drawn such that each circle passes through three of the five points?

5 In how many different ways can the individual letters of the word *PARABOLA* be arranged in a line such that the three letter *A*s appear consecutively?

LESSON

7-3 PROBABILITY PROBLEMS

OVERVIEW

*The **probability** of an event is a number from 0 to 1 that represents the likelihood that the event will occur. The closer a probability value for an event is to 1, the greater the likelihood that this event will occur.*

EXPRESSING PROBABILITY AS A FRACTION

If an event can happen in r of n equally likely ways, then the probability that this event will occur is $\frac{r}{n}$. For example, if a jar contains one red marble, two yellow marbles, and four green marbles, the probability that a marble picked from the jar at *random* (that is, without looking) will be yellow is

$$\frac{2}{1 + 2 + 4} = \frac{2}{7}$$

In the probability fraction above, 2 corresponds to the number of yellow marbles or "favorable outcomes," and 7 represents the total number of marbles or possible outcomes. In general,

$$\text{Probability of an event} = \frac{\text{Number of favorable outcomes}}{\text{Total number of outcomes}}$$

FINDING A PROBABILITY BY LISTING OUTCOMES

To figure out the number of favorable outcomes for an event, it may be helpful to first list all the different possible outcomes for that probability experiment.

EXAMPLE

A penny, a nickel, and a dime are placed in a hat. A coin is drawn, its value is noted, and then the coin is replaced in the same hat. If a second coin is drawn, what is the probability that the sum of the values of the two coins will be more than 10 cents?

Solution

- All possible outcomes can be listed as a set of ordered pairs:

(penny, penny)   (nickel, penny)   (dime, penny)
(penny, nickel)   (nickel, nickel)   (dime, nickel)
(penny, dime)   (nickel, dime)   (dime, dime)

- Five of the nine possible outcomes are favorable:

(penny, dime) = 11 cents (dime, penny) = 11 cents

(nickel, dime) = 15 cents (dime, nickel) = 15 cents

(dime, dime) = 20 cents

- Hence:

$$\text{Probability of an event} = \frac{\text{Number of favorable outcomes}}{\text{Total number of outcomes}} = \frac{5}{9}$$

MULTIPLYING PROBABILITIES

Multiplying the probability that one event will occur by the probability that a second event will occur gives the probability that both events will occur. For example, suppose a fair coin is tossed 2 times. Since the probability of getting a head on each toss is $\frac{1}{2}$, the probability of tossing two heads is $\frac{1}{2} \times \frac{1}{2}$ or $\frac{1}{4}$. Also, the probability of tossing three heads is $\frac{1}{2} \times \frac{1}{2} \times \frac{1}{2}$ or $\frac{1}{8}$.

SOME PROBABILITY FACTS

Since the numerator of a probability fraction must be less than or equal to the denominator, the probability of an event must be a number from 0 to 1.

- If an event is certain to happen, its probability is 1.
- If an event can never happen, its probability is 0.
- The probability that an event will *not* happen is 1 minus the probability that the event will happen. For example, if the probability that a basketball team will win tomorrow is $\frac{1}{5}$, then the probability that it will lose is $1 - \frac{1}{5} = \frac{4}{5}$.
- The sum of the probabilities of the events that comprise a probability experiment is 1. For example, if a fair coin is tossed, then

$$P(\text{head}) + P(\text{tail}) = \frac{1}{2} + \frac{1}{2} = 1$$

As another example, consider a jar that contains only white, blue, and orange marbles. If the probability of selecting a white marble is $\frac{1}{3}$ and the probability of selecting a blue marble is $\frac{1}{4}$, then the probability of selecting an orange marble must be the fraction that, when added to $\frac{1}{3}$ and $\frac{1}{4}$, gives 1. Hence, the probability of selecting an orange marble is

$$1 - \left(\frac{1}{3} + \frac{1}{4}\right) = 1 - \left(\frac{4}{12} + \frac{3}{12}\right) = \frac{5}{12}$$

EXAMPLE

A dart is thrown and lands in square *ABCD*. If *E* is the midpoint of $\overline{AD}$ and *F* is the midpoint of $\overline{CD}$, as shown in the accompanying diagram, what is the probability that the dart will land in the shaded region?

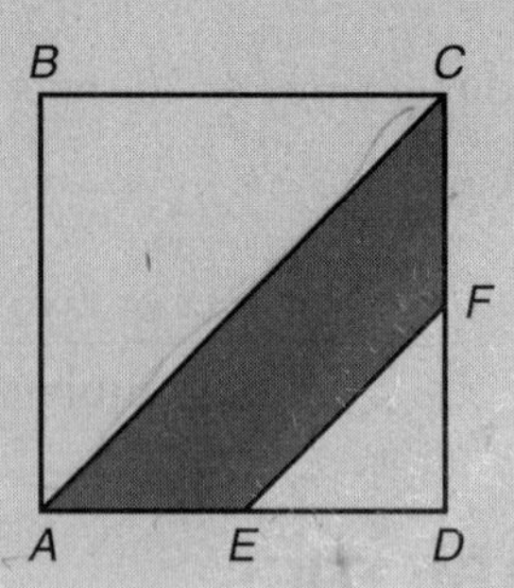

Solution

The required probability is the ratio of the area of the shaded region to the area of the square. Suppose the length of each side of the square is 2.

- Area of square $ABCD = 2 \times 2 = 4$.
- Area of right $\triangle ABC = \frac{1}{2} \times 2 \times 2 = 2$. Since *E* and *F* are midpoints, $DE = DF = 1$, so the area of right $\triangle EDF = \frac{1}{2} \times 1 \times 1 = \frac{1}{2}$.
- Area of the shaded region $= 4 - 2 - \frac{1}{2} = \frac{3}{2}$
- $P(\text{dart will land in shaded region}) = \dfrac{\text{Area of shaded region}}{\text{Area of square}}$

$$= \frac{3}{2} \div 4 = \frac{3}{8}$$

$$= \frac{3}{2} \times \frac{1}{4} = \frac{3}{8}$$

Lesson 7-3 Tune-Up Exercises

Multiple-Choice

1 A bag contains three green marbles, four blue marbles, and two orange marbles. If a marble is picked at random, what is the probability that an orange marble will NOT be picked?

(A) $\frac{1}{4}$
(B) $\frac{1}{3}$
(C) $\frac{4}{11}$
(D) $\frac{1}{2}$
(E) $\frac{7}{9}$

1, 2, 2, 3, 3, 3, 4, 4, 4, 4

2 What is the probability that a number selected at random from the set of numbers above will be the average of the set?

(A) 0
(B) $\frac{1}{10}$
(C) $\frac{1}{5}$
(D) $\frac{3}{10}$
(E) $\frac{2}{5}$

3 If the probability that an event will occur is $\frac{x}{4}$ and $x \neq 0$, what is the probability that this event will NOT occur ?

(A) $\frac{1 - x}{4}$
(B) $\frac{4 - x}{4}$
(C) $\frac{4 - x}{x}$
(D) $\frac{4}{x}$
(E) $\frac{x}{x - 4}$

Questions 4 and 5.

A jar contains only red, blue, and green marbles. The probability of picking a red marble is 0.25, and the probability of picking a blue marble is 0.40.

4 What is the probability of picking a green marble?

(A) 0.25
(B) 0.35
(C) 0.60
(D) 0.75
(E) 1.00

5 What is the LEAST number of marbles that could be in the jar?

(A) 4
(B) 5
(C) 10
(D) 20
(E) It cannot be determined from the information given.

6 A cube whose faces are numbered from 1 to 6 is rolled. Which of the following statements must be true?

I. The probability of getting a 5 on the top face is $\frac{1}{5}$.
II. The probability of getting an even number on the top face is $\frac{1}{2}$.
III. The probability of getting a prime number on the top face is $\frac{1}{2}$.

(A) I and II only
(B) I and III only
(C) II only
(D) II and III only
(E) All are true.

7 If the letters L, O, G, I, and C are randomly arranged to form a five-letter "word," what is the probability that the result will be the word LOGIC?

(A) $\frac{1}{120}$

(B) $\frac{1}{24}$

(C) $\frac{1}{5}$

(D) $\frac{1}{4}$

(E) $\frac{1}{2}$

8

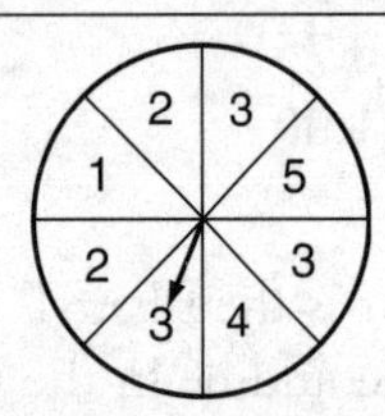

The circle above is divided into eight sectors of equal area. What is the probability that the spinner will land on an even-numbered region in each of two consecutive spins?

(A) $\frac{3}{4}$

(B) $\frac{25}{64}$

(C) $\frac{3}{8}$

(D) $\frac{1}{4}$

(E) $\frac{9}{64}$

9 Three fair coins are tossed at the same time. What is the probability that all three coins will come up heads *or* all will come up tails?

(A) $\frac{1}{8}$

(B) $\frac{1}{6}$

(C) $\frac{1}{4}$

(D) $\frac{1}{3}$

(E) $\frac{3}{8}$

10 The faces of a red cube and a yellow cube are numbered from 1 to 6. Both cubes are rolled. What is the probability that the top face of each cube will have the same number?

(A) $\frac{1}{8}$

(B) $\frac{1}{6}$

(C) $\frac{1}{4}$

(D) $\frac{1}{3}$

(E) $\frac{1}{2}$

11 A jar contains 54 marbles each of which is blue, green, or white. The probability of selecting a blue marble at random from the jar is $\frac{1}{3}$, and the probability of selecting a green marble at random is $\frac{4}{9}$. How many white marbles does the jar contain?

(A) 6
(B) 8
(C) 9
(D) 12
(E) 18

$$A = \{-2, -1, 0, 1, 2\}$$

12 If x represents a number picked at random from set A above, what is the probability that $x^2 < 2$?

(A) $\frac{1}{5}$

(B) $\frac{2}{5}$

(C) $\frac{3}{5}$

(D) $\frac{4}{5}$

(E) 1

---

13 If x represents a number picked at random from the set $\{-3, -2, -1, 0, 1, 2\}$, what is the probability that x will satisfy the inequality $4 - 3x < 6$?

(A) $\frac{1}{6}$

(B) $\frac{1}{3}$

(C) $\frac{1}{2}$

(D) $\frac{2}{3}$

(E) $\frac{5}{6}$

14

$$A = \{1, 2, 3\}$$
$$B = \{1, 4, 9\}$$

A number is randomly selected from set A above, and then a second number is randomly selected from set B. What is the probability that the product of the two numbers selected will be less than 9?

(A) $\frac{1}{3}$

(B) $\frac{5}{9}$

(C) $\frac{2}{3}$

(D) $\frac{7}{9}$

(E) $\frac{5}{6}$

15

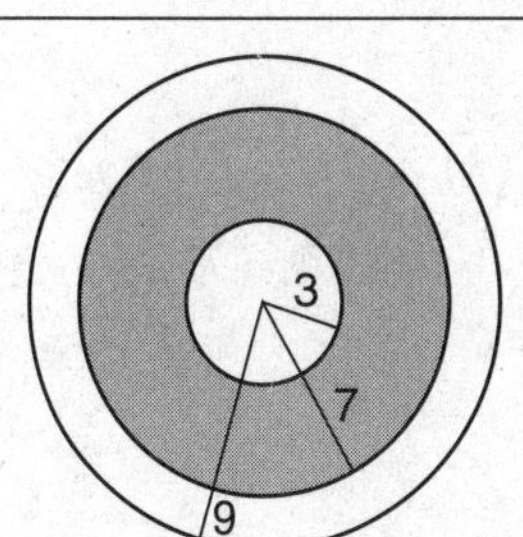

A target shown in the accompanying diagram consists of three circles with the same center. The radii of the circles are 3, 7, and 9. A dart is thrown and lands on the target. What is the probability that the dart will land on the shaded region?

(A) $\frac{4}{9}$

(B) $\frac{40}{81}$

(C) $\frac{49}{81}$

(D) $\frac{7}{10}$

(E) $\frac{40}{49}$

16 The six faces of a cube are numbered 1 through 6. If the cube will be rolled twice in succession, what is the probability that the number that faces up on the first roll will be greater than the number that faces up on the second roll?

(A) $\frac{1}{4}$

(B) $\frac{1}{3}$

(C) $\frac{5}{12}$

(D) $\frac{7}{12}$

(E) $\frac{2}{3}$

17

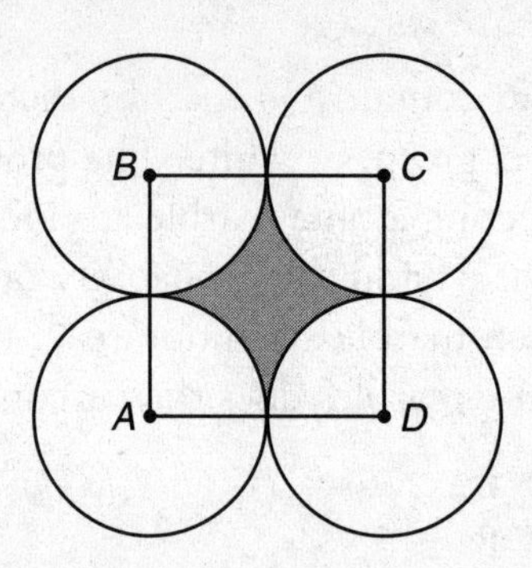

In the accompanying diagram, points *A*, *B*, *C*, and *D* are the centers of four circles that each have a radius length of 1. The circles are tangent at the points shown. If a point is selected at random from the interior of square *ABCD*, what is the probability that the point will be chosen from the shaded region?

(A) $1 - \frac{\pi}{4}$

(B) $1 - \frac{3\pi}{16}$

(C) $1 - \frac{\pi}{6}$

(D) $1 - \frac{\pi}{8}$

(E) $1 - \frac{\pi}{16}$

18

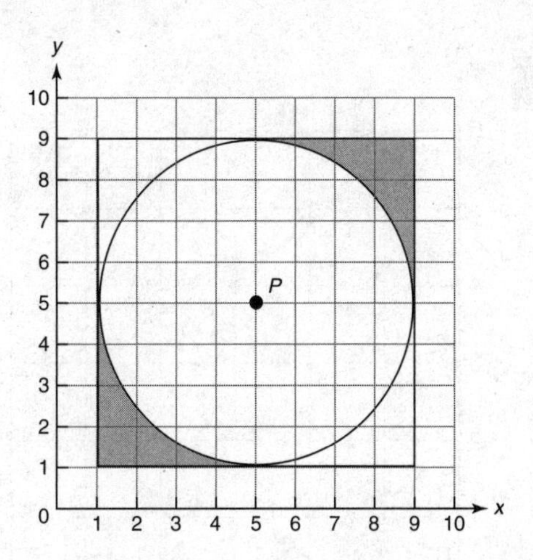

In the accompanying diagram, circle P is inscribed in a square. If a point is picked at random from the interior of the square, what is the probability that the point will lie in the shaded region?

(A) $\frac{4-\pi}{16}$

(B) $\frac{4-\pi}{8}$

(C) $\frac{8-\pi}{32}$

(D) $\frac{4-\pi}{4}$

(E) $\frac{9-2\pi}{9}$

19

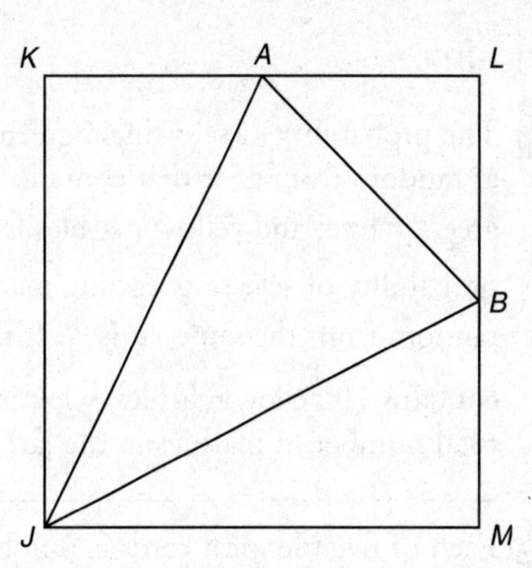

In the accompanying figure, $JKLM$ is a square with sides of length 6. Points A and B are the midpoints of sides $\overline{KL}$ and $\overline{LM}$, respectively. If a point is selected at random from the interior of the square, what is the probability that the point will be chosen from the interior of $\triangle JAB$?

(A) $\frac{3}{16}$

(B) $\frac{1}{4}$

(C) $\frac{5}{16}$

(D) $\frac{3}{8}$

(E) $\frac{1}{2}$

Grid-In

1 The probability of selecting a green marble at random from a jar that contains only green, white, and yellow marbles is $\frac{1}{4}$. The probability of selecting a white marble at random from the same jar is $\frac{1}{3}$. If this jar contains 10 yellow marbles, what is the total number of marbles in the jar?

2 Each of five identical cards is numbered on one of its sides with a different integer from 1 to 5. The cards are turned over and then shuffled so that they are not in a predictable order. If two cards are selected at random without replacement, what is the probability that their sum is at least 8?

3

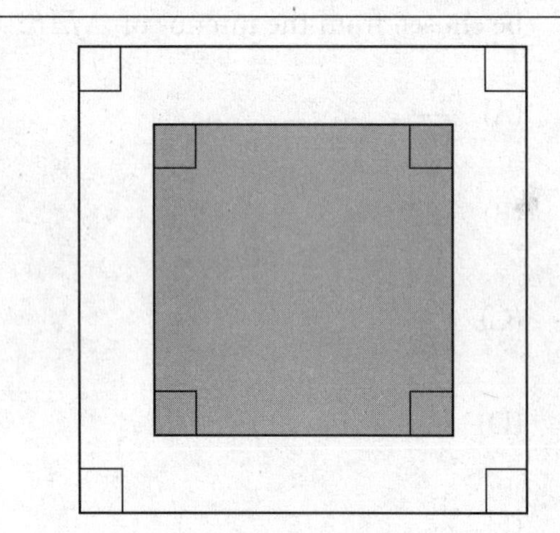

In the accompanying diagram of a square dartboard, the length of a side of the larger square is 1.5 times the length of a side of the smaller square. If a dart is thrown and lands on the larger square, what is the probability that it will land in the interior of the smaller square?

4

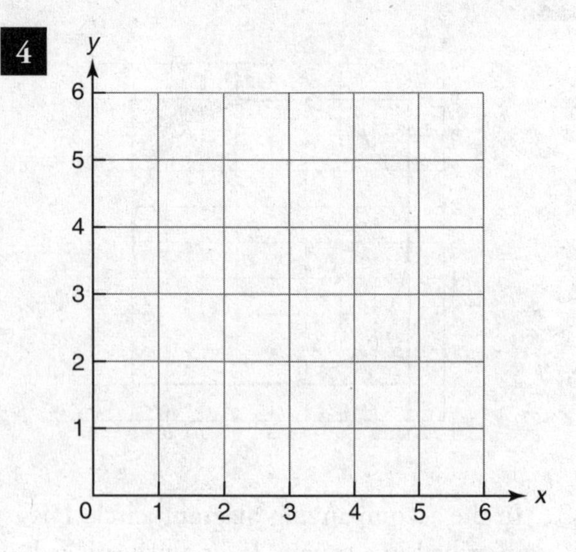

A square dartboard is placed in the first quadrant from $x = 0$ to 6 and $y = 0$ to 6, as shown in the accompanying figure. A triangular region on the dartboard is enclosed by the graphs of the equations $y = 2$, $x = 6$, and $y = x$ (not shown). Find the probability that a dart that randomly hits the dartboard will land in the triangular region formed by the three lines.

5 Five students, all of different heights, are to be randomly arranged in a line. What is the probability that the tallest student will be first in line and the shortest student will be last in line?

6 A jar contains four blue marbles and two green marbles. Without looking, two marbles are drawn from the jar. What is the probability that two marbles with the same color will be selected?

GRAPHS AND TABLES

OVERVIEW

Graphs and tables summarize information in a way that is easy to read. Each SAT *typically includes at least one or two questions that require you to read and analyze data from a graph or table. Before attempting to answer questions based on a graph, you should quickly scan the graph to get an overview of what data and information are being presented. Pay close attention to the type of graph and its title. Take note of any descriptive labels and units of measurement along horizontal and vertical sides of a bar or line graph. If a circle graph is given, look for the whole amount that the circle represents.*

COMMON TYPES OF GRAPHS

Recognizing the type of graph may help you answer questions about the graph. In general:

- Circle graphs or pie charts show how the parts that comprise a whole compare with the whole and to each other. Each "slice" of a pie chart is labeled with the percent that part is of the whole. The "slices" of a pie chart always add up to 100% of the whole.
- Bar graphs compare similar things where the height or length of each bar represents an amount of something.
- Broken-line graphs show how the amount of something changes over time. If a line segment slants down in a time interval, the amount is decreasing for that time interval. If a line segment rises in a time interval, the amount is increasing in that interval. A flat line segment indicates no change.

Lesson 7-4 Tune-Up Exercises

Multiple-Choice

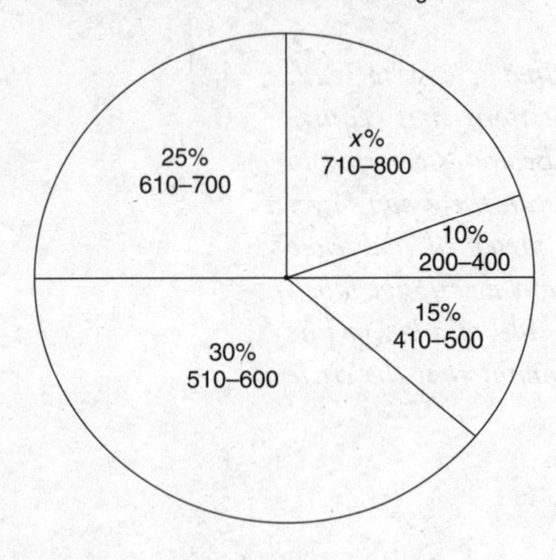

Questions 1 and 2 refer to the graph above.

1 If there are 72 SAT math scores between 510 and 600, how many SAT math scores are above 700?

(A) 40
(B) 48
(C) 56
(D) 64
(E) 72

2 If 20% of the students with SAT math scores from 610 to 700 received college scholarships, how many students with SAT math scores from 610 to 700 received college scholarships?

(A) 12
(B) 18
(C) 30
(D) 48
(E) 60

Minimum Age Requirement (years)	Number of States
14	7
15	12
16	27
17	2
18	2

3 The table above shows the minimum age requirement for obtaining a driver's license. In what percent of the states can a person obtain a driver's license before the age of 16?

(A) 94%
(B) 47%
(C) 38%
(D) 19%
(E) 6%

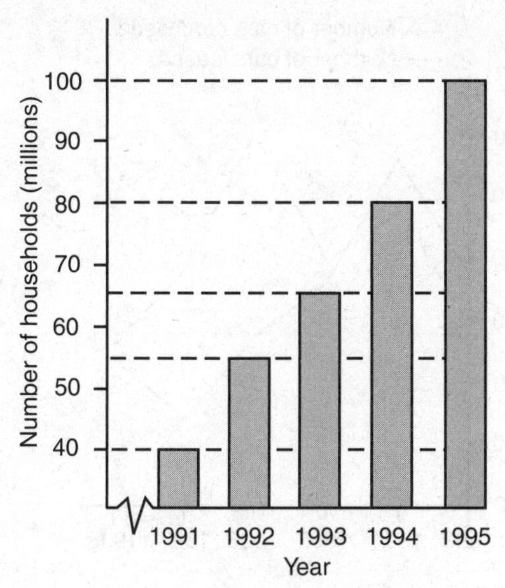

Growth of Computers in U.S. Households

Questions 4 and 5.

The graph above shows the number of U.S. households with computers for the years 1991 to 1995.

4 What was the percent of increase in the number of households with computers from 1991 to 1995?

(A) 60%
(B) 75%
(C) 80%
(D) 120%
(E) 150%

5 The greatest percent of increase in the number of households with computers occurred in which 2 consecutive years?

(A) 1991 to 1992
(B) 1992 to 1993
(C) 1993 to 1994
(D) 1994 to 1995
(E) It cannot be determined from the information given.

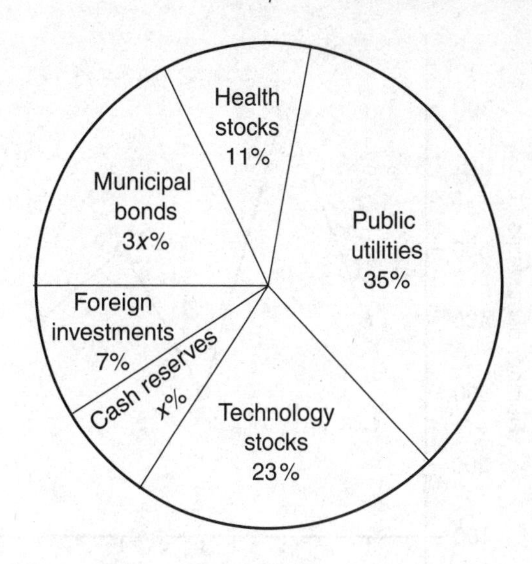

Investment Portfolio Valued at $250,000

Questions 6 and 7.

The graph above shows how $250,000 is invested.

6 How much money is invested in municipal bonds?

(A) $45,000
(B) $37,500
(C) $35,000
(D) $30,000
(E) $15,000

7 After 20% of the amount that is invested in technology stocks is reinvested in health stocks, how much money is invested in health stocks?

(A) $77,500
(B) $65,000
(C) $50,000
(D) $45,000
(E) $39,000

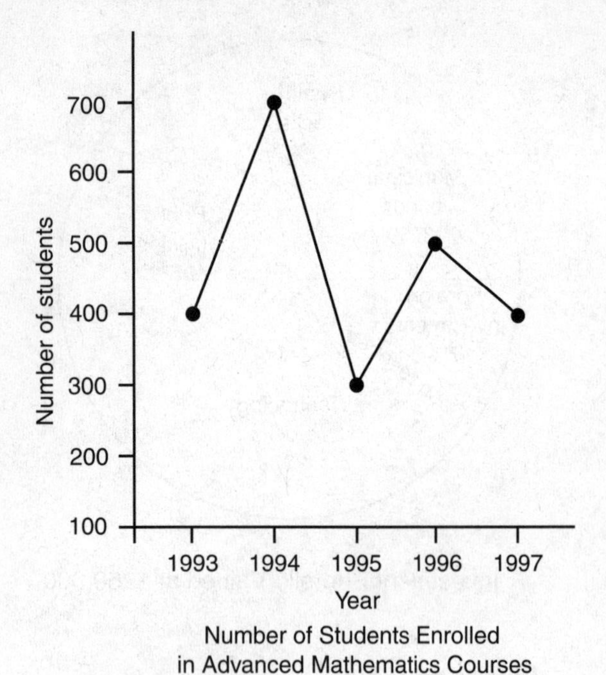

Number of Students Enrolled
in Advanced Mathematics Courses

Questions 8 and 9 refer to the graph above.

8 The percent increase in the number of students enrolled in advanced mathematics courses from 1993 to 1994 exceeded the percent increase from 1995 to 1996 by approximately what percent?

(A) 200
(B) 133
(C) 75
(D) 67
(E) 8

9 From 1997 to 1998 the number of students enrolled in advanced mathematics courses increased by the same percent that student enrollment in advanced mathematics courses dropped from 1996 to 1997. What was the approximate number of students enrolled in advanced mathematics courses in 1998?

(A) 420
(B) 440
(C) 450
(D) 460
(E) 480

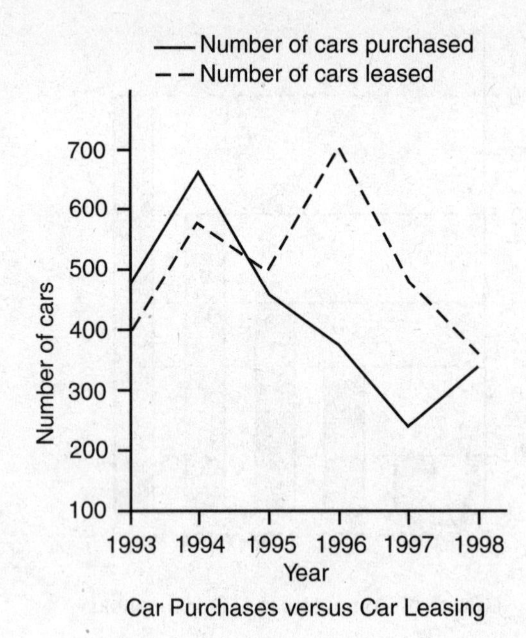

Car Purchases versus Car Leasing

Questions 10 and 11 refer to the graph above.

10 In 1995, the number of cars purchased was what percent of the number of cars leased?

(A) 86
(B) 88
(C) 90
(D) 92
(E) 94

11 Which of the following is the best approximation for the decrease in the number of cars purchased per year between 1994 and 1997?

(A) 105
(B) 140
(C) 300
(D) 420
(E) 480

LESSON 7-5

MISCELLANEOUS PROBLEM TYPES

OVERVIEW

Some SAT questions do not fall into a particular branch of mathematics. These questions may involve:

- *performing an unfamiliar arithmetic operation by following a given rule;*
- *finding a missing digit in a correctly worked out arithmetic problem; or*
- *figuring out a particular term in a list of numbers that follow a rule.*

NEW OPERATIONS THAT USE SPECIAL SYMBOLS

In some SAT math questions a new arithmetic operation is invented by giving a rule that uses an unfamiliar operation symbol. For example, if for all positive integer values of a and b, $\otimes$ is defined by the equation

$$a \otimes b = \left(\frac{a}{b}\right)^2$$

then the value of $6 \otimes 2$ can be obtained by substituting 6 for a and 2 for b in the given formula. Thus:

$$\underset{\downarrow}{a} \otimes \underset{\downarrow}{b} = \left(\frac{a}{b}\right)^2$$

$$6 \otimes 2 = \left(\frac{6}{2}\right)^2 = 3^2 = 9$$

MISSING-DIGIT PROBLEMS

In some SAT math questions a correctly worked out arithmetic problem is presented with one or more digits replaced by letters or symbols. You need to use mathematical reasoning, together with systematic trial and error, to figure out the missing digits.

EXAMPLE

$$\begin{array}{r} AB \\ \underline{\times B} \\ 1A9 \end{array}$$

In the correctly worked out multiplication problem above, A and B are single digits. What is a possible value of A?

Solution

- Note that since B times B contains the digit 9, the only possible choice for B is 3 or 7.
- Try $B = 3$ by systematically multiplying numbers of the form $A3$ by 3 to see whether any products of the form $1\,A\,9$ are possible. Since the product must be a three-digit number, A in this case must be greater than 3. The product on the left below is a three-digit number, but A is 4 in the multiplicand and 2 in the product. Try $A = 5$, and note that the product on the right has the desired form.

$$\begin{array}{r} 43 \\ \times 3 \\ \hline 129 \end{array} \qquad \begin{array}{r} 53 \\ \times 3 \\ \hline 159 \end{array}$$

- If no choices for A work with $B = 3$, repeat the process with $B = 7$.

NUMBER SEQUENCES

A list of numbers that follow a predictable pattern is called a **sequence.** For example, in the sequence

$$3, 6, 12, 24, 48, \ldots$$

each number after the first is obtained by multiplying the number that comes before it by 2. The three trailing dots (. . .) indicate that the pattern continues without ever ending. SAT test questions may ask you to find or compare the values of particular terms of a sequence.

EXAMPLE

In the sequence below, each term after the first term, x, is obtained by doubling the term that comes before it and then subtracting 3. What is the value of $x + z$?

$$x, y, 19, 35, 67, z, \ldots$$

Solution

- Since you need to find the terms that come *before* 19, reverse the given rule. *Add* 3 to 19 to get 22, and then take *one-half* of 22 to get $y = 11$.
- To find x, add 3 to 11 (the value of y) to get 14, and then take one-half of 14 to get $x = 7$.
- To find z, double the preceding term, 67, to get 134. Subtracting 3 from 134 gives $z = 131$.

Hence, $x + z = 7 + 131 = 138$.

ARITHMETIC SEQUENCES

In the *arithmetic* sequence

$$2, 5, 8, 11, 14, \ldots$$

5 is obtained by adding 3 to 2, 8 is obtained by adding 3 to 5, 11 is obtained by adding 3 to 8, and so forth. An **arithmetic sequence** is an ordered list of numbers in which each number after the first is obtained by adding the *same* number to the number that comes before it. The same number that is added to each term in order to get the next term is called the **common difference.** In the arithmetic sequence 2, 5, 8, 11, 14, . . . the common difference is 3.

Here are two formulas that are worth remembering for arithmetic sequences having n terms:

- Last term $= a_n = a_1 + (n - 1) \times d$
- Sum of terms $= S_n = \dfrac{n \times (a_1 + a_n)}{2}$

where:
a_1 = the first term of the arithmetic sequence.
a_n = the last term of the arithmetic sequence.
d = the common difference between consecutive terms.
S_n = the sum of all the terms from a_1 to a_n, inclusive.

EXAMPLE

The first row of an auditorium has 14 seats. If each row after the first has two more seats than the row in front of it, how many seats are in the 20th row?

Solution

Use the formula $a_n = a_1 + (n - 1) \times d$ where $a_1 = 14$, $n = 20$, and $d = 2$.

$$\begin{aligned} a_n &= a_1 + (n - 1) \times d \\ a_{20} &= 14 + (20 - 1) \times 2 \\ &= 14 + 19 \times 2 \\ &= 14 + 38 \\ &= 52 \end{aligned}$$

Thus, there are 52 seats in the 20th row of the auditorium.

EXAMPLE

If the auditorium in the previous example has 20 rows, what is the total number of seats in the auditorium?

Solution:

Use the formula $S_n = \frac{n \times (a_1 + a_n)}{2}$ where $a_1 = 14$, $n = 20$, and $a_{20} = 52$.

$$\begin{aligned} S_n &= \frac{n \times (a_1 + a_n)}{2} \\ S_{20} &= \frac{20 \times (14 + 52)}{2} \\ &= \frac{20 \times 66}{2} \\ &= 10 \times 66 \\ &= 660 \end{aligned}$$

Hence, the auditorium has a total of 660 seats.

Lesson 7-5 Tune-Up Exercises

Multiple-Choice

1 For all nonnegative numbers a, $\boxed{a} = \sqrt{a} + 1$. Which of the following is equal to the sum of $\boxed{36} + \boxed{64}$?

(A) 10
(B) 12
(C) 16
(D) 52
(E) 102

---

2 If ${}_{b}\overset{a}{\triangle}_{c} = \dfrac{a+b}{c} + \dfrac{a+c}{b} + \dfrac{b+c}{a}$ for all nonzero a, b, and c, then ${}_{2}\overset{1}{\triangle}_{3} =$

(A) 12
(B) 10
(C) 8
(D) 5
(E) 3

Questions 3 and 4 refer to the following definition:

For all positive integers k,

let $\boxed{k}\!\rhd = 2(k - 1)$ if k is even;

let $\boxed{k}\!\rhd = \dfrac{1}{2}(k + 1)$ if k is odd.

3 The product $\boxed{6}\!\rhd \times \boxed{11}\!\rhd =$

(A) 25
(B) 30
(C) 31
(D) 59
(E) 60

4 If $\boxed{N}\!\rhd$ is a multiple of 4, which of the following can be the value of N?

(A) 8
(B) 12
(C) 25
(D) 31
(E) 40

5

$$\begin{array}{r} A6 \\ B4 \\ +\ B8 \\ \hline 148 \end{array}$$

In the correctly worked out addition example above, what is the greatest possible value of digit B?

(A) 9
(B) 8
(C) 6
(D) 5
(E) 4

6

$$a_1, a_2, a_3, a_4, a_5, \ldots, a_n$$

In the sequence of positive integers above, $a_1 = a_2 = 1$, $a_3 = 2$, $a_4 = 3$, and $a_5 = 5$. If each term after the second is obtained by adding the two terms that come before it and if $a_n = 55$, what is the value of n?

(A) 12
(B) 10
(C) 9
(D) 8
(E) 5

7

$$\begin{array}{r} 63 \\ \times\ \ 4P \\ \hline 31P \\ 2P2 \\ \hline 2Q3P \end{array}$$

In the correctly worked out multiplication problem above, what digit does Q represent?

(A) 8
(B) 7
(C) 6
(D) 5
(E) 4

<u>**Questions 8 and 9**</u> refer to the following definition:

Let the operation Φ be defined for all positive integers a and b by the equation

$$a \Phi b = ab - a$$

8 For what value of a is $a \Phi 4 = 24$?

(A) 3
(B) 4
(C) 6
(D) $\frac{25}{4}$
(E) 8

9 Which of the following must be true?

I. $a \Phi b = b \Phi a$
II. $a \Phi (a + 1) = a^2$
III. $\frac{a}{2} \Phi \frac{b}{2} = \frac{a \Phi b}{2}$

(A) I only
(B) II only
(C) III only
(D) I and III only
(E) II and III only

10

$$\begin{array}{r} C5 \\ B7 \\ +1A9 \\ \hline 361 \end{array}$$

In the correctly worked out addition problem above, which of the following could be digit A?

I. 0
II. 4
III. 8

(A) I only
(B) III only
(C) I and II only
(D) II and III only
(E) I, II, and III

<u>**Questions 11 and 12**</u> refer to the following definition:

For all positive integers y,

$$\triangle_{y} = 2\sqrt{y}$$

11 Which of the following equals 8?

(A) $\triangle_{4}$
(B) $\triangle_{8}$
(C) $\triangle_{16}$
(D) $\triangle_{32}$
(E) $\triangle_{64}$

12 $\triangle_{y} \times \triangle_{y} =$

(A) 2
(B) 4
(C) $4\sqrt{y}$
(D) $4y$
(E) $4y^2$

13 Let $\frown x \frown$ be defined for all real values of x by the equation

$$\frown x \frown = 2x + 1$$

If $3 - \frown x \frown = x$, then $x =$

(A) $\frac{2}{3}$
(B) 1
(C) $\frac{3}{2}$
(D) 2
(E) 3

14

$$\begin{array}{r} 83A \\ +DBB \\ \hline CAC2 \end{array}$$

In the correctly worked out addition problem above, what digit does D represent?

(A) 2
(B) 4
(C) 5
(D) 6
(E) 7

Questions 15 and 16 refer to the following definition:

For all positive integers, let $\underset{\vee}{\hat{p}}$ be defined by the equation

$$\underset{\vee}{\hat{p}} = p + k$$

where k is the greatest divisor of p and $k < p$.

15 $\underset{\vee}{\overset{\wedge}{28}} =$

(A) 56
(B) 42
(C) 35
(D) 32
(E) 30

16 If $\underset{\vee}{\overset{\wedge}{15}} = n$, then $\underset{\vee}{\hat{n}} =$

(A) 20
(B) 24
(C) 25
(D) 30
(E) 40

17 $\frac{1}{2}, -\frac{1}{4}, \frac{1}{8}, -\frac{1}{16}, \ldots, a_n$

In the above sequence, $a_1 = \frac{1}{2}$, $a_2 = -\frac{1}{4}$, $a_3 = \frac{1}{8}$, $a_4 = -\frac{1}{16}$, and a_n represents the nth term. Which equation expresses a_n in terms of n?

(A) $(-1)^{n+1} \cdot \frac{1}{2^n}$

(B) $(-1)^n \cdot \frac{1}{2^n}$

(C) $(-1)^{n+1} \cdot \frac{1}{2^{-n}}$

(D) $\left(\frac{-1}{2}\right)^n$

(E) $\frac{-1}{2^n}$

Grid-In

1 If the operation Ω is defined by the equation

$$x \,\Omega\, y = 2x + 3y$$

what is the value of a in the equation $a \,\Omega\, 4 = 1 \,\Omega\, a$?

$$x, y, 22, 14, 10, \ldots$$

2 In the sequence above, each term after the first term, x, is obtained by halving the term that comes before it and then adding 3 to that number. What is the value of $x - y$?

3 In a certain sequence of positive integers, if m and n are consecutive terms, the next consecutive term is

$\frac{m+n}{2}$ if $m + n$ is even; and

$m+n-3$ if $m + n$ is odd

For example, if two consecutive terms are 11 and 17, the next two consecutive terms are 14 and 28. In a sequence that follows the same rule, if the first two terms are 15 and 19, what is the seventh term?

4 Let the operation ▲ be defined by the equation

$$a \blacktriangle b = ab - (a + b)$$

If $5 \blacktriangle b = 1$, what is the value of b?

ANSWERS TO CHAPTER 7 TUNE-UP EXERCISES

Lesson 7-1 (*Multiple-Choice*)

1. **(B)** If the average (arithmetic mean) of a set of seven numbers is 81, then the sum of these seven numbers is 7×81 or 567. Since, if one of the numbers is discarded, the average of the six remaining numbers is 78, the sum of these six numbers is 6×78 or 468. Since $567 - 468 = 99$, the value of the number that was discarded is 99.

2. **(E)** Since vertical angles are equal, the angle of the quadrilateral opposite the 105° angle also measures 105°. Since the sum of the measures of the angles of a quadrilateral is 360°,

$$a + b + 105^\circ + 65^\circ = 360^\circ$$
$$a + b = 360^\circ - 170^\circ$$
$$\frac{a+b}{2} = \frac{190}{2} = 95$$

3. **(C)** If the average of the two-digit numbers $1N$, $N7$, and NN is 35, then the sum of the three numbers is 3×35 or 105. Since $1N + N7 + NN = 105$, look for a value of N in the set of answer choices such that $N + 7 + N$ is 15 or 25. For choice (C), $N = 4$,

$$14 + 47 + 44 = 105$$

4. **(C)** The radii of circles with areas of 16π and 100π are 4 and 10, respectively. The average of 4 and 10 is

$$\frac{4+10}{2} = \frac{14}{2} = 7$$

The area of a circle of radius 7 is $\pi(7)^2$ or 49π.

5. **(B)** The average of 3^k and 3^{k+2} is their sum divided by 2:

$$\frac{3^k + 3^{k+2}}{2} = \frac{3^k + 3^2 \cdot 3^k}{2} = \frac{1 \cdot 3^k + 9 \cdot 3^k}{2} = \frac{10 \cdot 3^k}{2} = 5 \cdot 3^k$$

6. **(C)** When x is subtracted from $2y$ the difference is equal to the average of x and y, so $2y - x = \frac{x+y}{2}$ or $2(2y - x) = x + y$. Hence,

$$4y - 2x = x + y$$
$$4y - y = x + 2x$$
$$3y = 3x$$

Dividing both sides of $3y = 3x$ by $3y$ gives $\frac{x}{y} = 1$.

7. **(E)** If the average of x, y, and z is 32, then $\frac{x+y+z}{3} = 32$, so

$$x + y + z = 3 \times 32 = 96$$

Since the average of y and z is 27, then $\frac{y+z}{2} = 27$, so

$$y + z = 2 \times 27 = 54$$

Substituting the value of $y + z$ in the equation $x + y + z = 96$ gives $x + 54 = 96$, so

$$x = 96 - 54 = 42$$

and

$$2x = 84$$

Hence, the average of x and $2x$ equals

$$\frac{42+84}{2} = \frac{126}{2} = 63$$

8. **(B)** Eliminate choice (A) because the mode is 2 but the average (mean) of 2, 2, 2, 2, and 4 must be greater than 2. Eliminate choices (C), (D), and (E) because the mode in each case is 1 but the average (mean) in each of these answer choices must be greater than 1. In choice (B) the mode is 3, the average (mean) of 1, 3, 3, 3, and 5 is

$$\frac{1+3+3+3+5}{5} = \frac{15}{5} = 3$$

and the median is 3 since two values (1 and 3) are less than or equal to 3 and two other values (3 and 5) are equal to or greater than 3.

9. **(D)** Let x represent the average rate of speed in miles per hour for the second part of the trip, which lasts $7 - 3$ or 4 hours. Since rate multiplied by time equals distance, the man drives 45×3 or 135 miles during the first part of the trip and $4x$ miles during the second part. Total distance divided

by total time gives the average rate of speed for the entire trip, so

$$\frac{135 + 4x}{7} = 53$$

$$135 + 4x = 7 \times 53 = 371$$

$$4x = 371 - 135 = 236$$

$$x = \frac{236}{4} = 59$$

10. **(B)** For the set of numbers 2, 2, 4, 5, and 12, the mode is 2, the mean (or average) is

$$\frac{2 + 2 + 4 + 5 + 12}{5} = \frac{25}{5} = 5$$

and the median or middle value in the list is 4. Hence, the mean is greater than the mode.

11. **(A)** If x represents the lowest score Susan can receive on her next math exam and have an average of at least 85 on the five exams, then

$$\frac{78 + 93 + 82 + 76 + x}{5} = 85$$

$$329 + x = 5 \times 85$$

$$= 425$$

$$x = 425 - 329$$

$$= 96$$

12. **(E)** Since

$$(x + y)^2 = x^2 + 2xy + y^2$$

and

$$(x - y)^2 = x^2 - 2xy + y^2$$

the average of $(x + y)^2$ and $(x - y)^2$ is

$$\frac{\left(x^2 + 2xy + y^2\right) + \left(x^2 + 2xy + y^2\right)}{2}$$

or

$$\frac{2\left(x^2 + y^2\right)}{2} = x^2 + y^2$$

13. **(D)** Since the average of x and y is $\frac{11}{2}$, then,

$$\frac{x + y}{2} = \frac{11}{2} \text{ or } x + y = 11$$

If the average of $\frac{1}{x}$ and $\frac{1}{y}$ is $\frac{11}{24}$, then

$$\frac{1}{2}\left(\frac{1}{x} + \frac{1}{y}\right) = \frac{11}{24} \text{ or } \frac{1}{x} + \frac{1}{y} = \frac{11}{12}$$

Since

$$\frac{1}{x} + \frac{1}{y} = \frac{x + y}{xy} \text{ and } x + y = 11$$

then $\frac{11}{xy} = \frac{11}{12}$, so $xy = 12$.

14. **(C)** If the ratio of a to b is $\frac{1}{2}$, then $\frac{a}{b} = \frac{1}{2}$, so $2a = b$ or $a = \frac{b}{2}$. Since the ratio of c to b is $\frac{1}{3}$, then $\frac{c}{b} = \frac{1}{3}$, so $3c = b$ or $c = \frac{b}{3}$. Hence, the average of a and c in terms of b is

$$\frac{1}{2}(a + c) = \frac{1}{2}\left(\frac{b}{2} + \frac{b}{3}\right) = \frac{1}{2}\left(\frac{5b}{6}\right) = \frac{5b}{12}$$

15. **(D)** If the average of a, b, c, d, and e is 28, then

$$\frac{a + b + c + d + e}{5} = 28$$

or

$$a + b + c + d + e = 5 \times 28 = 140$$

If the average of a, c, and e is 24, then

$$\frac{a + c + e}{3} = 24$$

or

$$a + c + e = 3 \times 24 = 72$$

Substituting 72 for $a + c + e$ in $a + b + c + d + e = 140$ gives

$$b + d + 72 = 140$$

$$b + d = 140 - 72 = 68$$

$$\frac{b + d}{2} = \frac{68}{2} = 34$$

16. **(A)** Adding corresponding sides of the given equations, $2a + b = 7$ and $b + 2c = 23$, gives $2a + 2b + 2c = 30$ or $a + b + c = 15$. Hence:

$$\frac{a + b + c}{3} = \frac{15}{3} = 15$$

17. **(A)** If the average final exam grade of 10 students is x, then the sum of the final exam grades of the 10 students is $10x$. Similarly, the sum of the final exam grades of 15 students is $15y$. If the average final exam grade of the 25 students is t, then

$$\frac{10x + 15y}{25} = t$$

$$10x + 15y = 25t$$

$$2x + 3y = 5t$$

$$2x = 5t - 3y$$

$$x = \frac{5t - 3y}{2}$$

18. **(B)** The average of a, b, c, and d is p, so

$$\frac{a + b + c + d}{4} = p$$

or $a + b + c + d = 4p$. Similarly, since the average of a and c is q, then $a + c = 2q$, so $b + d + 2q = 4p$. Since $b + d = 4p - 2q$, then

$$\frac{b + d}{2} = \frac{4p}{2} - \frac{2q}{2} = 2p - q$$

(Grid-In)

1. **22** Let x, $x + 2$, $x + 4$, and $x + 6$ represent four consecutive even integers. If their average is 19, then

$$\frac{x + (x + 2) + (x + 4) + (x + 6)}{4} = 19$$

or

$$4x + 12 = 4 \times 19 = 76$$

Then

$$4x = 76 - 12 = 64$$
$$x = \frac{64}{4} = 16$$

Hence, $x + 6$, the greatest of the four consecutive even integers, is $16 + 6$ or 22.

2. **36** *Solution 1*: Since the average of x, y, and z is 12, then $x + y + z = 3 \times 12 = 36$. Hence,

$$3x + 3y + 3z = 3(36) = 108$$

The average of $3x$, $3y$, and $3z$ is their sum, 108, divided by 3 since three values are being added: $\frac{108}{3}$ or 36.

Solution 2: The average of x, y, and z is 12, so the average of any constant multiple of x, y, and z is 12 times that constant multiple, or $12(3) = 36$.

3. **28** If the average of three consecutive multiples of 4 is 32, then

$$\frac{4k + 4(k + 1) + 4(k + 2)}{3} = 32$$

where k is some positive integer. Hence,

$$12k + 12 = 3 \times 32$$
$$12k = 84$$
$$k = \frac{84}{12} = 7$$

The least of the three consecutive multiples is $4k = 4(7) = 28$.

4. **91** Let x represent the greatest possible test score. From the given information, at least four of the seven scores are known and two of the unknown scores must be below 82, which is given as the median:

$$70, ?, ?, 82, 87, 87, x$$

Since the average of the seven integer test scores is fixed at 80, the last test score will be greatest when the two unknown test scores below the average have the smallest possible value. The two unknown test scores below the median must be unequal since the mode is given as 87. Hence, these test scores must be 71 and 72. Since the average is 80,

$$\frac{70 + 71 + 72 + 82 + 87 + 87 + x}{7} = 80$$

$$70 + 71 + 72 + 82 + 87 + 87 + x = 80 \times 7$$
$$469 + x = 560$$
$$x = 560 - 469$$
$$= 91$$

Lesson 7-2 *(Multiple-Choice)*

1. **(A)** The total number of different sundaes that the ice cream parlor can make is the number of different flavors of ice cream times the number of different flavors of syrup times the number of different toppings: $6 \times 3 \times 4 = 72$.
2. **(D)** Although the red block must occupy the first position in the row, the four remaining blocks can be arranged in any order. Hence, there is 1 choice for the first position, 4 choices for the second position, 3 choices for the third position, 2 choices for the fourth position, and 1 choice for the last position. The total number of arrangements is $1 \times 4 \times 3 \times 2 \times 1$ or 24.
3. **(D)** The region in the diagram that represents the set of elements that belong to sets B and C but not to set A is the region in which circles B and C overlap outside of circle A. This region is labeled d.
4. **(D)** Any one of the five students can be seated in the first chair, any of the four remaining

students in the second chair, and any of the three remaining students in the third chair. Thus, five students can be seated in three chairs in $5 \times 4 \times 3$ or 60 different ways.

5. **(B)** Since the first coin is replaced before another coin is picked, the same coin may be drawn twice. There are seven possible ways in which two coins whose values add up to 25 or more cents can be picked: (quarter, quarter), (quarter, dime), (dime, quarter), (quarter, nickel), (nickel, quarter), (quarter, penny), and (penny, quarter).
6. **(C)** The number of students on one but not both of the two teams is the sum of the number of students on the basketball team plus the number on the tennis team minus the number on both teams: $15 + 18 - 11 = 22$. Since 22 students are on the basketball or the tennis team and 14 students are not on either team, there are $22 + 14$ or 36 students in the class.
7. **(D)** In forming a four-digit number, 0 cannot be used as the first digit, so only five out of the six possible digits can be used to fill the first decimal position of the number. Since the same digit cannot be used more than once, any of the five remaining digits can be used to fill the second position, any of the four remaining digits can be used to fill the third position, and any of the three remaining digits can be used to fill the last position. Thus, $5 \times 5 \times 4 \times 3$ or 300 four-digit numbers greater than 1000 can be formed from the digits 0, 1, 2, 3, 4, and 5 without repeating any digit.
8. **(C)** If the first letter is *T*, as the question states, and the second letter is *A*, then any one of the six other letters can fill the last position of each three-letter arrangement. Hence, there are $1 \times 1 \times 6$ or 6 possible three-letter arrangements. If the first letter is *T* and the third letter is *A*, then any one of the six other letters can fill the middle position of each three-letter arrangement. This possibility gives $1 \times 6 \times 1$ or 6 possible three-letter arrangements. Hence, the total number of three-letter arrangements in which the first letter is *T* and one of the other letters is *A* is $6 + 6$ or 12.
9. **(C)** If F students in the French class are in the French Club, S students are in the Spanish Club, and N of these students are in both the French and the Spanish clubs, then $F - N$ students belong only to the French Club and the total number of students in the class is $F + S - N$. Hence, the fraction of the class that belongs to the French Club but not to the Spanish Club is $\frac{F-N}{F+S-N}$.
10. **(D)** When the digits 0, 1, 3, 4, and 5 are used to form a three-digit number greater than 400, the first digit must be either 4 or 5. Hence, the number of *three*-digit numbers greater than 400 that can be formed from the digits 0, 1, 3, 4, and 5, without repeating any digit, is $2 \times 4 \times 3$ or 24. When the same set of digits is used to form a four-digit number, the first digit cannot be 0. Hence, the number of *four*-digit numbers that can be formed from the digits 0, 1, 3, 4, and 5, without repeating any digit, is $4 \times 4 \times 3 \times 2$ or 96. Hence, 120 (= 24 + 96) different numbers greater than 400 but less than 9999 can be formed.
11. **(A)** The shaded region in the diagram represents the set whose members are elements of sets X, Y, and Z. Look in the answer choices for a number that is odd (set X), has a square root that is an integer (set Y), and is a multiple of 5 (set Z). The only number that has all three properties is 25, choice (*A*).
12. **(C)** Given $X = \{1, 2, 4\}$ and $Y = \{1, 3, 4\}$, make a list of all the possible sums $x + y$:

    $1 + 1 = 2 \quad 2 + 1 = 3 \quad 4 + 1 = 5$
    $1 + 3 = 4 \quad 2 + 3 = 5 \quad 4 + 3 = 7$
    $1 + 4 = 5 \quad 2 + 4 = 6 \quad 4 + 4 = 8$

    Hence, seven different values of $x + y$ are possible: 2, 3, 4, 5, 6, 7, and 8.
13. **(B)** If in a music class of 30 students, 50% study woodwinds, 40% study strings, and 30% study both woodwinds and strings, then 50% of 30 or 15 students study woodwinds, 40% of 30 or 12 students study strings, and 30% of 30 or 9 students study both woodwinds and strings. Hence, $15 + 12 - 9$ or 18

students in the class study either woodwinds or strings. Thus, 30 − 18 or 12 students do not study either woodwind or string instruments. Since

$$\frac{12}{30} = \frac{2}{5} = 0.40$$

40% of the class do not study either woodwind or string instruments.

14. **(E)** Determine whether each Roman numeral statement is true or false, given that, of 25 city council members, 12 voted for proposition *A*, 19 voted for proposition *B*, and 10 voted for both propositions.
   - I. Since 19 members voted for proposition *B* and 10 voted for both propositions, 19 − 10 or 9 members voted only for proposition *B*. Hence, statement I is false.
   - II. Since 12 + 19 − 10 = 21, then 21 members voted for at least one of the two propositions. Since 25 members voted, 25 − 21 or 4 members voted against both propositions. Hence, statement II is true.
   - III. Since 10 members voted for both propositions, 12 − 10 or 2 members voted for proposition *A* but not for proposition *B*. Also, 19 − 10 = 9 members voted for proposition *B* but not for proposition *A*. Since 2 + 9 or 11 members voted for one proposition and against the other proposition, statement III is true.

   Only Roman numeral statements II and III are true.

15. **(C)** To count the number of elements in set *A* or *B* or *C*, find the sum of the members of each set, subtract from this sum the sum of the elements common to any two of these sets, and then add the number of elements common to all three sets. If x represents the number of people who subscribe to all three magazines, the number of people who subscribe to magazine *A* or *B* or *C* is
   $33 + 30 + 17 - (9 + 9 + 9) + x$ or $53 + x$
   Since 5 of the 63 people did not subscribe to any of the three magazines, 58 people subscribed to magazine *A* or *B* or *C*. Hence, $58 = 53 + x$, so $x = 5$.

16. **(C)** To find the number of five-member teams that can be made from a group of eight students, evaluate ${}_nC_r$ where $n = 8$ and $r = 5$:

$${}_nC_r = \frac{n!}{r!(n-r)!}$$

$${}_8C_5 = \frac{8!}{5!(8-5)!} = \frac{8 \times 7 \times 6 \times \cancel{5!}}{\cancel{5!} \times 3!}$$

$$= \frac{8 \times 7 \times \cancel{6}}{\cancel{3 \times 2} \times 1} = 8 \times 7$$

$$= 56$$

17. **(C)** Consider each of the voting choices in turn:
   - A voter may cast a ballot for two of the six candidates in ${}_6C_2 = 15$ ways.
   - A voter may cast a ballot for exactly one of the six candidates in 6 ways.
   - A voter may cast a ballot for none of the six candidates in 1 way.

   Hence, the total number of different voting choices is 15 + 6 + 1 = 22.

(Grid-In)

1. **84** To find the percentage of students enrolled in mathematics or computer science, calculate

$$\frac{\text{number of students in math or computer science}}{\text{total number of students}} \times 100\%$$

   First find the total number of students:
   - The number of students taking *only* mathematics = **15**
   - The number of students taking *only* computer science = **8**
   - The number of students taking *only* biology = **8**
   - The number of students taking mathematics *and* computer science but *not* biology = **7**
   - The number of students taking mathematics *and* biology but *not* computer science = **4**
   - The number of students taking computer science *and* biology but *not* mathematics = **6**

- The number of students taking mathematics *and* biology *and* computer science = **2**
- Hence, the total number of students is $15 + 8 + 8 + 7 + 4 + 6 + 2 = \mathbf{50}$

Since exactly 8 students are taking only biology, $50 - 8 = \mathbf{42}$ students must be taking mathematics, computer science, or both mathematics and computer science. Therefore:

$$\begin{aligned}\% \text{ in math or computer science} &= \frac{42}{50} \times 100\% \\ &= 84\%\end{aligned}$$

2. **6** If x represents the number of color choices, then $4 \cdot 3 \cdot x = 72$, so $x = \frac{72}{12} = 6$.
3. **8100** Since digits can be repeated and there are 9 digits excluding 0, there are 9 choices for the first and last digits:

$$\boxed{9} \times \square \times \square \times \boxed{9}$$

Any one of the 10 digits can be used to fill the second and third positions. Hence, the number of different telephone numbers that can be assigned when the last four digits of each telephone number does not begin or end with a zero is

$$\boxed{9} \times \boxed{10} \times \boxed{10} \times \boxed{9} = 8100$$

4. **10** To find the total number of different circles that can be drawn through any three of the five points, evaluate ${}_nC_r$ where $n = 5$ and $r = 3$:

$${}_nC_r = \frac{n!}{r!(n-r)!}$$

$$\begin{aligned}{}_5C_3 &= \frac{5!}{3!(5-3)!} = \frac{5 \times 4 \times \cancel{3!}}{\cancel{3!} \times 2!} \\ &= \frac{5 \times 4}{2 \times 1} \\ &= 10\end{aligned}$$

5. **720** The word *PARABOLA* contains 3 letter *A*s and 5 letters that are not *A*s. Since the three letter *A*s must appear consecutively, consider it one object and consider the remaining 5 letters to be 5 different objects. Hence, the 5 + 1 = 6 objects can be arranged in

$$6! = 6 \times 5 \times 4 \times 3 \times 2 \times 1 = 720 \text{ ways}$$

Lesson 7-3 (*Multiple-Choice*)

1. **(E)** The total number of marbles in the bag is $3 + 4 + 2$ or 9. Of these 9 marbles, $3 + 4$ or 7 marbles are not orange. Hence, the probability that an orange marble will NOT be picked is $\frac{7}{9}$.
2. **(D)** The average of the set 1, 2, 2, 3, 3, 3, 4, 4, 4, 4 is the sum of the 10 numbers divided by 10 or $\frac{30}{10} = 3$. Since three of the 10 numbers are 3, the probability that a number selected at random from the set of 10 numbers will be the average of the set is $\frac{3}{10}$.
3. **(B)** If the probability that an event will occur is $\frac{x}{4}$ and $x \neq 0$, then the probability that this event will NOT occur is

$$1 - \frac{x}{4} = \frac{4}{4} - \frac{x}{4} = \frac{4-x}{4}$$

4. **(B)** Since the probability of picking a red marble is 0.25 and the probability of picking a blue marble is 0.40, the probability of picking a green marble is $1 - (0.25 + 0.40)$ or 0.35.
5. **(D)** To find the LEAST number of marbles that could be in the jar, multiply each answer choice by 0.25, 0.40, and 0.35 until an integer is obtained for each product. Since 4 times 0.40 is 1.6, rule out choice (A); 5 times 0.25 is 1.25, so rule out choice (B); 10 times 0.35 is 3.5, so rule out choice (C). Since 20 times 0.25 is 5, 20 times 0.40 is 8, and 20 times 0.35 is 7, choice (D) is correct.
6. **(D)** Determine whether each Roman numeral statement is true or false when a cube whose faces are numbered from 1 to 6 is rolled.
   - I. Since any one of the six numbers may show up on the top face, the probability of getting a 5 on the top face is $\frac{1}{6}$. Hence, statement I is false.
   - II. Since there are three even numbers (2, 4, and 6), the probability of getting an even number on the top face is $\frac{3}{6}$ or $\frac{1}{2}$. Hence, statement II is true.
   - III. Since there are three prime numbers from 1 to 6 (2, 3, and 5), the probability of getting a prime number on the top face is $\frac{3}{6}$ or $\frac{1}{2}$. Hence, statement III is true.

Only Roman numeral statements II and III are true.

7. **(A)** The five letters *L, O, G, I,* and *C* can be arranged in $5 \times 4 \times 3 \times 2 \times 1$ or 120 different ways. Since only one of these arrangements is "LOGIC," the probability that a random arrangement of the five letters will form the word "LOGIC" is $\frac{1}{120}$.

8. **(E)** Since three of the eight sectors are even-numbered, the probability that the spinner will land on an even-numbered region is $\frac{3}{8}$. Hence, the probability that the spinner will land on an even-numbered region in each of two consecutive spins is $\frac{3}{8} \times \frac{3}{8}$ or $\frac{9}{64}$.

9. **(C)** If three fair coins are tossed at the same time, the probability that all three will come up heads is $\frac{1}{2} \times \frac{1}{2} \times \frac{1}{2}$ or $\frac{1}{8}$. The probability that all three coins will come up tails is also $\frac{1}{2} \times \frac{1}{2} \times \frac{1}{2}$ or $\frac{1}{8}$. Hence, the probability that all three coins will come up heads *or* all will come up tails is $\frac{1}{8} + \frac{1}{8}$ or $\frac{1}{4}$.

10. **(B)** The probability that the red and the yellow cube will each show a given number from 1 to 6 is $\frac{1}{6} \times \frac{1}{6} = \frac{1}{36}$. Hence, the probability that both cubes will show a 1 or a 2 or a 3 or a 4 or a 5 or a 6 is

$$\frac{1}{36} + \frac{1}{36} + \frac{1}{36} + \frac{1}{36} + \frac{1}{36} + \frac{1}{36} = \frac{6}{36} = \frac{1}{6}$$

11. **(D)** Since the probability of selecting a blue marble at random from the jar is $\frac{1}{3}$ and the probability of selecting a green marble at random from the jar is $\frac{4}{9}$, the probability of selecting a white marble is

$$1 - \left(\frac{1}{3} + \frac{4}{9}\right) = 1 - \frac{7}{9} = \frac{2}{9}$$

If x marbles of the 54 marbles in the jar are white, then

$$\frac{x}{54} = \frac{2}{9}$$
$$9x = 108$$
$$x = \frac{108}{9} = 12$$

12. **(C)** The squares of three of the five numbers in the set $A = \{-2, -1, 0, 1, 2\}$ are less than 2:
$(-1)^2 = 1 < 2, \quad 0^2 = 0 < 2, \quad 1^2 = 1 < 2$
Hence, the probability of picking a number x from set A such that $x^2 < 2$ is $\frac{3}{5}$.

13. **(C)** If $4 \; -3x < 6$, then $-3x < 2$, so $\frac{-3x}{-3} > \frac{2}{-3}$ and $x > -\frac{2}{3}$. Three of the six numbers in $\{-3, -2, -1, 0, 1, 2\}$ are greater than $-\frac{2}{3}$: 0, 1, and 2. Hence, the required probability is $\frac{3}{6}$ or $\frac{1}{2}$.

14. **(B)** If a number is randomly selected from set $A = \{1, 2, 3\}$ and then a second number is randomly selected from set $B = \{1, 4, 9\}$, the set of all possible products is as follows:
$1 \times 1 = 1 \quad 2 \times 1 = 2 \quad 3 \times 1 = 3$
$1 \times 4 = 4 \quad 2 \times 4 = 8 \quad 3 \times 4 = 12$
$1 \times 9 = 9 \quad 2 \times 9 = 18 \quad 3 \times 9 = 27$
Since five of the nine possible products are less than 9, the probability that the product of the two numbers selected will be less than 9 is $\frac{5}{9}$.

15. **(B)** To find the probability that the dart will land on the shaded region, find the ratio of the area of the shaded region to the area of the target:

$$P(\text{dart lands on target}) = \frac{\text{Area of shaded region}}{\text{Area of largest circle}}$$
$$= \frac{(\pi \times 7^2) - (\pi \times 3^2)}{\pi \times 9^2}$$
$$= \frac{49\pi - 9\pi}{81\pi}$$
$$= \frac{40\not{\pi}}{81\not{\pi}}$$
$$= \frac{40}{81}$$

16. **(C)** The total number of possible outcomes is $6 \times 6 = 36$. Count the number of favorable outcomes by making an organized list:

(6,5), (6,4) . . . , (6,1)	5 outcomes
(5,4), (5,3), . . . , (5,1)	4 outcomes
(4,3), (4,2), (4,1)	3 outcomes
(3,2),(3,1)	2 outcomes
(2,1)	1 outcome

Total favorable outcomes: 15

Calculate the probability ratio:

$$P(\text{first roll} > \text{second roll}) = \frac{15}{36} = \frac{5}{12}.$$

17. **(A)** To find the probability that the point will be chosen from the shaded region, find the ratio of the area of the shaded region to the area of square $ABCD$.
   - Since the circles have the same radius, the area of the four quarter circles is equal to the area of one circle, which is $\pi \times 1^2 = \pi$.
   - The length of a side of the square is $1 + 1 = 2$, so its area is $2 \times 2 = 4$.
   - The area of the shaded region is the difference between the areas of the square and the four quarter circles, which is $4 - \pi$.
   - Therefore:

$$P(\text{Point chosen from shaded region}) = \frac{\text{area of shaded region}}{\text{area of square } ABCD} = \frac{4-\pi}{4} = \frac{4}{4} - \frac{\pi}{4} = 1 - \frac{\pi}{4}$$

18. **(B)** To find the probability that the point will be chosen from the shaded region, find the ratio of the area of the shaded region to the area of the square.
   - The length of a side of the square is $9 - 1 = 8$, so its area is $8 \times 8 = 64$.
   - The diameter of the inscribed circle is equal to the length of the side of the square. Hence, the radius of the inscribed circle is $\frac{1}{2} \times 8 = 4$, so the area of the inscribed circle is $\pi \times 4^2 = 16\pi$.
   - The difference between the areas of the square and the inscribed circle is $64 - 16\pi$, which represents the sum of the areas of the four corners of the square bounded by the square and the circle. Since only two of these four corner regions are shaded, the area of the shaded region is $\frac{1}{2}(64 - 16\pi) = 32 - 8\pi$.
   - Therefore:

$$P(\text{Point chosen from shaded region}) = \frac{\text{area of shaded region}}{\text{area of square}} = \frac{32 - 8\pi}{64} = \frac{\cancel{8}(4-\pi)}{\underset{8}{\cancel{64}}} = \frac{4-\pi}{8}$$

19. **(D)** To find the probability that the point will be chosen from the interior of $\triangle JAB$, find the ratio of the area of $\triangle JAB$ to the area of square $JKLM$. Find the area of $\triangle JAB$ indirectly by subtracting the sum of the areas of right triangles JKA, BLA, and JMB from the area of square $JKLM$.

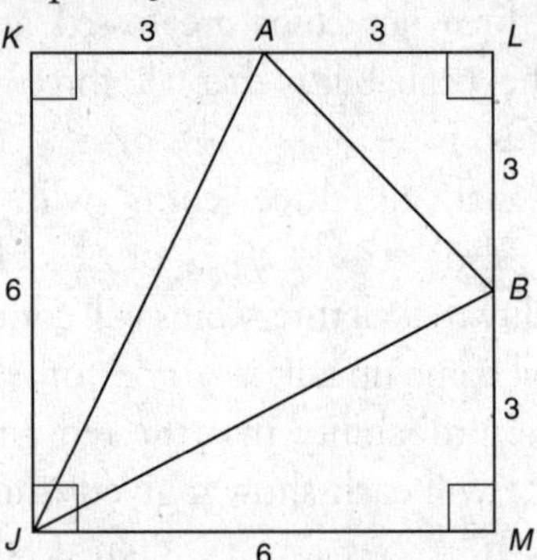

$$\text{Area of } \triangle JKA = \frac{1}{2} \times 6 \times 3 = 9$$

$$\text{Area of } \triangle BLA = \frac{1}{2} \times 3 \times 3 = 4.5$$

$$\text{Area of } \triangle JMB = \frac{1}{2} \times 6 \times 3 = 9$$

$$\text{Area of } \triangle JAB = (6 \times 6) - (9 + 4.5 + 9) = 13.5$$

Hence,

$$P(\text{Point chosen from } \triangle JAB) = \frac{\text{area of } \triangle JAB}{\text{area of square } JKLM} = \frac{13.5}{36} = \frac{135}{360} = \frac{135 \div 45}{360 \div 45} = \frac{3}{8}$$

(*Grid-In*)

1. **24** Since

$$1 - \left(\frac{1}{4} + \frac{1}{3}\right) = 1 - \frac{7}{12}$$

the probability of selecting a yellow marble is $\frac{5}{12}$. If 10 of the x marbles in the jar are yellow, then $\frac{5}{12} = \frac{10}{x}$. Since 10 is two times 5, x must be two times 12 or 24.

2. **4/20** If two cards are selected at random without replacement from a set of five identical cards each numbered on one side with a different integer from 1 to 5, the set of all possible sums is as follows:

$1 + 2 = 3$ $2 + 1 = 3$ $3 + 1 = 4$ $4 + 1 = 5$ $5 + 1 = 6$
$1 + 3 = 4$ $2 + 3 = 5$ $3 + 2 = 5$ $4 + 2 = 6$ $5 + 2 = 7$
$1 + 4 = 5$ $2 + 4 = 6$ $3 + 4 = 7$ $4 + 3 = 7$ $5 + 3 = 8$
$1 + 5 = 6$ $2 + 5 = 7$ $3 + 5 = 8$ $4 + 5 = 9$ $5 + 4 = 9$

Since four of the 20 possible sums are 8 or more, the probability of picking two cards whose sum will be at least 8 is $\frac{4}{20}$.
Grid in as 4/20.

3. **4/9** The probability that the dart will land in the interior of the inner square is equal to the ratio of the area of the inner square to the area of the larger square. It is given that the length of a side of the larger square is 1.5 times the length of a side of the smaller square. Suppose the length of a side of the smaller square is 2, then the length of a side of the larger square is 3 and the ratio of their areas is $\frac{2^2}{3^2} = \frac{4}{9}$.

4. **8/36** To find the probability that a dart will land in the triangular region formed by the lines $y = 2$, $x = 6$, and $y = x$, find the ratio of the area of the triangular region to the area of the square.
   - The area of the square dartboard is $6 \times 6 = 36$.

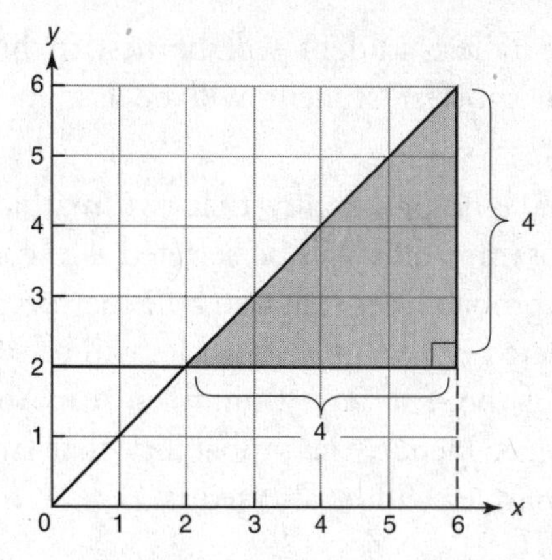

   - The area of the triangular region is $\frac{1}{2} \times 4 \times 4 = 8$.
   - Hence, the probability that the dart lands in the triangular region is $\frac{8}{36}$.

5. **1/20** To find the probability that the tallest student will be first in line and the shortest student will be last in line, find the ratio of the number of different ways in which the five students can be arranged in a line under these conditions to the number of different ways in which the five students can be arranged in a line when there are no conditions.
   - The tallest student must be in the first position and the shortest student must be in the last position. Hence, each of these positions can be filled in exactly one way:

$$\boxed{1} \times \boxed{} \times \boxed{} \times \boxed{} \times \boxed{1}$$

   The second position can be filled by any one of the remaining three students, the third position can be filled by any of the two remaining students, and the fourth position must be filled by the one remaining student:

$$\boxed{1} \times \boxed{3} \times \boxed{2} \times \boxed{1} \times \boxed{1} = 6$$

   Under the given conditions, there are six different ways in which the students can be arranged.
   - If there are no conditions, then the total number of ways in which the five students can be arranged in a line is $5! = 5 \times 4 \times 3 \times 2 \times 1 = 120$.
   - When the five students are randomly arranged in a line, the probability that

the tallest student will be first in line and the shortest student will be last in line is $\frac{6}{120} = \frac{1}{20}$.

6. **7/15** The probability that two marbles with the same color will be selected is the sum of the probabilities that two blue marbles will be selected or two green marbles will be selected.
   - Of the 4 + 2 = 6 marbles, 4 marbles are blue. Hence, the probability that two blue marbles will be selected is
   $$\frac{4}{6} \times \frac{3}{5} = \frac{12}{30}$$
   - Of the 6 marbles, 2 marbles are green. Hence, the probability that two green marbles will be selected is
   $$\frac{2}{6} \times \frac{1}{5} = \frac{2}{30}$$
   - Hence, the probability that two marbles with the same color will be selected is
   $$\frac{12}{30} + \frac{2}{30} = \frac{14}{30} = \frac{7}{15}$$

Lesson 7-4 (*Multiple-Choice*)

1. **(B)** Since the slices that comprise a pie chart must add up to 100%, 10 + 15 + 30 + 25 + x = 100, so x = 20.
   Solution I: If N represents the total number of SAT math scores, then 30% of N is 72 so $N = 72 \div 0.30 = 240$. Since 20% of 240 = 0.20 × 240 = 48, 48 SAT math scores are above 700.
   Solution II: Since 30% of the total number of SAT scores represented by the graph is 72, 10% of the total number of SAT scores represented by the graph is $\frac{1}{3} \times 72 = 24$. Since 10% of the circle represents 24 scores, 20% of the circle represents 2 × 24 or 48 SAT math scores.
2. **(A)** Since 25% of the 240 SAT math scores are from 610 to 700, 25% × 240 = $\frac{1}{4}$ × 240 = 60 students had scores from 610 to 700. If 20% of these students received scholarships, then 20% × 60 = 0.20 × 60 = 12 students with math scores from 610 to 700 received college scholarships.

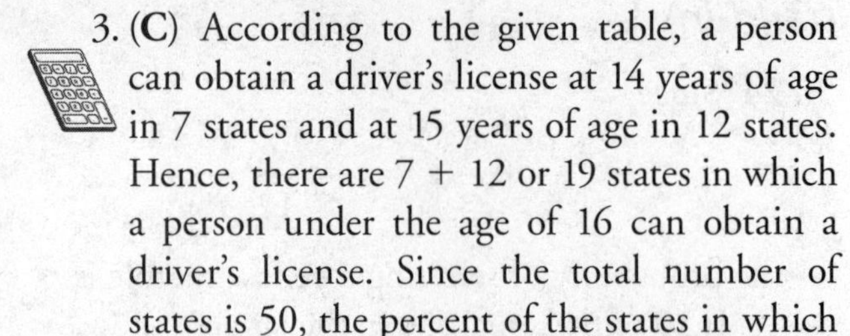

3. **(C)** According to the given table, a person can obtain a driver's license at 14 years of age in 7 states and at 15 years of age in 12 states. Hence, there are 7 + 12 or 19 states in which a person under the age of 16 can obtain a driver's license. Since the total number of states is 50, the percent of the states in which a person under the age of 16 can obtain a driver's license is $\frac{19}{50} \times 100\%$ or 38%.
4. **(E)** According to the height of the bars in the given graph, in 1991 the number of households with computers was 40 million and in 1995 the number was 100 million. Hence,
$$\begin{aligned}\text{Percent increase} &= \frac{\text{Amount of increase}}{\text{Original amount}} \times 100\% \\ &= \frac{100-40}{40} \times 100\% \\ &= \frac{60}{40} \times 100\% \\ &= 1.5 \times 100\% \\ &= 150\%\end{aligned}$$
5. **(A)** Calculate the percent of increase for each 2 consecutive years:
   - From 1991 to 1992:
   $$\begin{aligned}\text{Percent increase} &= \frac{\text{Amount of increase}}{\text{Original amount}} \times 100\% \\ &= \frac{55-40}{40} \times 100\% \\ &= \frac{15}{40} \times 100\% \\ &= 37.5\%\end{aligned}$$
   - From 1992 to 1993:
   $$\begin{aligned}\text{Percent increase} &= \frac{\text{Amount of increase}}{\text{Original amount}} \times 100\% \\ &= \frac{65-55}{55} \times 100\% \\ &= \frac{10}{55} \times 100\% \\ &= 18.18\%\end{aligned}$$
   - From 1993 to 1994:
   $$\begin{aligned}\text{Percent increase} &= \frac{\text{Amount of increase}}{\text{Original amount}} \times 100\% \\ &= \frac{80-65}{65} \times 100\% \\ &= \frac{15}{65} \times 100\% \\ &= 23.08\%\end{aligned}$$

- From 1994 to 1995:

$$\text{Percent increase} = \frac{\text{Amount of increase}}{\text{Original amount}} \times 100\% = \frac{100-80}{80} \times 100\% = \frac{20}{80} \times 100\% = 25\%$$

Hence, the greatest percent of increase occurred from 1991 to 1992.

6. **(A)** The sum of all of the sectors of a circle graph is 100%. Hence,

$$\begin{aligned} 11\% + 35\% + 23\% + x\% + 7\% + 3x\% &= 100\% \\ 76\% + x\% + 3x\% &= 100\% \\ 4x\% &= 100\% - 76\% \\ &= 24\% \\ x &= \frac{24\%}{4\%} = 6 \end{aligned}$$

Since $3x\% = 3(6)\% = 18\%$, municipal bonds make up 18% of the investment portfolio. To find the amount of money invested in municipal bonds, multiply the total value of the portfolio by 18%:
18% of \$250,000 = 0.18 × \$250,000 = \$45,000. Hence, \$45,000 is invested in municipal bonds.

7. **(E)** The original amount invested in health stocks is 11% of the total investment. You are told that 20% of the 23% that is invested in technology stocks is reinvested in health stocks. Since 20% of 23% = 0.2 × 15.6% is now invested in health stocks. Since

$$\begin{aligned} 15.6\% \text{ of } \$250{,}000 &= 0.156 \times \$250{,}000 \\ &= \$39{,}000 \end{aligned}$$

the amount of money now invested in health stocks is \$39,000.

8. **(E)** From data in the graph:
- Percent increase from 1993 to 1994

$$= \frac{700-400}{400} \times 100\% = \frac{300}{400} \times 100\% = 75\%$$

- Percent increase from 1995 to 1996

$$= \frac{500-300}{300} \times 100\% = \frac{200}{300} \times 100\% = 66.7\%$$

- Percent increase from 1993 to 1994 exceeds the percent increase from 1995 to 1996 by 75% − 66.7% or, approximately, 8%.

9. **(E)** From 1996 to 1997, the percent decrease was $\frac{500-400}{500} \times 100\%$ or 20%. If the percent increase from 1997 to 1998 was 20%, then the number of students enrolled in advanced mathematics courses in 1998 was 400 + 20% of 400 = 400 + 80 = 480.

10. **(D)** In 1995, 460 cars were purchased and 500 cars were leased. Since $\frac{460}{500} \times 100\% = 92\%$, the number of cars purchased in 1995 was 92% of the cars leased in that year.

11. **(B)** For 1994–1995, 1995–1996, and 1996–1997 the number of cars purchased decreased a total of 660 – 240 or 420 cars. Hence, the decrease in the number of cars per year was approximately $\frac{420}{3}$ or 140.

Lesson 7-5 (*Multiple-Choice*)

1. **(C)** Since $\boxed{a} = \sqrt{a} + 1$,

$$\begin{aligned} \boxed{36} + \boxed{64} &= \left(\sqrt{36}+1\right) + \left(\sqrt{64}+1\right) \\ &= (6+1) + (8+1) \\ &= 7 + 9 \\ &= 16 \end{aligned}$$

2. **(C)** To evaluate ${}_{2}\overset{1}{\triangle}_{3}$, let $a = 1$, $b = 2$, and $c = 3$ in the given formula:

$$\begin{aligned} {}_{b}\overset{a}{\triangle}_{c} &= \frac{a+b}{c} + \frac{a+c}{b} + \frac{b+c}{a} \\ &= \frac{1+2}{3} + \frac{1+3}{2} + \frac{2+3}{1} \\ &= \frac{3}{3} + \frac{4}{2} + \frac{5}{1} \\ &= 1 + 2 + 5 \\ &= 8 \end{aligned}$$

3. **(E)** Since the number 6 is even and the number 11 is odd,

$$\triangleright 6 = 2(6 - 1) = 2(5) = 10$$

and

$$\triangleright 11 = \frac{1}{2}(11 + 1) = \frac{1}{2}(12) = 6$$

Hence,

$$\triangleright 6 \times \triangleright 11 = 10 \times 6 = 60$$

4. **(D)** Evaluate $\triangleright N$ for each answer choice until you obtain a number that is a multiple of 4. For choice (D), if $N = 31$, then

$$\triangleright N = \frac{1}{2}(11 + 1) = \frac{1}{2}(12) = 6$$

Since 16 is a multiple of 4, (D) is the correct choice.

5. **(C)** Since 1 is carried over from adding the digits in the first column, in the second column $A + B + B + 1 = 14$, so $A + 2B = 13$. Since B must be an integer, A cannot be 0. Hence, the greatest possible value of digit B occurs when $A = 1$ and $B = 6$.

6. **(B)** Following the given rule for obtaining terms, write the terms of the sequence up to and including 55:

$$1, 1, 2, 3, 5, 8, 13, 21, 34, 55$$

Since 10 numbers are listed, $n = 10$.

7. **(A)** In the first row of multiplication, $63 \times P = 31P$. Hence, P must be equal to 5 or 6 since the product of each of these and 63 is a number between 300 and 400. Since $63 \times 5 = 315$ and $63 \times 6 = 378$, then $P = 5$, so

$$Q = 3 + P = 3 + 5 = 8$$

8. **(E)** You are told that $a \,\Phi\, b = ab - a$. If $a \,\Phi\, 4 = 24$, then $b = 4$ and $ab - a = 24$. Hence, $4a - a = 3a = 24$, so

$$a = \frac{24}{3} = 8$$

9. **(B)** Determine whether each Roman numeral equation is true or false when $a \,\Phi\, b = ab - a$.
   - I. Since $b \,\Phi\, a = ab - b$ and $ab - a$ is not necessarily equal to $ab - b$, $a \,\Phi\, b = b \,\Phi\, a$ is not always true. Equation I need not be true.
   - II. To evaluate $a \,\Phi\, (a + 1)$, let $b = a + 1$:

$$\begin{aligned} a \,\Phi\, (a + 1) &= a(a + 1) - a \\ &= a^2 + a \quad - a \\ &= a^2 \end{aligned}$$

   Hence, equation II must be true.
   - III. Since

$$\frac{a}{2} \Phi \frac{b}{2} = \left(\frac{a}{2}\right)\left(\frac{b}{2}\right) - a = \frac{ab}{4} - a$$

   and

$$\frac{a \Phi b}{2} = \frac{ab - a}{2} = \frac{ab}{2} - \frac{a}{2}$$

   $\frac{a}{2} \Phi \frac{b}{2}$ does not equal $\frac{a \Phi b}{2}$. Hence, equation III is false.

   Only Roman numeral equation II must be true.

10. **(B)** Since 2 is carried over from the addition in the first column to the second column and 2 is carried over from the addition in the second column to the third column, $A + B + C + 2 = 26$, so $A + B + C = 24$. Determine whether each Roman numeral value could be possible.
   - I. If $A = 0$, then $B + C = 24$, which is not possible since the maximum value of $B + C$ is $9 + 9$ or 18. Hence, value I could not be A.
   - II. If $A = 4$, then $B + C = 20$, which is not possible since the maximum value of $B + C$ is $9 + 9$ or 18. Hence, value II could not be A.
   - III. If $A = 8$, then $B + C = 16$, which is possible, so value III could be A.

   Only Roman numeral value III could be digit A.

11. **(C)** Substitute each of the answer choices for y into $\triangle y = 2\sqrt{y}$ until you get 8. Since

$$\triangle 16 = 2\sqrt{16} = 2(4) = 8$$

(C) is the correct choice.

12. **(D)**

$$\begin{aligned} \triangle y \times \triangle y &= \left(2\sqrt{y}\right) \times \left(2\sqrt{y}\right) \\ &= 4\left(\sqrt{y} \times \sqrt{y}\right) \\ &= 4y \end{aligned}$$

13. **(A)** If $\smile\!x\!\smile = 2x + 1$ and $3 - \smile\!x\!\smile = x$, then

$$\begin{aligned} 3 - (2x + 1) &= x \\ 3 - 2x - 1 &= x \\ 2 &= 3x \\ \frac{2}{3} &= x \end{aligned}$$

14. **(D)** In the correctly worked out addition problem, C must be equal to 1 since this is the only possible digit that can be carried over from the addition in column 3 to column 4. From the addition in the first column, there are two possibilities to consider: $A + B = 2$ or $A + B = 12$. If $A + B = 2$, then $A = B = 1$, $A = 0$ and $B = 2$, or $A = 2$ and $B = 0$. All of these combinations of values are impossible because they do not make $C = 1$ in the addition of the second column. Since, therefore, $A + B = 12$, 1 is carried from the addition in the first column to the second column. Hence, $3 + B + 1 = 11$ or $B = 7$, so $A = 5$. Also, since 1 is carried over from the addition in column 3 to column 4, $8 + D + 1 = A = 15$, so $D = 6$.

15. **(B)** Since the greatest divisor of 28 that is less than 28 is 14,

$$\overset{\wedge}{\underset{\vee}{28}} = 28 + 14 = 42$$

16. **(D)** The greatest divisor of 15 that is less than 15 is 5, so

$$\overset{\wedge}{\underset{\vee}{15}} = 15 + 5 = 20 = n$$

Since the greatest divisor of 20 that is less than 20 is 10,

$$\overset{\wedge}{\underset{\vee}{20}} = 20 + 10 = 30$$

17. **(A)** Since

$$\frac{1}{2^1} = \frac{1}{2}, \frac{1}{2^2} = \frac{1}{4}, \frac{1}{2^3} = \frac{1}{8}, \ldots,$$

the nth term, without regard to its sign, is $\frac{1}{2^n}$. Each odd-numbered term of the sequence is positive, while each even-numbered term is negative. When n is an odd integer, $(-1)^{n+1}$ is $+1$ because the exponent, $n + 1$, is even. Also, when n is an even integer, $(-1)^{n+1}$ is -1 since the exponent, $n + 1$, is odd. Hence, the nth term of the sequence is $(-1)^{n+1} \cdot \frac{1}{2^n}$.

(Grid-In)

1. **10** Since $x \,\Omega\, y = 2x + 3y$, evaluate $a \,\Omega\, 4$ by letting $x = a$ and $y = 4$:
$$a \,\Omega\, 4 = 2a + 3(4) = 2a + 12$$
Evaluate $1 \,\Omega\, a$ by letting $x = 1$ and $y = 4$:
$$1 \,\Omega\, a = 2(1) + 3a = 2 + 3a$$
Since $a \,\Omega\, 4 = 1 \,\Omega\, a$, then $2a + 12 = 2 + 3a$, or $12 - 2 = 3a - 2a$, so $10 = a$.

2. **32** In the sequence x, y, 22, 14, 10, . . . each term after the first term, x, is obtained by halving the term that comes before it and then adding 3 to that number. Hence, to obtain y do the opposite to 22: subtract 3 and then double the result, getting 38. To obtain x, subtract 3 from 38 and then double the result, getting 70. Thus, $x - y = 70 - 38 = 32$.

3. **54** Since the sum of the first two terms, $15 + 19 = 34$, and 34 is even, the next consecutive term is $\frac{15+19}{2} = 17$. The first three terms of this sequence are 15, 19, and 17. To obtain the next term, add 19 and 17, getting 36. Since 36 is even, the next consecutive term is $\frac{19+17}{2} = 18$. The first four terms of the sequence are 15, 19, 17, and 18. Since $17 + 18 = 35$ and 35 is odd, the next term of the sequence is $17 + 18 - 3$ or 32. Hence, the first five terms of the sequence are 15, 19, 17, 18, and 32. Since $18 + 32 = 50$ and 50 is even, the next term of the sequence is 25, so the first six terms are 15, 19, 17, 18, 32, and 25. Since $32 + 25 = 57$ and 57 is odd, the next term of the sequence is $32 + 25 - 3 = 54$. Hence, the seventh term of the sequence is 54.

4. **6/4** If $a \blacktriangle b = ab - (a + b)$ and $5 \blacktriangle b = 1$, then

$$\begin{aligned} 5 \blacktriangle b = 5b - (5 + b) &= 1 \\ 5b - 5 - b &= 1 \\ 4b &= 6 \\ b &= \frac{6}{4} \end{aligned}$$

Grid in as 6/4.

Algebra II Methods

CHAPTER 8

The math section of the SAT assumes some basic knowledge of skills and concepts typically studied in second-year, high-level algebra courses. This chapter reviews those Algebra II topics that you are expected to know.

LESSONS IN THIS CHAPTER

LESSON

8-1 ZERO, NEGATIVE, AND FRACTIONAL EXPONENTS

OVERVIEW

The rules that apply to positive integer exponents also work for negative integer and fractional exponents. A quantity raised to a fractional power may be written in radical form using the rule

$$x^{\frac{\text{power}}{\text{root}}} = \left(\sqrt[\text{root}]{x}\right)^{\text{power}}$$

RULE FOR ZERO EXPONENTS

MATH REFERENCE FACT

$x^0 = 1$, provided that $x \neq 0$

Any nonzero quantity raised to the 0 power is 1, as in $7^0 = 1$ and $(-2)^0 = 1$. The expression 0^0 is *not* defined. Keep in mind that $(2y)^0 = 1$ but $2y^0 = 2(1) = 2$.

RULE FOR NEGATIVE EXPONENTS

To change a negative exponent to a positive exponent, invert the base, as in $5^{-1} = \frac{1}{5}$ and $\frac{1}{y^{-2}} = y^2$. Here are more examples:

MATH REFERENCE FACT

- $x^{-n} = \frac{1}{x^n}$, provided that $x \neq 0$
- $\left(\frac{x}{y}\right)^{-n} = \left(\frac{y}{x}\right)^{n}$, provided that $x \neq 0$

- $2x^{-3} = \frac{2}{x^3}$
- $\left(\frac{3}{5}\right)^{-2} = \left(\frac{5}{3}\right)^{2} = \frac{25}{9}$
- $\frac{x^7 y^1}{x^3 y^4} = x^{7-3} \cdot y^{1-4} = x^4 y^{-3}$ *or* $\frac{x^4}{y^3}$

EXPONENT RULE FOR ROOTS

You already know that $\sqrt{9}$ multiplied by itself is 9. Using the product law of exponents,

$$9^{\frac{1}{2}} \times 9^{\frac{1}{2}} = 9^{\frac{1}{2}+\frac{1}{2}} = 9$$

MATH REFERENCE FACT

$x^{\frac{1}{2}} = \sqrt{x}$ and $x^{\frac{1}{n}} = \sqrt[n]{x}$

Since $9^{\frac{1}{2}}$ multiplied by itself is also 9, it must be the case that $9^{\frac{1}{2}}$ and $\sqrt{9}$ mean the same thing. Thus, $9^{\frac{1}{2}} = \sqrt{9}$.

Similarly, $8^{\frac{1}{3}}$ can be written as $\sqrt[3]{8}$. To represent the *n*th root of *x* in exponent form, write *x* raised to the $\frac{1}{n}$ power.

CHANGING FRACTIONAL EXPONENTS TO RADICAL FORM

Because $8^{\frac{4}{3}} = \left(8^{\frac{1}{3}}\right)^4$, $8^{\frac{4}{3}}$ is the cube root of 8 raised to the fourth power.

Thus,

$$8^{\frac{4}{3}} = \left(\sqrt[3]{8}\right)^4 = (2)^4 = 16$$

Since it is also true that $8^{\frac{4}{3}} = \left(8^4\right)^{\frac{1}{3}}$, the order in which the root and power are evaluated does not matter:

$$8^{\frac{4}{3}} = \sqrt[3]{\left(8^4\right)} = \sqrt[3]{4096} = 16$$

To ensure both speed and accuracy, learn how to use your calculator to evaluate roots and fractional powers. Using a typical graphing calculator,

- $\sqrt{121}$ can be evaluated by pressing

  [1] [2] [1] [^] [.] [5] [ENTER]

- $8^{\frac{4}{3}}$ can be evaluated by pressing

  [8] [^] [(] [4] [÷] [3] [)] [ENTER]

Here are some examples:

- $(xy)^{\frac{5}{3}} = \left(\sqrt[3]{xy}\right)^5$

- $81^{\frac{3}{4}} = \left(\sqrt[4]{81}\right)^3 = (3)^3 = 27$

- $8^{-\frac{2}{3}} = \dfrac{1}{8^{\frac{2}{3}}} = \dfrac{1}{\left(\sqrt[3]{8}\right)^2} = \dfrac{1}{2^2} = \dfrac{1}{4}$

- $\left(8x^6\right)^{\frac{4}{3}} = \left(8^{\frac{4}{3}}\right)\left(x^6\right)^{\frac{4}{3}}$

  $= \left(\sqrt[3]{8}\right)^4\left(x^{6\bullet\frac{4}{3}}\right)$

  $= \left(2^4\right)\left(x^8\right)$

  $= 16x^8$

MATH REFERENCE FACT

$$x^{\frac{y}{n}} = \left(\sqrt[n]{x}\right)^y = \sqrt[n]{x^y}$$

EXAMPLE

If $x > 1$ and $\frac{\sqrt{x^3}}{x^2} = x^n$, what is the value of n?

(A) $-\frac{3}{2}$

(B) -1

(C) $-\frac{1}{2}$

(D) $\frac{1}{2}$

(E) $\frac{3}{2}$

Solution

- Rewrite the radical using the exponent rule for square roots:

$$\frac{\sqrt{x^3}}{x^2} = x^n$$

$$\frac{\left(x^3\right)^{\frac{1}{2}}}{x^2} = x^n$$

$$\frac{x^{\frac{3}{2}}}{x^2} = x^n$$

- Use the quotient law of exponents:

$$x^{\frac{3}{2}-2} = x^n$$

- Simplify the exponent:

$$x^{\frac{3}{2}-\frac{4}{2}} = x^n$$

$$x^{-\frac{1}{2}} = x^n$$

$$n = -\frac{1}{2}$$

The correct choice is **(C)**.

Lesson 8-1 Tune-Up Exercises

Multiple-Choice

1 If x is a positive integer such that $x^{\frac{3}{4}} = c$, then $\sqrt{x} =$

(A) $c^{\frac{1}{3}}$

(B) $c^{\frac{1}{4}}$

(C) $c^{\frac{1}{2}}$

(D) $c^{\frac{2}{3}}$

(E) c

2 If $\sqrt{a} = x$ and $a^2 = y$, then $\sqrt{a^3} =$

(A) $\frac{y}{x}$

(B) $\frac{x}{y}$

(C) xy

(D) $\frac{x^2}{y}$

(E) xy^2

3 If $j = x^{-\frac{1}{2}}$ and $k = \frac{1}{x^2}$, then $\left(\frac{j}{k}\right)^3 =$

(A) $x^{-\frac{9}{2}}$

(B) $x^{-\frac{27}{8}}$

(C) $x^{-\frac{15}{2}}$

(D) $x^{\frac{27}{8}}$

(E) $x^{\frac{9}{2}}$

4 If $\sqrt{m} = 2p$, then $m^{\frac{3}{2}} =$

(A) $\frac{p}{3}$

(B) $2p^2$

(C) $6p^3$

(D) $8p^3$

(E) $\frac{1}{2}p^{\frac{2}{3}}$

5 If x, p, and q are positive numbers such that $x = p^2$, $x^3 = q$, and $\frac{p}{q} = x^n$, then $n =$

(A) $-\frac{7}{2}$

(B) $-\frac{5}{2}$

(C) $-\frac{3}{2}$

(D) $\frac{5}{2}$

(E) $\frac{7}{2}$

6 If x and y are positive and $40x = 10x^2y^{-1}$, what is x^{-1} in terms of y?

(A) $\frac{1}{4y}$
(B) $4y$
(C) $\frac{30}{y}$
(D) $\frac{y}{4}$
(E) $30\sqrt{y}$

7 If $\frac{x}{x^{1.5}} = 8x^{-1}$ and $x > 0$, then $x =$

(A) $\frac{\sqrt{2}}{4}$
(B) $2\sqrt{2}$
(C) 4
(D) 16
(E) 64

8 If $10^k = 64$, what is the value of $10^{\frac{k}{2}+1}$?
(A) 18
(B) 42
(C) 80
(D) 81
(E) 320

9 If x is a positive integer greater than 1, how much greater than x^2 is $x^{\frac{5}{2}}$

(A) $x^2\left(1 - x^{\frac{1}{2}}\right)$
(B) $x^{-\frac{1}{2}}$
(C) $x^2\left(x^{\frac{1}{2}} - 1\right)$
(D) $x^{\frac{1}{2}}$
(E) $\frac{1}{2}$

10 If $x, y, \neq 0$, $x^{-1} + x^{-1} = a$, and $y^{-1} + y^{-1} + y^{-1} = b$, then $\frac{a}{b} =$

(A) $\frac{3x}{2y}$
(B) $\frac{xy}{6}$
(C) $\frac{3y}{2x}$
(D) $\frac{6}{xy}$
(E) $\frac{2y}{3x}$

Grid-In

1 If $2x^{-1} = \frac{5}{3}$, what is the value of x?

2 If $x^{-\frac{1}{2}} = \frac{1}{8}$, what is the value of $x^{\frac{2}{3}}$?

3 If $x^{-1} = 3$ and $y^{-1} = 9$, what is the value of $(x + y)^{-\frac{1}{2}}$?

4 If $8(rs)^{-1} = (3rs^{-2})^2$, what is the value of $\frac{r}{s}$?

LESSON

8-2

EQUATIONS INVOLVING RADICALS AND EXPONENTS

OVERVIEW

The variable in an equation may be underneath a radical, have a fractional exponent, or be part of an exponent.

RADICAL EQUATIONS

If an equation contains the square root of a variable term, isolate the radical in the usual way. Eliminate the radical by raising both sides of the equation to the second power. To solve $\sqrt{3x + 1} - 2 = 3$ perform the following steps:

1. Isolate the radical:

$$\sqrt{3x + 1} = 5$$

2. Raise both sides of the equation to the second power:

$$\left(\sqrt{3x + 1}\right)^2 = (5)^2$$

3. Simplify:

$$3x + 1 = 25$$
$$\frac{3x}{3} = \frac{24}{3}$$
$$x = 8$$

MATH REFERENCE FACT

To solve an equation of the form $\sqrt[n]{x} = c$, raise both sides of the equation to the nth power:

$$\left(\sqrt[n]{x}\right)^n = (c)^n \text{ so } x = c^n$$

EQUATIONS WITH FRACTIONAL EXPONENTS

Since the reciprocal of $\frac{3}{2}$ is $\frac{2}{3}$, raising both sides of the equation $x^{\frac{3}{2}} = 8$ to the $\frac{2}{3}$ power creates an equivalent equation in which the power of x is 1:

$$x^{\frac{3}{2}\bullet\frac{2}{3}} = (8)^{\frac{2}{3}}$$
$$x^1 = \left(\sqrt[3]{8}\right)^2$$
$$x = (2)^2 = 4$$

MATH REFERENCE FACT

To solve an equation of the form $x^{\frac{m}{n}} = k$, raise both sides of the equation to the $\frac{n}{m}$ power so that the exponent of x becomes 1.

EXPONENTIAL EQUATIONS

If an equation contains a variable in an exponent, rewrite each side as a power of the same base. Then set the exponents equal to each other. For example, solve $8^{x-2} = 4^{x+1}$ as follows:

1. Rewrite each side as a power of 2: $(2^3)^{(x-2)} = (2^2)^{(x+1)}$
2. Simplify the exponents: $2^{3(x-2)} = 2^{2(x+1)}$
3. Set the exponents equal:

$$\begin{aligned} 3(x-2) &= 2(x+1) \\ 3x - 6 &= 2x + 2 \\ 3x - 2x &= 2 + 6 \\ x &= 8 \end{aligned}$$

Lesson 8-2 Tune-Up Exercises

Multiple-Choice

1 If $\dfrac{1}{\sqrt{x^2 - 2}} = 2$ and $x^2 > 2$, then x could be

(A) $\dfrac{3}{2}$
(B) $\dfrac{7}{4}$
(C) 2
(D) $\dfrac{9}{4}$
(E) $\dfrac{11}{4}$

2 If $z = x^3 = \sqrt{y}$ where $y > 0$, then $xy =$

(A) $z^{\frac{4}{3}}$
(B) $z^{\frac{5}{3}}$
(C) z^2
(D) $z^{\frac{7}{3}}$
(E) $z^{\frac{8}{3}}$

3 If $4^y + 4^y + 4^y + 4^y = 16^x$, then $y =$

(A) $2x - 1$
(B) $2x + 1$
(C) $x - 2$
(D) $x + 2$
(E) $4x$

4 If $t^9 = s^3$, what is s in terms of t?

(A) $\sqrt[3]{t}$
(B) $\sqrt{t}$
(C) t^2
(D) t^3
(E) t^6

5 If $3^x = 81$ and $2^{x+y} = 64$, then $\dfrac{x}{y} =$

(A) 1
(B) $\dfrac{3}{2}$
(C) 2
(D) $\dfrac{5}{2}$
(E) 3

6 If $64^{1-x} = \dfrac{1}{16^{2x}}$, then $x =$

(A) -3
(B) -2
(C) -1
(D) 2
(E) 3

7 If $4^{x-y} = 64$ and $3^{-2x+y} = \frac{1}{3}$, then $y =$

(A) -5
(B) -3
(C) -1
(D) 3
(E) 5

8 If $27^x = 9^{y-1}$, then

(A) $y = \frac{3}{2}x + 1$
(B) $y = \frac{3}{2}x + 2$
(C) $y = \frac{3}{2}x + \frac{1}{2}$
(D) $y = \frac{1}{2}x + \frac{2}{3}$
(E) $y = \frac{2}{3}x - 1$

9 If x and y are positive numbers that satisfy the equation $\sqrt{x^2 + y^2 + 1} = x + y$, then $xy =$

(A) $\frac{1}{2}$
(B) 1
(C) $\frac{3}{2}$
(D) 2
(E) 4

10 If w and y are positive integers such that $1000^y = 100^w$, what is the value of $\frac{y}{w}$?

(A) $\frac{1}{10}$
(B) $\frac{2}{3}$
(C) $\frac{4}{3}$
(D) $\frac{3}{2}$
(E) 10

11 If n and p are positive integers such that $8\left(2^p\right) = 4^n$, what is n in terms of p?

(A) $\frac{p+2}{3}$
(B) $\frac{2p}{3}$
(C) $\frac{p+3}{2}$
(D) $\frac{3p}{2}$
(E) $2p + 3$

12 If r and s are positive numbers and $\sqrt{rs} = 2r$, what is the value of $\frac{r}{s}$?

(A) $\frac{1}{4}$
(B) $\frac{1}{2}$
(C) $\frac{1}{\sqrt{2}}$
(D) 2
(E) 4

Grid-In

1 What is the value of x if $5\sqrt{x-1}+13=48$?

2 If $2\sqrt{x} = \sqrt{x+2}$, then what is the value of x?

3 If $\sqrt{x^y} = 9$ and $y > x$, what is the value of xy?

4 If m and p are positive integers and $\left(\sqrt{3}\right)^m = 27^p$, what is the value of $\frac{p}{m}$?

5 What is the value of x if $16^{x-2} = \frac{1}{2} \cdot 2^{3x+1}$?

6 If a, b, and c are positive numbers such that $\sqrt{\frac{a}{b}} = 8c$ and $ac = b$, what is the value of c?

LESSON

8-3 ABSOLUTE VALUE EQUATIONS AND INEQUALITIES

OVERVIEW

*The **absolute value** of x, written as |x|, refers to the number x without regard to its sign. Geometrically, |x| represents the distance from x to 0 on the number line. Since +2 and –2 are each 2 units from 0, $|+2| = 2$ and $|-2| = 2$. To solve absolute value equations and inequalities, you must remove the absolute value sign by considering two possibilities. The first possibility assumes the quantity inside the absolute value sign is nonnegative, in which case the absolute value sign is simply removed. The second possibility assumes the quantity inside the absolute value sign is negative. In this situation, the absolute value sign is removed by taking the opposite of the quantity inside the absolute value sign.*

SOLVING ABSOLUTE VALUE EQUATIONS

To solve an equation involving $|x|$, remove the absolute value sign by accounting for two possibilities:

1. If $x \geq 0$, then $|x| = x$. For example, $|+2| = 2$.
2. If $x < 0$, then $|x| = -x$. For example, $|-2| = -(-2) = 2$.

EXAMPLE

Solve for x: $|x - 1| = 4$

SOLUTION

Consider the two possibilities:

- If the quantity inside the absolute value sign is nonnegative, then

$$|x - 1| = x - 1 = 4, \text{ so } x = 5$$

- If the quantity inside the absolute value sign is negative, then

$$\begin{aligned} |x - 1| = -(x - 1) &= 4 \\ -x + 1 &= 4 \\ -x &= 3 \\ x &= -3 \end{aligned}$$

- The two possible solutions for x are **5** or **–3**. You should verify that both roots satisfy the original absolute value equation.

EXAMPLE

Solve for x: $|2x + 3| + 4 = 5$

Solution

If $|2x + 3| + 4 = 5$, then $|2x + 3| = 1$. Thus,

- $2x + 3 = 1$, so $2x = 1 - 3 = -2$ and $x = \frac{-2}{2} = -1$; or
- $2x + 3 = -1$, so $2x = -1 - 3 = -4$ and $x = \frac{-4}{2} = -2$

You should verify that both roots satisfy the original equation.

EXAMPLE

Solve and check: $|x - 3| = 2x$

Solution

If $|x - 3| = 2x$, then

- $x - 3 = 2x$ so $-3 = x$.

  Check: If $x = -3$, then
  $$|-3 - 3| = 2(-3)$$
  $$|-6| \neq -6$$

- $x - 3 = -2x$ so $3x - 3 = 0$, $3x = 3$ and $x = 1$.

  Check: If $x = 1$, then
  $$|x - 3| = 2x$$
  $$|1 - 3| = 2(1)$$
  $$|-2| = 2 \checkmark$$

Hence, $x = 1$ is the only root of the absolute value equation.

INTERPRETING ABSOLUTE VALUE INEQUALITIES

The absolute value inequality $|x - a| < d$ represents the set of all points x that are less than d units from a. For example,

- the inequality $|t - 68°| < 3°$ states that the temperature, t, is less than 3° from 68°, which means that t is between 65° and 71°, as shown in Figure 8.1.

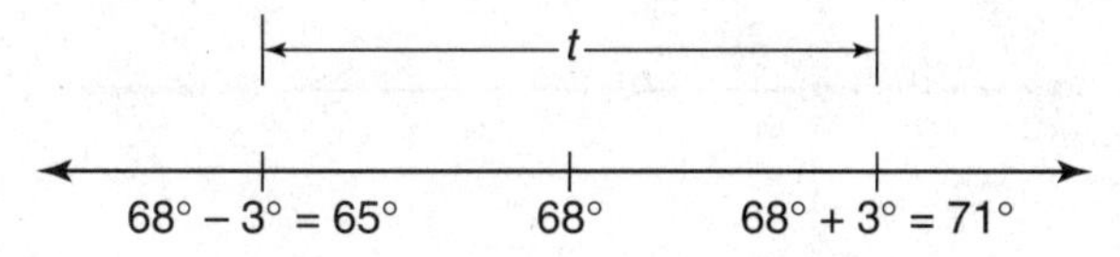

Figure 8.1 Solution of $|t - 68°| < 3°$

- the inequality $|t - 68°| > 3°$ states that the temperature, t, is more than 3° from 68°, which means that t is less than 65° or greater than 71°.

SOLVING ABSOLUTE VALUE INEQUALITIES

To solve an absolute value inequality algebraically, remove the absolute value sign according to the following rules where d is a positive number:

- if $|ax - b| < d$, then $-d < ax - b < d$.
- if $|ax - b| > d$, then $ax - b < -d$ or $ax - b > d$.

EXAMPLE

Solve and graph the solution set of $|2x - 1| \leq 7$.

Solution

- If $|2x - 1| \leq 7$, then $-7 \leq 2x - 1 \leq 7$.
- Add 1 to each member of the combined inequality:

$$\begin{array}{rcl} -7 \leq & 2x - 1 & \leq 7 \\ +1 & +1 & +1 \\ \hline -6 \leq & 2x & \leq 8 \end{array}$$

- Divide each member of the combined inequality by 2:

$$\frac{-6}{2} \leq \frac{2x}{2} \leq \frac{8}{2}$$
$$-2 \leq x \leq 4$$

- Graph the solution set:

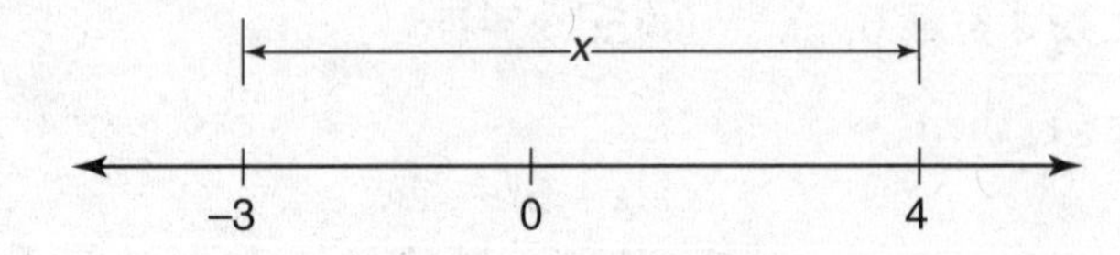

EXAMPLE

Solve and graph the solution set of $|3x - 1| > 5$.

Solution

If $|3x - 1| > 5$, then

- $3x - 1 > 5$, so $3x > 6$ and $x > 2$
- $3x - 1 < -5$, so $3x < -4$ and $x < -\frac{4}{3}$

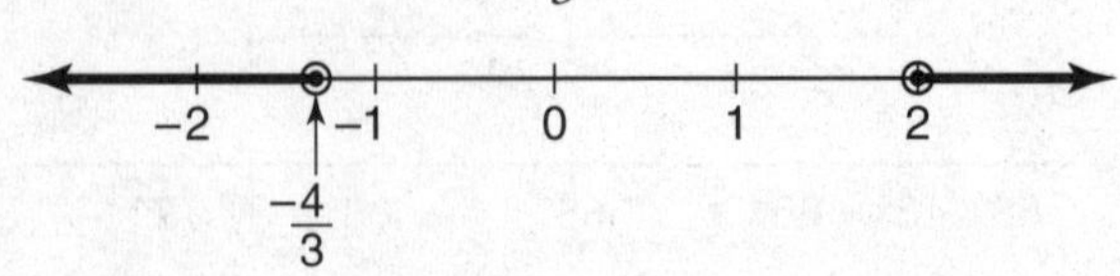

EXAMPLE

If $5 < |2 - x| < 6$ and $x > 0$, what is one possible value of x?

Solution

Remove the absolute value sign:

- If $2 - x \geq 0$, then $5 < 2 - x < 6$, so $3 < -x < 4$. Multiplying each term of the inequality by –1 gives $-3 > x > -4$. Since it is given that $x > 0$, disregard this solution.
- If $2 - x < 0$, then $5 < -(2 - x) < 6$ or, equivalently, $5 < x - 2 < 6$, so $5 + 2 < x < 6 + 2$ and $7 < x < 8$.
- Therefore, x can be any number between 7 and 8, such as **7.5**.

Lesson 8-3 Tune-Up Exercises

Multiple-Choice

1 If $|x| \leq 2$ and $|y| \leq 1$, then what is the least possible value of $x - y$?

(A) -3
(B) -2
(C) -1
(D) 0
(E) 1

2 If $\left|\frac{1}{2}x\right| \geq \frac{1}{2}$, then which statement must be true?

(A) $x \leq -2$ or $x \geq 2$
(B) $x \leq -1$ or $x \geq 1$
(C) $x \leq -\frac{1}{2}$ or $x \geq \frac{1}{2}$
(D) $-1 \leq x \leq 1$
(E) $-2 \leq x \leq 2$

3 If $\frac{1}{2}|x| = 1$ and $|y| = x + 1$, then y^2 could be

(A) 2
(B) 3
(C) 4
(D) 9
(E) 16

4 If $|1-x| < x$, then:

(A) $0 < x < \frac{1}{2}$
(B) $-\frac{1}{2} < x < 0$
(C) $x < \frac{1}{2}$
(D) $x < -1$ or $x > \frac{1}{2}$
(E) $x > \frac{1}{2}$

5 In a certain greenhouse for plants, the Fahrenheit temperature, F, is controlled so that it does *not* vary from 79° by more than 7°. Which of the following best expresses the possible range in Fahrenheit temperatures of the greenhouse?

(A) $|F - 79| \leq 7$
(B) $|F - 79| > 7$
(C) $|F - 7| \leq 79$
(D) $|F - 7| > 79$
(E) $F \leq 71$ or $F \geq 87$

6 If $\frac{|a + 3|}{2} = 1$ and $2|b + 1| = 6$, then $|a + b|$ could equal any of the following EXCEPT

(A) 1
(B) 3
(C) 5
(D) 7
(E) 9

Grid-In

$$|t - 7| = 4$$
$$|9 - t| = 2$$

1 What value of t satisfies both of the above equations?

2 If $5 < |k + 1| < 6$ and $k < 0$, then what is one possible value of $|k|$?

$$|3x - 4| = 2$$
$$|11 - 6x| = 7$$

3 What value of x satisfies both of the above equations?

4 If $|x - 16| \leq 4$ and $|y + 6| \leq 2$, what is the greatest possible value of $x - y$?

LESSON

8-4 WORKING WITH FUNCTIONS

OVERVIEW

A ***function*** *is a rule or formula that tells how to pair the elements of one set, called the* ***domain,*** *with exactly one element of another set, called the* ***range.*** *A function can be represented in different ways: numerically as a set of ordered pairs or table of values, algebraically as an equation, or visually as a graph.*

SETS OF ORDERED PAIRS AS FUNCTIONS

Suppose set X consists of 5 teenagers and set Y consists of their possible ages:

DOMAIN: X = {Alice, Barbara, Chris, Dennis, Enid}
RANGE: $Y = \{13, 14, 15, 16, 17, 18, 19\}$

If each teenager in set X is matched with his or her present age in set Y, then the result can be written as a set of ordered pairs:

{(Alice,17), (Barbara,13), (Chris,16), (Dennis,19), (Enid,15)}

or it can be presented in table form:

Name	Age
Alice	17
Barbara	13
Chris	16
Dennis	19
Enid	15

Because each teenager from set X is paired with exactly one age from set Y, the set of ordered pairs represents a function.

EQUATIONS AS FUNCTIONS

A function is usually named by a lowercase letter, such as f or g. The equation $y = 2x + 3$ describes a function, since it gives a rule for pairing any given x-value with one particular y-value: input any number x, multiply it by 2, add 3, and name the result y. When $x = 3$, $y = 2(3) + 3 = 9$. If this function is called f, then the ordered pair (3,9) belongs to function f. This fact can be abbreviated by writing $f(3) = 9$, which is read as "f of three equals nine":

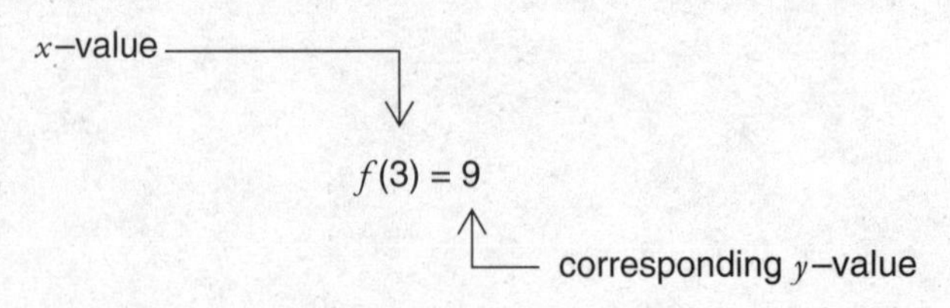

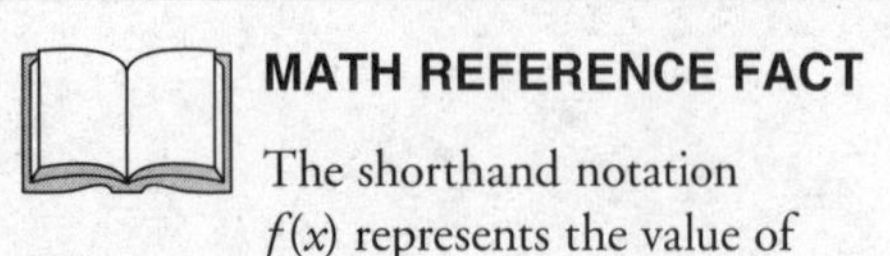

MATH REFERENCE FACT

The shorthand notation $f(x)$ represents the value of function f when x has the value inside the parentheses.

To understand functions further, study these examples:

- If $h(x) = \dfrac{x}{x^3 + 1}$, then to find $h(2)$, replace x with 2:

$$\begin{aligned} h(2) &= \frac{2}{(2)^3 + 1} \\ &= \frac{2}{8 + 1} \\ &= \frac{2}{9} \end{aligned}$$

- If $f(x) = (1 - x)^{\frac{5}{3}}$, then to find $f(-7)$, replace x with -7:

$$\begin{aligned} f(-7) &= (1 - (-7))^{\frac{5}{3}} \\ &= 8^{\frac{5}{3}} \\ &= \left(\sqrt[3]{8}\right)^5 \\ &= (2)^5 \\ &= 32 \end{aligned}$$

- Let functions f and g be defined by $f(x) = 2x - 1$ and $g(x) = x^2$, respectively. To find $f(g(2))$, start with the "inside" function: evaluate $g(2)$ and then use that result as the x-value for function f:

$$\begin{aligned} g(2) &= 2^2 = 4 \\ f(g(2)) &= f(4) \\ &= 2 \cdot 4 - 1 \\ &= 8 - 1 \\ &= 7 \end{aligned}$$

- If $g(x) = \dfrac{1}{2}x + 2$, then to find $4g(x) + 1$, multiply the equation that defines function g by 4 and then add 1:

$$\begin{aligned} 4g(x) + 1 &= 4\Bigg[\overbrace{\frac{1}{2}x + 2}^{g(x)}\Bigg] + 1 \\ &= (2x + 8) + 1 \\ &= 2x + 9 \end{aligned}$$

- If $f(x) = x^2 + x$, then to find $f(n - 1)$, replace each x with $n - 1$:

$$\begin{aligned} f(n-1) &= (n-1)^2 + (n-1) \\ &= (n-1)(n-1) + (n-1) \\ &= (n^2 - 2n + 1) + (n-1) \\ &= n^2 - n \end{aligned}$$

EXAMPLE

If function g is defined by $g(x) = 2x - 5$, what is the value of $2g(4) + 7$?

Solution

- Find the value of $g(4)$:

$$\begin{aligned} g(x) &= 2x - 5 \\ g(4) &= 2(4) - 5 \\ &= 8 - 5 \\ &= 3 \end{aligned}$$

- Evaluate $2g(4) + 7$:

$$\begin{aligned} 2g(4) + 7 &= 2(3) + 7 \\ &= 6 + 7 \\ &= 13 \end{aligned}$$

EXAMPLE

If $k(x) = \dfrac{2x - p}{5}$, then for what value of p is $k(7) = 3$?

Solution

Since $k(7) = 3$, replace x with 7 and set the result equal to 3:

$$\begin{aligned} k(7) &= \frac{2(7) - p}{5} \\ 3 &= \frac{14 - p}{5} \\ 15 &= 14 - p \\ p &= 14 - 15 \\ p &= -1 \end{aligned}$$

DETERMINING THE DOMAIN AND RANGE

Unless otherwise indicated, the *domain* of function f is the largest possible set of real numbers x for which $f(x)$ is a real number. There are two key rules to follow when finding the domain of a function:

1. **Do *not* divide by zero.** Exclude from the domain of a function any value of x that results in division by 0. If $f(x) = \dfrac{x + 2}{x - 1}$, then $f(1) = \dfrac{1 + 2}{1 - 1} = \dfrac{3}{0}$.

Since division by 0 is not allowed, x cannot be equal to 1. The domain of function f is the set of all real numbers *except* 1.

2. **Do *not* take the square root of a negative number.** Since the square root of a negative number is not a real number, the quantity underneath a square root sign must always evaluate to a number that is greater than or equal to 0. If $f(x) = \sqrt{x - 3}$, then x must be at least 3, since any lesser value of x will result in the square root of a negative number. For example, $f(1) = \sqrt{1 - 3} = \sqrt{-2}$, but $\sqrt{-2}$ is not a real number. Thus, the domain of function f is limited to the set of all real numbers greater than or equal to 3.

The *range* of the function $y = f(x)$ is the set of all values that y can have as x takes on each of its possible values. For example, if function f is defined by $f(x) = 1 + \sqrt{x}$, then the smallest possible function value is $f(0) = 1 + \sqrt{0} = 1$.

As x increases without bound, so does $1 + \sqrt{x}$. Therefore, the range of f is the set of all real numbers greater than or equal to 1.

GRAPHS AS FUNCTIONS

Since a graph is a set of ordered pairs located on a coordinate grid, a function may take the form of a graph. If Figure 8.2 shows the complete graph for function g, then you can tell the following from the graph:

- The *domain* of g is $-6 \leq x \leq 5$, since the greatest set of x-values over which the graph extends *horizontally* is from $x = -6$ to $x = 5$, inclusive.
- The *range* of g is $0 \leq y \leq 8$, because the greatest set of y-values over which the graph extends *vertically* is from $y = 0$ to $y = 8$, inclusive.

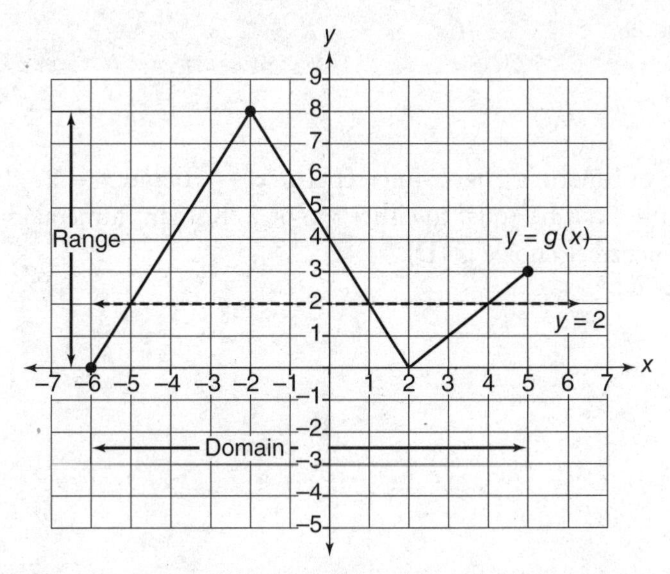

Figure 8.2 Domain and range of function g. The line $y = 2$ intersects the graph at points for which $g(x) = 2$.

Given the graph of function f, you can find those values of t that make $f(t) = k$ by drawing a horizontal line through k on the y-axis. The x-coordinates of the points at which the horizontal line intersects the graph, if any, represent the possible values of t.

You can also read specific function values from the graph of function g in Figure 8.2:

- To find $g(-1)$, determine the y-coordinate of the point on the graph whose x-coordinate is -1. Since the graph contains $(-1,6)$, $g(-1) = 6$.
- To find the values of t such that $g(t) = 2$, find all points on the graph whose y-coordinates are 2. Since the graph passes through $(-5,2)$, $(1,2)$, and $(4,2)$, $g(-5) = 2$, $g(1) = 2$, and $g(4) = 2$. Hence, the possible values of t are -5, 1, and 4.

FINDING THE ZEROS OF A FUNCTION

MATH REFERENCE FACT

The x-intercepts of the graph of function f, if any, correspond to those values of x for which $f(x) = 0$.

The zeros of a function f are those values of x, if any, for which $f(x) = 0$. You can determine the zeros of a function from its graph by locating the points at which the graph intersects the x-axis. At each of these points, the y-coordinate is 0, so $f(x) = 0$.

EXAMPLE

Referring to Figure 8.2, which could be the value of s when $g(s) = 0$?

I. -6
II. 2
III. 4

(A) I only
(B) II only
(C) III only
(D) I and II only
(E) I, II, and III

Solution

Since the graph of function g has x-intercepts at $x = -6$ and $x = 2$, $g(-6) = 0$ and $g(2) = 0$. Because s can be equal to either -6 or 2, Roman numeral choices I and II are correct. The correct choice is **(D)**.

Lesson 8-4 Tune-Up Exercises

Multiple-Choice

1 For which function is $f(-x) = f(x)$?

(A) $f(x) = \sqrt{x}$
(B) $f(x) = 2x$
(C) $f(x) = x^2$
(D) $f(x) = x^3$
(E) $f(x) = 2^x$

2 Let the function g be defined by $g(x) = x^{\frac{3}{4}} - 7$. If $g(k) = 57$, then $k =$

(A) 50
(B) 64
(C) 81
(D) 216
(E) 256

3

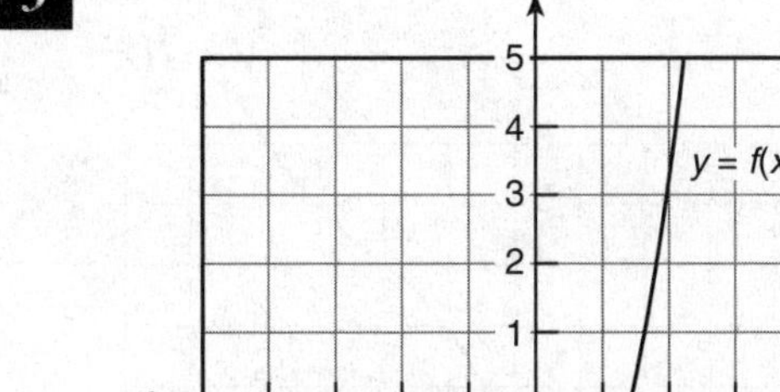
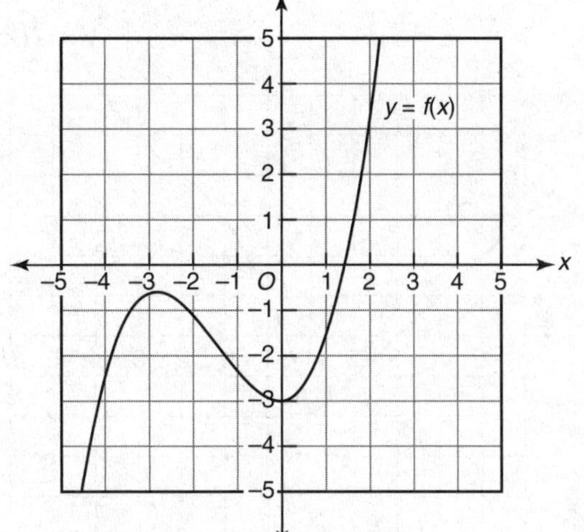

Let function f be defined by the accompanying graph. If $f(-4) + f(2) = f(p)$, what is the value of p?

(A) -3
(B) -2
(C) 0
(D) 1
(E) 1.5

4 If the function f is defined by $f(x) = 5x + 3$, then $2f(x) - 3 =$

(A) $10x$
(B) $10x + 3$
(C) $10x + 6$
(D) $7x + 3$
(E) $7x + 6$

5 If the function k is defined by $k(h) = (h + 1)^2$, then $k(x - 2) =$

(A) $x^2 - x$
(B) $x^2 - 2x$
(C) $x^2 - 2x + 1$
(D) $x^2 + 2x - 1$
(E) $x^2 - 1$

6

x	1	2	3	4	5
f(x)	3	4	5	6	7

x	3	4	5	6	7
g(x)	4	6	8	10	12

The accompanying tables define functions f and g. What is $g(f(3))$?

(A) 4
(B) 6
(C) 8
(D) 10
(E) 12

7 Function f is defined by $f(x) = \sqrt{x + 3}$. If $f(p) = 3$, what is the value of p?

(A) 0
(B) $\frac{1}{3}$
(C) 3
(D) 6
(E) 9

8 If function f is defined by $f(x) = 3x + 2$ and $\frac{1}{2}f\left(\sqrt{c}\right) = 3$, then $c =$

(A) $\frac{2}{\sqrt{3}}$

(B) $\frac{2}{3}$

(C) $\frac{4}{3}$

(D) $\frac{4}{9}$

(E) $\frac{16}{9}$

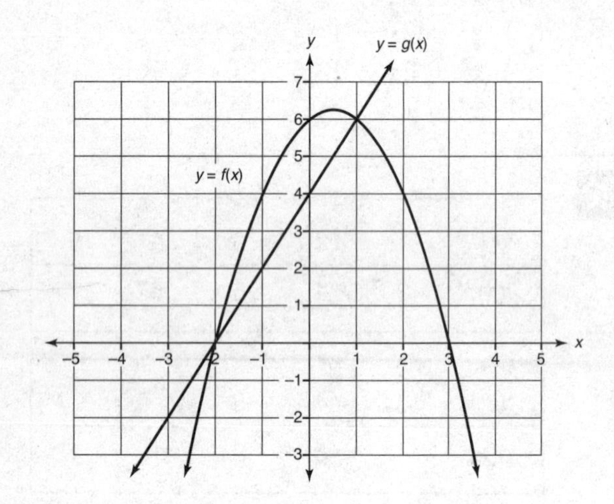

9 Based on the graphs of functions f and g shown in the accompanying figure, what are all values of x between -3 and 3 for which $f(x) \geq g(x)$?

I. $-2 \leq x \leq 0$
II. $0 \leq x \leq 1$
III. $1 \leq x \leq 3$

(A) I only
(B) II only
(C) III only
(D) I and II
(E) II and III

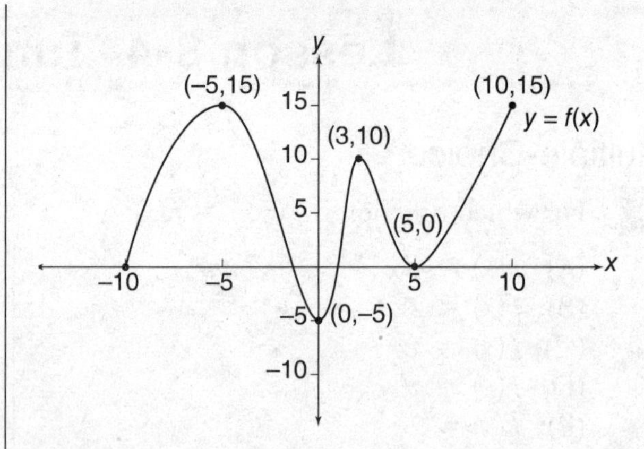

10 If in the accompanying figure (p,q) lies on the graph of $y = f(x)$ and $0 \leq p \leq 5$, which of the following represents the set of corresponding values of q?

(A) $-5 \leq q \leq 15$
(B) $-5 \leq q \leq 10$
(C) $-5 \leq q \leq 5$
(D) $5 \leq q \leq 10$
(E) $5 \leq q \leq 15$

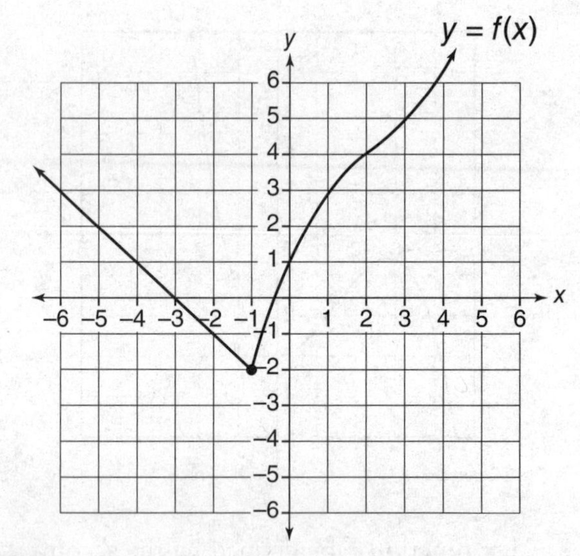

11 The accompanying figure shows the graph of $y = f(x)$. If function g is defined by $g(x) = f(x + 4)$, then $g(-1)$ could be

(A) -2
(B) 3
(C) 4
(D) 5
(E) 6

x	f(x)	g(x)
1	2	3
2	4	5
3	5	1
4	3	2
5	1	4

Questions 12–13 refer to the accompanying table, which gives the values of functions f and g for integer values of x from 1 to 5, inclusive.

12 According to the table, if $f(5) = p$, what is the value of $g(p)$?

(A) 1
(B) 2
(C) 3
(D) 4
(E) 5

13 Function h is defined by $h(x) = 2f(x) - 1$, where function f is defined in the accompanying table. What is the value of $g(k)$ when $h(k) = 5$?

(A) 1
(B) 2
(C) 3
(D) 4
(E) 5

14 If $f(x) = 1 + \sqrt{x}$ and $f(g(2)) = 4$, then $g(x)$ could be any of the following EXCEPT

(A) $5x - 1$
(B) $4x + 1$
(C) $3x + 2$
(D) $2x + 5$
(E) $x^3 + 1$

15 Let j be the function defined by $j(x) = x^2 + 1$ and k be the function defined by $k(x) = (x + 1)^2$. Which expression is equivalent to $k(p - 1)$?

(A) $j(p - 1)$
(B) $j(p) - 1$
(C) $j(p) + 1$
(D) $j(p + 1)$
(E) $j(p) - p$

16 According to market research, the number of magazine subscriptions that can be sold can be estimated using the function $n(p) = \dfrac{5000}{4p - k}$, where n is the number of thousands of subscriptions sold, p is the price in dollars for each individual subscription, and k is some constant. If 250,000 subscriptions were sold at \$15 for each subscription, how many subscriptions could be sold if the price were set at \$20 for each subscription?

(A) 40,000
(B) 50,000
(C) 75,000
(D) 100,000
(E) 125,000

17 If the function f is defined by $f(x) = 2x + k$ and the function g is defined by $g(x) = \dfrac{x - 5}{2}$, for what value of k is $f(g(x)) = g(f(x))$?

(A) -15
(B) -5
(C) 0
(D) 5
(E) 15

18 Let the function f be defined by $f(x) = 3x - 2$ and let the function g be defined by $g(x) = -x + 1$. If $f(x + y) = 4$ and $g(x - y) = -7$, then $x =$

(A) -3
(B) -1
(C) 1
(D) 3
(E) 5

Grid-In

1 Let h be the function defined by $h(x) = x + 4^x$. What is the value of $h\left(-\frac{1}{2}\right)$?

2 Let f be the function defined by $f(x) = 3^{2x-1}$. If $f(n) = 81$, what is the value of n?

3 Let the function f be defined by $f(x) = x^2 + 12$. If n is a positive number such that $f(3n) = 3f(n)$, what is the value of n?

4

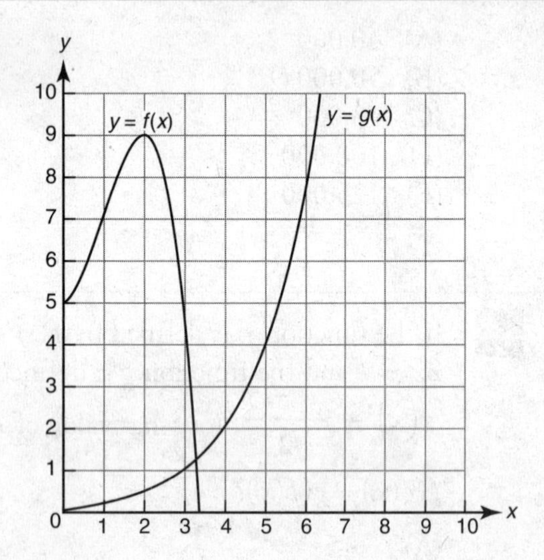

Let the function f and g be defined by the graphs in the accompanying diagram. What is the value of $f(g(3))$?

Questions 5 and 6. Let the function f be defined by the graph below.

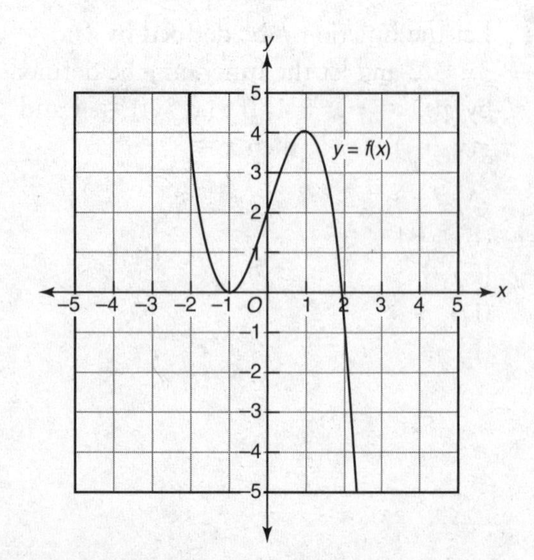

5 What is the integer value of $2f(-1) + 3f(1)$?

6 If n represents the number of different values of x for which $f(x) = 2$ and m represents the number of different values of x for which $f(x) = 4$, what is the value of mn?

7 Let g be the function defined by $g(x) = x - 1$. If $\frac{1}{2}g(c) = 4$, what is the value of $g(2c)$?

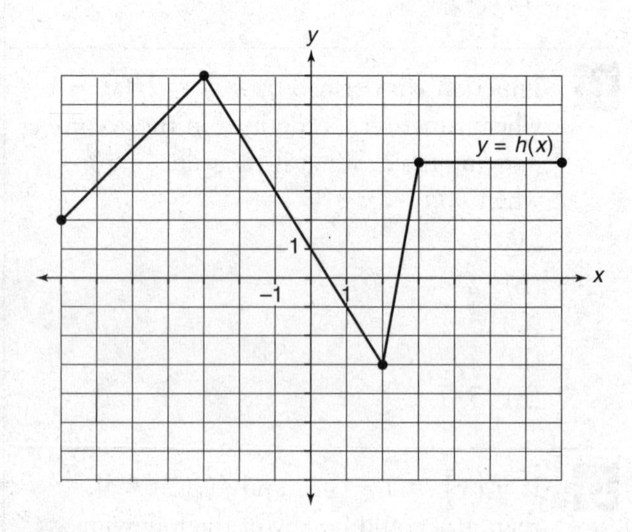

8 The figure above shows the graph of function h. If function f is defined by $f(x) = h(2x) + 1$, what is the value of $f(-1)$?

SOME SPECIAL FUNCTIONS

OVERVIEW

The SAT assumes that you are familiar with the key properties of linear, quadratic, and exponential functions.

- *In a* ***linear function,*** *the greatest power of the variable is 1, as in* $y = 3x + 2$.
- *In a* ***quadratic function,*** *the greatest power of the variable is 2, as in* $f(x) = 3x^2 - 5x + 1$.
- *In an* ***exponential function,*** *the variable is in an exponent, as in* $f(x) = 2^x$.

LINEAR FUNCTIONS AS MODELS

The graph of a linear function is a straight line. When a linear function is written in the form $f(x) = mx + b$, m is the slope of the line and b is its y-intercept. If $y = f(x) = 3x + 2$, then each time x increases by 1 unit, y increases by 3 units, so the rate of change is fixed at $\frac{3}{1}$ or 3. Thus, the slope of a line represents the rate at which a linear function changes in value with respect to x. Linear functions are used to represent, or "model," situations that involve *constant* rates of change.

EXAMPLE

After 2 hours of driving, Tyrone finds that 13 gallons of gas are left in his car's fuel tank, and after 3 hours of driving, 10.5 gallons are left.

a. Find a linear function that models the number of gallons g left in the tank as x function of x hours of driving.

b. At what rate is the car consuming gas?

c. How many gallons of gasoline were in the tank before the car was driven?

Solution

a. Find a linear function of the form $g(x) = mx + b$ that contains (2,13) and (3,10.5):
- Since the slope of the line is $\frac{10.5 - 13}{3 - 2} = -2.5$, the linear function has the form $g = -2.5x + b$.
- To find b, substitute (2,13) into $g(x) = -2.5x + b$, which makes $13 = -2.5(2) + b$, so $b = 18$.
- The linear function is $g = -2.5x + 18$.

b. Since the slope of $g(x) = -2.5x + 18$ is -2.5, the car is consuming gas at the constant rate of 2.5 gallons per hour of driving.

c. If $x = 0$, then $g = -2.5(0) + 18 = 18$ gallons.

QUADRATIC FUNCTIONS AND THEIR GRAPHS

A quadratic function has the form $f(x) = ax^2 + bx + c$, where a, b, and c are constants with a $\neq 0$. The graph of a quadratic function is a "U-shaped" curve called a **parabola,** as shown in Figures 8.3 and 8.4. A parabola function has a vertical line of symmetry that divides it into two "mirror image" parts. The constant c in $y = ax^2 + bx + c$ is the y-intercept of the parabola. The line of symmetry intersects the parabola at its **vertex,** or "turning point." The vertex is always the lowest point or the highest point on the parabola.

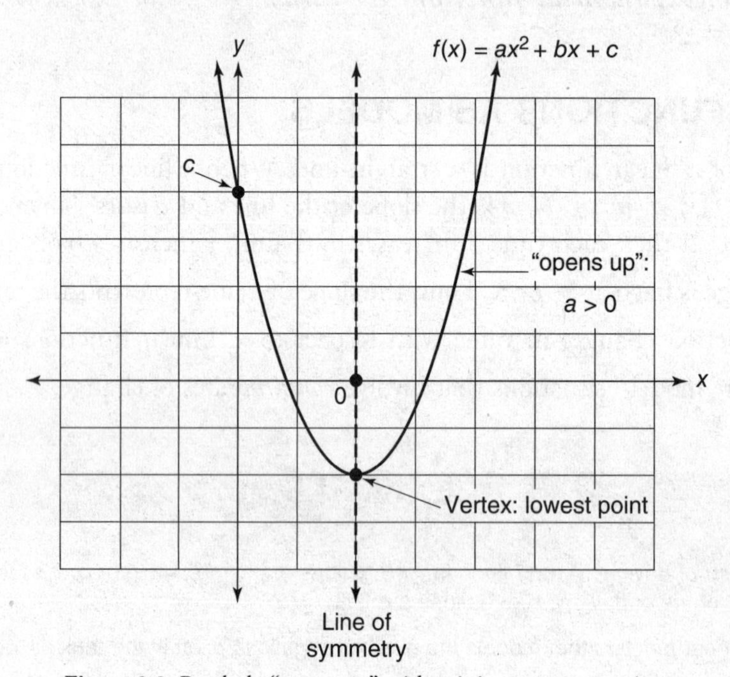

Figure 8.3 Parabola "opens up" with minimum vertex point.

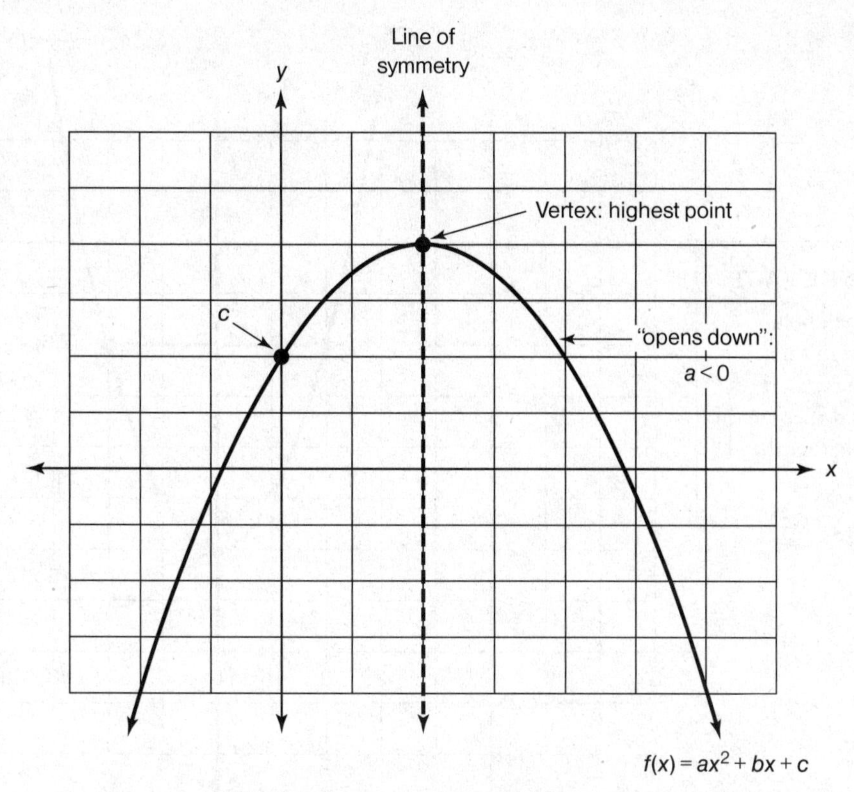

Figure 8.4 Parabola "opens down" with maximum vertex point.

Without graphing a parabola, you can tell from its equation whether it "opens up" or "opens down." If $f(x) = ax^2 + bx + c$ and

- $a > 0$, the parabola "opens up," and the vertex is the *lowest* point on the curve, as in Figure 8.3. At this point, has its *minimum* value. For example, the graph of $f(x) = 2x^2 - 6x + 3$ is a parabola that opens up with a minimum vertex point and a y-intercept of 3.

- $a < 0$, the parabola "opens down," and the vertex is the *highest* point on the curve, as in Figure 8.4. At this point, y reaches its *maximum* value. For example, the graph of $f(x) = -x^2 - 4x + 5$ is a parabola that opens down with a maximum vertex point and a y-intercept of 5.

MATCHING PAIRS OF PARABOLA POINTS

Each point on a parabola other than the vertex has a corresponding point on the opposite side of the line of symmetry and the same distance from it. Figure 8.5 shows the graph of $f(x) = x^2 - 10x + 27$ with $x = 5$ as its line of symmetry. The point (3,6) is on the parabola and 2 units from the line of symmetry. The matching point on the parabola is (7,6), which lies on the opposite side of the line of symmetry and is also 2 units from it. Using function notation, $f(3) = f(7)$. Similarly, $f(4) = f(6)$.

A parabola's line of symmetry bisects every horizontal segment that connects two points on the parabola.

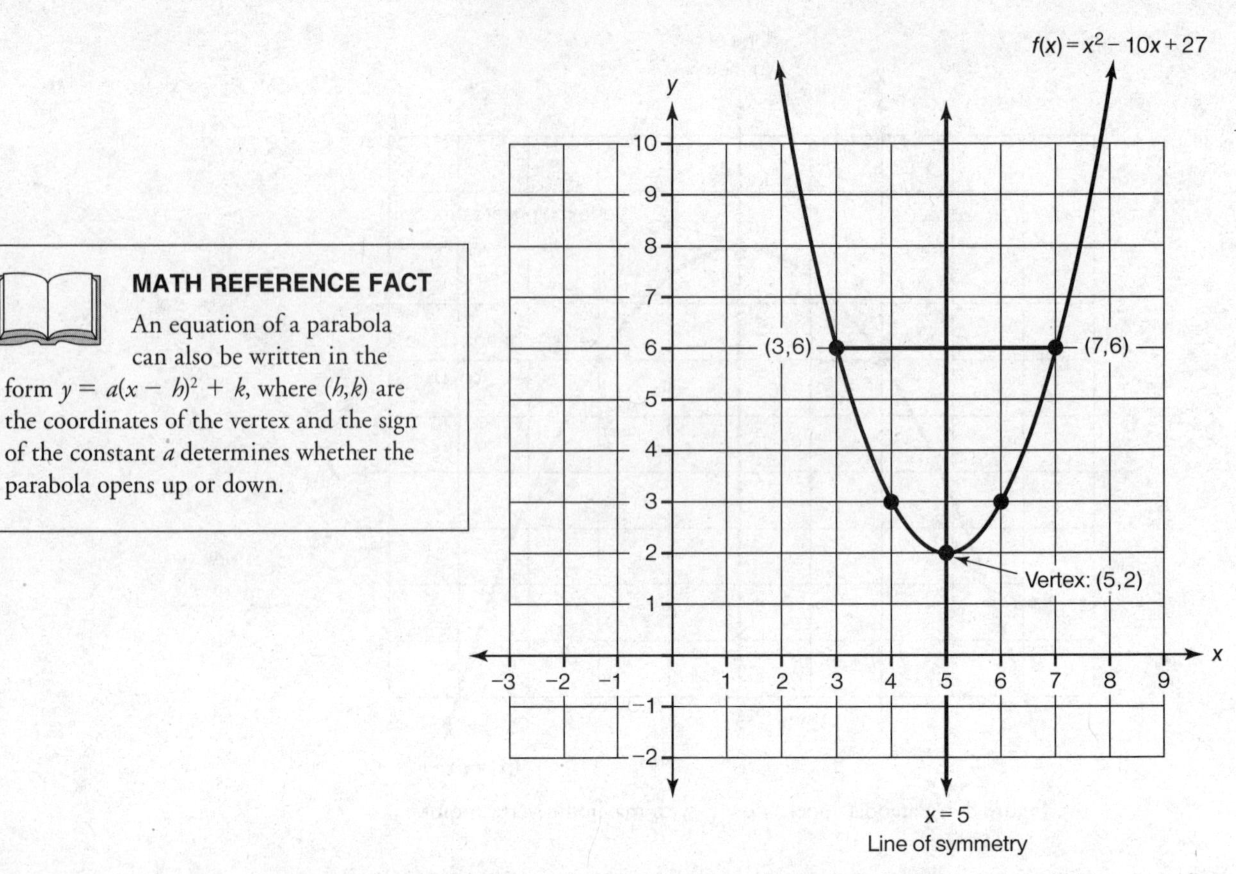

MATH REFERENCE FACT

An equation of a parabola can also be written in the form $y = a(x - h)^2 + k$, where (h,k) are the coordinates of the vertex and the sign of the constant a determines whether the parabola opens up or down.

Figure 8.5 Matching pairs of points on a parabola

"SIDEWAYS" PARABOLAS

The graph of $x = ay^2 + by + c$ $(a \neq 0)$ is a parabola with a *horizontal* line of symmetry, as shown in Figure 8.6. The equation $x = y^2 - 4y + 1$ does *not* represent a function, since matching points on the graph have the same x-value paired with different y-values, such as (1,0) and (1,4).

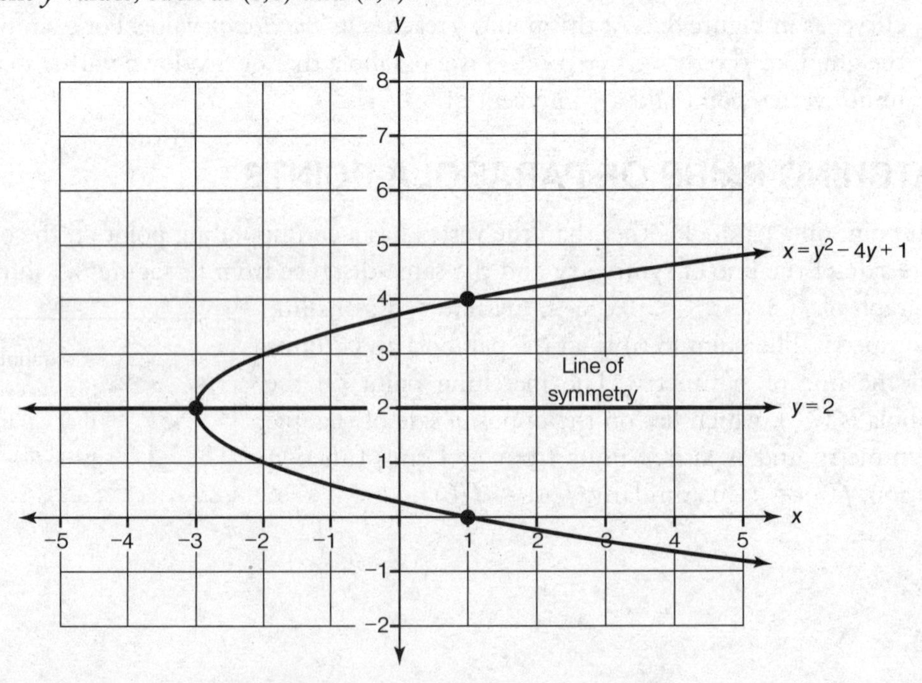

Figure 8.6 Graph of $x = y^2 - 4y + 1$

EXPONENTIAL CHANGE

An **exponential function** is a function that can be written in the form $f(x) = b^x$, where b stands for a positive constant other than 1. In an exponential function, y changes by a fixed multiplying factor of its previous value. The accompanying table illustrates this process using $y = f(x) = 2^x$. For each unit change in x, y doubles or, equivalently, increases by 100% of its previous value. The slope, or the *rate,* at which y changes varies along an exponential curve. The graph of $f(x) = 2^x$ in Figure 8.7 becomes steeper and rises more rapidly as x increases. As x decreases, the graph gets closer and closer to the x-axis without ever touching it.

x	$y = 2^x$
1	2
2	4
3	8
4	16

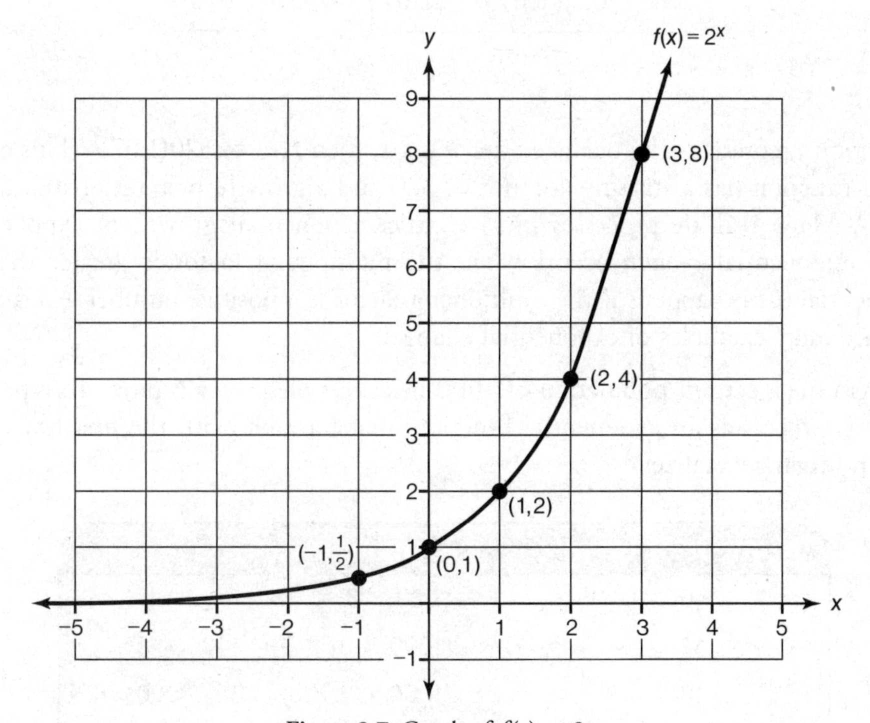

Figure 8.7 Graph of $f(x) = 2^x$

Examples of Exponential Change

Suppose \$320 is invested in an account that earns 7% interest compounded annually.

- After one year, the balance in the account is

$$\underbrace{320}_{\text{initial amount}} + \underbrace{320 \times 0.07}_{\text{interest}} = 320(1 + 0.07) = 320\underbrace{(1.07)}_{\substack{\uparrow \\ \text{multiplying factor}}}$$

- After the second year, the balance in the account is

$$\underbrace{320(1.07)}_{\text{old amount}} + \underbrace{[320(1.07)] \times (0.07)}_{\text{interest}} = 320(1.07)(1 + 0.07) = 320\underbrace{(1.07)}^{2}$$

- After n years, the balance, y, has increased by a factor of 1.07 for a total of n times. Thus,

$$y = \underbrace{320(1.07)(1.07)\ldots(1.07)}_{n \text{ factors}} = 320\underbrace{(1.07)}^{n}$$

If function f represents the balance after n years, then $f(n) = 320(1.07)^n$. This exponential function has a starting amount of 320 and a growth, or multiplying, factor of 1.07. Many real-life processes involve either exponential growth or exponential decay. Exponential *growth* occurs when the multiplying factor is greater than 1. Exponential *decay* happens if the multiplying factor is a positive number less than 1. Here are more examples of exponential change:

- Assume a certain population of 1000 insects triples every 6 days. To represent this process as an exponential function, make a table with the first few terms and then generalize:

Number of Days	Population
6	1000×3
12	$(1000 \times 3) \times 3 = 1000 \times 3^2$
18	$(1000 \times 3 \times 3) \times 3 = 1000 \times 3^3$
...	...
n	$1000 \times 3^{\frac{n}{6}}$

Thus, the exponential function $f(n) = 1000 \times 3^{\frac{n}{6}}$ describes this growth process, where n is the number of days that have elapsed.

- In the geometric sequence

$$400, 200, 100, 50, 25, \ldots$$

each term after the first is obtained by multiplying the term that comes before it by $\frac{1}{2}$. To represent this process as an exponential function, make a table with the first few terms and then generalize:

Term number	Term
1	400
2	$400 \times \frac{1}{2}$
3	$\left(400 \times \frac{1}{2}\right) \times \frac{1}{2} = 400 \times \left(\frac{1}{2}\right)^2$
4	$\left(400 \times \frac{1}{2} \times \frac{1}{2}\right) \times \frac{1}{2} = 400 \times \left(\frac{1}{2}\right)^3$
. . .	. . .
n	$400 \times \left(\frac{1}{2}\right)^{n-1}$

Thus, the exponential function $a(n) = 400 \times \left(\frac{1}{2}\right)^{n-1}$ describes this decay process, where $a(n)$ represents the nth term of this number sequence.

Lesson 8-5 Tune-Up Exercises

Multiple-Choice

1 If in the equation $y = 3^x$, x is increased by 2, then y is

(A) increased by 2
(B) increased by 6
(C) multiplied by 6
(D) increased by 9
(E) multiplied by 9

2 The graph of the quadratic function $f(x) = -x^2 - 4x + 1$ has a line of symmetry at $x = -2$. Which statement must be true?

(A) The maximum point is $(-2,5)$.
(B) The maximum point is $(-2,13)$.
(C) The minimum point is $(-2,5)$.
(D) The minimum point is $(-2,13)$.
(E) The maximum point is $(5,13)$.

3 A printer agrees to publish a school brochure. The charge for the first copy is \$300. For each additional copy, the printer charges \$0.15. If y represents the total cost of printing the brochure, which equation represents the function that gives the cost of printing x copies of the brochure.

(A) $y = 0.15(300x)$
(B) $y = \left(\dfrac{300}{0.15}\right)x$
(C) $y = 0.15 + 300x$
(D) $y = 0.15x + 300$
(E) $y = 300 + 0.15 + x$

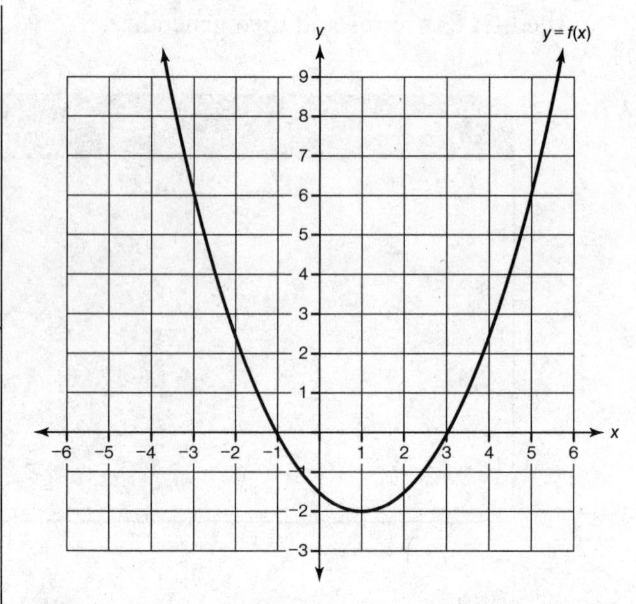

4 The figure above shows the graph of a quadratic function f with a minimum point at $(1,-2)$. If $f(5) = f(c)$, then which of the following could be the value of c?

(A) -5
(B) -3
(C) 0
(D) 6
(E) 25

5 The graph of a quadratic function f intersects the x-axis at $x = -2$ and $x = 6$. If $f(8) = f(p)$, which could be the value of p?

(A) -6
(B) -4
(C) -2
(D) 0
(E) 3

6 A certain population of insects starts at 16 and doubles every 6 days. What is the population after 60 days?

(A) 2^6
(B) 2^{10}
(C) 2^{14}
(D) 2^{32}
(E) 2^{40}

7 If in the quadratic function $f(x) = ax^2 + bx + c$, a and c are both negative constants, which of the following could be the graph of function f?

(A)

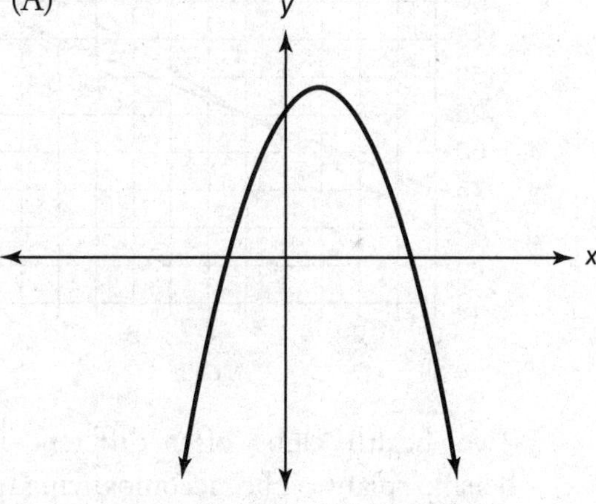

(B)

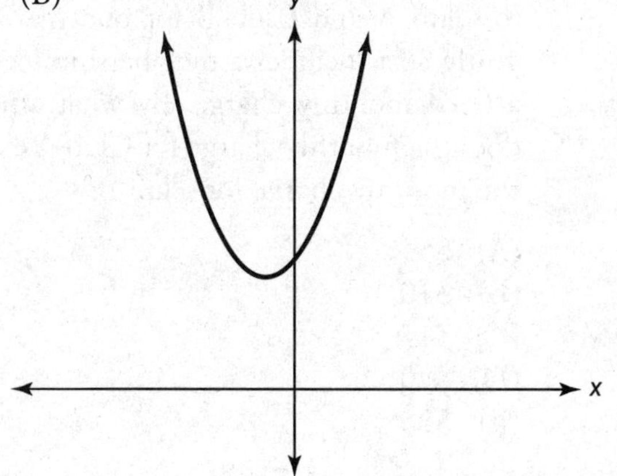

(C)

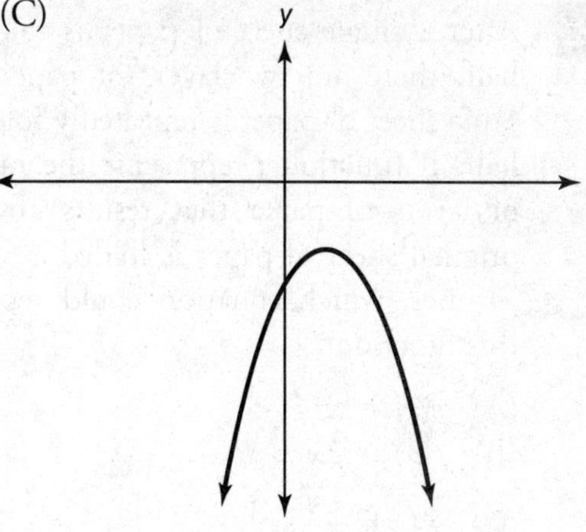

(D)

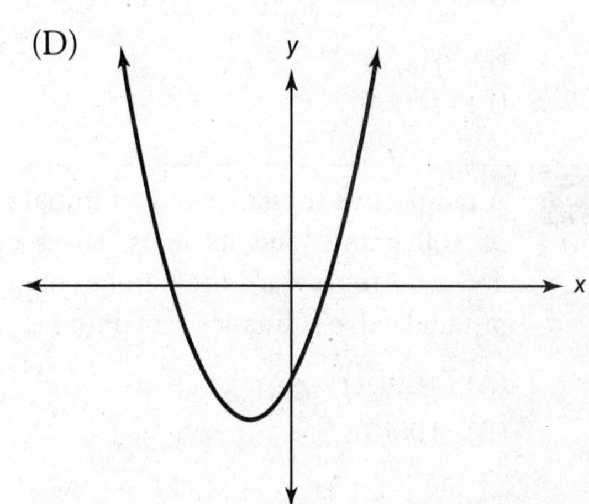

(E)

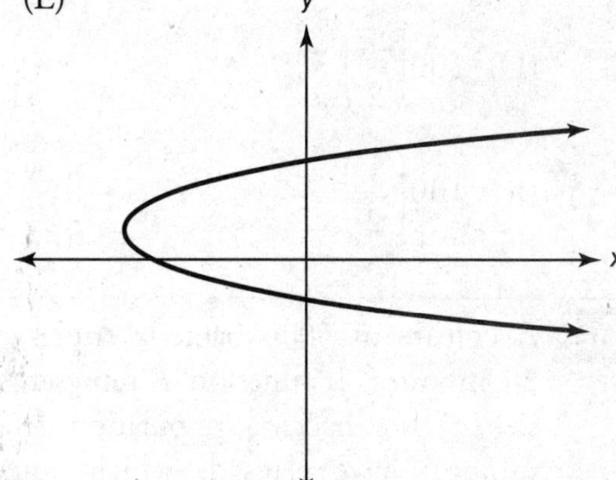

8 After a single sheet of paper is folded in half, there are two layers of paper. The same sheet of paper is repeatedly folded in half. If function f represents the number of layers of paper that results after the original sheet of paper is folded a total of x times, which equation could represent this function?

(A) $f(x) = x^2$
(B) $f(x) = 2x$
(C) $f(x) = \left(\frac{1}{2}\right)^x$
(D) $f(x) = 2^x$
(E) $f(x) = 2^{x-1}$

9 A radioactive substance has an initial mass of 100 grams, and its mass halves every 4 years. After t years, the number of grams of radioactive substance remaining is

(A) $100(4)^{\frac{t}{4}}$
(B) $100(4)^{-2t}$
(C) $100\left(\frac{1}{2}\right)^{\frac{t}{4}}$
(D) $100\left(\frac{1}{2}\right)^{4t}$
(E) $100\left(\frac{1}{2}\right)^{\frac{4}{t}}$

10 A culture of 5000 bacteria triples every 20 minutes. If function P represents the size of the bacteria population after m minutes have elapsed, which equation could represent $P(m)$?

(A) $P(m) = 5000(20)^{3m}$
(B) $P(m) = 5000(3)^{20m}$
(C) $P(m) = 5000(3)^{\frac{m}{20}}$
(D) $P(m) = 5000\left(20^{\frac{m}{3}}\right)$
(E) $P(m) = \left(\frac{5000}{3}\right)^{20m}$

11 In the number sequence $8, 24, 72, 216, \ldots,$ each term after the first is three times as great as the term that comes before it. If n is a positive integer, what is the $n + 1$ term of this sequence?

(A) $8 \times 3^{n-2}$
(B) $8 \times 3^{n-1}$
(C) 8×3^{n}
(D) $8 \times 3^{n+1}$
(E) $8 \times 3^{n+2}$

12

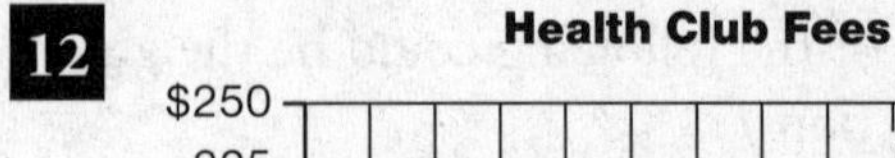

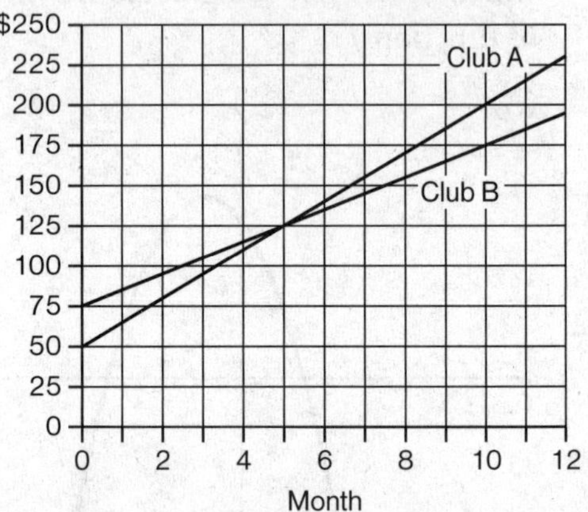

Two health clubs offer different membership plans. The accompanying graph represents the total yearly cost of belonging to Club A and Club B for one year. The yearly cost includes a membership fee plus a fixed monthly charge. By what amount does the monthly charge for Club A exceed the monthly charge for Club B?

(A) \$5
(B) \$10
(C) \$15
(D) \$20
(E) \$25

13 A parabola passes through the points (0, 0) and (6, 0). If the turning point of the parabola is $T(h,4)$, which statement must be true?

I $h = 2$.
II If the parabola passes through (1,2), then it must all pass through (5,2).
III Point T is the highest point of the parabola.

(A) II only
(B) III only
(C) I and II only
(D) I and III only
(E) II and III only

Grid-In

1 Ari begins painting at 12:00 noon. At 12:30 PM he estimates that 15.75 gallons of paint are left, and at 2:00 PM he estimates that 10.5 gallons of paint remain. If the paint is being used at constant rate, how many gallons of paint did Ari have when he started the job?

2 The number of hours, H, needed to manufacture X computer monitors is given by the linear function defined by $H = kX + q$, where k and q are constants. If it takes 270 hours to manufacture 100 computer monitors and 410 hours to manufacture 160 computer monitors, how many *minutes* are required to manufacture one computer monitor?

3

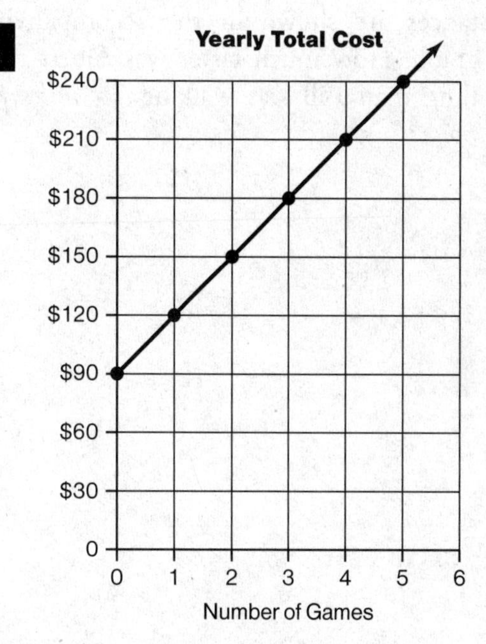

The accompanying graph represents the yearly cost of playing 0 to 5 games of golf at the Shadybrook Golf Course. The yearly cost includes a membership fee plus the cost of the 5 games. If the cost of each game is the same, what would be the total cost of membership and playing 12 games during the year?

4

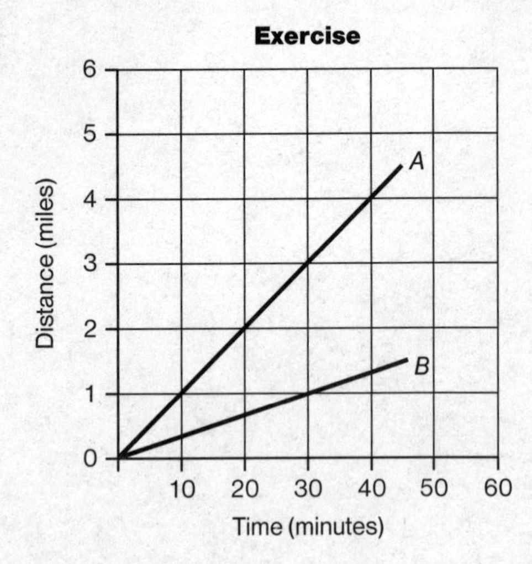

During a 45-minute lunch period, Albert (A) went running, and Bill (B) walked for exercise. Their times and distances are shown in the accompanying graph. How much faster was Albert running than Bill was walking, in *miles per hour*?

LESSON

8-6

REFLECTING AND TRANSLATING FUNCTION GRAPHS

OVERVIEW

*Making certain types of simple changes in the equation of a function can change the position of its graph without changing its shape or size. A **reflection** of a graph in a line flips the graph over that line so that the original and reflected graphs are exact "mirror images." A **translation** of a graph shifts every point of the graph in the same way so that its shape and size are not affected.*

REFLECTING GRAPHS OF FUNCTIONS

Changing the x-coordinate or the y-coordinate of a point to its opposite reflects, or "flips," that point over a coordinate axis. Referring to Figure 8.8 we find the following:

- (2,5) becomes (–2,5) after it is reflected over the y-axis.
- (2,5) becomes (2,–5) after it is reflected over the x-axis.

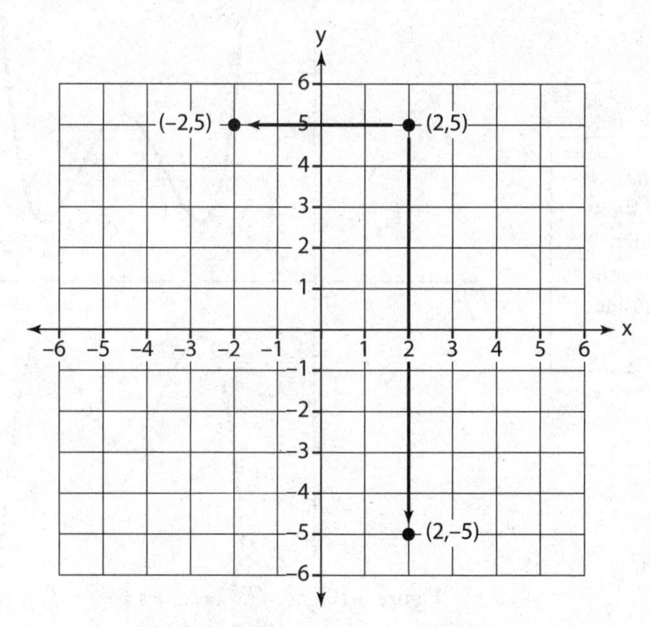

Figure 8.8 Reflecting (2,5) in the coordinate axes

A graph may also be reflected in a coordinate axis. In Figure 8.9, corresponding points on the two graphs have the same y-coordinates but x-coordinates with opposite signs. The graph of $y = f(-x)$ is the graph of $y = f(x)$ reflected over the y-axis.

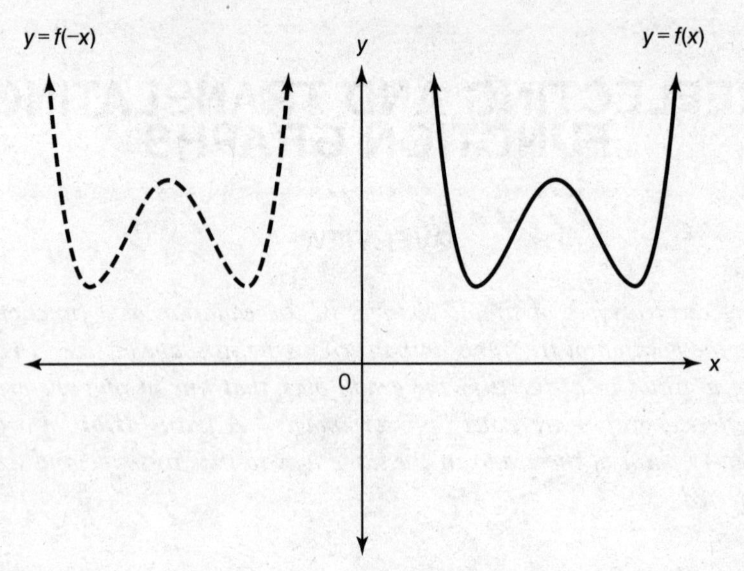

Figure 8.9 $y = f(x)$ becomes $y = f(-x)$ when it is reflected in the y-axis.

In Figure 8.10, corresponding points on the two graphs have the same x-coordinates but y-coordinates with opposite signs. The graph of $y = -f(x)$ is the graph of $y = f(x)$ reflected over the x-axis.

Writing a negative sign *inside* the parentheses of $y = f(x)$ flips its graph over the y-axis, while writing a negative sign *outside* the parentheses flips its graph upside down over the x-axis.

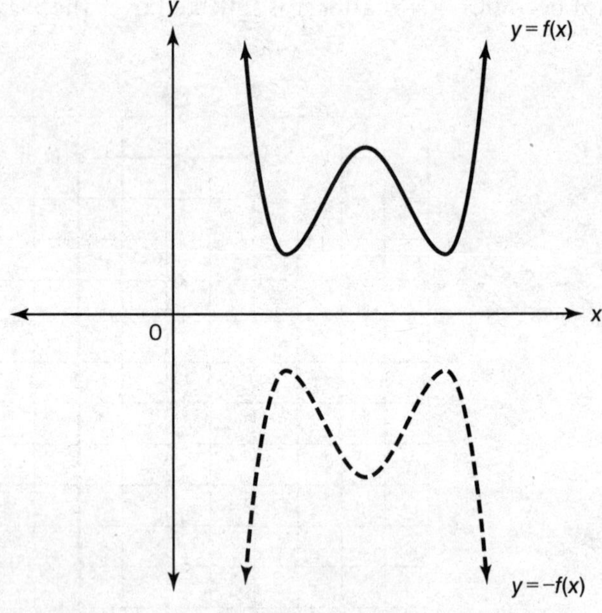

Figure 8.10 $y = f(x)$ becomes $y = -f(x)$ when it is reflected in the x-axis.

RULES FOR REFLECTING GRAPHS

- To flip $y = f(x)$ upside down over the x-axis, write the negative sign *outside* the parentheses: $y = -f(x)$ is the graph of $y = f(x)$ reflected over the x-axis.
- To flip $y = f(x)$ over the y-axis, write the negative sign *inside* the parentheses: $y = f(-x)$ is the graph of $y = f(x)$ reflected over the y-axis.

TRANSLATING GRAPHS UP AND DOWN

If c is a positive number, the graph of $y = f(x)$ can be shifted up c units by graphing $y = f(x) + c$ or shifted down c units by graphing $y = f(x) - c$. Here are some examples:

- The graph of $y = f(x) + 2$ is the graph of $y = f(x)$ shifted *up* 2 units, as in Figure 8.11.
- The graph of $y = f(x) -2$ is the graph of $f(x) = x^2$ shifted *down* 2 units, as in Figure 8.12.

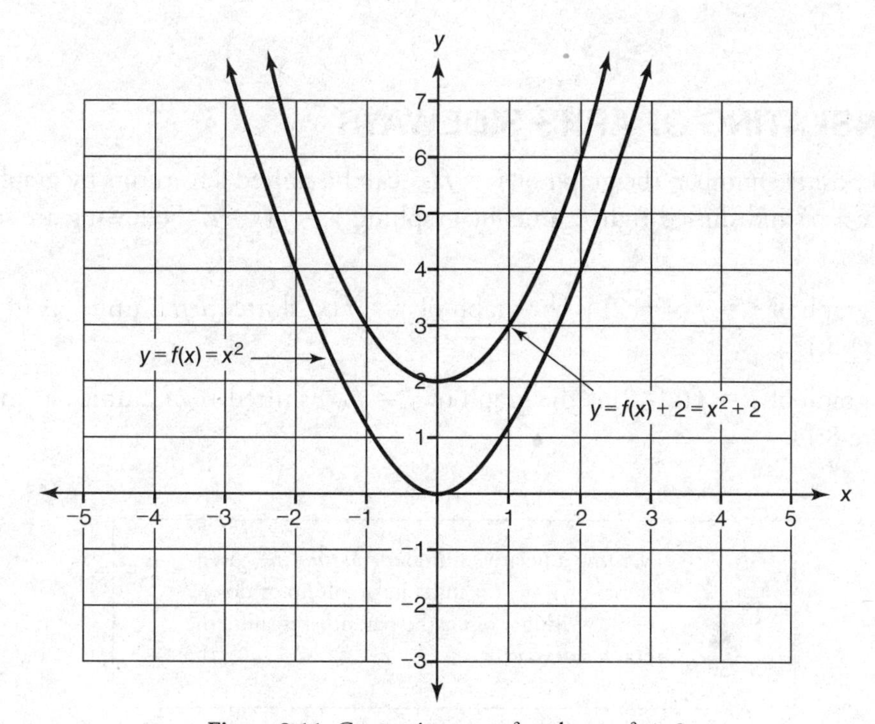

Figure 8.11 Comparing $y = x^2$ and $y = x^2 + 2$

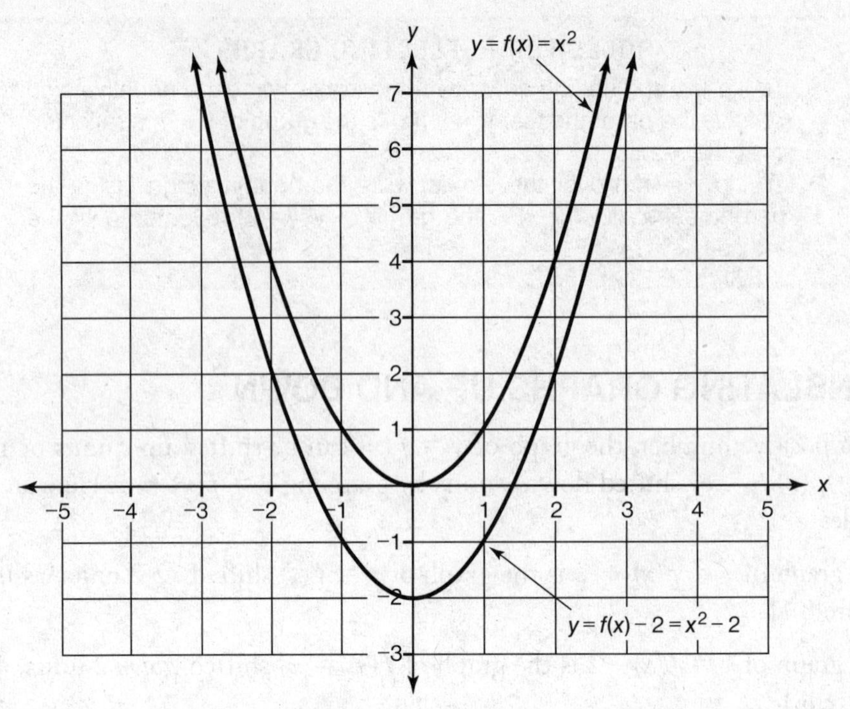

Figure 8.12 Comparing $y = x^2$ and $y = x^2 - 2$

TRANSLATING GRAPHS SIDEWAYS

If c is a positive number, the graph of $y = f(x)$ can be shifted left c units by graphing $y = f(x + c)$ and shifted right c units by graphing $y = f(x - c)$. Following are some examples:

- The graph of $y = f(x + 2)$ is the graph of $y = f(x)$ shifted *left* 2 units, as in Figure 8.13.
- The graph of $y = f(x - 2)$ is the graph of $y = f(x)$ shifted *right* 2 units, as in Figure 8.14.

Writing a positive number c *outside* the parentheses of $y = f(x)$ shifts its graph up or down, while writing c *inside* the parentheses shifts the graph sideways.

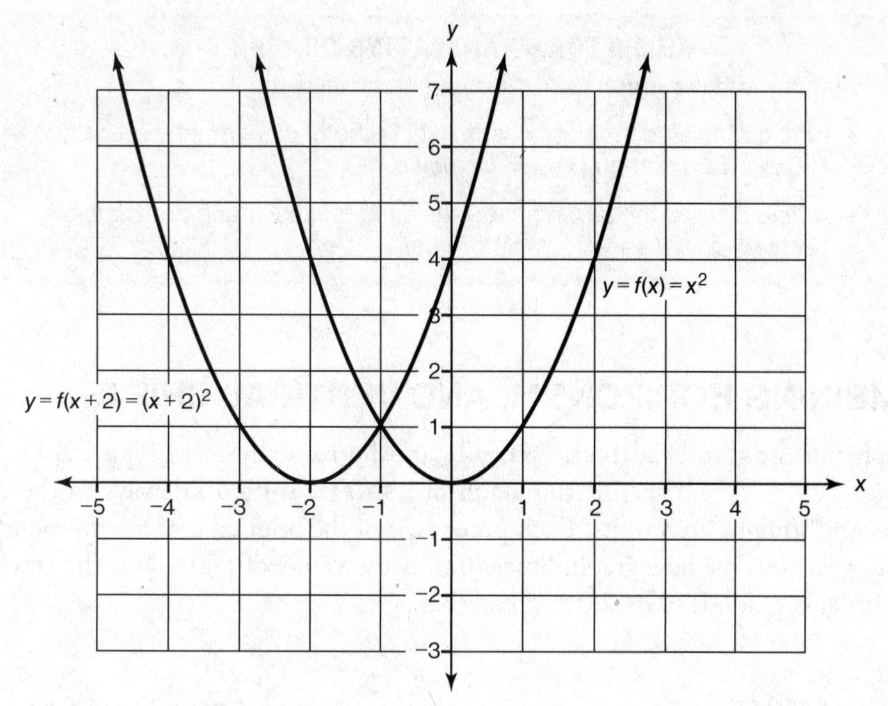

Figure 8.13 Comparing $y = x^2$ and $y = (x + 2)^2$

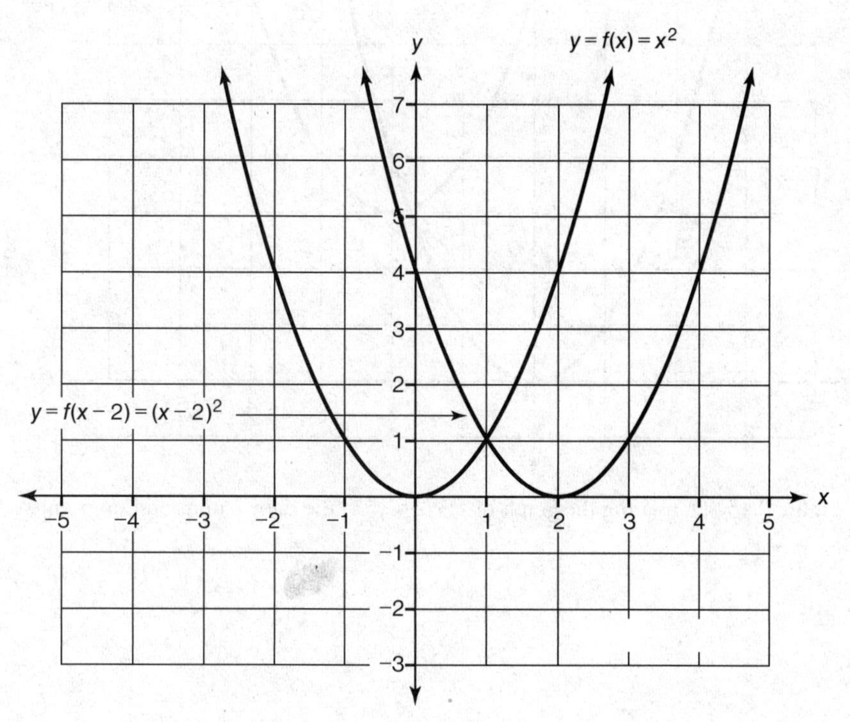

Figure 8.14 Comparing $y = x^2$ and $y = (x - 2)^2$

RULES FOR TRANSLATING GRAPHS

Starting with the graph of $y = f(x)$ and a positive number c:

- writing c *outside* the parentheses shifts the original graph vertically: $f(x) + c$ for *up* and $f(x) - c$ for *down*.
- writing c *inside* the parentheses shifts the original graph horizontally: $f(x + c)$ for *left* and $f(x - c)$ for *right*.

COMBINING HORIZONTAL AND VERTICAL SHIFTS

A graph may be translated both vertically and horizontally as in Figure 8.15. The graph of $y = f(x - 2) + 3$ is the graph of $y = f(x)$ shifted sideways to the right 2 units and straight up 3 units. Each point (x,y) of the original graph corresponds to $(x + 2, y + 3)$ of the new graph. Since (0,0) is the vertex of $f(x) = x^2$, the vertex of the translated graph is $(0 + 2, 0 + 3) = (2,3)$.

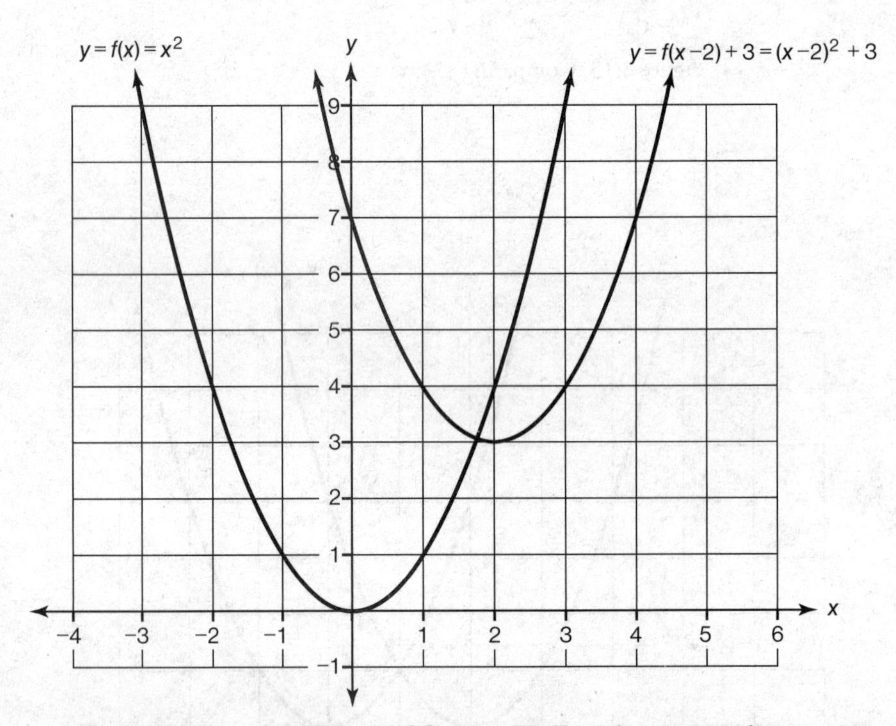

Figure 8.15 Translating the graph of $f(x) = x^2$ to the right 2 units and up 3 units

Lesson 8-6 Tune-Up Exercises

Multiple-Choice

1 The graph of $y = 2^{x-3}$ can be obtained by shifting the graph of $y = 2^x$

(A) 3 units to the right
(B) 3 units to the left
(C) 3 units up
(D) 3 units down
(E) 2 units to the right and 3 units up

2 Which equation represents the line that is the reflection of the line $y = 2x - 3$ in the x-axis?

(A) $y = -2x - 3$
(B) $y = -2x + 3$
(C) $y = 2x + 3$
(D) $y = 3x - 2$
(E) $y = -\frac{1}{2}x - 3$

3 The endpoints of $\overline{AB}$ are $A(0,0)$ and $B(9,-6)$. What is an equation of the line that contains the reflection of $\overline{AB}$ in the y-axis?

(A) $y = -\frac{3}{2}x$

(B) $y = -\frac{2}{3}x$

(C) $y = -x + 3$

(D) $y = \frac{2}{3}x$

(E) $y = \frac{3}{2}x$

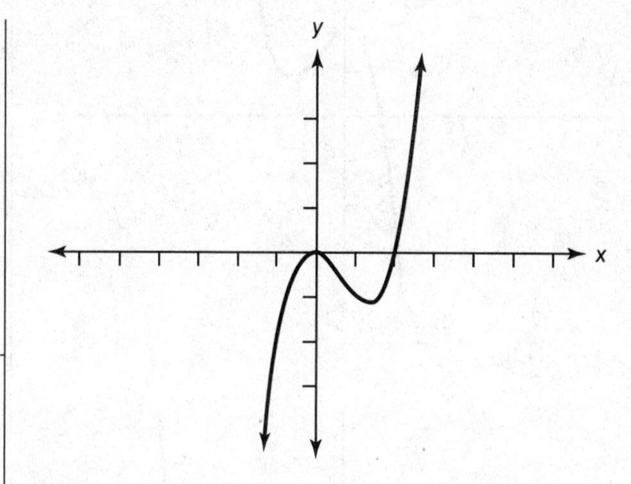

4 The figure above shows the graph of function f. If $g(x) = -f(x)$, which graph represents function g?

(A)

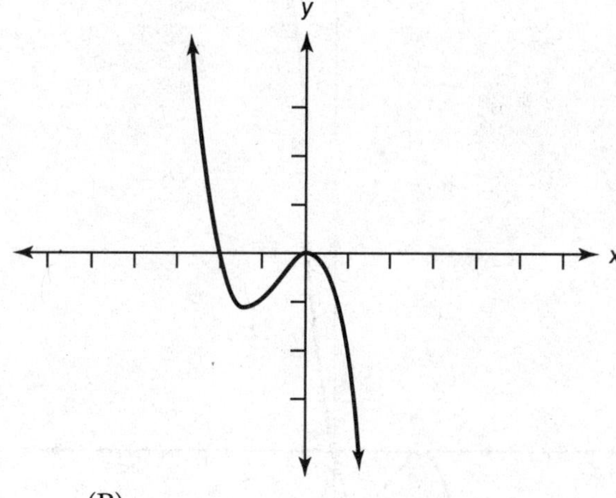

(B)

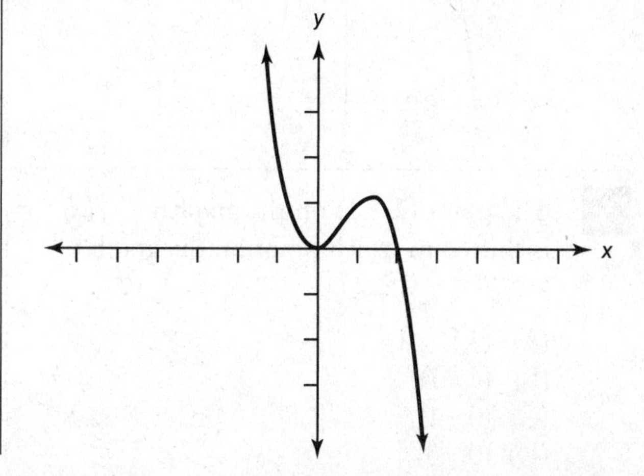

(C)

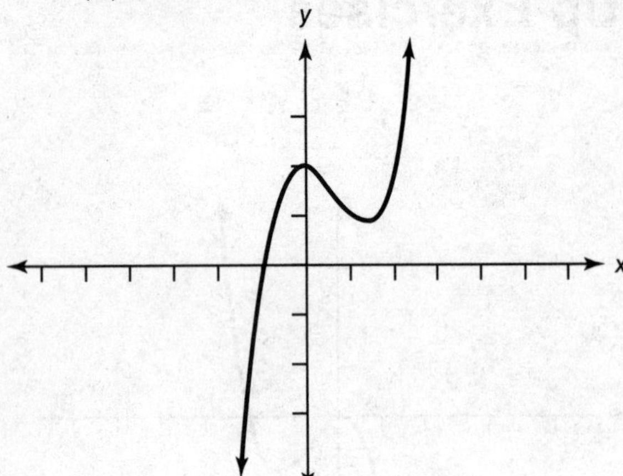

(D)

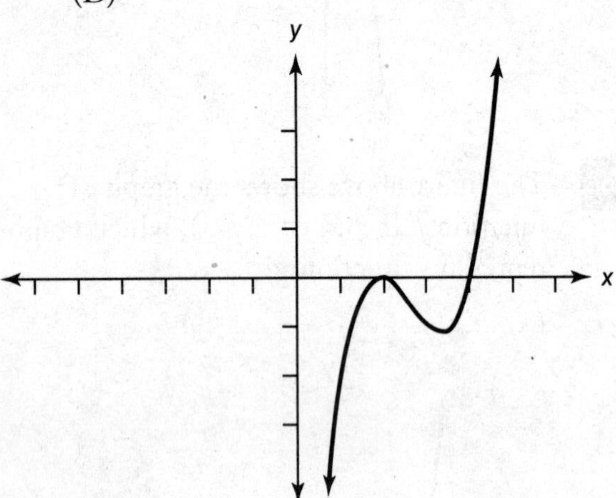

(E)

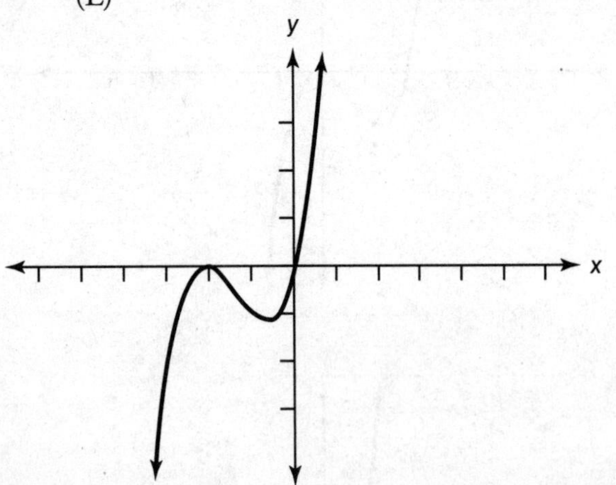

5 The point $(2,-1)$ on the graph $y = f(x)$ is shifted to which point on the graph of $y = f(x + 2)$?

(A) $(4,1)$
(B) $(4,-1)$
(C) $(0,-1)$
(D) $(0,-3)$
(E) $(2,1)$

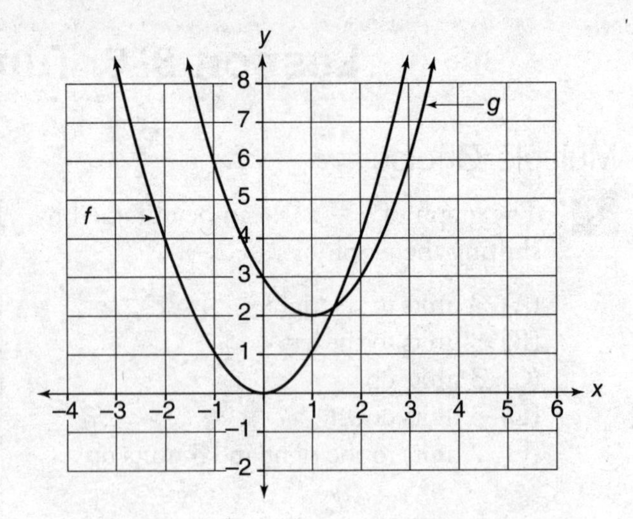

6 The accompanying figure shows the graphs of functions f and g. If f is defined by $f(x) = x^2$ and g is defined by $g(x) = f(x + h) + k$, where h and k are constants, what is the value of $h + k$?

(A) −3
(B) −2
(C) −1
(D) 1
(E) 2

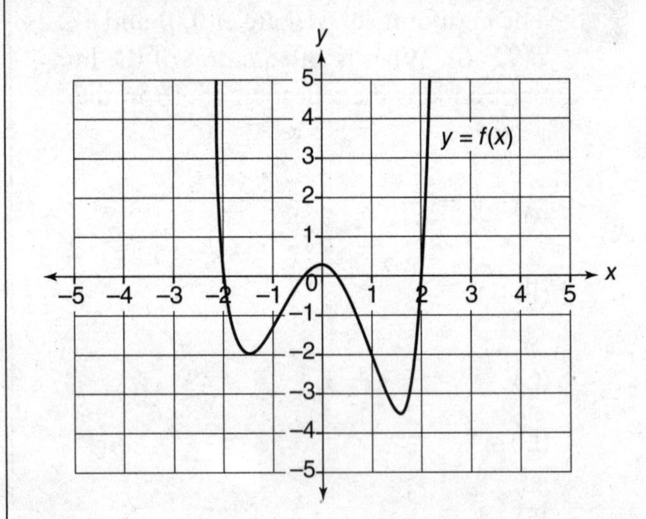

7 If $g(x) = -2$ intersects the graph of $y = f(x) + k$ at one point, which of these choices could be the value of k?

(A) −1.5
(B) −0.5
(C) 0
(D) 0.5
(E) 1.5

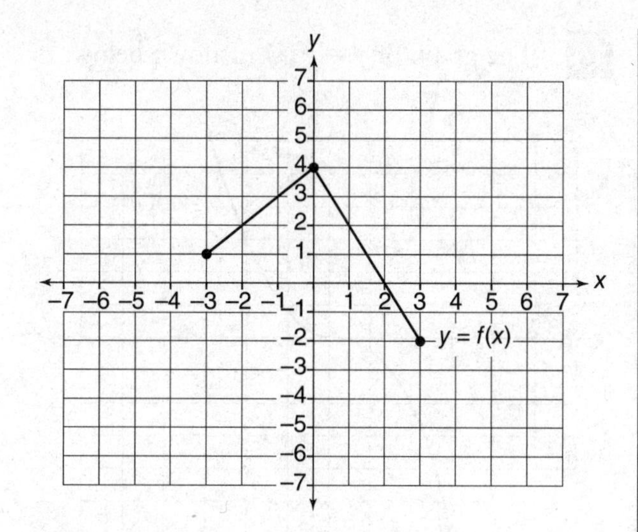

8 If the accompanying figure shows the graph of function f, which of the following could represent the graph of $y = f(x + 1)$?

(A)

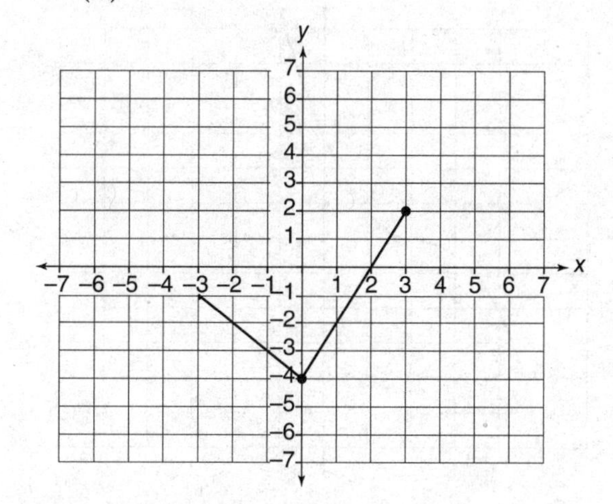

(B)

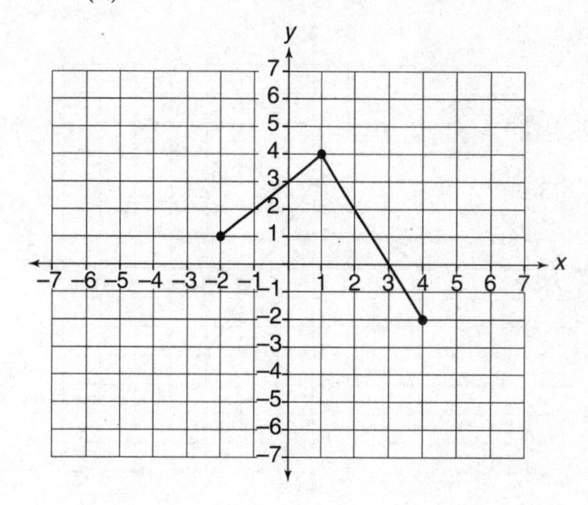

(C)

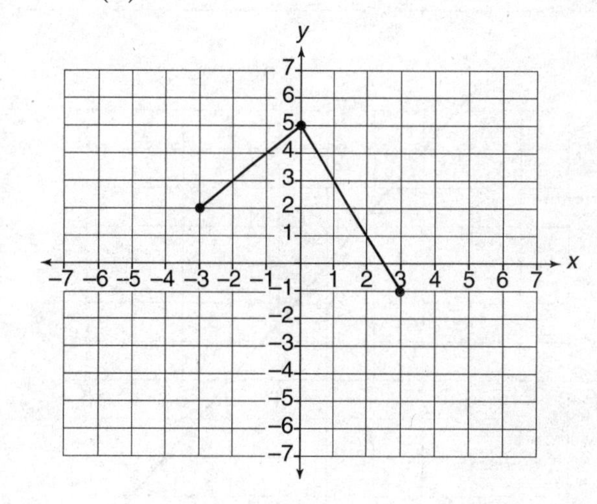

(D)

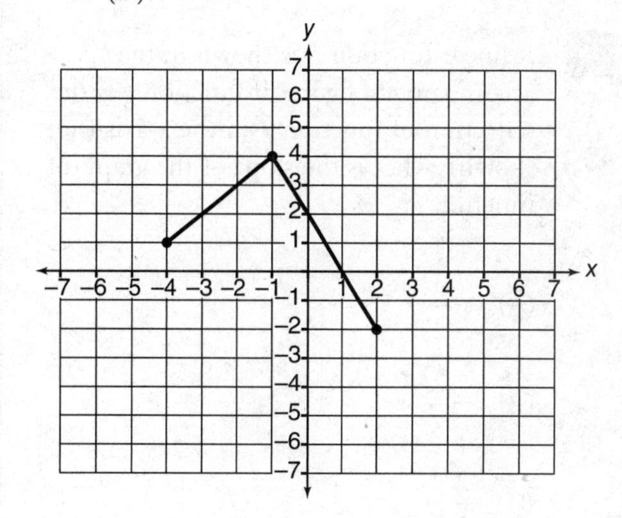

(E)

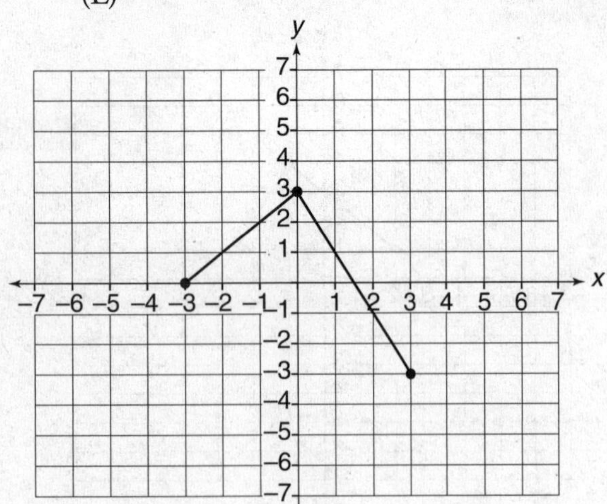

9 A linear function f is shown in the accompanying figure. If function g is the reflection of function f in the x-axis (not shown), what is the slope of the graph of function g?

(A) $-\frac{3}{2}$

(B) $-\frac{2}{3}$

(C) $\frac{2}{3}$

(D) $\frac{3}{2}$

(E) 3

10 The graph of $y = f(x)$ is shown below.

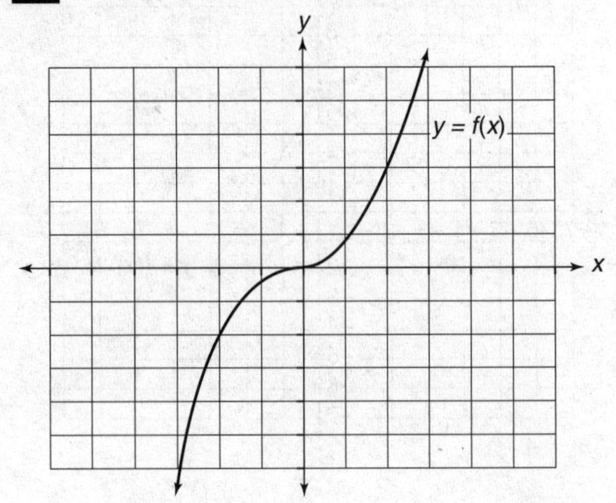

Which of the following could represent the graph of $y = f(x - 2) + 1$?

(A)

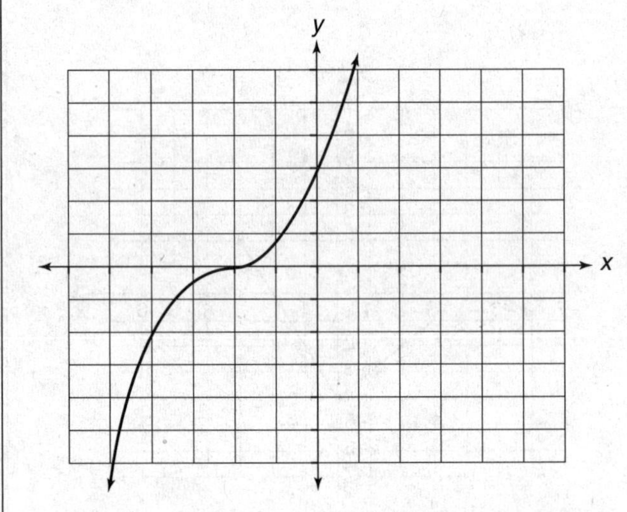

(B)

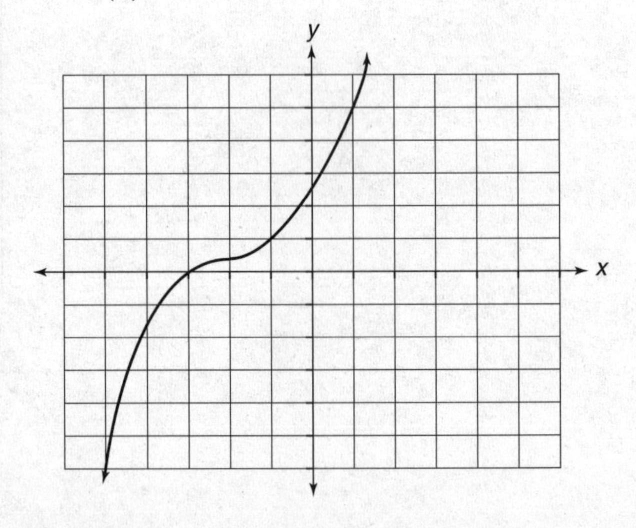

(C)

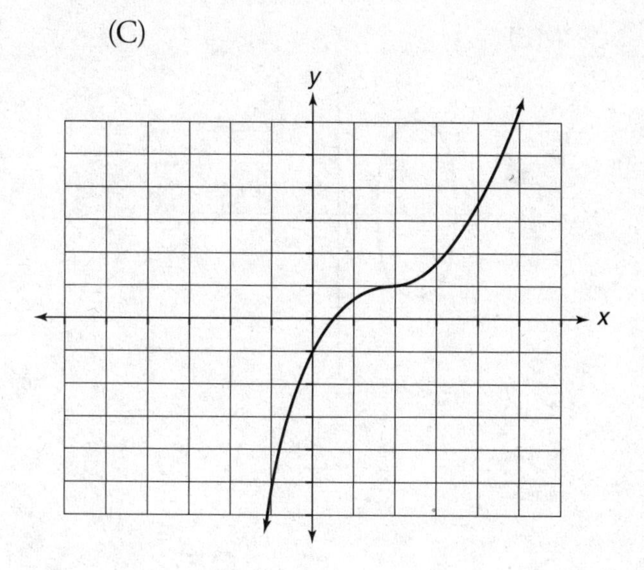

(D)

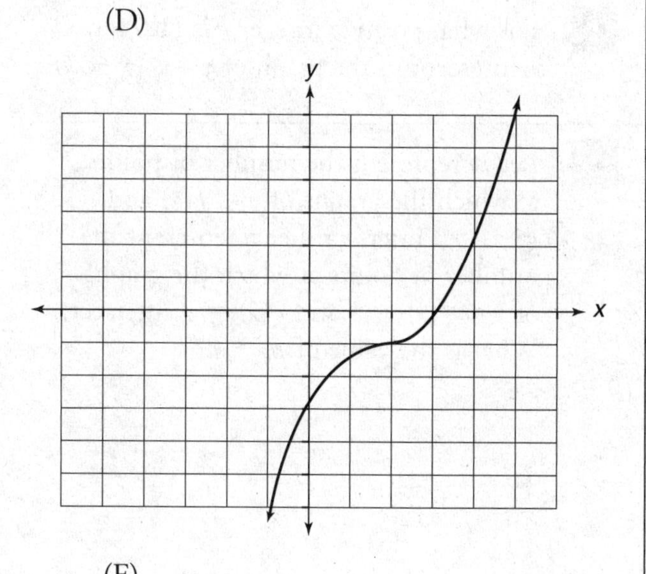

(E)

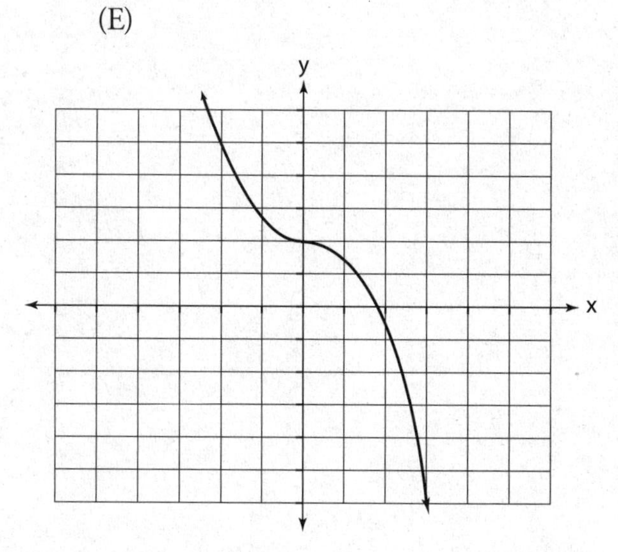

11

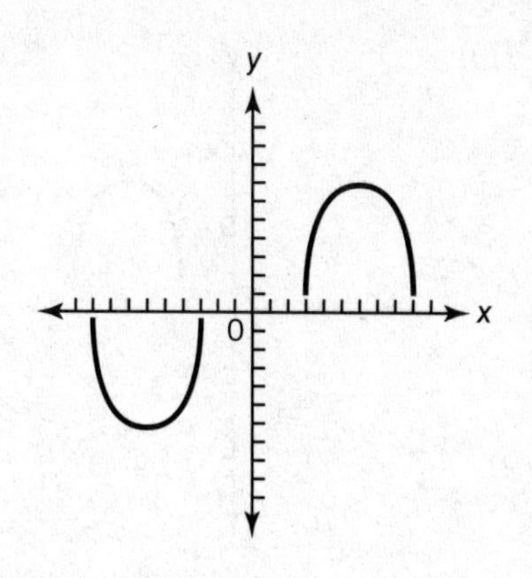

The graph of the function f is shown above. Which of the following could represent the graph of $y = |f(x)|$?

(A)

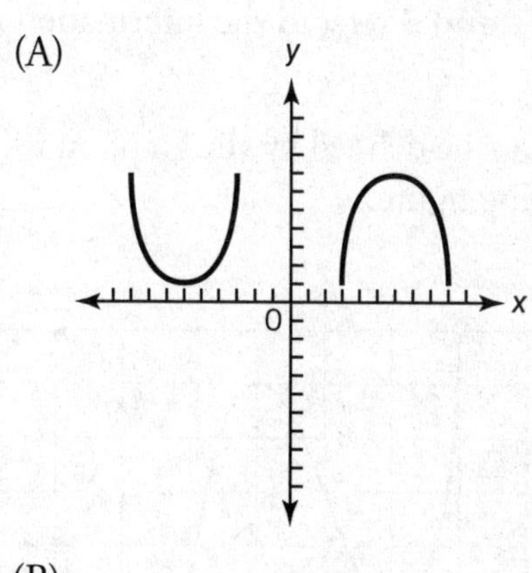

(B)

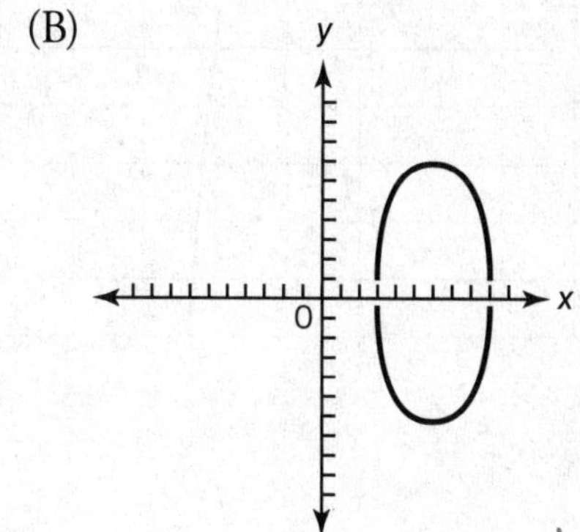

(C)

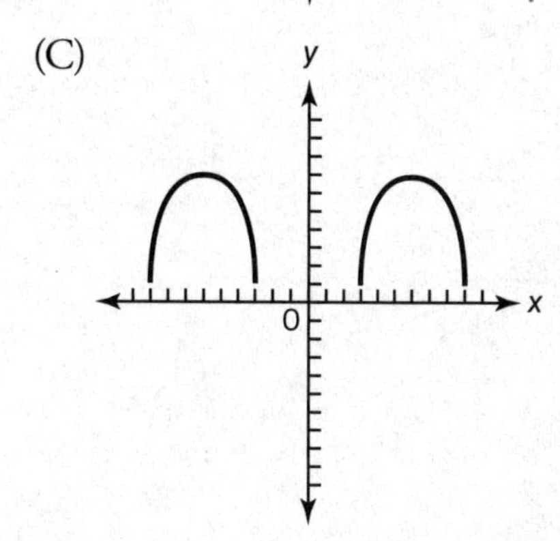

(D)

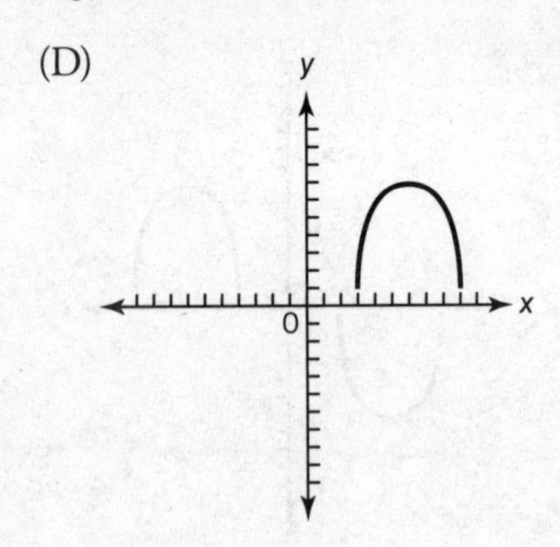

(E)

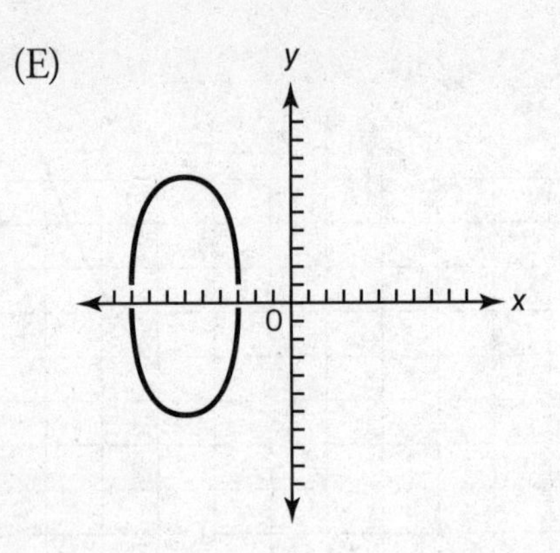

Grid-In

Questions 1 and 2 refer to the information and graph below.

Let function f be defined by the graph in the accompanying figure.

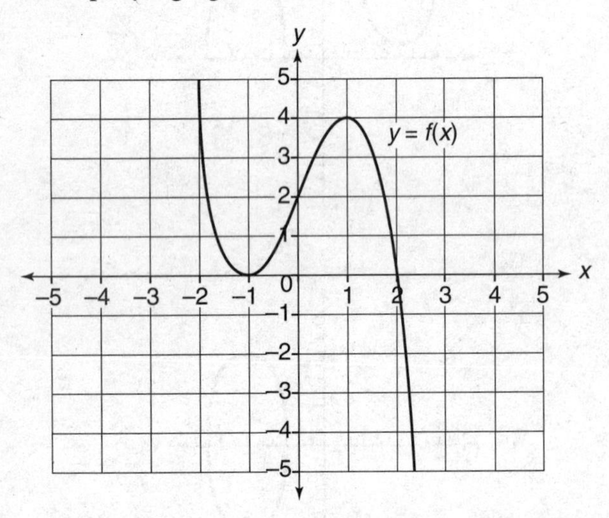

1 For what positive integer k is (1,0) an x-intercept of the graph of $y = f(x - k)$?

2 Let m represent the number of points at which the graphs of $y = f(x)$ and $g(x) = 3$ intersect. Let n represent the number of points at which the graphs of $y = f(x) - 1$ and $g(x) = 3$ intersect. What is the value of $m + n$?

ANSWERS TO CHAPTER 8 TUNE-UP EXERCISES

Lesson 8-1 (*Multiple-Choice*)

1. **(D)** If $x^{\frac{3}{4}} = c$, then $\left(x^{\frac{3}{4}}\right)^{\frac{4}{3}} = x = c^{\frac{4}{3}}$. Then $\sqrt{x} = (c)^{\frac{1}{2}} = \left(c^{\frac{4}{3}}\right)^{\frac{1}{2}} = c^{\frac{4}{6}} = c^{\frac{2}{3}}$.

2. **(A)** Since $\sqrt{a^3} = a^{\frac{3}{2}}$, we need to use the given facts that $\sqrt{a} = x$ and $a^2 = y$ to find $a^{\frac{3}{2}}$ in terms of x and y.
   - Using the quotient law,
   $$\frac{a^2}{\sqrt{a}} = \frac{a^2}{a^{\frac{1}{2}}} = a^{2-\frac{1}{2}} = a^{\frac{3}{2}}$$
   - From the given, $\dfrac{a^2}{\sqrt{a}} = \dfrac{y}{x}$
   - Hence, $\dfrac{a^2}{\sqrt{a}} = a^{\frac{3}{2}} = \dfrac{y}{x}$

3. **(E)** If $j = x^{-\frac{1}{2}}$ and $k = \dfrac{1}{x^2}$, then $j = \dfrac{1}{x^{\frac{1}{2}}}$ and $\dfrac{1}{k} = x^2$, so
$$\left(\frac{j}{k}\right)^3 = \left(\frac{x^2}{x^{\frac{1}{2}}}\right)^3 = \left(x^{\frac{3}{2}}\right)^3 = x^{\frac{3}{2}\times 3} = x^{\frac{9}{2}}$$

4. **(D)** If $\sqrt{m} = 2p$, then $m^{\frac{1}{2}} = 2p$, so
$$\left(m^{\frac{1}{2}}\right)^3 = (2p)^3$$
$$m^{\frac{3}{2}} = 8p^3$$

5. **(B)** If $x = p^2$, $x^3 = q$, and $\dfrac{p}{q} = x^n$, then $p = \sqrt{x} = x^{\frac{1}{2}}$. Thus,
$$\frac{p}{q} = x^n$$
$$\frac{x^{\frac{1}{2}}}{x^3} = x^n$$
$$x^{\frac{1}{2}-3} = x^n$$
$$x^{-\frac{5}{2}} = x^n$$
Therefore, $n = -\dfrac{5}{2}$.

6. **(A)** If $40x = 10x^2y^{-1}$, then
$$\frac{40x}{40x^2} = \frac{10x^2y^{-1}}{40x^2}$$
$$\frac{1}{x} = \frac{y^{-1}}{4}$$
$$x^{-1} = \frac{1}{4y}$$

7. **(E)** If $\dfrac{x}{x^{1.5}} = 8x^{-1}$, then $\dfrac{x}{x^{1.5}} = \dfrac{8}{x}$, so $x \cdot x = 8x^{1.5}$.
Since $x^2 = 8x^{1.5}$, $\dfrac{x^2}{x^{1.5}} = x^{2-1.5} = x^{0.5} = \sqrt{x} = 8\cdot$
Hence, $x = 8^2 = 64$.

8. **(C)** If $10^k = 64$, then $\left(10^k\right)^{\frac{1}{2}} = 64^{\frac{1}{2}} = \sqrt{64} = 8$.
Hence, $10^{\frac{k}{2}} = 8$. Multiply both sides of $10^{\frac{k}{2}} = 8$ by 10:
$$10^1 \times \left(10^{\frac{k}{2}}\right) = 10^1 \times 8$$
$$10^{\frac{k}{2}+1} = 80$$

9. **(C)** To find how much greater $x^{\frac{5}{2}}$ is than x^2, subtract x^2 from $x^{\frac{5}{2}}$: $x^{\frac{5}{2}} - x^2$. Then factor out the GCF of x^2 from the difference:
$$x^{\frac{5}{2}} - x^2 = x^2\left(x^{\frac{1}{2}} - 1\right)$$

10. **(E)** $a = x^{-1} + x^{-1} = \frac{1}{x} + \frac{1}{x} = \frac{2}{x}$, and $b = y^{-1} + y^{-1} + y^{-1} = \frac{1}{y} + \frac{1}{y} + \frac{1}{y} = \frac{3}{y}$.
Hence,
$$\frac{a}{b} = a \cdot \frac{1}{b} = \frac{2}{x} \cdot \frac{y}{3} = \frac{2y}{3x}$$

(*Grid-In*)

1. **6/5** If $2x^{-1} = \frac{5}{3}$, then $\frac{2}{x} = \frac{5}{3}$, so $5x = 6$ and $x = \frac{6}{5}$.

2. **16** If $x^{-\frac{1}{2}} = \frac{1}{8}$, then $x^{\frac{1}{2}} = 8$. Since $\sqrt{x} = 8$, $x = 64$. Hence,
$$x^{\frac{2}{3}} = 64^{\frac{2}{3}} = \left(\sqrt[3]{64}\right)^2 = 4^2 = 16.$$

3. **3/2** If $x^{-1} = 3$ and $y^{-1} = 9$, then $x = \frac{1}{3}$ and $y = \frac{1}{9}$. Thus,

$$\begin{aligned}(x + y)^{-\frac{1}{2}} &= \left(\frac{1}{3} + \frac{1}{9}\right)^{-\frac{1}{2}} \\ &= \left(\frac{1}{3} + \frac{1}{9}\right)^{-\frac{1}{2}} \\ &= \left(\frac{3}{9} + \frac{1}{9}\right)^{-\frac{1}{2}} \\ &= \left(\frac{4}{9}\right)^{-\frac{1}{2}} \\ &= \left(\frac{9}{4}\right)^{\frac{1}{2}} = \sqrt{\frac{9}{4}} = \frac{3}{2}\end{aligned}$$

4. **2/3** If $8(rs)^{-1} = (3rs^{-2})^2$, then

$$\begin{aligned}\frac{8}{rs} &= 9r^2s^{-4} \\ \frac{8}{rs} &= \frac{9r^2}{s^4} \\ \frac{9r^3s}{9s^4} &= \frac{8s^4}{9s^4} \\ \frac{r^3}{s^3} &= \frac{8}{9} \\ \left(\frac{r}{s}\right)^3 &= \frac{8}{9} \\ \frac{r}{s} &= \sqrt[3]{\frac{8}{9}} = \frac{2}{3}\end{aligned}$$

Lesson 8-2 (*Multiple-Choice*)

1. **(A)** If $\frac{1}{\sqrt{x^2-2}} = 2$ and $x^2 > 2$, then

$$\begin{aligned}\sqrt{x^2 - 2} &= \frac{1}{2} \\ \left(\sqrt{x^2 - 2}\right)^2 &= \left(\frac{1}{2}\right)^2 \\ x^2 - 2 &= \frac{1}{4} \\ x^2 &= 2 + \frac{1}{4} \\ &= \frac{8}{4} + \frac{1}{4} \\ &= \frac{9}{4}\end{aligned}$$

$$\begin{aligned}\sqrt{x^2} &= \sqrt{\frac{9}{4}} \\ x &= \frac{3}{2}\end{aligned}$$

2. **(D)** If $z = x^3 = \sqrt{y}$, where $y > 0$, then $x = z^{\frac{1}{3}}$ and $y = z^2$. Hence,

$$xy = z^{\frac{1}{3}} \cdot z^2 = z^{\frac{7}{3}}$$

3. **(A)** It is given that $4^y + 4^y + 4^y + 4^y = 16^x$. On the left side of the equation, four identical terms are being added together. Thus, $4 \cdot 4^y = (4^2)^x$, so $4^{y+1} = (4^{2x})$.
Hence, $y + 1 = 2x$ and $y = 2x - 1$.

4. **(D)** Since $t^9 = x^3$, $\left(t^9\right)^{\frac{1}{3}} = \left(s^3\right)^{\frac{1}{3}}$, so $s = t^3$.

5. **(C)** Solve each exponential equation

- Since $3^x = 81 = 3^4$, $x = 4$
- Since $2^{x+y} = 64 = 2^6$, $x + y = 6$. Because $x = 4$, $4 + y = 6$, so $y = 2$.
- Hence, $\frac{x}{y} = \frac{4}{2} = 2$.

6. **(A)** If $64^{1-x} = \frac{1}{16^{2x}}$, then $64^{1-x} = 16^{-2x}$. Writing each side of the equation as a power of 2 gives $2^{6(1-x)} = 2^{4(-2x)}$. Therefore,

$$\begin{aligned}6(1 - x) &= 4(-2x) \\ 6 - 6x &= -8x \\ 8x - 6x &= -6 \\ 2x &= -6 \\ x &= \frac{-6}{2} = -3\end{aligned}$$

7. **(A)** Use the two exponential equations to obtain a pair of linear equations in x and y:

- If $4^{x-y} = 64$, then $2^{2(x-y)} = 2^6$, so $2(x - y) = 6$ or, equivalently, $x - y = 3$.
- If $3^{-2x+y} = \frac{1}{3}$, then $3^{-2x+y} = 3^{-1}$, so $-2x + y = -1$.
- Solve the linear system consisting of $x - y = 3$ and $-2 + y = -1$ by adding corresponding sides of the two equations to eliminate y:

$$\begin{aligned} x - y &= 3 \\ -2x + y &= -1 \\ \hline -x + 0 &= 2 \\ \text{so } x &= -2. \end{aligned}$$

Then $-2 - y = 3$, so $y = 5$.

8. **(A)** If $27^x = 9^{y-1}$, then $3^{3x} = 3^{2(y-1)}$. Hence, $3x = 2(y - 1)$, so $3x = 2y - 2$ and $3x + 2 = 2y$. Solving for y:

$$\begin{aligned} y &= \frac{3x + 2}{2} \\ &= \frac{3}{2}x + \frac{2}{2} \\ &= \frac{3}{2}x + 1 \end{aligned}$$

9. **(A)** Eliminate the radical by raising both sides of the given equation to the second power:

$$\begin{aligned} \left(\sqrt{x^2 + y^2 + 1}\right)^2 &= (x + y)^2 = (x + y)(x + y) \\ \cancel{x^2} + \cancel{y^2} + 1 &= \cancel{x^2} + 2xy + \cancel{y^2} \\ 1 &= 2xy \\ \frac{1}{2} &= \frac{2xy}{2} \\ \frac{1}{2} &= xy \end{aligned}$$

10. **(B)** If $1000^y = 100^w$, then $(10^3)^y = (10^2)^w$. Because $10^{3y} = 10^{2w}$, $3y = 2w$. Thus,

$$\frac{\cancel{3}y}{\cancel{3}w} = \frac{2\cancel{w}}{3\cancel{w}} \text{ and } \frac{y}{w} = \frac{2}{3}.$$

11. **(C)** If $8(2^p) = 4^n$, then $(2^3)(2^p) = (2^2)^n$, and $2^{p+3} = 2^{2n}$. Thus, $p + 3 = 2n$, and $n = \dfrac{p + 3}{2}$.

12. **(A)** If $\frac{1}{2}\sqrt{rs} = r$, then

$$\begin{aligned} \sqrt{rs} &= 2r \\ \left(\sqrt{rs}\right)^2 &= (2r)^2 \\ \frac{rs}{r^2} &= \frac{4\cancel{r^2}}{\cancel{r^2}} \\ \frac{s}{r} &= 4 \\ \frac{r}{s} &= \frac{1}{4} \end{aligned}$$

(Grid-In)

1. **50** If $5\sqrt{x - 1} + 13 = 48$, then $5\sqrt{x - 1} = 35$, so $\sqrt{x - 1} = \frac{35}{5} = 7$. Hence, $(\sqrt{x - 1})^2 = 7^2$, so $x - 1 = 49$ and $x = 50$.

2. **2/3** If $2\sqrt{x} = \sqrt{x + 2}$, then $\left(2\sqrt{x}\right)^2 = \left(\sqrt{x + 2}\right)^2$. Thus, $4x = x + 2$, $3x = 2$, and $x = \frac{2}{3}$.

3. **12** If $\sqrt{x^y} = 9$ and $y > x$, then

$$\begin{aligned} \left(\sqrt{x^y}\right)^2 &= 9^2 \\ x^y &= 81 \end{aligned}$$

Since it is also given that $y > x$, $3^4 = 81$ where $x = 3$ and $y = 4$. Hence, $xy = 12$.

4. **1/6** If m and p are positive integers and $\left(\sqrt{3}\right)^m = 27^p$, then

$$\begin{aligned} \left(3^{\frac{1}{2}}\right)^m &= \left(3^3\right)^p \\ 3^{\frac{m}{2}} &= 3^{3p} \\ \frac{m}{2} &= 3p \\ \frac{m}{6m} &= \frac{6p}{6m} \\ \frac{1}{6} &= \frac{p}{m} \end{aligned}$$

5. **8** Since $16^{x-2} = \frac{1}{2} \cdot 2^{3x+1}$, rewrite both sides of the equation as a power of 2:

$$2^{4(x-2)} = 2^{-1} \cdot 2^{3x+1} = 2^{3x+1-1} = 2^{3x}$$

Setting the exponents equal makes $4(x - 2) = 3$, so $4x - 8 = 3x$ and $x = 8$.

6. **1/4** If a, b, and c are positive numbers such that $\sqrt{\frac{a}{b}} = 8c$, then

$$\begin{aligned} \left(\sqrt{\frac{a}{b}}\right)^2 &= (8c)^2 \\ \frac{a}{b} &= 64c^2 \end{aligned}$$

Since it is also given that $ac = b$,

$$\frac{a}{ac} = 64c^2$$
$$\frac{1}{c} = 64c^2$$
$$64c^3 = 1$$
$$c^3 = \frac{1}{64}$$
$$\left(c^3\right)^{\frac{1}{3}} = \left(\frac{1}{64}\right)^{\frac{1}{3}}$$
$$c = \frac{1}{4}$$

Lesson 8-3 (*Multiple-Choice*)

1. **(A)** If $|x| \leq 2$ and $|y| \leq 1$, then $-2 \leq x \leq 2$ and $-1 \leq y \leq 1$. The least possible value of $x - y$ is $-2 - 1 = -3$
2. **(B)** If $\left|\frac{1}{2}x\right| \geq \frac{1}{2}$, then either $\frac{1}{2}x \geq \frac{1}{2}$ so $x \geq 1$ or $\frac{1}{2}x \leq -\frac{1}{2}$ so $x \leq -1$.
3. **(D)** If $\frac{1}{2}|x| = 1$, then $|x| = 2$, so $x = \pm 2$. If $x = -2$, then $|y| = x + 1 = -1$, which is impossible. If $x = 2$, then $|y| = x + 1 = 3$, so $y = \pm 3$ and $y^2 = (\pm 3)^2 = 9$.
4. **(E)** Since $|1 - x| < x$, then $-x < 1 - x < x$, which means $-x < 1 - x$ or $1 - x < x$.
   - If $1 - x < x$, then $1 < 2x$ or, equivalently, $2x > 1$, so $x > \frac{1}{2}$. Pick a test value greater than $\frac{1}{2}$, such as $x = 1$. Since $|1 - 1| < 1$ is true, $x > \frac{1}{2}$ is a true statement.
   - If $-x < 1 - x$, then the inequality simplifies to another inequality that does not involve x; this inequality does not produce a solution.
5. **(A)** Since the temperature, F, can range from 7° below 79° to 7° above 79°, the positive difference between F and 79° is always less than or equal to 7°, which is expressed by the inequality $|F - 79| \leq 7$.
6. **(D)** If $\frac{|a + 3|}{2} = 1$, then $|a + 3| = 2$, so $a + 3 = 2$ or $a + 3 = -2$. Hence, $a = -1$ or $a = -5$.

If $2|b + 1| = 6$, then $|b + 1| = 3$, so $b + 1 = 3$ or $b + 1 = -3$. Hence, $b = 2$ or $b = -4$.
Then $|a + b|$ could equal the following:
- $|-1 + 2| = 1$
- $|-1 + (-4)| = 5$
- $|-5 + 2| = 3$
- $|-5 + (-4)| = 9$

Thus, $|a + b|$ could not equal 7.

(*Grid-In*)

$$|t - 7| = 4$$
$$|9 - t| = 2$$

1. **11** If $|t - 7| = 4$, then $t - 7 = 4$ or $t - 7 = -4$. Then $t = \mathbf{11}$ or $t = 3$. If $|9 - t| = 2$, then $9 - t = 2$ or $9 - t = -2$. Then $t = 7$ or $t = \mathbf{11}$.
2. **6.5** It is given that $5 < |k + 1| < 6$ and $k < 0$.
   - If $k + 1 \geq 0$, then $5 < k + 1 < 6$, so $4 < k < 5$. Since it is given that $k < 0$, disregard this solution.
   - If $k + 1 < 0$, then $5 < -(k + 1) < 6$ or, equivalently, $-5 > k + 1 > -6$. Thus, $-6 > k > -7$.

   Since k is between -6 and -7, the value of $|k|$ can be *any* decimal or fraction between 6 and 7, such as 6.5.
3. **2/3** Solve each of the given absolute value inequalities:
   - If $|3x - 4| = 2$, then

   $3x - 4 = 2$ or $3x - 4 = -2$
   $3x = 6$ | $3x = 2$
   $x = 2$ | $x = \frac{2}{3}$

   - If $|11 - 6x| = 7$, then

   $11 - 6x = 7$ or $11 - 6x = 7$
   $-6x = -4$ | $-6x = -18$
   $x = \frac{2}{3}$ | $x = 3$

   Hence, $x = \frac{2}{3}$ satisfies both equations.
4. **16** It is given that $|x - 16| \leq 4$ and $|y - 6| \leq 2$. The greatest possible value of $x - y$ occurs when x takes on its greatest value and, at the same time, y takes on its smallest value.

- If $|x - 16| \leq 4$, then $-4 \leq x - 16 \leq 4$. Adding 16 to each member of the inequality makes $12 \leq x \leq 20$. Hence, the greatest value of x is 20.
- If $|y - 6| \leq 2$, then $-2 \leq y - 6 \leq 2$. Adding 6 to each member of the inequality makes $4 \leq y \leq 8$. Hence, the smallest value of y is 4.
- The greatest value of $x - y$ is $20 - (4) = 16$.

Lesson 8-4 (*Multiple-Choice*)

1. **(C)** Test each answer choice in turn until you find the function for which replacing x with $-x$ results in the same function. For choice (C), $f(-x) = (-x)^2 = x^2 = f(x)$.
2. **(E)** If $g(x) = x^{\frac{3}{4}} - 7$, then $g(k) = k^{\frac{3}{4}} - 7 = 57$. Thus, $k^{\frac{3}{4}} = 57 + 7 = 64$, and
$$\left(k^{\frac{3}{4}}\right)^{\frac{4}{3}} = 64^{\frac{4}{3}}.$$
Hence, $k = \left(\sqrt[3]{64}\right)^4 = 4^4 = 256$.
3. **(E)** From the given graph, since $(-4,-3)$ is a point on the graph, $f(-4) = -3$, and since $(2,3)$ is a point on the graph, $f(2) = 3$. Hence, $f(p) = f(-4) + f(2) = -3 + 3 = 0$. Since $f(p) = 0$, find the x-coordinate of the point on the graph at which $y = 0$, which is 1.5.
4. **(B)** If the function f is defined by $f(x) = 5x + 3$, then
$$\begin{aligned} 2f(x) - 3 &= 2(5x + 3) - 3 \\ &= 10x + 6 - 3 \\ &= 10x + 3 \end{aligned}$$
5. **(C)** If $k(h) = (h + 1)^2$, then
$$\begin{aligned} k(x - 2) &= ((x - 2) + 1)^2 \\ &= (x - 1)^2 \\ &= (x - 1)(x - 1) \\ &= x^2 - 2x + 1 \end{aligned}$$
6. **(C)** The accompanying tables define functions f and g. Since $f(3) = 5$, $g(f(3)) = g(5) = 8$.

x	1	2	3	4	5
$f(x)$	3	4	5	6	7

x	3	4	5	6	7
$g(x)$	4	6	8	10	12

7. **(D)** Function f is defined by $f(x) = \sqrt{x + 3}$. If $f(p) = 3$, then
$$\begin{aligned} f(p) &= \sqrt{p + 3} = 3 \\ \left(\sqrt{p + 3}\right) &= 3^2 \\ p + 3 &= 9 \\ p &= 6 \end{aligned}$$
8. **(C)** If function f is defined by $f(x) = 3x + 2$ and $\frac{1}{2}f\left(\sqrt{c}\right) = 3$, then $f(c) = 3c + 2$ and
$$\begin{aligned} \frac{1}{2}f\left(\sqrt{c}\right) &= \frac{3c + 2}{2} = 3 \\ 3c + 2 &= 6 \\ 3c &= 4 \\ c &= \frac{4}{3} \end{aligned}$$
9. **(D)** The graphs of functions f and g, shown in the accompanying figure, are given. The region in which $f(x) \geq g(x)$ is the region where the graph of $y = f(x)$ is above the graph of $y = g(x)$, as indicated by the shaded region. Thus, $f(x) \geq g(x)$ for $-2 \leq x \leq 1$. Roman numeral choices I and II are included in this interval, which corresponds to choice (D).

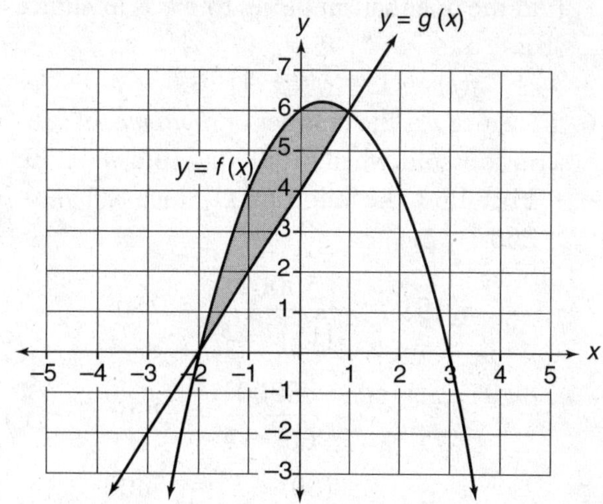

10. **(B)** Reading from the graph, when p in (p,q) is between 0 and 5, the y-coordinates of all such points on the graph range from a minimum of -5 to a maximum of 10, which is represented by the inequality $-5 \leq q \leq 10$.
11. **(D)** If $x = -1$, $g(-1) = f(-1 + 4) = f(3)$. Reading from the given graph of function f, $f(3) = 5 = g(-1)$.
12. **(C)** According to the table, $f(5) = p = 1$, so $g(p) = g(1) = 3$.

x	f(x)	g(x)
1	2	3
2	4	5
3	5	1
4	3	2
5	1	4

13. **(B)** It is given that function h is defined by $h(x) = 2f(x) - 1$, where function f is defined in the accompanying table. If $h(k) = 5$, then $h(k) = 2f(k) - 1 = 5$, $2f(k) = 6$ so $f(k) = 3$. From the table, $f(4) = 3$, so $k = 4$. Again reading from the table, $g(k) = g(4) = 2$.
14. **(C)** If $f(x) = 1 + \sqrt{x} = 4$, then $\sqrt{x} = 3$ so $x = 9$. Function g must have the property that $g(2) = 9$. Substitute 2 for x in each of the answer choices until you find the expression for which $g(2) \neq 9$. Hence, $g(x)$ could not be $3x + 2$, since $3(2) + 2 = 8 \neq 9$.
15. **(B)** It is given that function j is defined by $j(x) = x^2 + 1$ and function k is defined by $k(x) = (x + 1)^2$. Thus,
$$k(p - 1) = (p - 1 + 1)^2 = p^2$$
Evaluate each of the answer choices until you find the one that simplifies to p^2, as in choice (B):
$$j(p) - 1 = (p^2 + 1) - 1 = p^2$$
16. **(E)** Since n is the number of *thousands* of subscriptions sold, when 250,000 are sold, $n = 250$.
   - First find the value of k. Since $n(15) = 250$,
$$n(15) = \frac{5000}{4(15) - k} = 250$$
$$\frac{5000}{60 - k} = 250$$
$$250(60 - k) = 5000$$
$$\frac{250(60 - k)}{250} = \frac{5000}{250}$$
$$60 - k = 20$$
$$\frac{k}{n} = 40$$
   - Rewrite the original function with $k = 40$:
$$n(p) = \frac{5000}{4p - 40}$$
   - Find $n(20)$:
$$n(20) = \frac{5000}{4(20) - 40} = \frac{5000}{40} = 125$$

Because n is the number of *thousands* of subscriptions sold, the number of subscriptions that could be sold at \$20 for each subscription is 125,000.

17. **(D)** Since $f(x) = 2x + k$ and $g(x) = \dfrac{x - 5}{2}$,
   - $$f(g(x)) = f\left(\frac{x - 5}{2}\right) = 2\left(\frac{x - 5}{2}\right) + k = x = x - 5 + k$$
   - $$g(f(x)) = g(2x + k) = \frac{(2x + k) - 5}{2} = \frac{2x + k - 5}{2}$$
   - If $f(g(x)) = g(f(x))$, then $x - 5 + k = \dfrac{2x + k - 5}{2}$. Solve for k by first multiplying both sides of the equation by 2 in order to eliminate the fractional term:
$$2(x - 5 + k) = 2x + k - 5$$
$$2x - 10 + 2k = 2x + k - 5$$
$$2k = k - 5 + 10$$
$$k = 5$$
18. **(E)** Since $f(x) = 3x - 2$ and $g(x) = -x + 1$,
   - $f(x + y) = 3(x + y) - 2 = 4$, so $3(x + y) = 6$ and $x + y = \dfrac{6}{3} = 2$.
   - $g(x - y) = -(x - y) + 1 = -7$, so $-(x - y) = -8$ or, equivalently, $x - y = 8$.

Solve the system of equations by adding corresponding sides to eliminate y:
$$\begin{aligned} x + y &= 2 \\ x - y &= 8 \\ \hline 2x &= 10 \\ x &= \frac{10}{2} = 5 \end{aligned}$$

(*Grid-In*)

1. **0** If $h(x) = x + 4^x$, then

$$h\left(-\frac{1}{2}\right) = -\frac{1}{2} + 4^{-\frac{1}{2}}$$
$$= -\frac{1}{2} + \frac{1}{\sqrt{4}}$$
$$= -\frac{1}{2} + \frac{1}{2}$$
$$= 0$$

2. **5/2** or **2.5** If $f(n) = 3^{2n-1} = 81$, then $3^{2n-1} = 3^4$. Hence, $2n - 1 = 4$ and $2n = 5$, so $n = \frac{5}{2}$.

3. **2** Since, $f(x) = x^2 + 12$, $f(n) = n^2 + 12$, and $f(3n) = (3n)^2 + 12 = 9n^2 + 12$. If $f(3n) = 3f(n)$, then,

$$9n^2 + 12 = 3\left(n^2 + 12\right)$$
$$9n^2 + 12 = 3n^2 + 36$$
$$9n^2 - 3n^2 = 36 - 12$$
$$6n^2 = 24$$
$$n^2 = \frac{24}{6}$$
$$n = \pm\sqrt{4} = \pm 2$$

Since n must be a positive number, $n = 2$.

4. **7** Use the graph of each function to find the function values you need.
   - Since (3,1) is a point on the graph of function g, $g(3) = 1$. Hence, $f(g(3)) = f(1)$.
   - Because (1,7) is a point on the graph of function f, $f(1) = 7$.
   - Therefore, $f(g(3)) = f(1) = 7$.

5. **12** From the graph, $f(-1) = 0$ and $f(1) = 4$. Hence, $2f(-1) + 3f(1) = 2 \cdot 0 + 3 \cdot 4 = 12$.

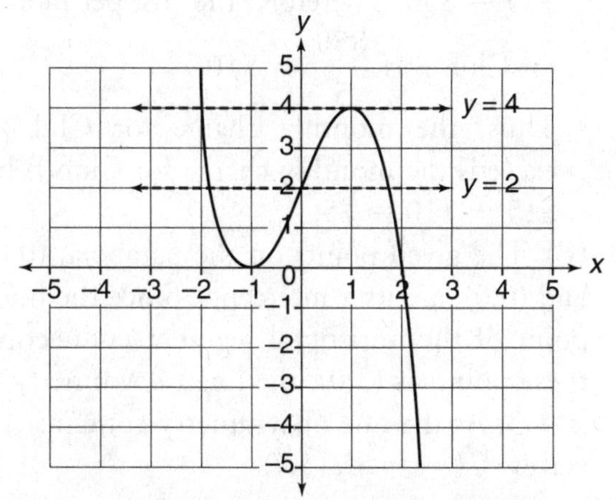

Graph for Exercises 5 and 6

6. **6** Use the accompanying graph.
   - To find n, the number of values of x for which $f(x) = 2$, count the number of points at which the horizontal line $y = 2$ intersects the graph. As shown in the accompanying graph, $n = 3$.
   - To find m, number of values of x for which $f(x) = 4$, count the number of points at which the horizontal line $y = 4$ intersects the graph. As shown in the accompanying graph, $m = 2$.

   Hence, $mn = 2 \times 3 = 6$.

7. **17** It is given that $g(x) = x - 1$ and $\frac{1}{2}g(c) = 4$. Hence,

$$\frac{1}{2}g(c) = \frac{1}{2}(c - 1) = 4$$
$$c - 1 = 8$$
$$c = 9$$

Now find the value of $g(2c)$ where $c = 9$:
$$g(2c) = g(18) = 18 - 1 = 17$$

8. **6** Since $f(x) = h(2x) + 1$, $f(-1) = h(-2) + 1$. According to the given graph of function h, $h(-2) = 5$, so $f(-1) = 5 + 1 = 6$.

Lesson 8-5 (*Multiple-Choice*)

1. **(E)** If in the equation $y = 3^x$, x is increased by 2, then $y = 3^{x+2} = 3^2 \cdot 3^x = 9(3^x)$.
2. **(A)** Since graph of the quadratic function $f(x) = -x^2 - 4x + 1$ has a line of symmetry at $x = -2$, the x-coordinate of the vertex is -2. When $x = -2$, the corresponding value of y is $f(-2) = -(2)^2 - 4(-2) + 1 = -4 + 8 + 1 = 5$. Because the coefficient of the x^2 term is negative, the vertex $(-2,5)$ is a maximum point.
3. **(D)** It is given that y represents the total cost of printing a total of x brochures with an initial cost of \$300 and a charge of \$0.15 for each additional copy. Since the first copy costs \$300, the remaining $x - 1$ copies cost \$0.15 each. Hence, $y = 0.15(x - 1) + 300$.
4. **(B)** Because of symmetry, $f(5) = f(-3) = f(c)$, so $c = -3$.
5. **(B)** It is given that the graph of a quadratic function f intersects the x-axis at $x = -2$ and

$x = 6$. Because the parabola's line of symmetry bisects every horizontal segment that joins two points on the parabola, the equation of the line of symmetry is $x = \dfrac{-2+6}{2} = 2$. Since $x = 8$ is 6 units to the right of $x = 2$, the corresponding point on the parabola is 6 units to the left of $x = 2$, which makes its x-coordinate $2 - 6 = -4$. Thus, $f(8) = f(-4)$.

6. **(C)** In exponential growth, if the starting population of 16 doubles every 6 days, then the population after 60 days is $16 \times 2^{\frac{60}{6}} = 2^4 \times 2^{10} = 2^{14}$.
7. **(C)** If in the quadratic function $f(x) = ax^2 + bx + c$, a and c are both negative, the graph is a parabola that opens down ($a < 0$) and intersects the negative y-axis at a negative value ($c < 0$), as shown in choice (C).
8. **(D)** After a single sheet of paper is folded in half, there are two layers of paper. When the same sheet of paper is folded in half for the second time, there are $4(=2^2)$ layers of paper. If this folding is performed a total of x times, then the number of layers of paper that results is 2^x.
9. **(C)** After 4 years, the amount of radioactive substance that remains is $100 \times \frac{1}{2}$; after 8 years, $\left(100 \times \frac{1}{2}\right) \times \frac{1}{2} = 100 \times \left(\frac{1}{2}\right)^2$ remains; and after t years, $100 \times \left(\frac{1}{2}\right)^{\frac{t}{4}}$.
10. **(C)** A culture of 5000 bacteria triples every 20 minutes.
    - After 20 minutes there are 5000×3^1 bacteria, after 40 minutes there are $(5000 \times 3^2$ bacteria, after 60 minutes there are 5000×3^3 bacteria, and so on.
    - In each case, the power of 3 can be obtained by dividing the number of minutes that elapses by 20.
    - Hence, the number of bacteria present after a total of m minutes is $5000 \times 3^{\frac{m}{20}}$ bacteria.
11. **(C)** Given the sequence 8, 24, 72, 216, . . . , each term after the first is obtained by multiplying the term before it by 3:

$$8,\ \underbrace{8 \times 3}_{\text{2nd term}},\ \underbrace{8 \times 3^2}_{\text{3rd term}},\ \underbrace{8 \times 3^3}_{\text{4th term}},\ldots,\underbrace{8 \times 3^{n-1}}_{n\text{th term}}$$

Hence, the $n + 1$ term of the sequence is $(8 \times 3^{n-1}) \times 3 = 8 \times 3^n$.

12. **(A)**

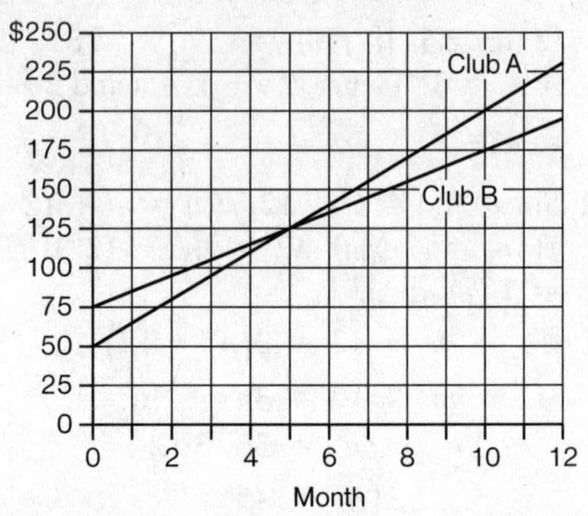

According to the graph:

- At month 0, the cost of Club A is \$50, and the cost at month 5 is \$125. This means that the monthly membership fee is \$50 and the cost of 5 months without the membership fee included is \$125 − \$50 = \$75. Therefore, the cost per month for Club A is $\frac{\$75}{5} = \$15$.
- At month 0, the cost of Club B is \$75, and the cost at month 5 is \$125. This means that the monthly membership fee is \$75 and the cost of 5 months without the membership fee included is \$125 − \$75 = \$50. Therefore, the cost per month for Club B is $\frac{\$50}{5} = \$10$.
- Thus, the monthly charge for Club A exceeds the monthly charge for Club B by \$15 − \$10 = \$5.

13. **(E)** The given points on the parabola, (0,0) and (6,0), are its x-intercepts. Since the midpoint of the horizontal segment connecting these points is (3,0), the line of symmetry is $x = 3$. As the line of symmetry contains the vertex, $T(h,4) = T(3,4)$.

- Since $h = 3$, Roman numeral choice I is false.
- Because (1,2) is 2 units to the left of the line of symmetry, its matching point on the opposite side of the line of symmetry is $(3 + 2, 2) = (5,2)$. So Roman numeral choice II is true.
- Vertex T is either the highest or lowest point of the parabola. If T were the lowest point on the parabola, then the parabola would *not* intersect the x-axis. Since it is given that the parabola intersects the x-axis, T must be the highest point on the parabola. Roman numeral choice III is true. Because Roman numeral choices II and III are true, the correct answer choice is (E).

(*Grid-In*)

1. **13.4** Find the number of gallons used per hour:

$$\frac{\Delta \text{ gallons}}{\Delta \text{ time}} = \frac{5 - 12}{3{:}00 - 12{:}30} = \frac{7 \text{ gallons}}{2.5 \text{ hours}} = -2.8\,\frac{\text{gallons}}{\text{hour}}$$

Since Ari uses $2.8\,\frac{\text{gallons}}{\text{hour}}$, at the same rate, he uses $\frac{1}{2} \times 2.8 = 14$ gallons during one-half hour. Hence, at 12:00 PM, one-half hour before 12:30, Ari started the job with $12 + 1.4 = 13.4$ gallons of paint.

2. **140** It is given that the number of hours H needed to manufacture X computer monitors is defined by $H = kX + q$. In the function $H = kX + q$, k represents the change in the number of hours, H, when X is increased by 1 unit. Hence, we need to use the given information to solve for k.
   - If it takes 270 hours to manufacture 100 computer monitors, then $270 = 100k + q$.
   - If it takes 410 hours to manufacture 160 computer monitors, then $410 = 160k + q$.
   - Solve the system $270 = 100k + q$ and $410 = 100k + q$ for k by subtracting corresponding sides of the equations to eliminate q:

$$\begin{aligned} 160k + q &= 410 \\ 100k + q &= 270 \\ \hline 60k &= 140 \\ k &= \frac{140}{60} = \frac{7}{3} \end{aligned}$$

   - Hence, each additional monitor after the first requires $\frac{7}{3}$ hours to manufacture. Since the question asks for the time in minutes, convert $\frac{7}{3}$ hours to minutes by multiplying by 60:

$$\frac{7}{\not{3}} \times \overset{20}{\not{60}} = 140 \text{ minutes}$$

3. **450** The total cost of membership and playing 12 games is the sum of the membership fee and the cost of each game times 12.
   - To figure out the cost of membership from the graph, find the value of y when x, the number of games, is 0. Since the graph intersects the y-axis at (0,90), the cost of membership is \$90.
   - When x increases from 0 to 1, y increases from \$90 to \$120. Hence, the cost of one game is $\$120 - \$90 = \$30$, so the cost of 12 games is $12 \times \$30 = \$360$.
   - The total cost of membership and playing 12 games is $\$90 + \$360 = \$450$.

4. **4**

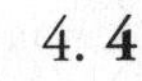

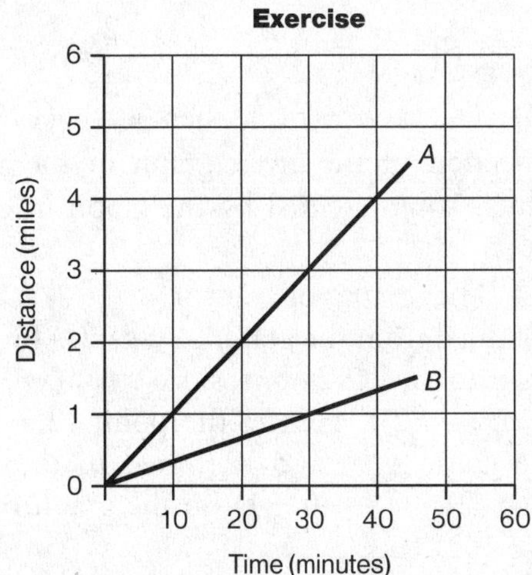

According to the graph,
   - Albert runs 1 mile in 10 minutes. At this constant rate, he runs 6 miles in 60 minutes, or 1 hour.

- Bill walks 1 mile in 30 minutes. At this constant rate, he walks 2 miles in 60 minutes, or 1 hour.

Since Albert runs at a rate of 6 miles per hour and Bill walks at a rate of 2 miles per hour, Albert was running $6 - 2 = 4$ miles per hour faster than Bill was walking.

Lesson 8-6 (*Multiple-Choice*)

1. **(A)** If c is a positive number, the graph of $y = f(x)$ can be shifted left c units by graphing $y = f(x + c)$ and shifted right c units by graphing $y = f(x - c)$. If $y = f(x) = 2^x$, then graphing $y = f(x - 3) = 2^{x-3}$ shifts the graph 3 units to the right.
2. **(A)** In general, $y = -f(x)$ is the graph of $y = f(x)$ reflected over the x-axis. Since it is given that $y = f(x) = 2x - 3$, $f(-x) = 2(-x) - 3 = -2x - 3$.
3. **(D)** Given that the endpoints of $\overline{AB}$ are $A(0,0)$ and $B(9, -6)$, the slope of $\overline{AB}$ is $\frac{-6}{9} = -\frac{2}{3}$, its equation is $y = -\frac{2}{3}x$. In general, $y = f(-x)$ is the graph of $y = f(x)$ reflected over the y-axis. Hence, the equation of the line that contains the reflection of $\overline{AB}$ in the y-axis is $y = f(-x) = -\frac{2}{3}(-x) = \frac{2}{3}x$.
4. **(B)** If $g(x) = -f(x)$, then $y = g(x)$ is the reflection of the given graph in the y-axis, which is represented by the graph in choice (B).
5. **(C)** The graph of $y = f(x + 2)$ can be obtained from the graph of $y = f(x)$ by shifting it to the left 2 units, since $y = f(x + 2) = f(x - (-2))$. Hence, the point $(2,-1)$ on the graph $y = f(x)$ is shifted to the point $(2 - 2,-1) = (0,-1)$ on the graph of $y = f(x + 2)$.
6. **(D)** Compare the coordinates of the vertex of the two function graphs. The vertex of function f is at $(0,0)$, and the vertex of function g is at $(1,2)$. This means that function g is the graph of function f shifted 1 unit to the right and 2 units up. Hence, an equation of function g is $g(x) = f(x - 1) + 2$. By comparing this equation to $g(x) = f(x + h) + k$, you know that $h = -1$ and $k = 2$. Thus, $h + k = -1 + 2 = 1$.
7. **(E)** The graph of $g(x) = -2$ is a horizontal line 2 units below the x-axis. If the graph of $g(x) = -2$ intersects the graph of $y = f(x) + k$ at one point, then the point of intersection must be one of the turning points. Consider the turning point located between 1 and 2. Since the y-coordinate of that turning point is approximately -3.5, the graph of $y = f(x)$ would need to be shifted up approximately 1.5 units. Hence, a possible value of k is 1.5.

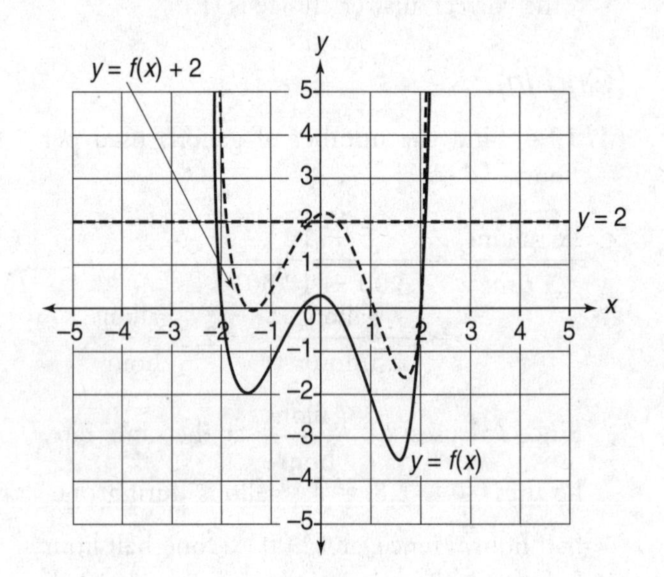

8. **(D)** The graph of $y = f(x + 1)$ is the graph of $y = f(x)$ shifted horizontally to the left 1 unit, shown by the graph in choice (D). You should confirm that this is the correct graph by checking a few specific points. The original graph includes the points $(-3,1)$, $(0,4)$ and $(3,-2)$. Since the original graph is shifted to the left 1 unit, the matching points on the graph of $y = f(x + 1)$ must have their x-coordinates reduced by 1: $(-4,1)$, $(-1,4)$, and $(2,-2)$. Only the graph in choice (D) contains all three of these points.
9. **(D)** METHOD 1: Since the line contains the points $(0,3)$ and $(2,0)$, its slope is $\frac{3 - 0}{0 - 2} = -\frac{3}{2}$. The y-intercept of the line is

3, so its equation is $y = f(x) = -\frac{3}{2}x + 3$. In general, $y = -f(x)$ is the graph of $y = f(x)$ reflected over the x-axis. Since $y = f(-x) = -\frac{3}{2}(-x) + 3 = \frac{3}{2}x + 3$, the slope of the reflected line is $\frac{3}{2}$.

METHOD 2: Find that the slope of the given line is $-\frac{3}{2}$. By sketching the reflected line, it is easy to calculate its slope directly or to see that the slopes of the two lines must be opposite in sign:

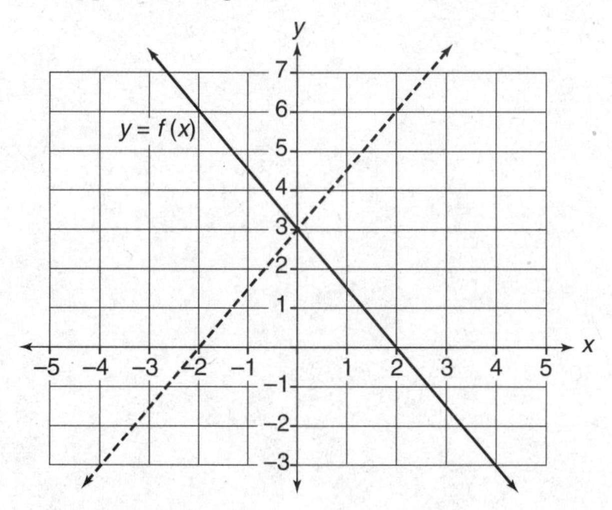

10. **(C)** The graph of $y = f(x - 2) + 1$ is the graph of $y = f(x)$ shifted two units to the right and 1 unit up. Since the original graph contains (0,0) and (2,2), the translated graph must contain $(0 + 2, 0 + 1) = (2,1)$ and $(2 + 2, 2 + 1) = (4,3)$. Only the graph in choice (C) contains both of these points.
11. **(C)** According to the definition of absolute value, when $f(x) \geq 0$, $y = f(x)$, and when $f(x) < 0$, $y = -f(x)$. Hence, the graph of $y = |f(x)|$ includes the portion of the graph of function f that is in quadrant I, where $f(x) > 0$, and the reflection in the x-axis of the portion of the original graph of function f that is in quadrant III, where $f(x) < 0$. This is the graph in choice (C).

(Grid-In)

In the answer solutions to questions 1 and 2, function f is defined by the graph in the figure below.

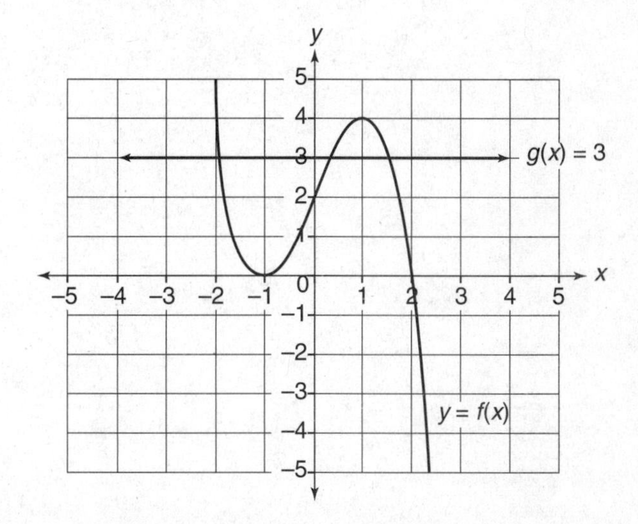

1. **2** When $k > 0$, the graph of $y = f(x - k)$ can be obtained from the graph of $y = f(x)$ by shifting it to the right k units. Hence, we need to find k so that $(-1,0)$, the x-intercept of $y = f(x)$, is mapped onto (1,0). This means that $-1 + k = 1$, so $k = 2$.
2. **5** The graph of $g(x) = 3$ is a horizontal line 3 units above the x-axis.
   - Since $g(x) = 3$ and $y = f(x)$ intersect at 3 points, $m = 3$.
   - The graph of $y = f(x) - 1$ is the graph of $y = f(x)$ shifted down 1 unit, which shifts the turning point at (1,4) to (1,3). Hence, $g(x) = 3$ intersects $y = f(x) - 1$ at 2 points, so $n = 2$.
   - Thus, $m + n = 3 + 2 = 5$.

PART 3

TAKING PRACTICE TESTS

Practice Tests 1 and 2

CHAPTER 9

This chapter provides two full-length practice SAT math tests. Each test consists of three timed math sections that should be taken under testlike conditions, using the sample SAT answer forms at the beginning of each test.

At the end of each practice test you will find an Answer Key, as well as detailed solutions for all questions. If you think you need additional practice on any type of question you missed or skipped over, you can quickly locate the lesson on which the question was based by referring to the number that follows each answer in the Answer Key.

Answer Sheet

PRACTICE TEST 1

Start with number 1 for each new section. If a section has fewer questions than answer spaces, leave the extra answer spaces blank.

SECTION 1

1 Ⓐ Ⓑ Ⓒ Ⓓ Ⓔ 6 Ⓐ Ⓑ Ⓒ Ⓓ Ⓔ 11 Ⓐ Ⓑ Ⓒ Ⓓ Ⓔ 16 Ⓐ Ⓑ Ⓒ Ⓓ Ⓔ
2 Ⓐ Ⓑ Ⓒ Ⓓ Ⓔ 7 Ⓐ Ⓑ Ⓒ Ⓓ Ⓔ 12 Ⓐ Ⓑ Ⓒ Ⓓ Ⓔ 17 Ⓐ Ⓑ Ⓒ Ⓓ Ⓔ
3 Ⓐ Ⓑ Ⓒ Ⓓ Ⓔ 8 Ⓐ Ⓑ Ⓒ Ⓓ Ⓔ 13 Ⓐ Ⓑ Ⓒ Ⓓ Ⓔ 18 Ⓐ Ⓑ Ⓒ Ⓓ Ⓔ
4 Ⓐ Ⓑ Ⓒ Ⓓ Ⓔ 9 Ⓐ Ⓑ Ⓒ Ⓓ Ⓔ 14 Ⓐ Ⓑ Ⓒ Ⓓ Ⓔ 19 Ⓐ Ⓑ Ⓒ Ⓓ Ⓔ
5 Ⓐ Ⓑ Ⓒ Ⓓ Ⓔ 10 Ⓐ Ⓑ Ⓒ Ⓓ Ⓔ 15 Ⓐ Ⓑ Ⓒ Ⓓ Ⓔ 20 Ⓐ Ⓑ Ⓒ Ⓓ Ⓔ

SECTION 2

1 Ⓐ Ⓑ Ⓒ Ⓓ Ⓔ 5 Ⓐ Ⓑ Ⓒ Ⓓ Ⓔ
2 Ⓐ Ⓑ Ⓒ Ⓓ Ⓔ 6 Ⓐ Ⓑ Ⓒ Ⓓ Ⓔ
3 Ⓐ Ⓑ Ⓒ Ⓓ Ⓔ 7 Ⓐ Ⓑ Ⓒ Ⓓ Ⓔ
4 Ⓐ Ⓑ Ⓒ Ⓓ Ⓔ 8 Ⓐ Ⓑ Ⓒ Ⓓ Ⓔ

ONLY ANSWERS ENTERED IN THE OVALS IN EACH GRID AREA WILL BE SCORED. YOU WILL NOT RECEIVE CREDIT FOR ANYTHING WRITTEN IN THE BOXES ABOVE THE OVALS.

9 10 11 12 13

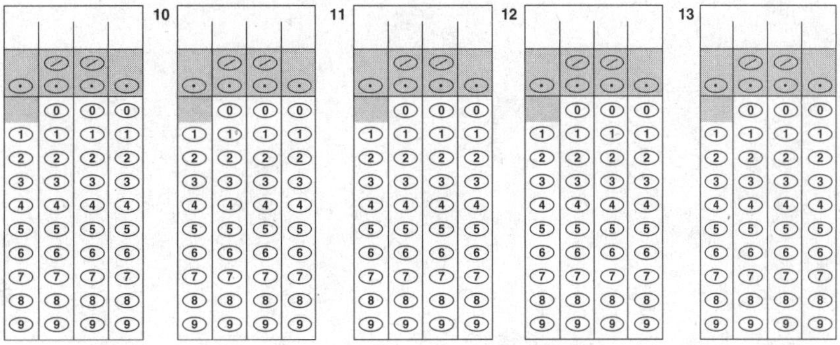

14 15 16 17 18

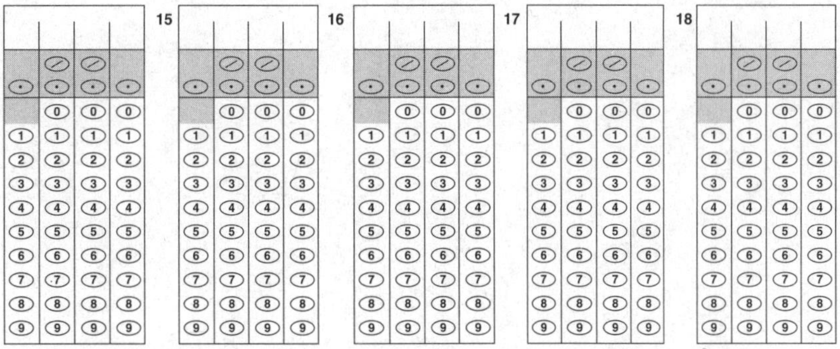

SECTION 3

1 Ⓐ Ⓑ Ⓒ Ⓓ Ⓔ 6 Ⓐ Ⓑ Ⓒ Ⓓ Ⓔ 11 Ⓐ Ⓑ Ⓒ Ⓓ Ⓔ 16 Ⓐ Ⓑ Ⓒ Ⓓ Ⓔ
2 Ⓐ Ⓑ Ⓒ Ⓓ Ⓔ 7 Ⓐ Ⓑ Ⓒ Ⓓ Ⓔ 12 Ⓐ Ⓑ Ⓒ Ⓓ Ⓔ 17 Ⓐ Ⓑ Ⓒ Ⓓ Ⓔ
3 Ⓐ Ⓑ Ⓒ Ⓓ Ⓔ 8 Ⓐ Ⓑ Ⓒ Ⓓ Ⓔ 13 Ⓐ Ⓑ Ⓒ Ⓓ Ⓔ 18 Ⓐ Ⓑ Ⓒ Ⓓ Ⓔ
4 Ⓐ Ⓑ Ⓒ Ⓓ Ⓔ 9 Ⓐ Ⓑ Ⓒ Ⓓ Ⓔ 14 Ⓐ Ⓑ Ⓒ Ⓓ Ⓔ 19 Ⓐ Ⓑ Ⓒ Ⓓ Ⓔ
5 Ⓐ Ⓑ Ⓒ Ⓓ Ⓔ 10 Ⓐ Ⓑ Ⓒ Ⓓ Ⓔ 15 Ⓐ Ⓑ Ⓒ Ⓓ Ⓔ 20 Ⓐ Ⓑ Ⓒ Ⓓ Ⓔ

BE SURE TO ERASE ANY ERRORS OR STRAY MARKS COMPLETELY.

Practice Test 1

Section 1 TIME: 25 MINUTES, 20 QUESTIONS

In this section solve each problem, using any available space on the page for scratchwork. Then decide which is the best of the choices given and fill in the corresponding oval on the answer sheet.

Notes:

1. The use of a calculator is permitted. All numbers used are real numbers.
2. Figures that accompany problems in this test are intended to provide information useful in solving the problems. They are drawn as accurately as possible EXCEPT when it is stated in a specific problem that the figure is not drawn to scale. All figures lie in a plane unless otherwise indicated.
3. Unless otherwise indicated, the domain of any function f is assumed to be the set of all real numbers for which $f(x)$ is a real number.

Reference Information

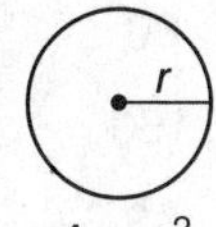

$A = \pi r^2$
$C = 2\pi r$

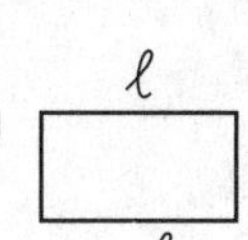

$A = \ell w$

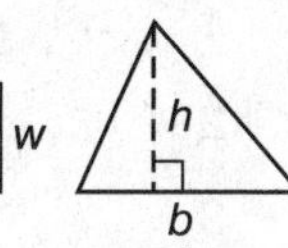

$A = \frac{1}{2}bh$

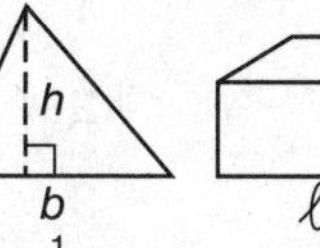

$V = \ell wh$

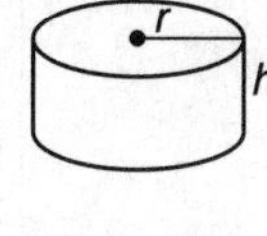

$V = \pi r^2 h$

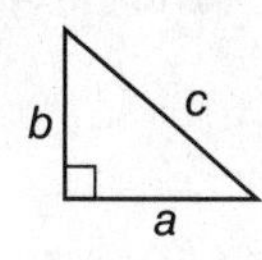

$c^2 = a^2 + b^2$

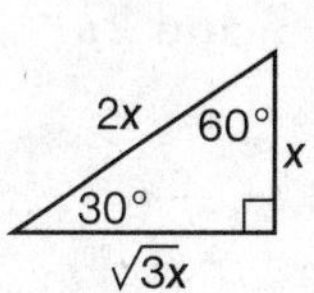

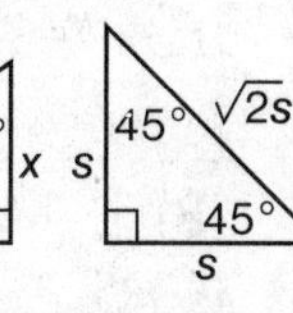

Special Right Triangles

The number of degrees of arc in a circle is 360.
The sum of the measures in degrees of the angles of a triangle is 180.

1 How many positive integers less than 36 are equal to 4 times an *odd* integer?

(A) Two
(B) Three
(C) Four
(D) Five
(E) Six

2

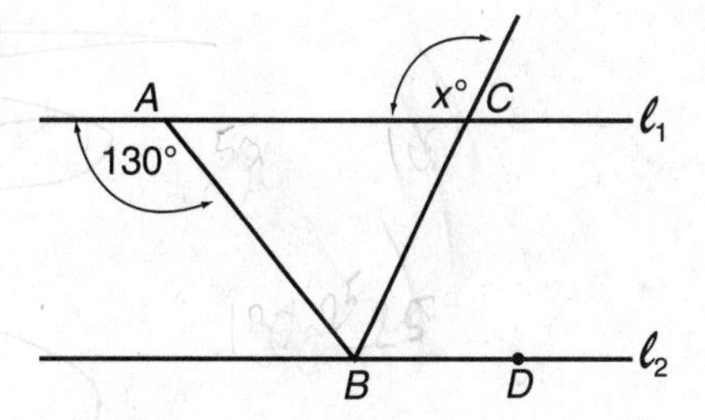

If, in the figure above, $\ell_1 \parallel \ell_2$ and BC bisects $\angle ABD$, then $x =$

(A) 75
(B) 95
(C) 105
(D) 115
(E) 150

GO ON TO THE NEXT PAGE

3 A long-distance telephone call costs \$1.80 for the first 3 minutes and \$0.40 for each additional minute. If the charge for an x-minute long-distance call at this rate was \$4.20, then $x =$

(A) 7
(B) 8
(C) 9
(D) 10
(E) 12

4 What is the LEAST number of squares each of side length 2 inches that is needed to cover completely, without overlap, a larger square with side length 8 inches?

(A) 4
(B) 8
(C) 9
(D) 16
(E) 32

5 If $3a = 2$ and $2b = 15$, then $ab =$

(A) 1
(B) 2
(C) 5
(D) 7
(E) 12

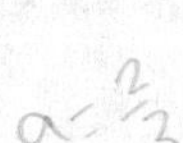

6

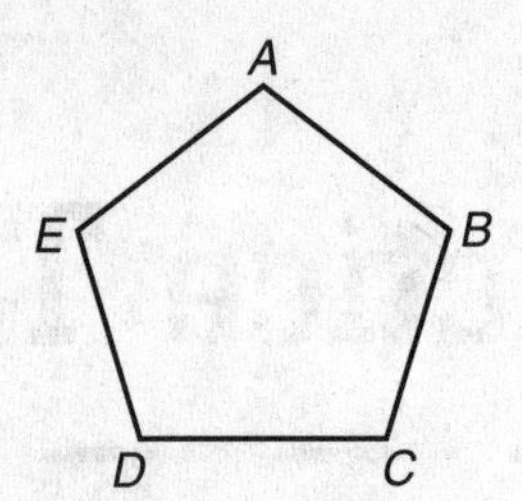

In figure *ABCDE* above, the length of each side is 1 unit. If a point starts at vertex A and moves along each side in a clockwise direction, at which vertex will the point be when it has traveled a distance of exactly 715 units?

(A) A
(B) B
(C) C
(D) D
(E) E

7

Number of Television Sets	Number of Families
0	32
1	80
2	160
3	83
More than 3	45

The table above summarizes the results of a survey in which families reported the number of television sets they have in their homes. What percent of the families surveyed have two or fewer television sets?

(A) 75%
(B) 68%
(C) 60%
(D) 40%
(E) 33%

GO ON TO THE NEXT PAGE

8 If $0 < x < 1$, which of the following expressions must decrease in value as x increases?

I. $\frac{1}{1-x}$

II. $\frac{1}{x^2}$

III. $1 - \sqrt{x}$

(A) I only
(B) II only
(C) I and II only
(D) I and III only
(E) II and III only

9 After $\frac{1}{8}$ of a ribbon is thrown away, the remaining part is cut into two pieces whose lengths are in the ratio of 4 : 5. If 9 inches of the original ribbon was thrown away, how many inches long is the shorter of the two remaining pieces of ribbon?

(A) 21
(B) 28
(C) 30
(D) 35
(E) 42

10 If a cube with edge length π has the same volume as a cylinder with a height of π, then the radius of the base of the cylinder is

(A) π

(B) $\sqrt{\pi}$

(C) $\frac{1}{\sqrt{\pi}}$

(D) $\frac{1}{\pi}$

(E) $\frac{1}{\pi^2}$

11 If $\frac{5}{a} = b$ and $a = 3$, then $a(b + 1) =$

(A) 8

(B) $7\frac{1}{3}$

(C) $6\frac{2}{3}$

(D) 4
(E) 3

12 A person spent a total of $720 for dress shirts and sport shirts, each priced at $35 and $20, respectively. If the person purchased two $35 dress shirts for each $20 sport shirt, what is the total number of shirts purchased?

(A) 16
(B) 21
(C) 24
(D) 28
(E) 32

13 If 10 cubic centimeters of blood contains 1.2 grams of hemoglobin, how many grams of hemoglobin are contained in 35 cubic centimeters of the same blood?

(A) 2.7
(B) 3.0
(C) 3.6
(D) 4.2
(E) 4.8

GO ON TO THE NEXT PAGE

14

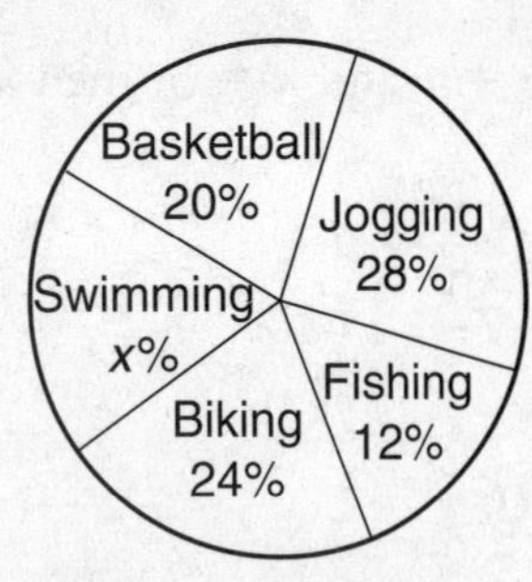

Favorite Sports Activities

The circle graph above summarizes the results of a survey of 2500 students who named their favorite sports activities. If each student in the survey named exactly one activity, what is the total number of students who named either swimming or biking as their favorite sports activity?

(A) 1000
(B) 875
(C) 750
(D) 600
(E) 400

15 If $x - 3$ is 1 less than $y + 3$, then $x + 2$ exceeds y by what amount?

(A) 4
(B) 5
(C) 6
(D) 7
(E) 9

16

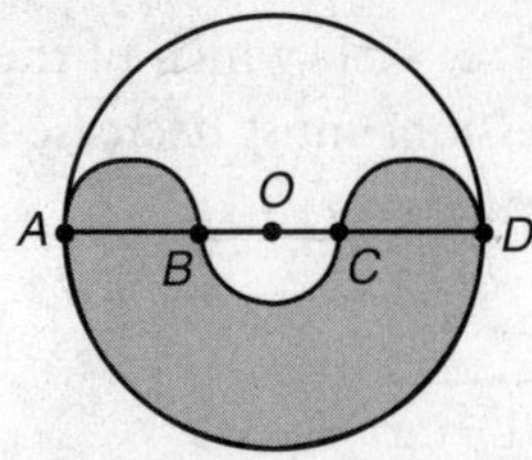

In the figure above, arcs *AB*, *BC*, and *CD* are semicircles with diameters $AB = BC = CD$. If the diameter of the largest circle with center *O* is 12, what is the area of the shaded region?

(A) 20π
(B) 24π
(C) 30π
(D) 36π
(E) 40π

17 For all positive integers *p*, let $\bigtriangledown p$ be defined by the equation

$$\bigtriangledown p = p + \frac{p}{k}$$

where *k* is the largest prime number that is a factor of *p*, and $k < p$. If $\bigtriangledown 110 = x$, what is the value of $\bigtriangledown x$?

(A) 120
(B) 133
(C) 144
(D) 148
(E) 160

GO ON TO THE NEXT PAGE

18

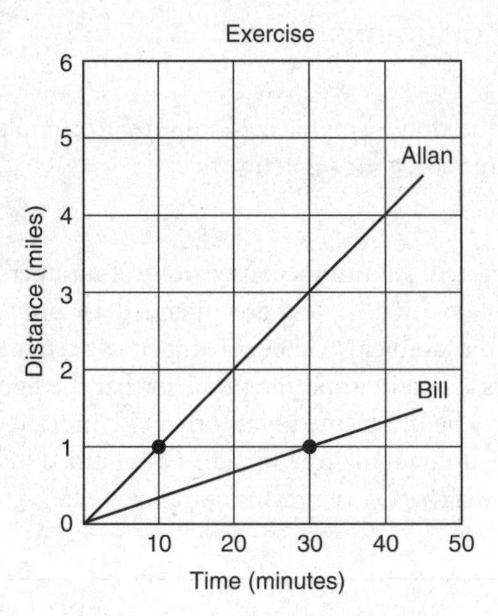

At 9:00 A.M. Allan began jogging and Bill began walking at constant rates around the same circular $\frac{1}{4}$ mile track. The figure above compares their times in minutes and corresponding distances in miles. Which statement or statements must be true?

I. Bill's average rate of walking was 2 miles per hour.
II. At 9:10 A.M., Allan had jogged $\frac{3}{5}$ mile more than Bill had walked.
III. At 9:30 A.M., Allan had completed 8 more laps around the track than Bill.

(A) I only
(B) II only
(C) I and II only
(D) I and III only
(E) I, II, and III

19 A group of p people plan to contribute equally to the purchase of a gift that costs d dollars. If n of the p people decide not to contribute, by what amount in dollars does the contribution needed from each of the remaining people increase?

(A) $\frac{d}{p-n}$

(B) $\frac{pd}{p-n}$

(C) $\frac{pd}{n(p-n)}$

(D) $\frac{nd}{p(p-n)}$

(E) $\frac{d}{n(pd-1)}$

20 In a certain class, exactly $\frac{2}{3}$ of the students applied for admission to a 4-year college and, of these, $\frac{1}{8}$ applied also to a 2-year junior college. If more than one student applied to both 4-year and 2-year colleges, what is the LEAST number of students who could be in this class?

(A) 12
(B) 24
(C) 28
(D) 30
(E) 36

STOP

If you finish before time is called, you may check your work on this section only. Do not turn to any other section in the test.

Section 2 TIME: 25 MINUTES, 18 QUESTIONS

This section contains two types of questions. You have 25 minutes to complete both types. You may use any available space for scratchwork.

Notes:

1. The use of a calculator is permitted. All numbers used are real numbers.
2. Figures that accompany problems in this test are intended to provide information useful in solving the problems. They are drawn as accurately as possible EXCEPT when it is stated in a specific problem that the figure is not drawn to scale. All figures lie in a plane unless otherwise indicated.
3. Unless otherwise indicated, the domain of any function f is assumed to be the set of all real numbers for which $f(x)$ is a real number.

Reference Information

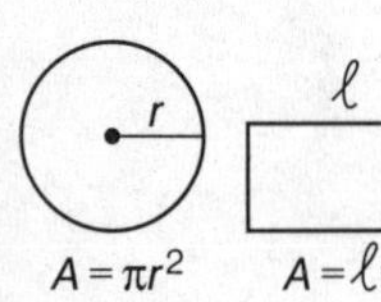

$A = \pi r^2$
$C = 2\pi r$

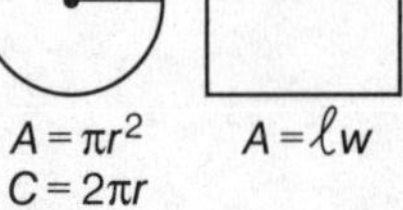

$A = \ell w$

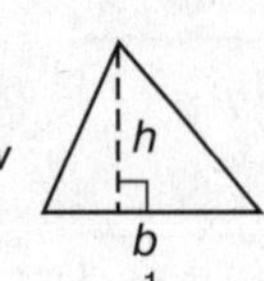

$A = \frac{1}{2}bh$

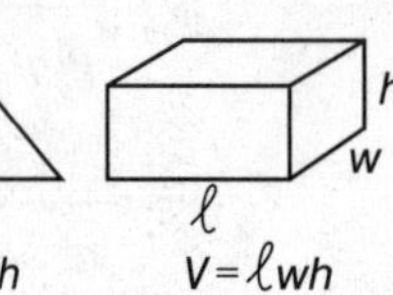

$V = \ell wh$

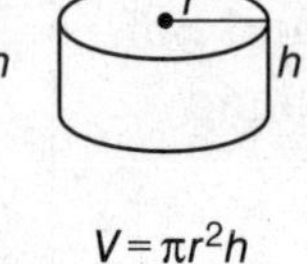

$V = \pi r^2 h$

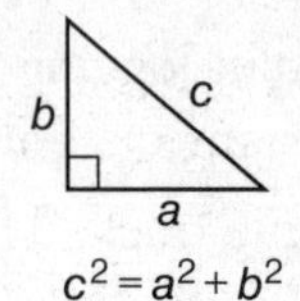

$c^2 = a^2 + b^2$

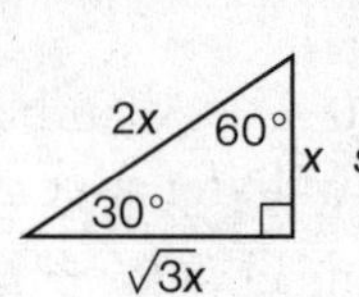

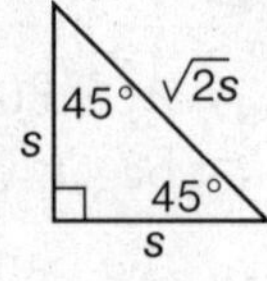

Special Right Triangles

The number of degrees of arc in a circle is 360.
The sum of the measures in degrees of the angles of a triangle is 180.

1 If $x^{-2} = 64$, what is the value of $x^{\frac{1}{3}}$?

(A) $\frac{1}{8}$
(B) $\frac{1}{4}$
(C) $\frac{1}{2}$
(D) 2
(E) 4

2 Set $X = \{21, 22, 23, 24, 25\}$
Set $Y = \{18, 20, 22, 24, 26, 28\}$

Sets X and Y are shown above. If a number is picked at random from set X, what is the probability that the number selected is also in set Y?

(A) 0.2
(B) 0.25
(C) 0.4
(D) 0.6
(E) 0.8

GO ON TO THE NEXT PAGE

3

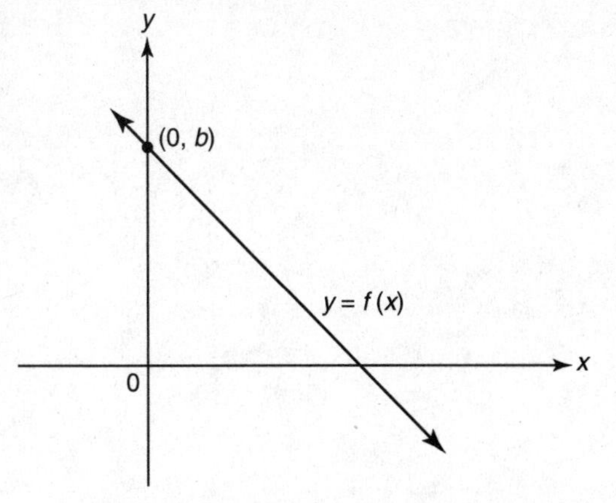

Note: Figure is not drawn to scale.

The figure above shows the graph of the linear function $y = f(x)$. If the slope of line is -2 and $f(3) = 4$, what is the value of b?

(A) 8
(B) 9
(C) 10
(D) 11
(E) 12

4 If $j - k = 9$ and $k = 6m$, and $3m = 7$, what is the value of j?

(A) 3
(B) 6
(C) 15
(D) 21
(E) 23

5 If $3\sqrt{x - 2} - 4 = 11$ and $y^3 = x^2$, what is the value of y?

(A) $\sqrt[3]{5}$
(B) 9
(C) 27
(D) 81
(E) 729

6 Function f is a linear function such that $f(2) = 3$ and $f(3) = 2$. What is a possible equation for function f?

I. $y = x + 1$
II. $y = x - 1$
III. $y = -x + 5$

(A) I only
(B) II only
(C) III only
(D) I and II only
(E) none

7 If $7^k = 100$, what is the value of $7^{\frac{k}{2}+1}$?

(A) 18
(B) 51
(C) 57
(D) 70
(E) 107

GO ON TO THE NEXT PAGE

Practice Test 1

8

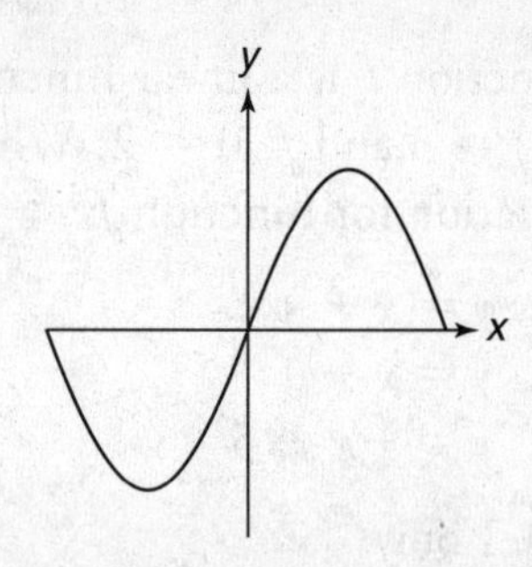

The graph of $y = f(x)$ is shown in the figure above. Which graph could represent the graph of $y = |f(x)|$?

(A)

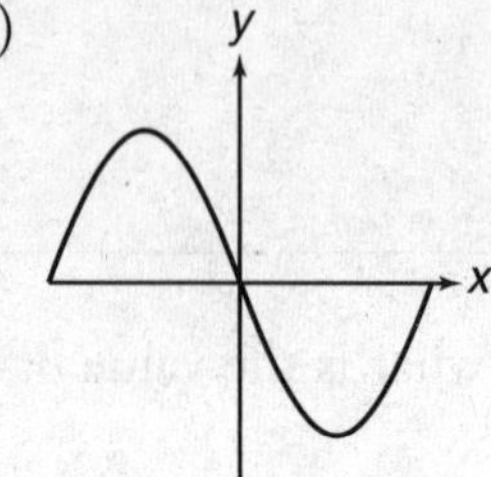

(B)

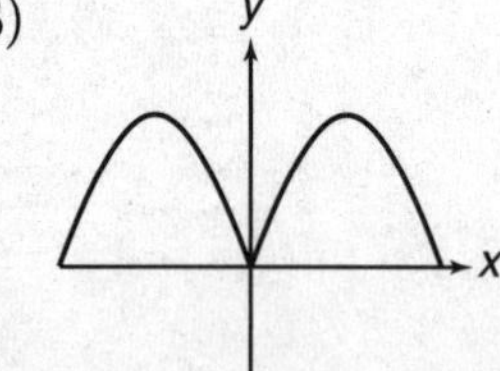

(C)

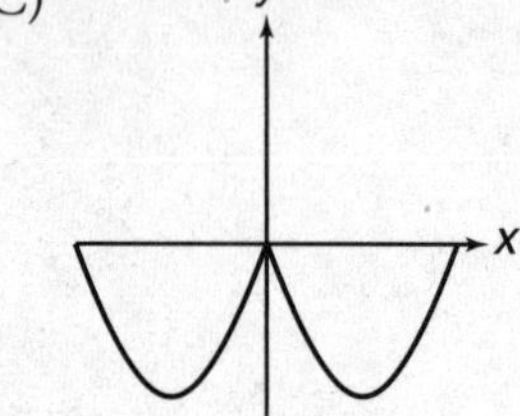

(D)

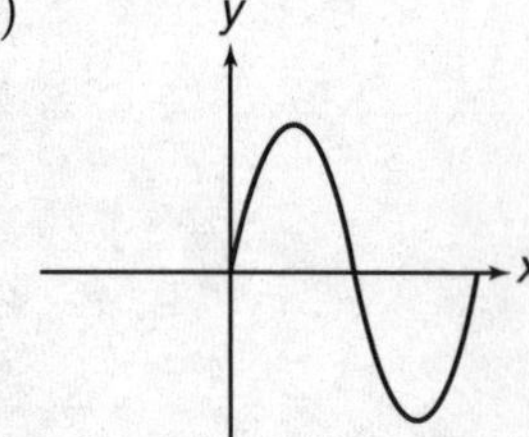

(E)

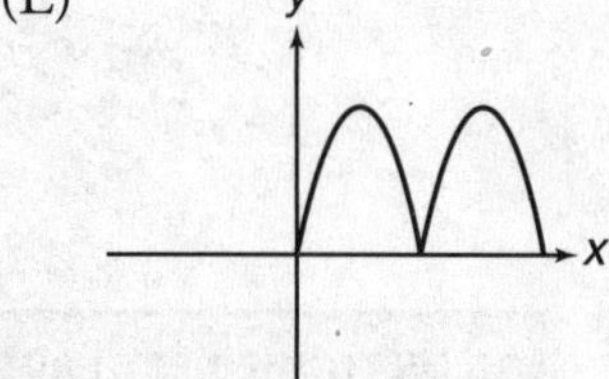

GO ON TO THE NEXT PAGE

Directions for Student-Produced Response Questions

Each of the remaining 10 questions (9–18) requires you to solve the problem and enter your answer by marking the ovals in the special grid, as shown in the examples below.

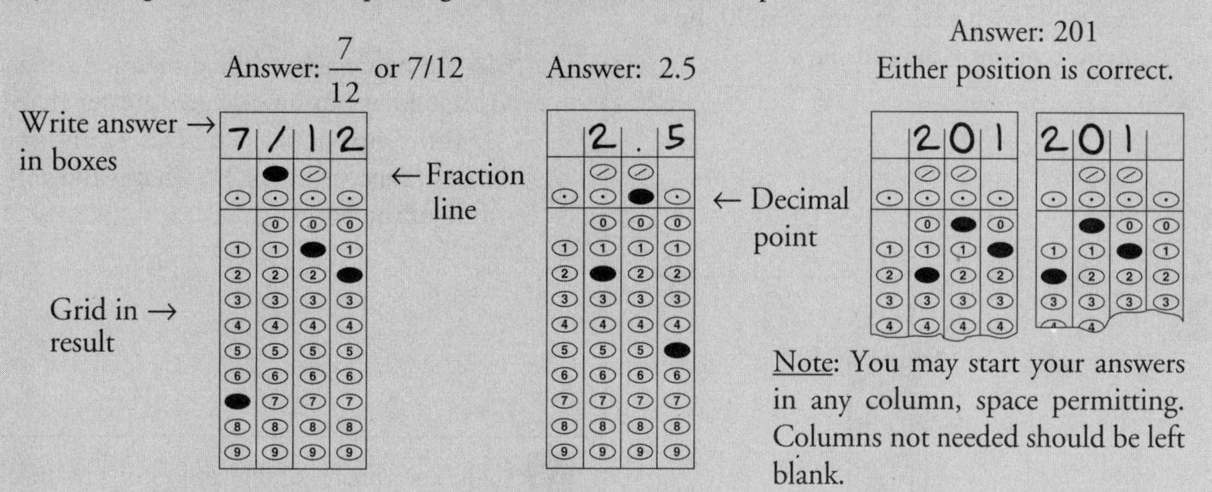

Note: You may start your answers in any column, space permitting. Columns not needed should be left blank.

- Mark no more than one oval in any column.
- Because the answer sheet will be machine-scored, **you will receive credit only if the ovals are filled in correctly.**
- Although not required, it is suggested that you write your answer in the boxes at the top of the columns to help you fill in the ovals accurately.
- Some problems may have more than one correct answer. In such cases, grid only one answer.
- No question has a negative answer.
- **Mixed numbers** such as $2\frac{1}{2}$ must be gridded as 2.5 or 5/2. (If 2 1 / 2 is gridded, it will be interpreted as $\frac{21}{2}$, not $2\frac{1}{2}$.)

- Decimal Accuracy: If you obtain a decimal answer, **enter the most accurate value the grid will accommodate.** For example, if you obtain an answer such as 0.6666 . . . , you should record the result as .666 or .667. **Less accurate values such as .66 or .67 are not acceptable.**

Acceptable ways to grid $\frac{2}{3} = .6666\ldots$

2/3 .666 .667

9 If $x + 2x + 3x + 4x = 1$, then what is the value of x^2?

10 What is the least positive integer p for which $441p$ is the cube of an integer?

11 During a certain month, a car salesperson sold three economy cars for every two luxury cars. If during that month the total number of economy and luxury cars sold by this salesperson was 90, how many economy cars did she sell?

12

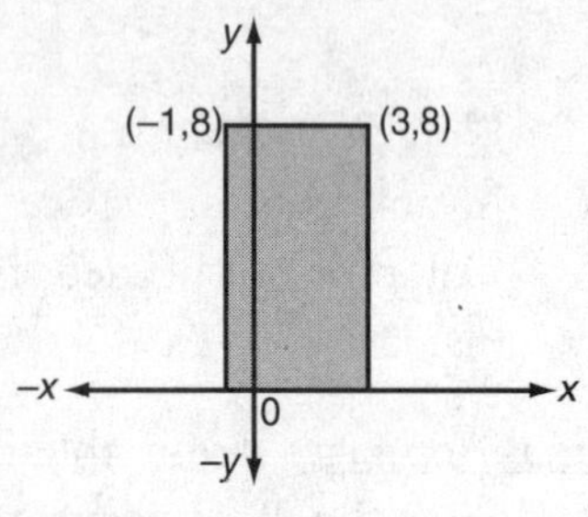

In the figure above, what is the perimeter of the shaded rectangle?

13 A committee of 43 people is to be divided into subcommittees so that each person serves on exactly one subcommittee. Each subcommittee must have at least three members but not more than five members. If M represents the maximum number of subcommittees that can be formed and m represents the least number of subcommittees that can be formed, what is the value of $M - m$?

14

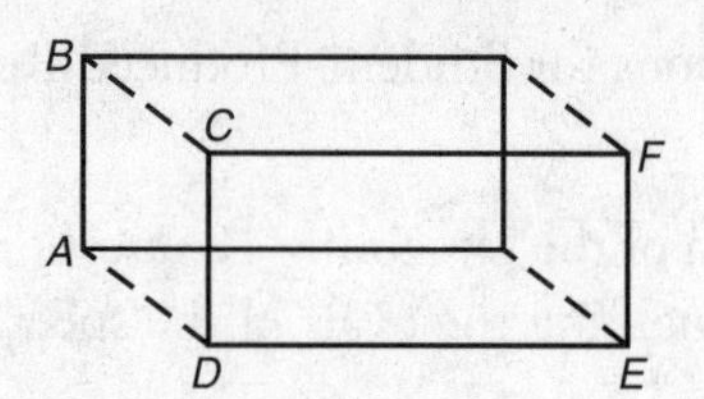

In the figure above, the dimensions of the rectangular box are integers greater than 1. If the area of face $ABCD$ is 12 and the area of face $CDEF$ is 21, what is the volume of the box?

15 The coordinates of the vertices of a triangle are $(1,-2)$, $(9,-2)$, and (h,k). If the area of the triangle is 40, what is a possible value for k?

16 In the games that the Bengals played against the Lions, the Bengals won $\frac{2}{3}$ of the games and lost $\frac{3}{4}$ of the other games. If the teams tied in two games, how many games did the Bengals play against the Lions?

GO ON TO THE NEXT PAGE

17

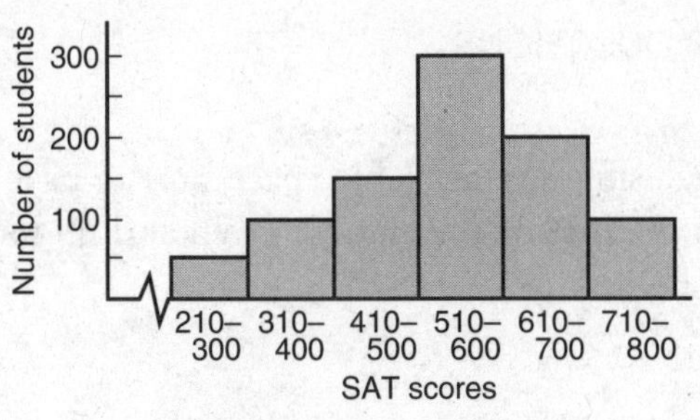

SAT Scores of Students
at Oceanview High School

If the graph above shows the distribution of SAT scores at Oceanview High School, what percent of the students in this school scored at least 610? (Omit the percent sign in your answer.)

18 Let f be the function defined by $f(x) = 8^x - [x]$, where the symbol $[x]$ represents the greatest integer that is *less than or equal to x*. For example, $[4.2] = 4$. What is the numerical value of $f\left(-\frac{2}{3}\right)$?

STOP

If you finish before time is called, you may check your work on this section only. Do not turn to any other section in the test.

Section 3 TIME: 20 MINUTES, 16 QUESTIONS

In this section solve each problem, using any available space on the page for scratchwork. Then decide which is the best of the choices given and fill in the corresponding oval on the answer sheet.

Notes:

1. The use of a calculator is permitted. All numbers used are real numbers.
2. Figures that accompany problems in this test are intended to provide information useful in solving the problems. They are drawn as accurately as possible EXCEPT when it is stated in a specific problem that the figure is not drawn to scale. All figures lie in a plane unless otherwise indicated.
3. Unless otherwise indicated, the domain of any function f is assumed to be the set of all real numbers for which $f(x)$ is a real number.

Reference Information

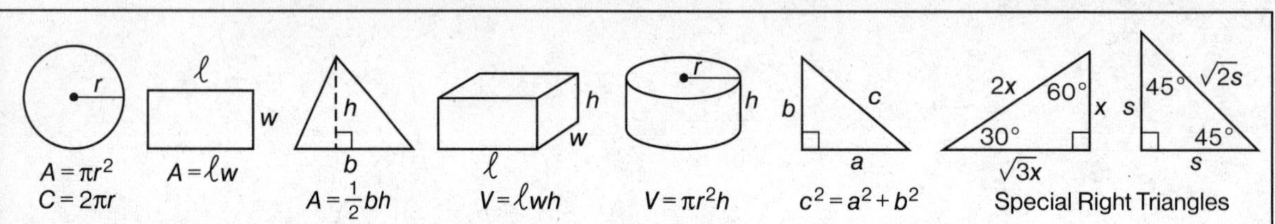

The number of degrees of arc in a circle is 360.
The sum of the measures in degrees of the angles of a triangle is 180.

1 A bell that rings every m minutes and another bell that rings every n minutes, where m and n are prime numbers, ring at the same time. What is the LEAST number of minutes that must elapse before the two bells again ring at the same time?

(A) $m + n$
(B) $\frac{m}{n}$
(C) mn
(D) $60mn$
(E) $\frac{60}{m + n}$

2 Pens that usually sell at two for \$3 are on sale at three for \$4. How much money is saved when 18 pens are purchased at the sale price rather than at the usual price?

(A) \$2
(B) \$3
(C) \$4
(D) \$5
(E) \$6

GO ON TO THE NEXT PAGE

5, 10, 14, 16, 20, 23

If each of the six numbers above is increased by x, the average of the resulting set of six numbers is 15. The value of x is

(A) $\frac{1}{6}$
(B) $\frac{1}{3}$
(C) $\frac{1}{2}$
(D) $\frac{2}{3}$
(E) $\frac{5}{6}$

In the figure above, if the slope of line ℓ is $\frac{1}{4}$, what are the coordinates (a, b) of point P?

(A) (12,5)
(B) (12,8)
(C) (8,4)
(D) (5,20)
(E) $\left(\frac{3}{4},5\right)$

If x is an odd integer, for how many values of x is $3 \leq 2x \leq 56$?

(A) 11
(B) 12
(C) 13
(D) 14
(E) 15

6 In the figure above, semicircle AD is inscribed in rectangle $ABCD$. What is the area of the shaded region in terms of w?

(A) $w^3\left(4 - \frac{\pi}{2}\right)$
(B) $w^2\left(2 - \frac{\pi}{4}\right)$
(C) $w^2\left(2 - \frac{\pi}{2}\right)$
(D) $\pi w^2 - 2$
(E) $2\pi w^2 - 1$

7 If $\frac{z}{2b} = 4$, $\frac{z}{2c} = 6$, and $2b + 3c = 12$, what is the value of z?

(A) 8
(B) 16
(C) 20
(D) 24
(E) 30

8 If $3a + b = 24$ and a is a positive even integer, which of the following statements must be true?

I. b is divisible by 3.
II. b is an even integer.
III. a is less than 8.

(A) II only
(B) I and II only
(C) I and III only
(D) II and III only
(E) I, II, and III

GO ON TO THE NEXT PAGE

Practice Test 1

9 After the length and width of a rectangle are each reduced by 40%, the length and width of the new rectangle are each increased by $\frac{1}{3}$ of their new values. In producing the final rectangle, by what percent is the area of the original rectangle reduced?

(A) 20%
(B) 24%
(C) 30%
(D) 36%
(E) 40%

10 If 1 blip + 4 beeps = 1 glitch and 3 blips + 1 beep = 2 glitches, how many beeps equal a glitch?

(A) 3
(B) 5
(C) 8
(D) 11
(E) 13

11 Tom loaded $\frac{1}{2}$ of the cartons at a shipping dock on a truck. When Tom went to lunch, George loaded $\frac{1}{3}$ of the remaining cartons on the truck. After George finished, 40 cartons remained to be loaded on the truck. How many cartons were on the shipping dock before any were loaded on the truck?

(A) 60
(B) 80
(C) 120
(D) 150
(E) 180

12 If $k^{-\frac{1}{2}} = 4$ and $j^{-3} = 8$, then $\frac{j}{k} =$

(A) $\frac{1}{8}$
(B) $\frac{1}{2}$
(C) 2
(D) 4
(E) 8

13

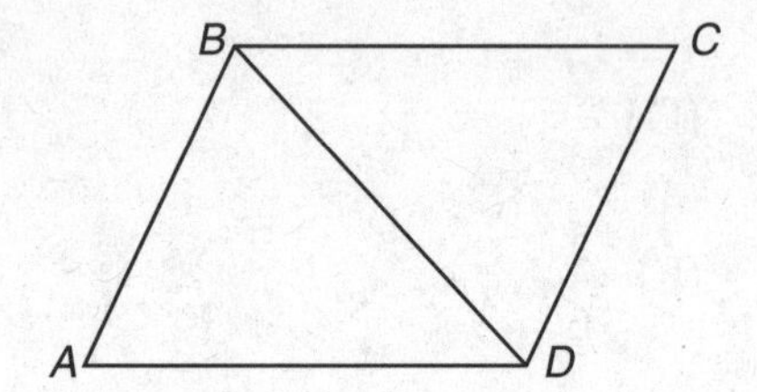

Note: Figure is not drawn to scale.

In the figure above, $AB = BC = CD = AD = BD$. If the perimeter of $ABCD$ is 8, what is the length of diagonal $\overline{AC}$ (not shown)?

(A) $\sqrt{2}$
(B) $\sqrt{3}$
(C) $2\sqrt{2}$
(D) $2\sqrt{3}$
(E) 3

14 If $2^x \cdot 4^y = 8^{x+y}$, then $\frac{x}{y} =$

(A) $-\frac{3}{2}$
(B) $-\frac{1}{2}$
(C) $\frac{3}{4}$
(D) 2
(E) $\frac{5}{2}$

GO ON TO THE NEXT PAGE

15

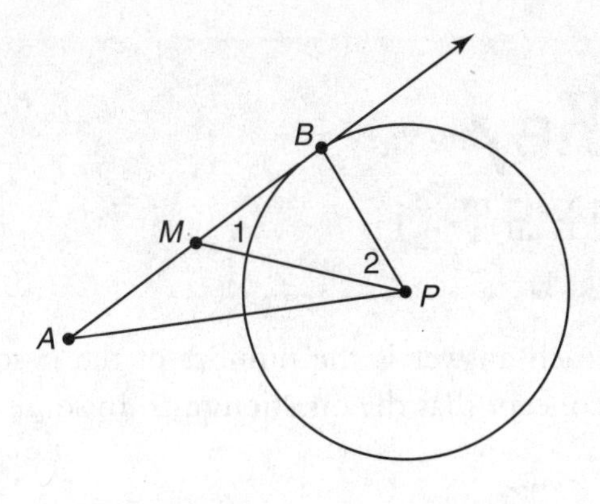

Note: Figure is not drawn to scale.

In the figure above, $\overline{AB}$ is tangent to circle P at point B. Line segment PM bisects $\overline{AB}$, and the measure of angle 1 is equal to the measure of angle 2. If r represents the radius length of the circle, what is the length of $\overline{AP}$ in terms of r?

(A) $\sqrt{2}r$
(B) $\sqrt{r^2 + 2}$
(C) $\sqrt{3}r$
(D) $2r$
(E) $\sqrt{5}r$

16

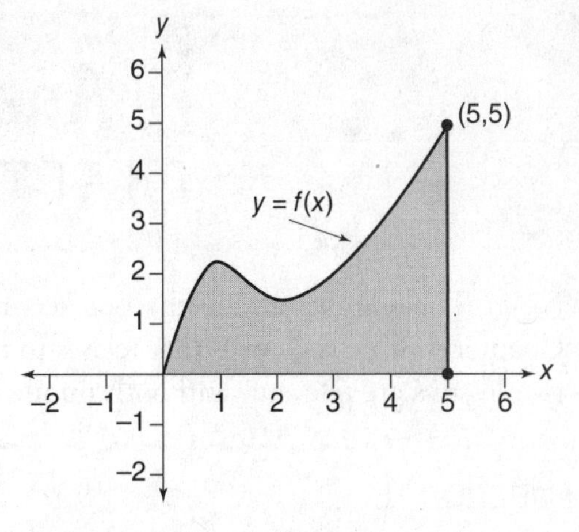

Function f is defined for $0 \leq x \leq 5$, as shown in the accompanying figure. If (r,s) is a point inside the shaded region bounded by the x-axis, the line $x = 5$, and $y = f(x)$, which statement must be true?

I. $r + s \leq 5$
II. $s \leq f(r)$
III. $r \neq s$

(A) I only
(B) II only
(C) III only
(D) I and III only
(E) II and III only

STOP

If you finish before time is called, you may check your work on this section only. Do not turn to any other section in the test.

Answer Key

PRACTICE TEST 1

Note: The number inside the brackets that follows each answer is the number of the lesson in Chapter 3, 4, 5, 6, 7, or 8 that relates to the topic or concept that the question tests. In some cases two lessons are relevant, and both numbers are given.

Section 1

1. C [3-1]
2. D [6-1]
3. C [3-5]
4. D [6-5]
5. C [4-2]
6. A [3-3]
7. B [7-4]
8. E [3-5]
9. B [5-4]
10. B [6-7]
11. A [4-2]
12. C [5-1]
13. D [5-4]
14. A [7-4]
15. D [5-1]
16. A [6-6]
17. C [7-5]
18. D [8-5]
19. D [5-1]
20. B [3-5]

Section 2

1. C [8-1]
2. C [7-3]
3. C [8-5]
4. E [4-1]
5. B [8-2]
6. C [8-5]
7. D [8-1]
8. B [8-3, 8-4]
9. .01 [4-1]
10. 21 [3-2]
11. 54 [5-1]
12. 24 [6-5]
13. 5 [7-2]
14. 84 [6-7]
15. 8 [6-8]
16. 24 [3-7]
17. 33.3 [7-4]
18. 5/4 [8-4]

Section 3

1. C [3-3]
2. B [3-1]
3. B [7-1]
4. A [6-8]
5. C [3-1]
6. C [6-5, 6-6]
7. D [4-2]
8. B [3-3]
9. D [7-4, 6-5]
10. D [4-6]
11. C [3-7]
12. E [8-1]
13. D [6-4]
14. B [8-2]
15. E [6-6]
16. B [8-4]

Self-Scoring Chart

SECTION 1 *(20 Questions)*

Number Correct ________ (a)
Number Incorrect ________ (b)
Number Omitted ________ (c)
(a) minus $\frac{1}{4}$ (b) = **Raw Score I** ______

SECTION 2 *(18 Questions)*

Part 1 Questions 1–8

Number Correct ________ (d)
Number Incorrect ________ (e)
(d) minus $\frac{1}{4}$ (e) = **Raw Score II** ______

Part 2 Questions 9–18

Number Correct **Raw Score III** ______

SECTION 3 *(16 Questions)*

Number Correct ________ (f)
Number Incorrect ________ (g)
Number Omitted ________ (h)
(f) minus $\frac{1}{4}$ (g) = **Raw Score IV** ______

Total Score = Raw Score I + II + III + IV ______

Evaluation Chart

50–54	Superior
45–49	Very Good
40–44	Good
35–39	Above Average
30–34	Average
20–29	Below Average
0–19	Below Average; needs intensive study

ANSWER EXPLANATIONS FOR PRACTICE TEST 1

Section 1

1. **(C)** Multiply odd integers beginning with 1 by 4 until the product is more than 36:
$$4 \times 1 = 4 \quad 4 \times 3 = 12$$
$$4 \times 5 = 20 \quad 4 \times 7 = 28$$
$$4 \times 9 = 36$$
Hence, there are four positive integers less than 36 (4, 12, 20, and 28) that are equal to 4 times an *odd* integer.

2. **(D)** Since pairs of obtuse angles formed by parallel lines are equal, the obtuse angles at A and B are equal, so $\angle ABD = 130°$. You are told that BC bisects $\angle ABD$, so
$$\angle DBC = \frac{1}{2}(130°) = 65°$$
An acute and an obtuse angle formed by parallel lines are supplementary. Hence, the obtuse angle at C, which is labeled $x°$, is supplementary to $\angle DBC$ and
$$x = 180 - 65 = 115$$

3. **(C)** If a call lasts x minutes and x is greater than 3, then the charge for the first 3 minutes is \$1.80 and the charge for the next $x - 3$ minutes is $0.40(x - 3)$. Since the total charge was \$4.20,
$$0.40(x - 3) + 1.80 = 4.20$$
$$40(x - 3) + 180 = 420$$
$$40x - 120 + 180 = 420$$
$$40x = 360$$
$$x = \frac{360}{40} = 9$$

4. **(D)** The area of a square with a side length of 8 inches is 8×8 or 64 square inches. Since the area of a smaller square with a side length of 2 inches is 2×2 or 4 square inches, $\frac{64}{4}$ or 16 of the smaller squares are needed to cover completely, without overlap, the larger square.

5. **(C)** If $3a = 2$ and $2b = 15$, then
$$(3a)(2b) = (2)(15)$$
$$6ab = 30$$
$$ab = \frac{30}{6} = 5$$

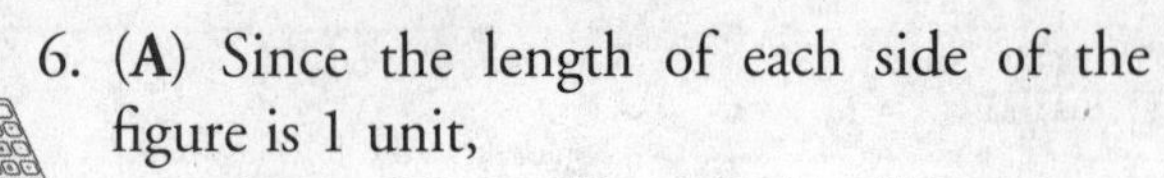

6. **(A)** Since the length of each side of the figure is 1 unit,

$$AB + BC + CD + DE + EA = 5$$

Therefore, each time a point starts and ends at vertex A, the point moves 5 units or a multiple of 5 units. Since $\frac{715}{5} = 143$ with no remainder, a point that starts at vertex A and moves a distance of 715 units along each side of the figure will go around the entire figure exactly 143 times, so it will end up at vertex A.

7. **(B)** According to the table, $32 + 80 + 160 + 83 + 45$ or 400 families were surveyed. Since the number of families with two or fewer television sets is $32 + 80 + 160$ or 272, the percent of families surveyed with two or fewer television sets is $\frac{272}{400} \times 100\%$ or 68%.

8. **(E)** Determine whether each Roman numeral expression decreases when x increases, where $0 < x < 1$.
   - I. As x increases, the denominator of the fraction $\frac{1}{1 - x}$ gets smaller, so the value of the fraction increases. Hence, expression I does not increase.
   - II. As x increases, x^2 also increases. Since the denominator of the fraction $\frac{1}{x^2}$ increases, the value of the fraction decreases. Hence, expression II decreases.
   - III. As x increases but remains less than 1, $\sqrt{x}$ increases, so the difference $1 - \sqrt{x}$ decreases. Hence, expression III decreases.

   Only Roman numeral expressions II and III decrease.

9. **(B)** Since the length of the discarded ribbon is 9 inches, and this length is $\frac{1}{8}$ of the length of the original ribbon, the original ribbon was 8×9 or 72 inches long. Hence, the remaining ribbon is $72 - 9$ or 63 inches long. You are given that the 63-inch ribbon is divided into two pieces whose lengths are in the ratio of 4 : 5. If x represents the length of the shorter ribbon, then $63 - x$ is the length of the longer ribbon. Hence:

$$\frac{\text{shorter ribbon}}{\text{longer ribbon}} = \frac{x}{63 - x} = \frac{4}{5}$$

Cross-multiplying gives

$$5x = 4(63 - x)$$
$$= 252 - 4x$$
$$9x = 252$$
$$x = \frac{252}{9} = 28$$

10. **(B)** If the edge length of a cube is π, then the volume of the cube is the edge length raised to the third power or π^3. If the height of a cylinder is π and the radius of its base is represented by r, then the volume of the cylinder is the area of the base times its height or $(\pi r^2) \times \pi = \pi^2 r^2$. Since you are told that the volumes of the cube and cylinder are equal,

$$\pi^3 = \pi^2 r^2$$
$$r^2 = \frac{\pi^3}{\pi^2} = \pi$$
$$r = \sqrt{\pi}$$

11. **(A)** By the distributive law, $a(b + 1) = ab + a$. If $\frac{5}{a} = b$, then $5 = ab$. Since $ab = 5$ and $a = 3$,

$$a(b + 1) = ab + a = 5 + 3 = 8$$

12. **(C)** If x represents the number of \$20 sport shirts purchased, then $2x$ is the number of \$35 dress shirts purchased. Since a total of \$720 was spent on the shirts, $20(x) + 35(2x) = 720$ or $20x + 70x = 720$, so $90x = 720$. Hence,

$$x = \frac{720}{90} = 8 \quad \text{and} \quad 2x = 2(8) = 16$$

The total number of shirts purchased is $x + 2x$ or $8 + 16 = 24$.

13. **(D)** If there are 1.2 grams of hemoglobin in 10 cubic centimeters of blood and x represents the number of grams of hemoglobin contained in 35 cubic centimeters of the same blood, then

$$\frac{\text{Blood}}{\text{Hemoglobin}} = \frac{10}{1.2} = \frac{35}{x}$$
$$10x = 42$$
$$x = \frac{42}{10} = 4.2$$

14. **(A)** Since the sum of the sectors of a circle graph must add up to 100%,

$$x\% + 20\% + 28\% + 12\% + 24\% = 100\%$$
$$x\% + 84\% = 100\%$$
$$x = 16$$

Since $16\% + 24\% = 40\%$, then 0.40×2500 or 1000 students named either swimming or biking as their favorite sports activity.

15. **(D)** If $x - 3$ is 1 less than $y + 3$, then $x - 3 = (y + 3) - 1$, so $x = y + 5$. Since
$$x + 2 = y + 5 + 2 = y + 7$$
then $x + 2$ exceeds y by 7.

16. **(A)** Since diameter $AD = 12$, $AB = BC = CD = 4$. The areas of semicircles BC and CD are equal, so the area of the shaded region is simply the sum of the areas of semicircles AD and AB. The area of semicircle AD is $\frac{1}{2}(6^2\pi)$ or 18π, and the area of semicircle AB is $\frac{1}{2}(2^2)\pi$ or 2π. Hence, the area of the shaded region is $18\pi + 2\pi$ or 20π.

17. **(C)** Since $110 = 11 \times 10$, the largest prime factor of 110 is 11. Hence,
$$\nabla 110 = 110 + \frac{110}{11} = 110 + 10$$
$$= 120 = x$$
Since $120 = 3 \times 8 \times 5$, the largest prime factor of 120 is 5. Hence,
$$\nabla x = \nabla 120 = 120 + \frac{120}{5}$$
$$= 120 + 24 = 144$$

18. **(D)** Determine whether each Roman numeral choice is true or false:
- I. From the graph, Bill walks 1 mile in 30 minutes. Since he is walking at a constant rate, he walks 2 miles in 60 minutes. Hence, Bill's average rate of walking was 2 miles per hour. This choice is correct.
- II. At 9:10 A.M., Allan jogged 1 mile. Since Bill was walking at a constant rate of 2 miles per hour, he walked
$$2 \text{ miles} \times \frac{1}{6} \text{ hr}(= 10 \text{ min}) = \frac{1}{3} \text{ mile.}$$
Hence, at 9:10 A.M. Allan had jogged $1 - \frac{1}{3} = \frac{2}{3}$ mile more than Bill had walked. This choice is not correct.
- III. At 9:30 A.M., Allan had jogged 3 miles and Bill had walked 1 mile. Hence, Allan jogged 2 miles more than Bill had walked. Four laps around a $\frac{1}{4}$-mile track equals 1 mile. Hence, Allan completed 8 more laps around the track than Bill. This choice is correct.

Since Roman numeral choices I and III are correct, the answer is choice (D).

19. **(D)** If a group of p people plan to contribute equally to the purchase of a gift that costs d dollars, then each person must contribute $\frac{d}{p}$ dollars. If n of the p people decide not to contribute, then each of the $p - n$ people who are left must contribute $\frac{d}{p-n}$ dollars. The difference between the two contribution rates represents the amount of increase for each person who contributes:
$$\frac{d}{p-n} - \frac{d}{p} = \frac{d}{p-n}\left(\frac{p}{p}\right) - \frac{d}{p}\left(\frac{p-n}{p-n}\right)$$
$$= \frac{dp - d(p-n)}{p(p-n)}$$
$$= \frac{dp - dp + nd}{p(p-n)}$$
$$= \frac{nd}{p(p-n)}$$

20. **(B)** In a certain class, exactly $\frac{2}{3}$ of the students applied for admission to a 4-year college. Of these, $\frac{1}{8}$ or $\frac{1}{8} \times \frac{2}{3} = \frac{2}{24} = \frac{1}{12}$ of the class also applied to a 2-year junior college. Hence, the number of students who could be in this class must be divisible by the least common denominator of $\frac{2}{3}$ and $\frac{1}{12}$, which is 12, or a multiple of the least common denominator. Suppose there are 12 students in the class. Then $\frac{1}{12}$ of 12 or

one student applied to both a 4-year and a 2-year college. Since you are told that more than one student applied to both 4-year and 2-year colleges, the LEAST number of students in the class is the next consecutive multiple of 12, which is 24.

Section 2

1. **(C)** If $x^{-2} = 64$, then $x^2 = \frac{1}{64}$ so $x = \frac{1}{\sqrt{64}} = \frac{1}{8}$. Hence, $x^{\frac{1}{3}} = \left(\frac{1}{8}\right)^{\frac{1}{3}} = \frac{1}{\sqrt[3]{8}} = \frac{1}{2}$.
2. **(C)** Since exactly two of the five numbers from set *X*, 22 and 24, are also in set *Y*, the probability that the number selected from set *X* is also in set *Y* is $\frac{2}{5} = 0.4$.
3. **(C)** It is given that the slope of the graph of the linear function is -2 and $f(3) = 4$, which means that the point (3,4) is on the line. The equation of the linear function has the form $y = mx + b$. To find the value of b, let $m = -2$, $x = 3$, and $y = 4$:
$$4 = -2(3) + b$$
$$4 = -6 + b$$
$$10 = b$$
4. **(E)** Since $3m = 7$, $k = 6m = 2(3m) = 2(7) = 14$. Hence, $j - 14 = 9$, so $j = 9 + 14 = 23$.
5. **(B)** First solve for x:
   - If $3\sqrt{x-2} = 4 = 11$, then $3\sqrt{x-2} = 15$ and $\sqrt{x-2} = \frac{15}{3} = 5$.
   - Eliminate the radical: $\left(\sqrt{x-2}\right)^2 = 5^2$, so $x - 2 = 25$ and $x = 27$.
   - $y^3 = x^2 = 27^2$, so $y = 27^{\frac{2}{3}} = \left(\sqrt[3]{27}\right)^2 = (3)^2 = 9$.
6. **(C)** Since $f(2) = 3$ and $f(3) = 2$, the graph of the linear function contains (2,3) and (3,2). Check each Roman numeral choice:
   - I. This choice is not correct, since (3,2) does not satisfy the equation $y = x + 1$.
   - II. This choice is not correct, since (2,3) does not satisfy the equation $y = x - 1$.
   - III. This choice is correct, since (3,2) and (2,3) both satisfy the equation $y = -x + 5$.

   Hence, the correct answer is (C).
7. **(D)** If $7^k = 100$, then
$$\left(7^k\right)^{\frac{1}{2}} = (100)^{\frac{1}{2}}$$
$$7^{\frac{k}{2}} = \sqrt{100} = 10$$
Hence, $7^{\frac{k}{2}+1} = 7^1 \cdot 7^{\frac{k}{2}} = 7(10) = 70$.
8. **(B)** Since the range (set of y-values) of the graph of $y = |f(x)|$ must be nonnegative, eliminate choices (A), (C), and (D). Among the remaining two choices, look for the graph for which the section of the graph that falls below the y-axis has been reflected in the x-axis so that all the y-values of the graph are now nonnegative, as in the graph in choice (B).
9. **.01** If $x + 2x + 3x + 4x = 1$, then $10x = 1$ so $x = \frac{1}{10}$ and $x^2 = \frac{1}{10} \times \frac{1}{10} = \frac{1}{100}$. Since 1/100 does not fit the grid, grid in .01 instead.
10. **(21)** Since $441p = 9 \times 49 \times p = 3^2 \times 7^2 \times p$, let $p = 3 \times 7$, which makes $441p = 3^3 \times 7^3 = (3 \times 7)^3 = 21^3$.
11. **(54)** If $2x$ represents the number of luxury cars sold, then $3x$ is the number of economy cars sold. Since the total number of economy and luxury cars sold was 90,
$$2x + 3x = 90$$
$$5x = 90$$
Hence, the number of economy cars sold was $3x = 3(18) = 54$.
12. **(24)** One side of the shaded rectangle lies on the x-axis, and the side opposite has endpoints $(-1,8)$ and $(3,8)$. Since this side is 8 units above the x-axis, the height of the rectangle is 8. The width of this rectangle is the difference between the x-coordinates of the points $(-1,8)$ and $(3,8)$ which is $3 - (-1)$ or 4. Hence, the perimeter of the rectangle is $2(4 + 8) = 2(12) = 24$.

13. **(5)** You are told that each subcommittee formed from a committee of 43 people must have at least three but not more than five members. Since each person serves on exactly one subcommittee, the maximum number of subcommittees is the integer part of $43 \div 3$ or 14. To find the minimum number of subcommittees, divide 43 by 5 to get 8 with a remainder of 3. Hence, the 43 people can be divided into eight subcommittees with five members each plus one subcommittee with three members, for a total of nine subcommittees. Thus, $M - m = 14 - 9 = 5$.
14. **(84)** Since the area of face *ABCD* is 12, then $AD \times CD = 12$. The area of face *CDEF* is 21, so $CD \times DE = 21$. Since the dimensions of the rectangular box are integers greater than 1, *CD* must be a common factor, other than 1, of 12 and 21. Since $12 = 3 \times 4$ and $21 = 3 \times 7$, then $CD = 3$. Since $CD = 3$, then $AD = 4$ and $DE = 7$, so the volume of the box is $3 \times 4 \times 7$ or 84.
15. **(8)** Sketch a possible triangle, as shown below.

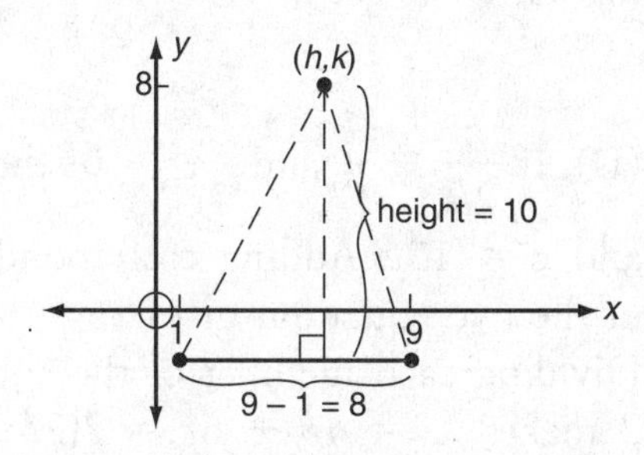

The length of the horizontal base of the triangle is the difference between the *x*-coordinates of the endpoints of the base, which is $9 - 1$ or 8. The area of a triangle is $\frac{1}{2}$ of the product of the base and the height. Since the area of the triangle is given as 40 and $\frac{1}{2} \times 8 \times 10 = 40$, the height of the triangle is 10 units. Since the *y*-coordinate of a point 10 units above the base is $-2 + 10$ or 8, a possible value of *k* is 8.

16. **(24)** The Bengals won $\frac{2}{3}$ of the games played, so the number of games that ended in a loss or tie is $\frac{1}{3}$ of the total number of games played. The Bengals lost $\frac{3}{4}$ of the number of games that remained, so they lost $\frac{3}{4}$ of $\frac{1}{3}$ or $\frac{1}{4}$ of the games they played. Since the Bengals won $\frac{2}{3}\left(=\frac{8}{12}\right)$ and lost $\frac{1}{4}\left(=\frac{3}{12}\right)$ of the games played, $\frac{1}{12}$ of the games played ended in a tie. You are told that the teams tied in two games. Therefore, the total number of games played is 24, since $\frac{1}{12}$ of 24 is 2.
17. **(33.3)** The total number of students who took the SAT at Oceanview High School is the sum of the heights of all of the bars on the graph. Thus,

$$50 + 100 + 150 + 300 + 200 + 100$$
or 900 students

took the test. The number of students who scored at least 610 is the sum of the heights of the last two bars, which is $200 + 100$ or 300. The percent of the students in this school who scored at least 610 is

$$\frac{300}{900} \times 100\% = 33.3\%$$

18. **(5/4)** If $f(x) = 8^x - [x]$, then

$$f\left(-\frac{2}{3}\right) = 8^{-\frac{2}{3}} - \left[\!\left[-\frac{2}{3}\right]\!\right]$$
$$= \frac{1}{\left(\sqrt[3]{8}\right)^2} - (-1)$$
$$= \frac{1}{2^2} + 1$$
$$= \frac{1}{4} + 1$$
$$= \frac{5}{4}$$

Section 3

1. **(C)** If a bell rings every m minutes and another bell rings every n minutes, then the number of minutes that must elapse before the bells again ring at the same time must be divisible by both m and n. Since m and n are prime numbers, the least number of minutes that must elapse before the two bells ring again is the lowest common multiple of m and n, which is m times n or mn.
2. **(B)** When 18 pens are purchased at two for \$3, the cost is 9 × \$3 or \$27. When 18 pens are purchased at three for \$4, the cost is 6 × \$4 or \$24. The amount saved is \$27 − \$24 or \$3.
3. **(B)** The sum of numbers 5, 10, 14, 16, 20, 23 is 88. If each of the six numbers is increased by x, their sum is $88 + 6x$. If the average of this new set of six numbers is 15, the sum of these six numbers is 6 × 15 or 90. Hence:

$$88 + 6x = 90$$
$$6x = 2$$
$$x = \frac{2}{6} = \frac{1}{3}$$

4. **(A)** Since point P lies on the same horizontal line as $(-1,5)$, its y-coordinate is also 5. Hence,

$$\text{Slope of line } \ell = \frac{\text{change in } y}{\text{change in } x} = \frac{5-2}{a-0} = \frac{1}{4}$$
$$\frac{3}{a} = \frac{1}{4}$$
$$a = 12$$

Since $a = 12$ and $b = 5$, the coordinates of point P are (12,5).

5. **(C)** If $3 \leq 2x \leq 56$, then dividing each member of the inequality gives

$$\frac{3}{2} \leq \frac{2x}{2} \leq \frac{56}{2} \text{ or } \frac{3}{2} \leq x \leq 28$$

Count the number of odd integers from $\frac{3}{2}$ to 28. There are 13 odd integers in this interval: 3, 5, 7, 9, 11, 13, 15, 17, 19, 21, 23, 25, and 27.

6. **(C)** Since $AD = BC = 2w$, the diameter of semicircle AD is $2w$, so its radius is w. Hence:

$$\underbrace{\text{Area shaded region}} = \underbrace{\text{Area rectangle } ABCD} - \underbrace{\text{Area semicircle } AD}$$
$$= 2w(w) - \frac{1}{2}\pi w^2$$
$$= 2w^2 - \frac{1}{2}\pi w^2$$

Since $2w^2 - \frac{1}{2}\pi w^2$ is not among the answer choices, write it in a different form by factoring out w^2 to get $w^2\left(2 - \frac{\pi}{2}\right)$.

7. **(D)** If $\frac{z}{2b} = 4$ and $\frac{z}{2c} = 6$, then $z = 8b$ and $z = 12c$. Adding corresponding sides of the two equations gives $2z = 8b + 12c$. Dividing each member of the equation by 2 makes $z = 4b + 6c = 2(2b + 3c) = 2(12) = 24$.
8. **(B)** Determine whether each Roman numeral statement is true or false when $3a + b = 24$ and a is a positive even integer.
   - I. Since $3a + b = 24$, then $b = 24 - 3a = 3(8 - a)$. Since $3(8 - a)$ is divisible by 3, then b is divisible by 3, so statement I is true.
   - II. Since $b = 3(8 - a)$ and a is a positive even integer, $8 - a$ is an even integer. Since the product of 3 and an even integer is always an even integer, b is an even integer, so statement II is true.

- III. Since there is no restriction on whether b is positive, negative, or 0, a could be less than 8, greater than 8, or equal to 8. Hence, statement III is false.

Only Roman numeral choices I and II are true.

9. **(D)** Suppose the length and width of the original rectangle are each 10; then the area of the original rectangle is 10 times 10 or 100. After the length and width of this rectangle are each reduced by 40%, the length and width of the new rectangle are each $10 - (0.40 \times 10)$ or 6. If the length and width of this rectangle are then each increased by $\frac{1}{3}$ of their value, then the length and width of the final rectangle is $6 + (\frac{1}{3} \times 6)$ or 8. The area of the final rectangle is 8×8 or 64. Hence,

$$\text{Percent of decrease in area} = \frac{\text{Decrease in area}}{\text{Original area}} \times 100\% = \frac{100 - 64}{100} \times 100\% = \frac{36}{100} \times 100\% = 36\%$$

10. **(D)** Multiply the first equation by 3 and then subtract:

1 blip + 4 beeps = 1 glitch → 3 blips + 12 beeps = 3 glitches
3 blips + 1 beep = 2 glitches → −3 blips + 1 beep = 2 glitches
11 beeps = 1 glitch

Hence, 11 beeps equal a glitch.

11. **(C)** Draw a rectangle and subdivide it into $2 \times 3 = 6$ equal parts:

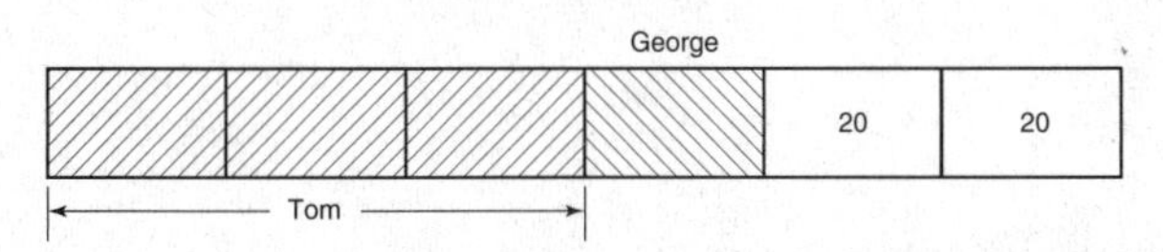

- Since Tom loaded $\frac{1}{2}$ of the cartons, shade in the first three rectangles.
- Because George loaded $\frac{1}{3}$ of the remaining cartons, shade in only one of the three remaining rectangles.
- Since 40 cartons remain, fill in 20 cartons in each of the remaining two rectangles. This means that the original number of cartons was $6 \times 20 = 120$ cartons.

12. **(E)**

- If $k^{-\frac{1}{2}} = 4$, then

$$\frac{1}{k^{\frac{1}{2}}} = \frac{1}{\sqrt{k}} = 4, \text{ so}$$

$$\left(\frac{1}{\sqrt{k}}\right)^2 = 4^2, \text{ which makes } \frac{1}{k} = 16.$$

- If $j^{-3} = 8$, then $j^3 = \frac{1}{8}$, so

$$j = \frac{1}{\sqrt[3]{8}} = \frac{1}{2}.$$

- Hence, $\frac{j}{k} = \frac{1}{k} \times j = 16 \times \frac{1}{2} = 8.$

13. **(D)**

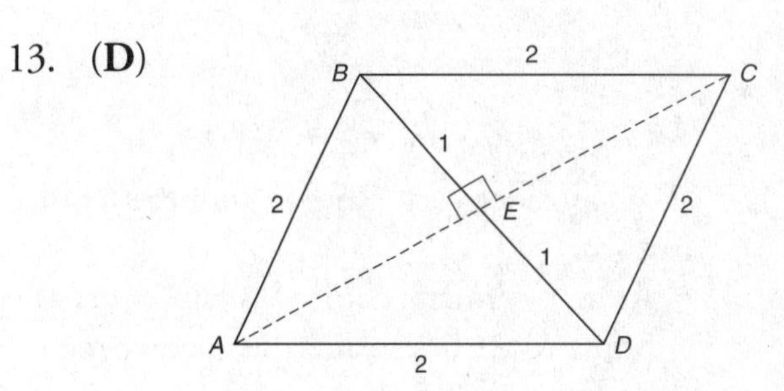

Note: Figure is not drawn to scale.

Since it is given that, in the accompanying figure, $AB = BC = CD = AD = BD$, quadrilateral $ABCD$ is a rhombus.

- Since the perimeter of the figure is 8, the length of each of the four equal sides is 2.
- Draw diagonal $\overline{AC}$. Since the diagonals of a rhombus intersect each other at right angles and at their midpoints, $\triangle AEB$ is a right triangle with hypotenuse $AB = 2$ and leg $BE = 1$. Use the Pythagorean theorem to find AE:

$$(AE)^2 + 1^2 = 2^2$$
$$(AE)^2 + 1 = 4$$
$$AE = \sqrt{3}$$

- Thus, $AC = 2 \times AE = 2\sqrt{3}$.

14. **(B)** First change to a common base:

$$2^x \cdot 4^y = 8^{x+y}$$
$$2^x \cdot 2^{2y} = 2^{3(x+y)}$$
$$2^{x+2y} = 2^{3(x+y)}$$

Set the exponents equal and solve for $\frac{x}{y}$:

$$x + 2y = 3x + 3y$$
$$-2x = y$$
$$\frac{-2x}{-2y} = \frac{y}{-2y}$$
$$\frac{x}{y} = -\frac{1}{2}$$

15. **(E)** Let $PB = r$.
    - Since $m\angle 1 = m\angle 2$, $MB = PB = r$.
    - Because $\overline{PM}$ bisects $\overline{AB}$, $AM = MB = r$, which means $AB = r + r = 2r$.
    - Since a radius is perpendicular to a tangent at the point of contact with the circle, $\triangle ABP$ is a right triangle. Use the Pythagorean theorem to find the length of hypotenuse $\overline{AP}$:

$$(AP)^2 = (PB)^2 + (AB)^2$$
$$= (r)^2 + (2r)^2$$
$$= r^2 + 4r^2$$
$$= 5r^2$$
$$AP = \sqrt{5r^2} = \sqrt{5}\,r$$

16. **(B)** Consider each Roman numeral choice in turn:
    - Roman numeral choice I is false, since the point (4,2) lies inside the shaded region and $4 + 2 > 5$.
    - For any given point (r,s) inside the shaded region, $f(r)$ represents the y-value of the point on the graph directly above it, so $s \leq f(r)$. For example, if $f(r,s) = (3,1)$, then $1 < f(3)$ since $f(3)$, according to the graph, has a value between 1 and 2. Roman numeral choice II must be true.
    - Because points such as (1,1) are contained in the shaded region, it is not always the case that $r \neq s$. Roman numeral choice III is false.

    Since only Roman numeral choice II must be true, the correct answer is choice **(B)**.

Answer Sheet

PRACTICE TEST 2

Start with number 1 for each new section. If a section has fewer questions than answer spaces, leave the extra answer spaces blank.

SECTION 1

1 Ⓐ Ⓑ Ⓒ Ⓓ Ⓔ	6 Ⓐ Ⓑ Ⓒ Ⓓ Ⓔ	11 Ⓐ Ⓑ Ⓒ Ⓓ Ⓔ	16 Ⓐ Ⓑ Ⓒ Ⓓ Ⓔ
2 Ⓐ Ⓑ Ⓒ Ⓓ Ⓔ	7 Ⓐ Ⓑ Ⓒ Ⓓ Ⓔ	12 Ⓐ Ⓑ Ⓒ Ⓓ Ⓔ	17 Ⓐ Ⓑ Ⓒ Ⓓ Ⓔ
3 Ⓐ Ⓑ Ⓒ Ⓓ Ⓔ	8 Ⓐ Ⓑ Ⓒ Ⓓ Ⓔ	13 Ⓐ Ⓑ Ⓒ Ⓓ Ⓔ	18 Ⓐ Ⓑ Ⓒ Ⓓ Ⓔ
4 Ⓐ Ⓑ Ⓒ Ⓓ Ⓔ	9 Ⓐ Ⓑ Ⓒ Ⓓ Ⓔ	14 Ⓐ Ⓑ Ⓒ Ⓓ Ⓔ	19 Ⓐ Ⓑ Ⓒ Ⓓ Ⓔ
5 Ⓐ Ⓑ Ⓒ Ⓓ Ⓔ	10 Ⓐ Ⓑ Ⓒ Ⓓ Ⓔ	15 Ⓐ Ⓑ Ⓒ Ⓓ Ⓔ	20 Ⓐ Ⓑ Ⓒ Ⓓ Ⓔ

SECTION 2

1 Ⓐ Ⓑ Ⓒ Ⓓ Ⓔ	5 Ⓐ Ⓑ Ⓒ Ⓓ Ⓔ
2 Ⓐ Ⓑ Ⓒ Ⓓ Ⓔ	6 Ⓐ Ⓑ Ⓒ Ⓓ Ⓔ
3 Ⓐ Ⓑ Ⓒ Ⓓ Ⓔ	7 Ⓐ Ⓑ Ⓒ Ⓓ Ⓔ
4 Ⓐ Ⓑ Ⓒ Ⓓ Ⓔ	8 Ⓐ Ⓑ Ⓒ Ⓓ Ⓔ

ONLY ANSWERS ENTERED IN THE OVALS IN EACH GRID AREA WILL BE SCORED.
YOU WILL NOT RECEIVE CREDIT FOR ANYTHING WRITTEN IN THE BOXES ABOVE THE OVALS.

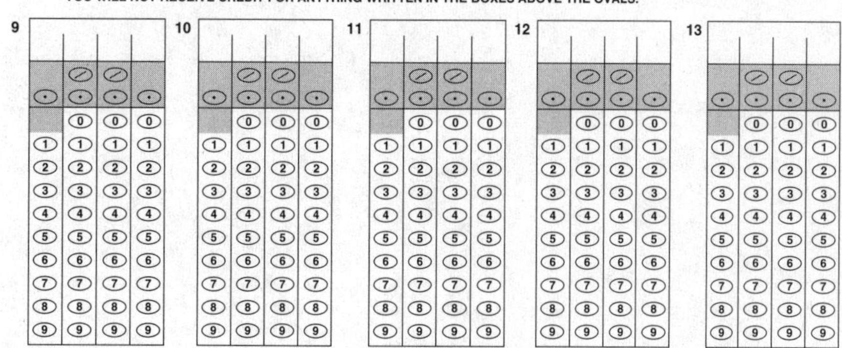

14 15 16 17 18

SECTION

1 Ⓐ Ⓑ Ⓒ Ⓓ Ⓔ	6 Ⓐ Ⓑ Ⓒ Ⓓ Ⓔ	11 Ⓐ Ⓑ Ⓒ Ⓓ Ⓔ	16 Ⓐ Ⓑ Ⓒ Ⓓ Ⓔ
2 Ⓐ Ⓑ Ⓒ Ⓓ Ⓔ	7 Ⓐ Ⓑ Ⓒ Ⓓ Ⓔ	12 Ⓐ Ⓑ Ⓒ Ⓓ Ⓔ	17 Ⓐ Ⓑ Ⓒ Ⓓ Ⓔ
3 Ⓐ Ⓑ Ⓒ Ⓓ Ⓔ	8 Ⓐ Ⓑ Ⓒ Ⓓ Ⓔ	13 Ⓐ Ⓑ Ⓒ Ⓓ Ⓔ	18 Ⓐ Ⓑ Ⓒ Ⓓ Ⓔ
4 Ⓐ Ⓑ Ⓒ Ⓓ Ⓔ	9 Ⓐ Ⓑ Ⓒ Ⓓ Ⓔ	14 Ⓐ Ⓑ Ⓒ Ⓓ Ⓔ	19 Ⓐ Ⓑ Ⓒ Ⓓ Ⓔ
5 Ⓐ Ⓑ Ⓒ Ⓓ Ⓔ	10 Ⓐ Ⓑ Ⓒ Ⓓ Ⓔ	15 Ⓐ Ⓑ Ⓒ Ⓓ Ⓔ	20 Ⓐ Ⓑ Ⓒ Ⓓ Ⓔ

BE SURE TO ERASE ANY ERRORS OR STRAY MARKS COMPLETELY.

Practice Test 2

Section 1 TIME: 25 MINUTES, 20 QUESTIONS

In this section solve each problem, using any available space on the page for scratchwork. Then decide which is the best of the choices given and fill in the corresponding oval on the answer sheet.

Notes:

1. The use of a calculator is permitted. All numbers used are real numbers.
2. Figures that accompany problems in this test are intended to provide information useful in solving the problems. They are drawn as accurately as possible EXCEPT when it is stated in a specific problem that the figure is not drawn to scale. All figures lie in a plane unless otherwise indicated.
3. Unless otherwise indicated, the domain of any function f is assumed to be the set of all real numbers for which $f(x)$ is a real number.

Reference Information

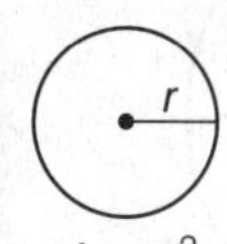
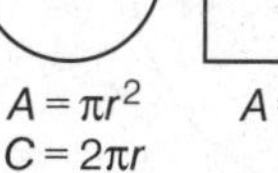
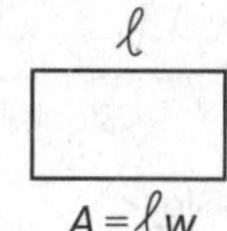
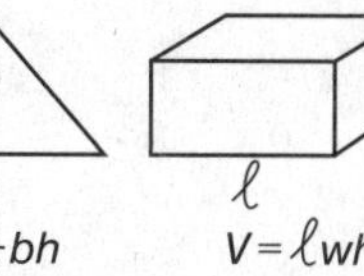
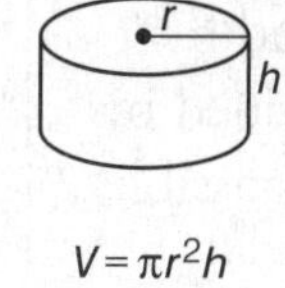

$A = \pi r^2$, $C = 2\pi r$ $A = \ell w$ $A = \frac{1}{2}bh$ $V = \ell wh$ $V = \pi r^2 h$ $c^2 = a^2 + b^2$ Special Right Triangles (30°-60°: x, $\sqrt{3}x$, $2x$; 45°-45°: s, s, $\sqrt{2}s$)

The number of degrees of arc in a circle is 360.
The sum of the measures in degrees of the angles of a triangle is 180.

1 If $(y - 1)(1 - 3) = (5 - 13)$, then $y =$

(A) -5
(B) $-\frac{5}{2}$
(C) $\frac{3}{2}$
(D) 5
(E) 6

2 The number of chirps a cricket makes increases with the temperature at a constant rate. If at a temperature of 12° Fahrenheit a cricket chirps 30 times per minute, how many times per minute will the cricket chirp when the temperature is 20° Fahrenheit?

(A) 28
(B) 32
(C) 36
(D) 48
(E) 50

GO ON TO THE NEXT PAGE

3 If one of the angles of a quadrilateral measures 75° and the degree measures of the other angles are in the ratio of 2 : 4 : 9, what is the degree measure of the largest angle of the quadrilateral?

(A) 171
(B) 168
(C) 135
(D) 105
(E) 63

4 If the average of y, $2y$, and $3y$ is 12, then $y =$

(A) 2
(B) 4
(C) 6
(D) 8
(E) 12

5 When a whole number N is divided by 2, the remainder is 0. When N is divided by 3, the remainder is 2. Which is a possible value of N?

(A) 17
(B) 30
(C) 49
(D) 53
(E) 74

6 Which of the following expressions is NOT equal to $\frac{1}{3}$?

(A) $\frac{1}{2} - \frac{1}{6}$

(B) $\frac{1}{6} \div 2$

(C) $\frac{4}{9} \times \frac{3}{4}$

(D) $\sqrt{\frac{1}{9}}$

(E) $\frac{\frac{1}{12}}{\frac{1}{4}}$

7

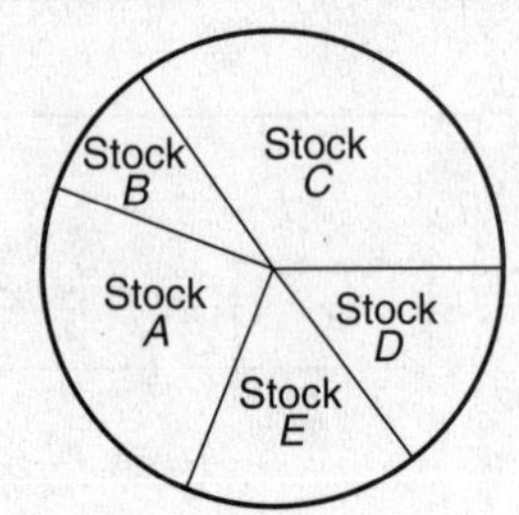

Investment of $4800 in Five Stocks

The circle graph above shows the distribution of an investment of $4800 among five different stocks. In which stock is the investment closest in value to $1200?

(A) A
(B) B
(C) C
(D) D
(E) E

GO ON TO THE NEXT PAGE

8 A class consists of 15 boys and 12 girls. On Tuesday, all the boys were present and the number of girls who were present represented 40% of the class attendance for that day. How many girls were ABSENT on Tuesday?

(A) Two
(B) Three
(C) Four
(D) Five
(E) Six

9 If $2^x = 16$ and $2x + y = 16$, then $y =$

(A) 2
(B) 4
(C) 6
(D) 8
(E) 12

10 A car averages 20 miles per gallon in city driving and 25 miles per gallon in highway driving. If gas costs $2 per gallon, how much money must be spent on gas for a 500-mile trip if $\frac{3}{5}$ of the distance consists of city driving and the remainder consists entirely of highway driving?

(A) $36
(B) $40
(C) $46
(D) $48
(E) $50

Questions 11–12.

A yellow marble, a blue marble, a red marble, a white marble, and a green marble are dropped into an empty jar, in the order given. This process is repeated until there are 88 marbles in the jar.

11 How many red marbles are in the jar?

(A) 17
(B) 18
(C) 19
(D) 20
(E) 21

12 What is the LEAST number of marbles that must be added to the jar so the jar will contain the same number of marbles of each color?

(A) 0
(B) 1
(C) 2
(D) 3
(E) 4

GO ON TO THE NEXT PAGE

13

Note: Figure is not drawn to scale.

In the figure above, what is the value of $\frac{a}{b+c}$?

(A) $\frac{1}{4}$

(B) $\frac{1}{3}$

(C) $\frac{1}{2}$

(D) $\frac{5}{8}$

(E) It cannot be determined from the information given.

14

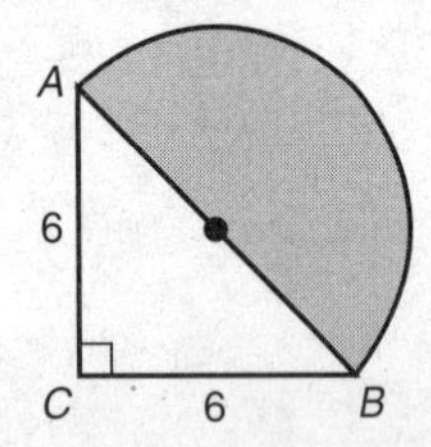

In the figure above, if arc AB is a semicircle with diameter AB, what is the area of the shaded region?

(A) 4π
(B) 9π
(C) 18π
(D) $18\sqrt{2}\,\pi$
(E) 36π

15 If $r = s^{t+1}$ and $s \neq 0$, then $\frac{r}{s} =$

(A) t
(B) s^t
(C) $t + 1$
(D) $s^t + 1$
(E) $\frac{1}{s^{t+1}}$

16 Which of the following statements must be true when $0 < a < 1$?

I. $\frac{\sqrt{a}}{a} > 1$
II. $2a < 1$
III. $a^2 - a^3 < 0$

(A) I only
(B) III only
(C) I and III only
(D) II and III only
(E) I, II, and III

17 A rectangular container with dimensions of 3 inches by 4 inches by 10 inches is filled to the top with lemonade. The lemonade is then poured into an empty cylindrical pitcher with a diameter of 8 inches. If there is no overflow, what is the height in inches of the lemonade in the pitcher?

(A) $\frac{15}{8\pi}$

(B) $\frac{15}{2\pi}$

(C) $\frac{30}{\pi}$

(D) 15

(E) $\frac{45-\pi}{15}$

GO ON TO THE NEXT PAGE

18 If $\frac{a+b}{a} = 3$ and $\frac{a+c}{c} = 2$, then $\frac{c}{b} =$

(A) $\frac{1}{2}$
(B) $\frac{2}{3}$
(C) 1
(D) $\frac{3}{2}$
(E) 2

19 In a certain sequence of numbers, each number after the first is 7 less than twice the preceding number. If the third number in the sequence is 23, what is the first number in the sequence?

(A) 9
(B) 11
(C) 13
(D) 15
(E) 17

20 Let

$$\textcircled{x} = \frac{x^2}{2} \quad \text{and} \quad \boxed{y} = y + 3$$

If $\textcircled{6} \div \boxed{6} = k$, then $\boxed{\textcircled{k}} =$

(A) 5
(B) 8
(C) 9
(D) 12
(E) 16

STOP

If you finish before time is called, you may check your work on this section only. Do not turn to any other section in the test.

Section 2 TIME: 25 MINUTES, 18 QUESTIONS

This section contains two types of questions. You have 25 minutes to complete both types. You may use any available space for scratchwork.

Notes:

1. The use of a calculator is permitted. All numbers used are real numbers.
2. Figures that accompany problems in this test are intended to provide information useful in solving the problems. They are drawn as accurately as possible EXCEPT when it is stated in a specific problem that the figure is not drawn to scale. All figures lie in a plane unless otherwise indicated.
3. Unless otherwise indicated, the domain of any function f is assumed to be the set of all real numbers for which $f(x)$ is a real number.

Reference Information

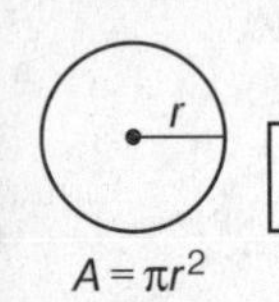

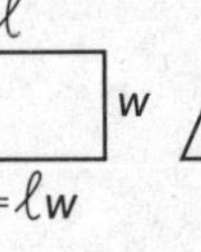

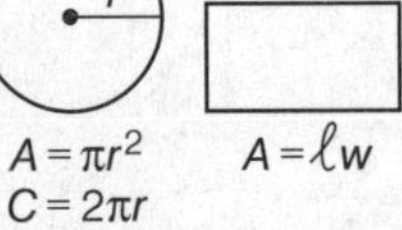

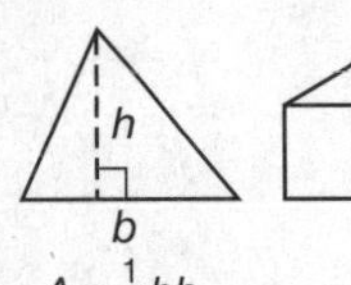

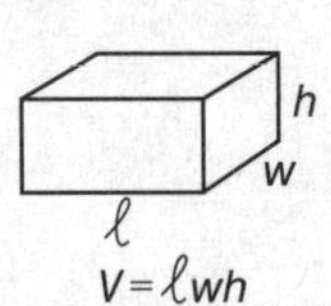

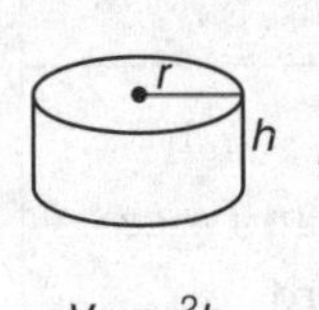

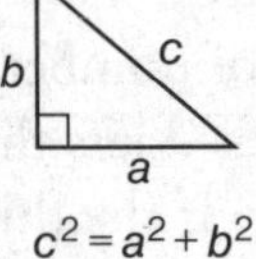
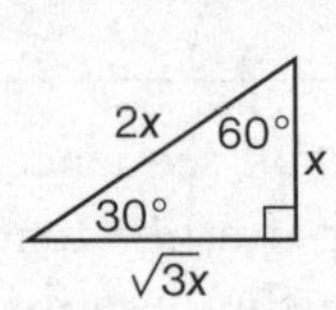

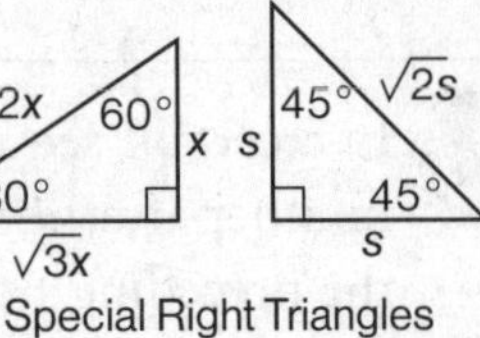

$A = \pi r^2$
$C = 2\pi r$

$A = \ell w$

$A = \frac{1}{2}bh$

$V = \ell wh$

$V = \pi r^2 h$

$c^2 = a^2 + b^2$

Special Right Triangles

The number of degrees of arc in a circle is 360.
The sum of the measures in degrees of the angles of a triangle is 180.

1 Valerie's age plus Emily's age equals 40 years. Michael's age plus Jacob's age equals 25 years. Valerie's age plus Michael's age equals 29 years. If Emily's age is 21 years, how old is Jacob?

(A) 8
(B) 10
(C) 14
(D) 15
(E) 18

2 If $2^x = \sqrt{p}$, what is 16^x in terms of p?

(A) p^2
(B) p^4
(C) $4\sqrt{p}$
(D) $8\sqrt{p}$
(E) $2p$

3 If $x^{-1} > x^{-2}$, then a possible value of x is

(A) -1
(B) $-\frac{1}{2}$
(C) $\frac{3}{4}$
(D) 1
(E) $\frac{3}{2}$

GO ON TO THE NEXT PAGE

4 A rubber ball is dropped from a height of 8 feet above the floor and is allowed to bounce up and down. At each bounce it rises 75% of the height from which it fell. What is the number of feet in the height of the ball after n bounces?

(A) $0.75(8^n)$
(B) $8(0.75)^n$
(C) $8 - (0.75)^n$
(D) $(8 \times 0.75)^n$
(E) $(8 - 0.75)^n$

5 Let x and y be numbers such that $0 < x < y < 1$, and let $d = x - y$. Which graph could represent the location of d on the number line?

(A)
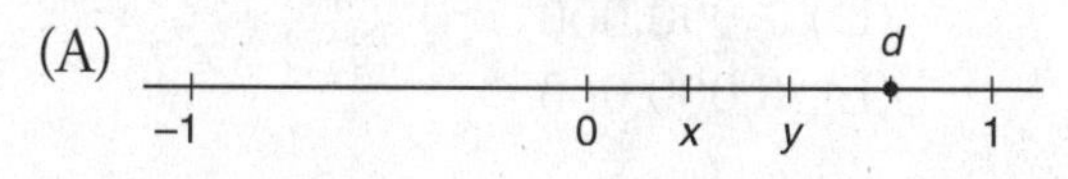

(B)
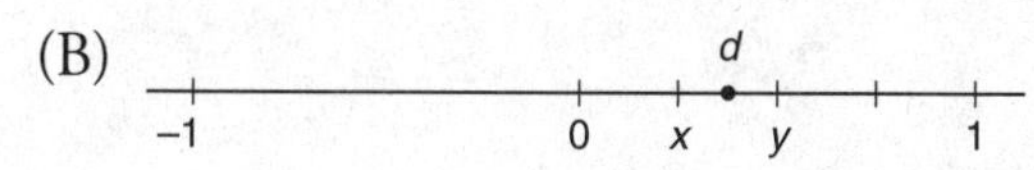

(C)
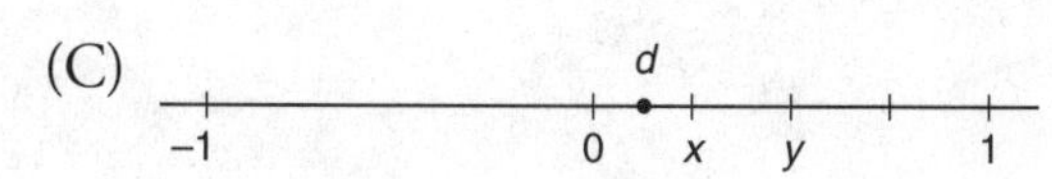

(D)
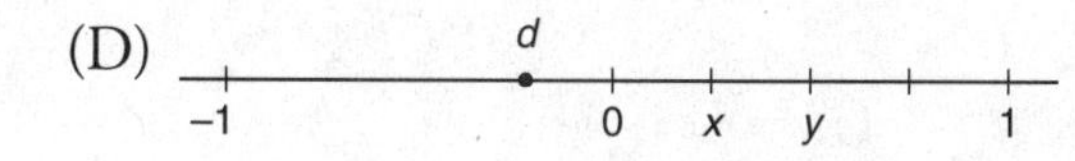

(E)
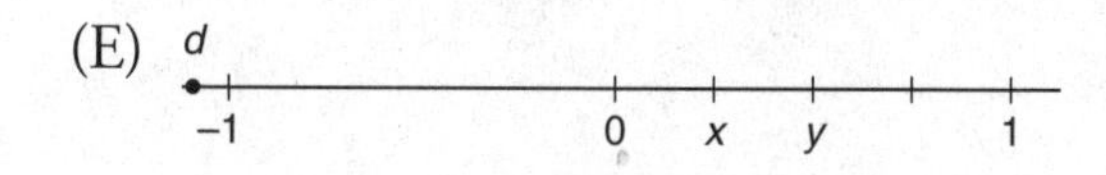

6
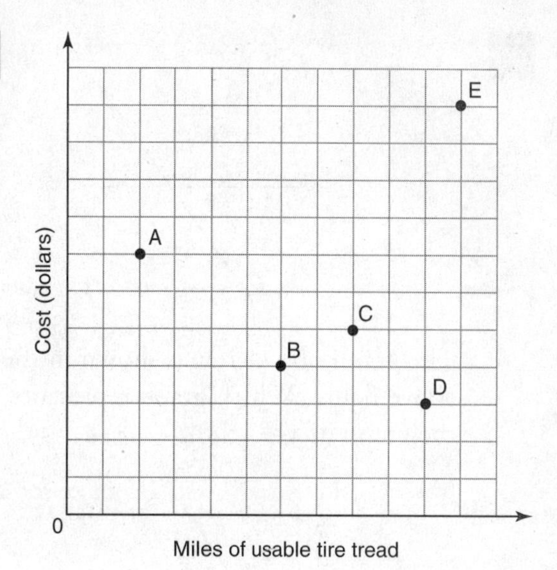

A consumer laboratory tested five competing automobile tires. For each tire brand, the number of miles the tire lasted before the tire tread became worn was plotted against the cost of the tire, as shown in the figure above. Of the five labeled data points, which one corresponds to the tire that cost the least amount per mile of useable tire tread?

(A) Brand *A*
(B) Brand *B*
(C) Brand *C*
(D) Brand *D*
(E) Brand *E*

GO ON TO THE NEXT PAGE

7

The graph of $y = f(x)$ is shown in the above figure. Which graph represents the graph of $y = |f(x)|$?

(A)

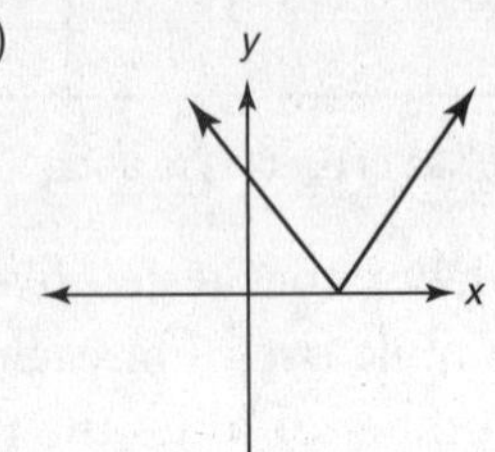

(B)

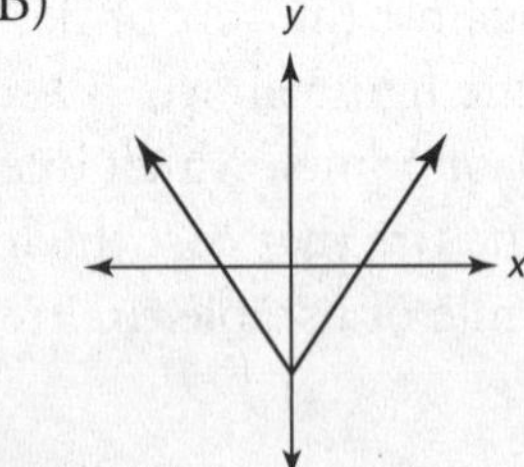

(C)

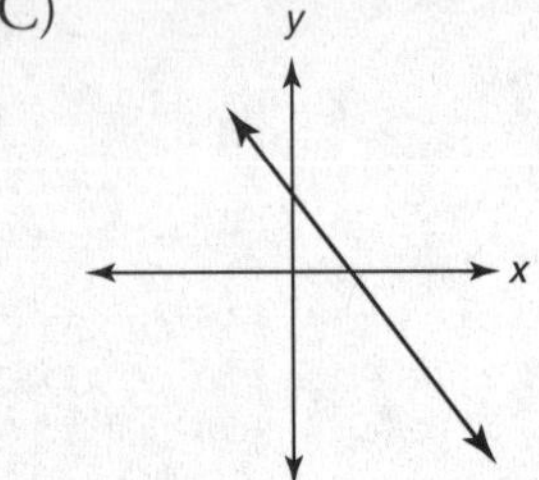

(D)

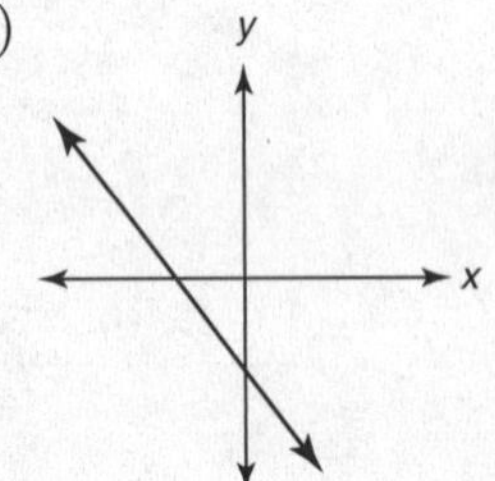

(E)

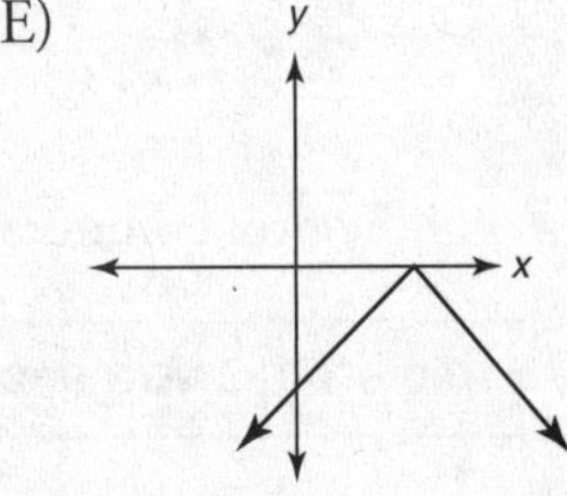

8 A telephone company has run out of seven-digit telephone numbers for a certain area code. To fix the problem, the telephone company will introduce a new area code. What is the maximum number of new seven-digit telephone numbers that will be generated for the new area code if both of the following conditions must be met?

Condition 1: The first digit cannot be a zero or a one.

Condition 2: The first three digits cannot be the emergency number 911 or the information number 411.

(A) 7,940,000
(B) 7,960,000
(C) 7,980,000
(D) 7,990,000
(E) 8,000,000

GO ON TO THE NEXT PAGE

Directions for Student-Produced Response Questions

Each of the remaining 10 questions (9–18) requires you to solve the problem and enter your answer by marking the ovals in the special grid, as shown in the examples below.

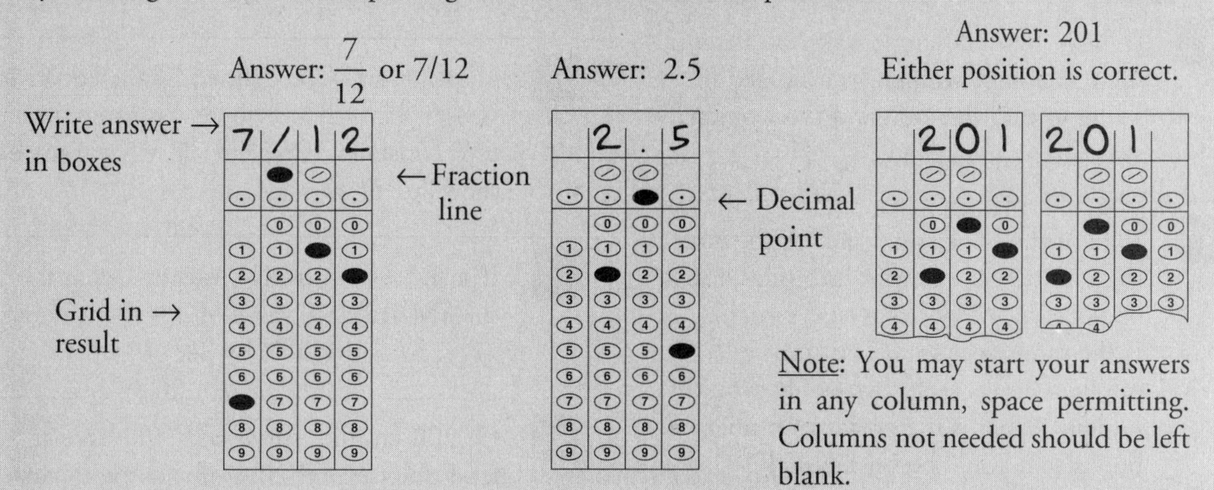

Note: You may start your answers in any column, space permitting. Columns not needed should be left blank.

- Mark no more than one oval in any column.
- Because the answer sheet will be machine-scored, **you will receive credit only if the ovals are filled in correctly.**
- Although not required, it is suggested that you write your answer in the boxes at the top of the columns to help you fill in the ovals accurately.
- Some problems may have more than one correct answer. In such cases, grid only one answer.
- No question has a negative answer.
- **Mixed numbers** such as $2\frac{1}{2}$ must be gridded as 2.5 or 5/2. (If 2 1 / 2 is gridded, it will be interpreted as $\frac{21}{2}$ not $2\frac{1}{2}$.)
- Decimal Accuracy: If you obtain a decimal answer, **enter the most accurate value the grid will accommodate.** For example, if you obtain an answer such as 0.6666 . . . , you should record the result as .666 or .667. **Less accurate values such as .66 or .67 are not acceptable.**

Acceptable ways to grid $\frac{2}{3}$ = .6666 . . .

2 / 3   . 6 6 6   . 6 6 7

9

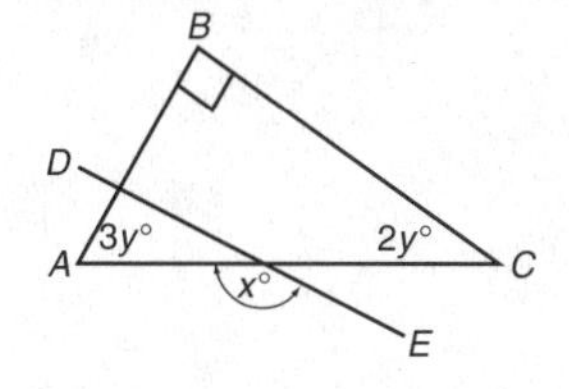

In the figure above, if $BC \parallel DE$, what is the value of x?

10

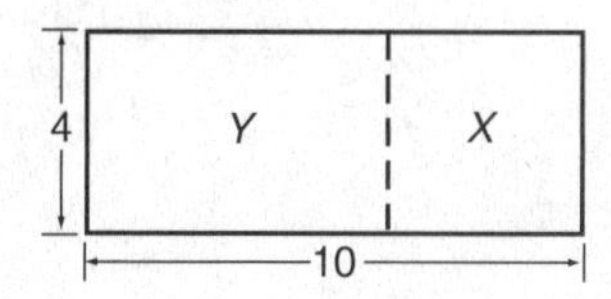

In the figure above, when square X is folded back along the broken line, what will be the area of the part of rectangle Y that is NOT covered by square X?

GO ON TO THE NEXT PAGE

11 The average of 14 scores is 80. If the average of four of these scores is 75, what is the average of the remaining 10 scores?

12 In how many different ways can three men and three women be arranged in a line so that the men and the women alternate?

13 If all of the books on a shelf with fewer than 45 books were put into piles of five books each, no books would remain. If the same books were put into piles of seven books each, two books would remain. What is the greatest number of books that could be on the shelf?

14

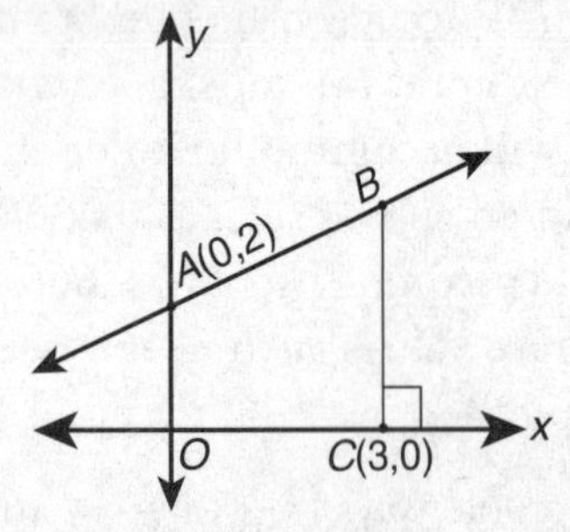

If the slope of line AB in the figure above is $\frac{1}{3}$, what is the area of quadrilateral $OABC$?

15 The ratio of boys to girls in a room is 2 to 3. After three boys leave the room, the ratio of boys to girls becomes 1 to 2. How many girls are in the room?

16 Let function f be defined by the equation $f(x) = kx + 2^{x-1}$, where k is some nonzero constant. If $f(5) = 31$, what must be the value of k?

17 If a and b are positive integers, what is the SMALLEST value of $a + b$ for which $2a + 3b$ is divisible by 24?

18 Of 200 families surveyed, 75% have at least one car and 20% of those with cars have more than two cars. If 50 families each have exactly two cars, how many families have exactly one car?

STOP

If you finish before time is called, you may check your work on this section only. Do not turn to any other section in the test.

Section 3 TIME: 20 MINUTES, 16 QUESTIONS

In this section solve each problem, using any available space on the page for scratchwork. Then decide which is the best of the choices given and fill in the corresponding oval on the answer sheet.

Notes:

1. The use of a calculator is permitted. All numbers used are real numbers.
2. Figures that accompany problems in this test are intended to provide information useful in solving the problems. They are drawn as accurately as possible EXCEPT when it is stated in a specific problem that the figure is not drawn to scale. All figures lie in a plane unless otherwise indicated.
3. Unless otherwise indicated, the domain of any function f is assumed to be the set of all real numbers for which $f(x)$ is a real number.

Reference Information

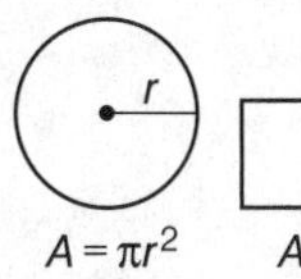

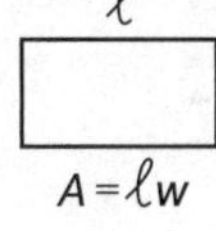

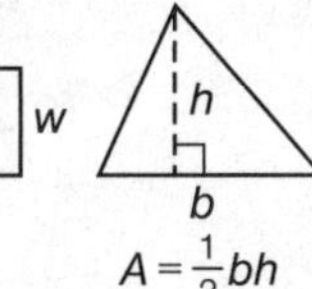

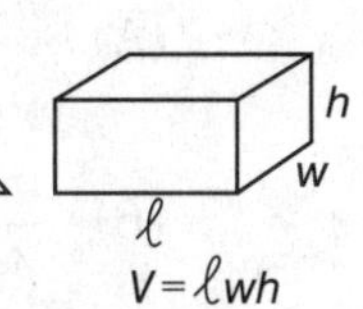

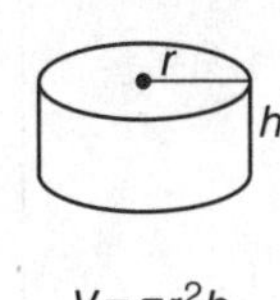

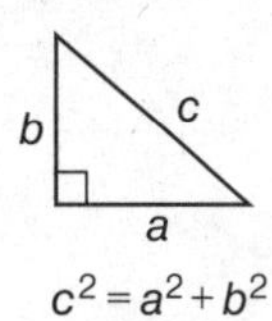

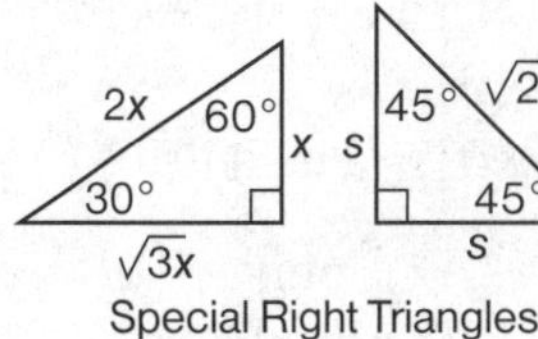

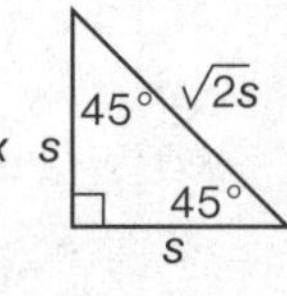

Special Right Triangles

The number of degrees of arc in a circle is 360.
The sum of the measures in degrees of the angles of a triangle is 180.

1

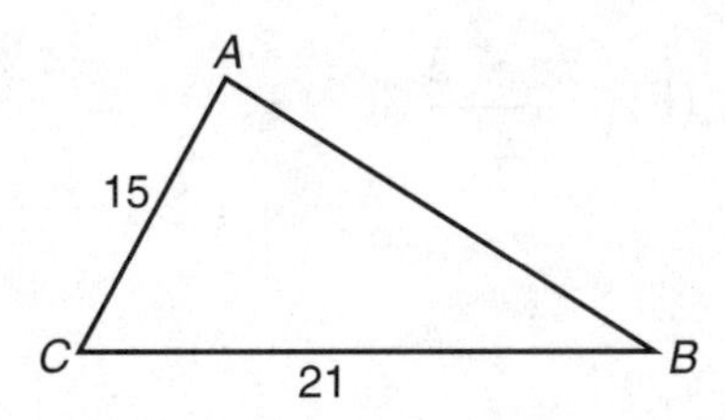

In the figure above, if the length of line segment AB is $\frac{1}{4}$ of the perimeter of $\triangle ABC$, what is the length of AB?

(A) 6
(B) 8
(C) 9
(D) 12
(E) 18

2 If $x^2 + 3y = x^2 - 3$, then $y =$

(A) -1
(B) 0
(C) 1
(D) $\sqrt{x}$
(E) $\frac{-\sqrt{x}}{3}$

3 If $a = b \div c$, where $b, c \neq 0$, then, when b is multiplied by 2 and c is divided by 2, a is

(A) unchanged
(B) decreased by 4
(C) increased by 4
(D) multiplied by 4
(E) divided by 4

GO ON TO THE NEXT PAGE

4 The average of a, b, and c is 3 times the median. If $0 < a < b < c$, what is the average of a and c in terms of b?

(A) $2b$
(B) $\frac{5}{2}b$
(C) $3b$
(D) $\frac{7}{2}b$
(E) $4b$

5 If $x - 1$ represents an odd integer, then which of the following must represent an even integer?

I. $x + 1$
II. $3x$
III. $(x - 1)^2$

(A) I only
(B) II only
(C) III only
(D) I and III only
(E) II and III only

6

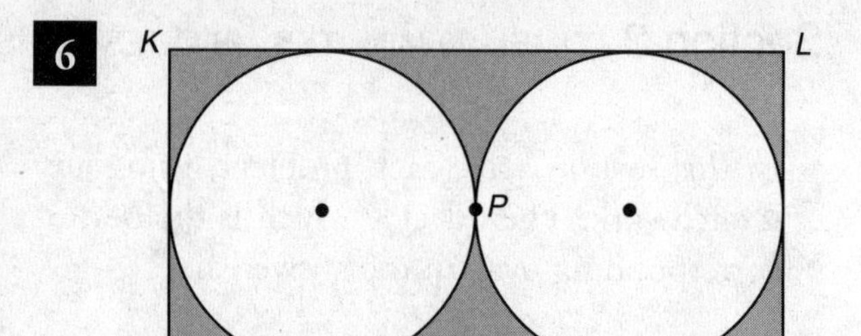

In the figure above, rectangle *JKLM* is circumscribed about two circles tangent to each other at point *P*. The radius length of each circle is 2. If a point is picked at random from the interior of the rectangle, what is the probability that the point chosen will lie in the shaded region?

(A) $\frac{\pi}{8}$
(B) $\frac{\pi}{4}$
(C) $1 - \frac{\pi}{4}$
(D) $1 - \frac{\pi}{8}$
(E) $\frac{\pi - 2}{4}$

GO ON TO THE NEXT PAGE

7

In the figure above, if $AB = AC$ and $AC \parallel BD$, what is the value of x?

(A) 45
(B) 50
(C) 60
(D) 65
(E) 70

8

$$\begin{array}{r} 3A.B5 \\ +\quad 3.AB \\ \hline C1.C1 \end{array}$$

In the correctly worked out addition problem above, if A, B, and C represent nonzero digits, what digit does A represent?

(A) 3
(B) 5
(C) 6
(D) 7
(E) 8

9

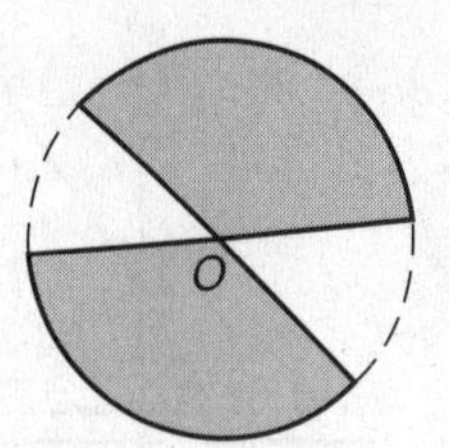

In the figure above, the sum of the lengths of the broken arcs of circle O is 2π. If the diameter of the circle is 18, what is the area of the shaded region?

(A) 9π
(B) 18π
(C) 45π
(D) 63π
(E) 72π

10 George spent 25% of the money he had on lunch and 60% of the remaining money on dinner. If he then had \$9.00 left, how much money did he spend on lunch?

(A) \$2.50
(B) \$4.00
(C) \$5.00
(D) \$7.50
(E) \$10.00

11 In the figure above, which of the following ordered pairs could be the coordinates (a, b) of point S?

(A) (2,9)
(B) (2,11)
(C) (3,9)
(D) (4,8)
(E) (5,10)

---

12 If 3 times 1 less than a number n is the same as twice the number increased by 14, what is n?

(A) 15
(B) 17
(C) 19
(D) 21
(E) 23

---

13 If the length of line segment AB is 9 inches, the total number of points 4 inches from point A and 6 inches from point B is

(A) one
(B) two
(C) three
(D) four
(E) It cannot be determined from the information given.

14

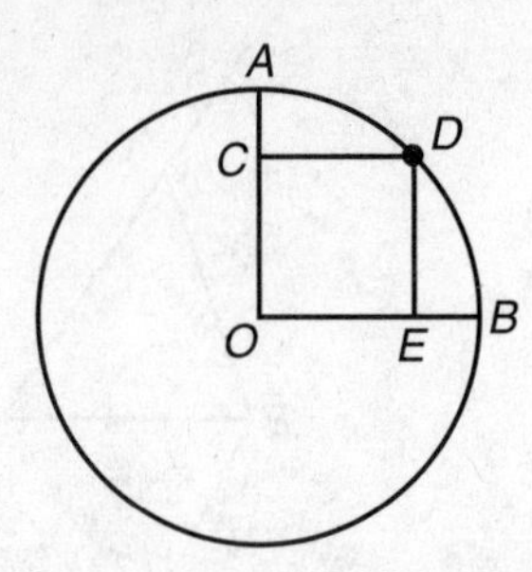

The circle in the figure above has center O. If the area of square $OCDE$ is 2, what is the length of arc ADB?

(A) 4π
(B) 2π
(C) $\sqrt{2\pi}$
(D) π
(E) It cannot be determined from the information given.

---

15 When report cards were distributed in a certain homeroom class, it was found that 13 students received at least one grade of A, 15 students received at least one grade of B, and 6 students received both As and Bs. If one-third of the students in the homeroom class received no As and no Bs, how many students are in the homeroom class?

(A) 27
(B) 30
(C) 33
(D) 36
(E) 39

---

16 If $g(x) = x \cdot 2^x$, then $g(a + 1) - g(a) =$

(A) $(a + 2)2^a$
(B) $(2a + 1)2^a$
(C) $(2a - 1)2^a$
(D) $(a + 1)2^{a+1}$
(E) $(a)2^{a+1}$

STOP

If you finish before time is called, you may check your work on this section only. Do not turn to any other section in the test.

Answer Key

PRACTICE TEST 2

Note: The number inside the brackets that follows each answer is the number of the lesson in Chapter 3, 4, 5, 6, 7, or 8 that relates to the topic or concept that the question tests. In some cases two lessons are relevant, and both numbers are given.

Section 1

1. D [4-1]
2. E [5-4]
3. A [6-4]
4. C [7-1]
5. E [3-3]
6. B [3-5]
7. A [7-4]
8. A [3-8]
9. D [4-1]
10. C [3-7]
11. B [3-3]
12. C [3-3]
13. C [6-2]
14. B [6-2, 6-6]
15. B [4-2]
16. A [3-5, 4-4]
17. B [6-7]
18. A [4-3]
19. B [7-5]
20. A [7-5]

Section 2

1. D [4-1]
2. A [8-2]
3. E [8-1]
4. B [8-5]
5. D [3-4]
6. D [8-5]
7. A [8-4]
8. C [7-2]
9. 144 [6-1]
10. 8 [6-5]
11. 82 [7-1]
12. 72 [7-2]
13. 30 [3-3]
14. 7.5 [6-8]
15. 18 [5-4]
16. 3 [8-4]
17. 9 [3-3]
18. 70 [3-8]

Section 3

1. D [6-5]
2. A [4-1]
3. D [3-6, 4-3]
4. E [7-1]
5. B [3-1]
6. C [7-3]
7. D [6-1]
8. D [7-5]
9. E [6-6]
10. D [5-2]
11. C [6-8]
12. B [5-1]
13. B [6-6]
14. D [6-6]
15. C [7-2]
16. A [8-4]

Self-Scoring Chart

SECTION 1 *(20 Questions)*

Number Correct ________ (*a*)
Number Incorrect ________ (*b*)
Number Omitted ________ (*c*)
(*a*) minus $\frac{1}{4}$ (*b*) = **Raw Score I** ________

SECTION 2 *(18 Questions)*

Part 1 Questions 1–8

Number Correct ________ (*d*)
Number Incorrect ________ (*e*)
(*d*) minus $\frac{1}{4}$ (*e*) = **Raw Score II** ________

Part 2 Questions 9–18

Number Correct **Raw Score III** ________

SECTION 3 *(16 Questions)*

Number Correct ________ (*f*)
Number Incorrect ________ (*g*)
Number Omitted ________ (*h*)
(*f*) minus $\frac{1}{4}$ (*g*) = **Raw Score IV** ________

Total Score = Raw Score I + II + III + IV ________

Evaluation Chart

50–54	Superior
45–49	Very Good
40–44	Good
35–39	Above Average
30–34	Average
20–29	Below Average
0–19	Below Average; needs intensive study

ANSWER EXPLANATIONS FOR PRACTICE TEST 2

Section 1

1. **(D)** If $(y-1)(1-3) = (5-13)$, then
$$(y-1)(-2) = (-8)$$
$$y - 1 = \frac{-8}{-2} = 4$$
$$y = 4 + 1 = 5$$

2. **(E)** If x represents the number of times per minute that the cricket will chirp when the temperature is 20° Fahrenheit, then
$$\frac{\text{Temperature}}{\text{Chirps}} = \frac{12}{30} = \frac{20}{x}$$
$$12x = 20(30) = 600$$
$$x = \frac{600}{12} = 50$$

3. **(A)** If three angles are in the ratio of 2 : 4 : 9, their measures can be represented as $2x$, $4x$, and $9x$. Since the measures of the four angles of a quadrilateral must add up to 360°,
$$2x + 4x + 9x + 75 = 360$$
$$15x + 75 = 360$$
$$15x = 285$$
$$x = \frac{285}{15} = 19$$
Since $9x = 9(19) = 171$, the largest angle of the quadrilateral measures 171°.

4. **(C)** Since the average of the three quantities y, $2y$, and $3y$ is 12,
$$\frac{y + 2y + 3y}{3} = 12$$
$$\frac{6y}{3} = 12$$
$$2y = 12$$
$$y = \frac{12}{2} = 6$$

5. **(E)** The question states that, when a whole number N is divided by 2, the remainder is 0. Therefore, N must be even, and you can eliminate choices (A), (C), and (D). You are also told that, when N is divided by 3, the remainder is 2. Test each of the remaining choices:
   - (B) $30 \div 3 = 10$ remainder of 0
   - (E) $74 \div 3 = 24$ remainder of 2

   A possible value of N is 74.

6. **(B)** To determine which choice is not equal to $\frac{1}{3}$, do the arithmetic for each choice until you find a result that is not $\frac{1}{3}$.
   - (A) $\frac{1}{2} - \frac{1}{6} = \frac{3}{6} - \frac{1}{6} = \frac{2}{6} = \frac{1}{3}$
   - (B) $\frac{1}{6} \div 2 = \frac{1}{6} \times \frac{1}{2} = \frac{1}{12}$

7. **(A)** The stock investment that is closest in value to \$1200 will correspond to the sector of the circle graph that is $\frac{\$1200}{\$4800}$ or $\frac{1}{4}$ of the circle. Since all the sectors of a circle must add up to 360°, look for the sector whose sides form an angle that measures approximately $\frac{1}{4} \times 360°$ or 90°. This sector is labeled stock A.

8. **(A)** If x girls were present, then
$$x = 0.40(x + 15)$$
$$= 0.40x + 6$$
$$0.60x = 6$$
$$x = \frac{6}{0.60} = 10$$
Since 10 of the 12 girls were present on Tuesday, 12 − 10 or 2 girls were absent on that day.

9. **(D)** Since $2^x = 16 = 2^4$, then $x = 4$. Hence, $2x + y = 16$ becomes
$$2(4) + y = 16$$
$$8 + y = 16$$
$$y = 16 - 8 = 8$$

10. **(C)** Since
$$\frac{3}{5} \text{ of } 500 = \frac{3}{\cancel{5}} \times \overset{100}{\cancel{500}} = 3 \times 100 = 300$$
there are 300 miles of city driving and 500 − 300 or 200 miles of highway driving. In city

Practice Test 2

driving, the car averages 20 miles per gallon, so the car uses $\frac{300}{20}$ or 15 gallons of gas during city driving. You are also given that the car averages 25 miles per gallon in highway driving, so the car uses $\frac{200}{25}$ or 8 gallons of gas during this part of the trip. Thus, a total of 15 + 8 or 23 gallons of gas are used for the 500-mile trip. Since gas costs $2 per gallon, the cost of the gas for the whole trip is 23 × $2 or $46.

11. **(B)** Marbles of five different colors are dropped into the jar in a fixed order: yellow, blue, red, white, and then green. Since 88 ÷ 5 = 17 with a remainder of 3, the jar contains 17 of each color marble *plus* three additional marbles—one yellow, one blue, and one red marble. Hence, there are 17 + 1 or 18 red marbles in the jar.

12. **(C)** Since the jar contains 18 yellow, blue, and red marbles, but only 17 white and green, one white and one green marble or two marbles must be added so the jar will contain the same number of marbles of each color.

13. **(C)** Since the three angles of the triangle have equal measures, the three sides of the triangle have equal lengths. Suppose $a = b = c = 2$:

$$\frac{a}{b+c} = \frac{2}{2+2} = \frac{2}{4} = \frac{1}{2}$$

14. **(B)** In any isosceles right triangle the hypotenuse is $\sqrt{2}$ times the length of a leg. Since, in the figure, hypotenuse $AB = 6\sqrt{2}$, the radius of semicircle AB is $3\sqrt{2}$. Hence:

$$\begin{aligned}\text{Area semicircle } AB &= \frac{1}{2}\left[\pi\left(3\sqrt{2}\right)^2\right]\\ &= \frac{1}{2}\left[\pi\left(3\sqrt{2}\right)\times\left(3\sqrt{2}\right)\right]\\ &= \frac{1}{2}[\pi(9\times 2)]\\ &= \frac{1}{2}(18\pi)\\ &= 9\pi\end{aligned}$$

15. **(B)** If $r = s^{t+1}$ and $s \neq 0$, then

$$\frac{r}{s} = \frac{s^{t+1}}{s^1} = s^{(t+1)-1} = s^t$$

16. **(A)** Determine whether each Roman numeral statement is always true when $0 < a < 1$.
    - I. Since $0 < a < 1$, then $\sqrt{a} > a$. For example, if $a = 0.25$, then $\sqrt{0.25} = 0.5$ and $0.5 > 0.25$. Dividing the inequality by a produces the equivalent inequality $\frac{\sqrt{a}}{a} > 1$. Statement I is always true.
    - II. If $2a < 1$, then $a < \frac{1}{2}$, which is not necessarily true since you are told only that $0 < a < 1$. Statement II is not always true.
    - III. Since $0 < a < 1$, $a^2 > a^3$, so $a^2 - a^3 > 0$. Statement III is always false.

    Only Roman numeral statement I is always true.

17. **(B)** A rectangular container with dimensions of 3 inches by 4 inches by 10 inches can hold 3 × 4 × 10 or 120 cubic inches of lemonade. If h represents the number of inches in the height of the lemonade after it is poured into a cylindrical pitcher with a diameter of 8 inches, then $\pi r^2 h = 120$, where $r = \frac{8}{2} = 4$. Since the volume of the lemonade remains the same,

$$\begin{aligned}\pi(4)^2 h &= 120\\ 16\pi h &= 120\\ h &= \frac{120}{16\pi} = \frac{15}{2\pi}\end{aligned}$$

18. **(A)** Simplify each of the given fractions:

$$\frac{a+b}{a} = \frac{a}{a} + \frac{b}{a} = 1 + \frac{b}{a} = 3, \text{ so } \frac{b}{a} = 2$$

$$\frac{a+c}{c} = \frac{a}{c} + \frac{c}{c} = \frac{a}{c} + 1 = 2, \text{ so } \frac{a}{c} = 1$$

Eliminate a by multiplying corresponding sides of $\frac{b}{a} = 2$ and $\frac{a}{c} = 1$ together:

$$\frac{b}{a} \times \frac{a}{c} = 2 \times 1$$

$$\frac{b}{c} = 2$$

Since $\frac{b}{c} = 2$, then $\frac{c}{b} = \frac{1}{2}$.

19. **(B)** You are told that, in a certain sequence of numbers, each number after the first is 7 less than twice the preceding number, and that the third number in the sequence is 23. For each answer choice, find the third term of the sequence until you get 23 as the third term.
    - (A) The second term $= 2 \times 9 - 7 = 18 - 7 = 11$, and the third term $= 2 \times 11 - 7 = 22 - 7 = 15$.
    - (B) The second term $= 2 \times 11 - 7 = 22 - 7 = 15$, and the third term $= 2 \times 15 - 7 = 30 - 7 = 23$.

20. **(A)** Since

$$\textcircled{x} = \frac{x^2}{2} \quad \text{and} \quad \boxed{y} = y + 3$$

then

$$k = \textcircled{6} \div \boxed{6} = \frac{6^2}{2} \div (6 + 3)$$

$$= 18 \div 9$$

$$= 2$$

Thus,

$$\textcircled{k} = \boxed{\frac{2^2}{2}} = \boxed{2} = 2 + 3 = 5$$

Section 2

1. **(D)** Valerie + Emily = 40, Michael + Jacob = 25, and Valerie + Michael = 29.
    - Since Emily is 21 years old, Valerie is 40 − 21 = 19 years old.
    - Hence, Michael is 29 − 19 = 10 years old.
    - Thus, Jacob is 25 − 10 = 15 years old.

2. **(A)** $16^x = (2^4)^x = (2^x)^4 = \left(\sqrt{p}\right)^4 = \left(p^{\frac{1}{2}}\right)^4 = p^2$.

3. **(E)** If $x^{-1} > x^{-2}$, then $\frac{1}{x} > \frac{1}{x^2}$. This inequality is false when x is negative or when $x = 1$, so eliminate choices (A), (B), and (D). When $x > 1$, $x^2 > x$, so $\frac{1}{x} > \frac{1}{x^2}$, which means choice (C) is incorrect and choice (E) is correct. You can verify that choice (E) is correct since, if $x = \frac{3}{2}$:
    - the reciprocal of x is $\frac{2}{3}$
    - $x^2 = \frac{9}{4}$ and the reciprocal of x^2 is $\frac{4}{9}$
    - $\frac{2}{3}\left(=\frac{6}{9}\right) > \frac{4}{9}$

4. **(B)** The height of the ball after:
    - one bounce is 0.75(8)
    - two bounces is $0.75[0.75(8)] = 8(0.75)^2$
    - three bounces is $0.75[8(0.75)^2] = 8(0.75)^3$
    - n bounces is $8(0.75)^n$

5. **(D)** Since $0 < x < y < 1$, x is less than y and both are positive, so $d = x - y < 0$. Eliminate choices (A), (B), and (C). Pick easy numbers for x and y. If $x = \frac{1}{3}$ and $y = \frac{2}{3}$, $d = x - y = -\frac{1}{3}$, which means that d must lie between −1 and 0.

6. **(D)**

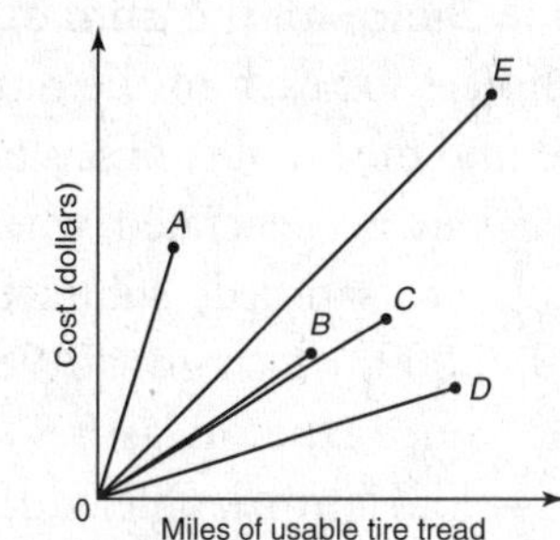

Draw line segments $\overline{OA}$, $\overline{OB}$, $\overline{OC}$, $\overline{OD}$, and $\overline{OE}$. The slope of each of these segments represents the cost of each tire per

Practice Test 2

mile of useable tire tread. Line segment $\overline{OD}$ has the slope with the smallest value.

7. **(A)** The graph of $y = |f(x)|$ can be obtained from the graph of $y = f(x)$ by reflecting in the x-axis the portion of the graph of $y = f(x)$ that falls below the x-axis, where y is negative. As a result, no points on the new graph will have points with negative y-values. The reflected graph is shown in choice (A).
8. **(C)** There are 10 digits from 0 to 9.
   - Find the number of seven-digit telephone numbers that can be generated that satisfy the first condition. If the first digit of a seven-digit telephone number cannot be a zero or a one, then there are only 8 digits that can be used for the first digit. Without considering the second condition, each of the remaining digits can be filled with any of the 10 digits. Hence, a seven-digit number that satisfies condition 1 can be formed by

$$\boxed{8} \times \boxed{10} \times \boxed{10} \times \boxed{10} \times \boxed{10} \times \boxed{10} \times \boxed{10} =$$

$$8 \times 10^6 = 8{,}000{,}000.$$

   - Now consider the second condition. The 8,000,000 numbers include numbers that begin with 911 or 411. The number of seven-digit numbers that begin with 911 or 411 is

$$\boxed{2} \times \boxed{1} \times \boxed{1} \times \boxed{10} \times \boxed{10} \times \boxed{10} \times \boxed{10} =$$

$$2 \times 10^4 = 20{,}000.$$

   because there are exactly two choices for the first digit (9 or 4), exactly one choice for the second digit (1), exactly one choice for the third digit (1), and 10 choices for each of the remaining digits.
   - To find the number of seven-digit numbers that can be generated when both conditions are satisfied, subtract 20,000 from 8,000,000, which gives 7,980,000.
9. **(144)** Since the measures of the acute angles of a right triangle add up to 90,

$$3y + 2y = 90$$
$$5y = 90$$
$$y = \frac{90}{5} = 18$$

Hence, in right triangle ABC, $\angle C = 2y = 2(18) = 36$. An acute and an obtuse angle formed by parallel lines are supplementary. Since $BC \parallel DE$, $x + C = 180$, so

$$x = 180 - 36 = 144$$

10. **(8)** Since the height of rectangle Y is 4, the length of each side of square X is also 4. Then the base of rectangle Y is $10 - 4$ or 6. Thus, when square X is folded back along the broken line, $6 - 4$ or 2 units of the base of rectangle Y will not be covered by square X. The area of the part of rectangle Y that is not covered by square X is 2×4 or 8.
11. **(82)** If the average of 14 scores is 80, the sum of the 14 scores is 14×80 or 1120. If the average of four of these scores is 75, the sum of these four scores is 300, so the sum of the remaining 10 scores is $1120 - 300$ or 820. The average of these 10 scores is $\frac{820}{10}$ or 82.
12. **(72)** In the six places on the line, the three men can be arranged in the first, third, and fifth positions in $3 \times 2 \times 1$ or 6 different ways. The three women can be arranged in the second, fourth, and sixth positions in $3 \times 2 \times 1$ or 6 different ways. Hence, the three men and the three women can be arranged so that they alternate, with a man in the first position, in 6×6 or 36 different ways. Similarly, the three men and three women can be arranged so that they alternate, with a woman in the first position, in 6×6 or 36 different ways. Hence, the total number of different ways three men and three women can be arranged in a line so that they alternate is $36 + 36$ or 72.
13. **(30)** You are looking for the greatest positive integer less than 45 that, when divided by 5, gives a remainder of 0 and that, when divided by 7, gives a remainder of 2. List all multiples of 5 less than 45, and pick the largest one that, when divided by 7, gives a remainder of 2.
   - The multiples of 5 less than 45 are 40, 35, 30, 25, 20, 15, 10, and 5.
   - $40 \div 7$ gives a remainder of 5.
     $35 \div 7$ gives a remainder of 0.
     $30 \div 7$ gives a remainder of 2.

   Hence, the greatest number of books that could be on the shelf is 30.

14. (**7.5**) *Solution 1:* Quadrilateral $OABC$ is a trapezoid so its area is equal to the height of the trapezoid times half the sum of the bases (parallel sides). Before you can find the area of trapezoid $OABC$, you need to know the coordinates of point B.
   - Since segment BC is perpendicular to the x-axis, points B and C have the same x-coordinate, 3.
   - The slope of the line through $A(0,2)$ and $B(3, y)$ is given as $\frac{1}{3}$. Hence:

$$\text{Slope of } AB = \frac{\text{Change in } y}{\text{Change in } x}$$

$$\frac{1}{3} = \frac{y-2}{3-0}$$

$$\frac{1}{3} = \frac{y-2}{3} \quad \text{so } y - 2 = 1 \text{ and } y = 3$$

   - Thus, the coordinates of B are (3,3) so $OC = 3$, $AO = 2$, and $BC = 3$.

$$\text{Area trapezoid } OABC = OC \times \frac{AO + BC}{2}$$

$$= 3 \times \frac{2+3}{2}$$

$$= \frac{15}{2} \text{ or } 7.5$$

Solution 2: Divide quadrilateral $OABC$ into a right triangle and a rectangle by drawing a horizontal line from A to side BC:

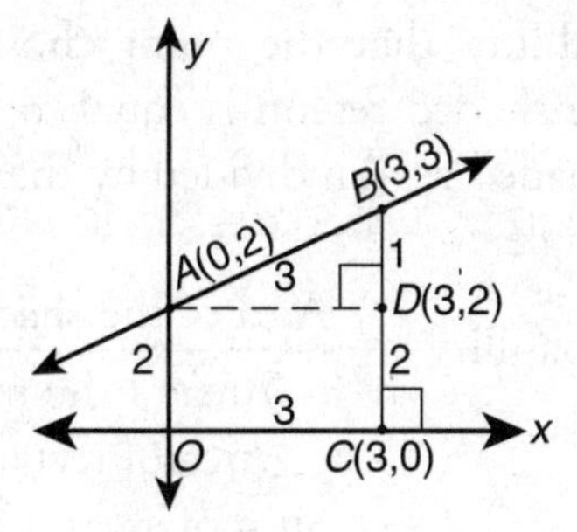

Then:

$$\text{Area of quad } OABC = \text{Area of right } \triangle ADB + \text{Area of rectangle } OADC$$

$$= \frac{1}{2}(BD)(AD) + (OC \times CD)$$

$$= \frac{1}{2}(1)(3) + (3 \times 2)$$

$$= 1.5 + 6$$

$$= 7.5$$

15. (**18**) Since the ratio of boys to girls in a room is 2 to 3, let $2x$ and $3x$ equal the numbers of boys and girls, respectively, in the room. After three boys leave the room, the ratio of boys to girls becomes 1 to 2. Set up a proportion:

$$\frac{\text{Boys}}{\text{Girls}} = \frac{2x-3}{3x} = \frac{1}{2}$$

$$3x = 2(2x-3)$$
$$= 4x - 6$$
$$6 = x$$

Since $3x = 3(6) = 18$, there are 18 girls in the room.

16. (**3**) If $f(x) = kx + 2^{x-1}$ and $f(5) = 31$, then let $x = 5$ and set the corresponding function value equal to 31:

$$f(5) = k(5) + 2^{5-1} = 31$$
$$5k + 2^4 = 31$$
$$5k + 16 = 31$$
$$5k = 31 - 16$$
$$k = \frac{15}{5} = 3$$

17. (**9**) List all positive integer values for a and b that make $2a + 3b = 24$. Then pick the pair of values for which $a + b$ has the smallest value.

a	$b = \frac{24-2a}{3}$	$a + b$
1	$\frac{22}{3}$	
2	$\frac{20}{3}$	
3	$\frac{18}{3}$ or 6	$3 + 6 = 9$
4	$\frac{16}{3}$	
5	$\frac{14}{3}$	
6	$\frac{12}{3}$ or 4	$6 + 4 = 10$
7	$\frac{10}{3}$	
8	$\frac{8}{3}$	

There is no need to go further since the next value of a is 9, which is the smallest value of $a + b$ obtained thus far.

18. (**70**) You are given that 75% of the 200 families surveyed have at least one car, so 0.75×200 or 150 families have at least one car. Since 20% of those with cars have more than two cars, 0.20×150 or 30 families have more than two cars. If 50 families each have exactly two cars, then $150 - (50 + 30)$ or 70 families have exactly one car.

Section 3

1. (**D**) Let x represent the length of AB. Since AB is $\frac{1}{4}$ of the perimeter of $\triangle ABC$,

$$x = \frac{1}{4}(15 + 21 + x) = \frac{1}{4}(36 + x)$$
$$4x = 36 + x$$
$$3x = 36$$
$$x = \frac{36}{6} = 12$$

2. (**A**) If $x^2 + 3y = x^2 - 3$, then subtracting x^2 from each side of the equation gives

$$3y = -3 \quad \text{or} \quad y = \frac{-3}{3} = -1$$

3. (**D**) *Solution 1*: If $a = b \div c$, then $a = \frac{b}{c}$. When b is multiplied by 2 and c is divided by 2,

$$a_{\text{new}} = 2b \div \frac{c}{2} = 2b \times \frac{2}{c} = 4\frac{b}{c} = 4a$$

Hence, a is multiplied by 4.
Solution 2: Plug in easy-to-work-with numbers for b and c. When $b = 8$ and $c = 2$, then $a = 8 \div 2 = 4$. Multiplying b by 2 gives 16, and dividing c by 2 gives 1 so, $a_{\text{new}} = 16 \div 1 = 16$. Since 16 is 4 times as great as 4, a is multiplied by 4.

4. (**E**) If $0 < a < b < c$, then the median or middle value of a, b, and c is b. Since the average of a, b, and c is 3 times the median,

$$\frac{a + b + c}{3} = 3b$$
$$a + b + c = 9b$$
$$a + c = 8b$$

Since $\frac{a + c}{2} = \frac{8b}{2} = 4b$, the average of a and c in terms of b is $4b$.

5. (**B**) Determine whether each Roman numeral expression always represents an even integer when $x - 1$ represents an odd integer.
   - I. Since $x - 1$ is an odd integer, $x - 1 + 2$ or $x + 1$ is the next consecutive odd integer. Hence, expression I does not represent an even integer.
   - II. Since $x - 1$ represents an odd integer, $x - 1 + 1$ or x represents the next even integer. Since any multiple of an even integer is also even, $3x$ is an even integer. Hence, expression II represents an even integer.
   - III. The square of an odd integer is always odd. For example, $3^2 = 9$. Since $(x - 1)^2$ represents the square of an odd integer, expression III does not represent an even integer.

   Hence, only Roman numeral expression II is an even integer.

6. (**C**)

The area of each circle is $\pi \times (radius)^2 = \pi \times (2)^2 = 4\pi$. The length of the rectangle is equal to $4 \times$ radius length $= 4 \times 2 = 8$. The width of the rectangle is equal to $2 \times$ radius length $= 2 \times 2 = 4$. Hence, the area of the rectangle is $8 \times 4 = 32$. The probability that the point chosen will lie in the shaded region is equal to the area of the shaded region divided by the area of the rectangle:

$$\text{Probability} = \frac{\text{Area of the shaded region}}{\text{Area of the rectangle}}$$
$$= \frac{\text{Area of rectangle} - \text{Sum of areas of the circles}}{\text{Area of rectangle}}$$
$$= \frac{32 - (4\pi + 4\pi)}{32}$$
$$= \frac{32 - 8\pi}{32}$$
$$= \frac{32}{32} - \frac{8\pi}{32}$$
$$= 1 - \frac{\pi}{4}$$

7. **(D)** An exterior angle of a triangle is equal to the sum of the two nonadjacent interior angles of the triangle, so $\angle B + \angle C = 130$. Since $AB = AC$, the angles opposite these sides, angles B and C, must be equal. Hence, $\angle B = \angle C = \frac{1}{2}(130) = 65$. You are also told that $AC \parallel BD$. Since acute angles formed by parallel lines are equal, $x = \angle C = 65$.
8. **(D)** In the given addition problem, consider the rightmost column, that is, the first column. In the first column, B must be 6 since $5 + 6 = 11$. One is then carried over to the second column, so $1 + 6 + A = 1C$. When 1 is carried over to the third column, $1 + A + 3 = 11$, so A is 7.
9. **(E)** Since the diameter of the circle is 18, its circumference is 18π. The sum of the lengths of the broken arcs of circle O is 2π, so the sum of the lengths of the unbroken arcs that bound the shaded region is $18\pi - 2\pi$ or 16π. This represents $\frac{16\pi}{18\pi}$ or $\frac{8}{9}$ of the circle. Since the radius of the circle is $\frac{18}{2}$ or 9, the area of the circle is $9^2 \times \pi$ or 81π. The area of the shaded region is $\frac{8}{9}$ of the area of the circle. Since $\frac{8}{9} \times 81\pi = 8 \times 9\pi = 72\pi$, the area of the shaded region is 72π.
10. **(D)** Suppose George started with x dollars. After he spent 25% of x dollars on lunch, he had $0.75x$ dollars left. Since he spent 60% of the amount of money that was left on dinner, he spent $0.60(0.75x)$ or $0.45x$ dollars on dinner. If he then had \$9.00 left,

$$x - 0.25x - 0.45x = \$9$$
$$x - 0.70x = \$9$$
$$0.30x = \$9$$
$$x = \frac{\$9}{0.30} = \$30$$

Hence, George spent 25% of \$30 or \$7.50 on lunch.
11. **(C)** The angles at R and S have the same measure, so $RS = TS$. Vertex S must lie on a vertical line that contains the midpoint a of horizontal base RT, as shown in the figure below:

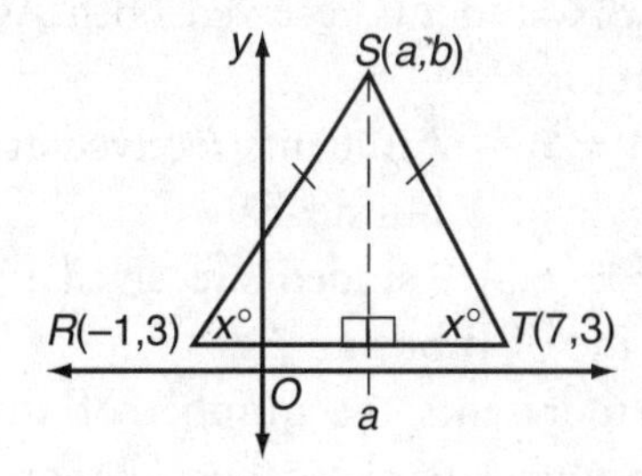

The x-coordinate of the midpoint of RT is $\frac{-1 + 7}{2}$ or 3, so $a = 3$. Hence, you can rule out choices (A), (B), (D), and (E).
12. **(B)** If 3 times 1 less than a number n is the same as twice the number increased by 14, then $3(n - 1) = 2n + 14$. Hence:

$$3n - 3 = 2n + 14$$
$$3n - 2n = 14 + 3$$
$$n = 17$$

13. **(B)** All points 4 inches from point A lie on a circle that has point A as its center and that has a radius of 4 inches. Similarly, all points 6 inches from point B lie on a circle that has point B as its center and that has a radius of 6 inches. Since the length of line segment AB is 9 inches, the two circles intersect at two points that are 4 inches from A and 6 inches from B, as shown in the figure below.

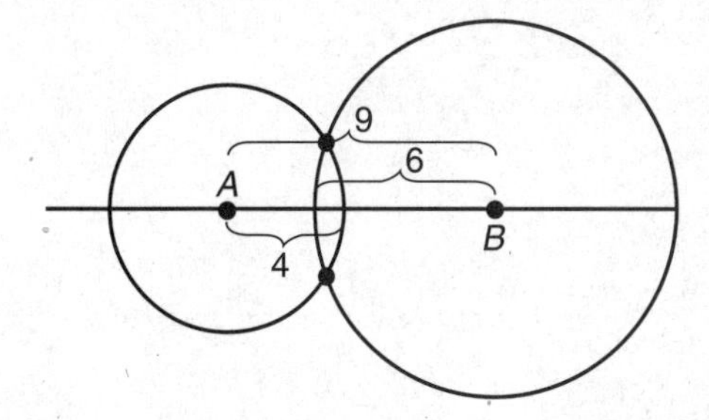

14. **(D)** Draw radius OD. Since the area of a square is equal to one-half the square of a diagonal, $\frac{1}{2}(OD)^2 = 2$ or $(OD)^2 = 2 \times 2 = 4$ so $OD = \sqrt{4} = 2$. The circumference of circle $O = 2\pi r = 2\pi(2) = 4\pi$. Since angle AOB is a right angle, arc ADB is $\frac{90}{360}$ or

$\frac{1}{4}$ of the circumference of circle *O*. Hence, the length of arc $ADB = \frac{1}{4} \times 4\pi = \pi$.

15. **(C)** Six students received both As and Bs. Therefore:
   - $13 - 6 = 7$ students received at least one grade of A but no Bs.
   - $15 - 6 = 9$ students received at least one grade of B but no As.

   If x represents the number of students in the homeroom class, then $\frac{1}{3}x$ represents the number of students who received no As or no Bs. This means that $\frac{2}{3}x$ represents the number of students who received As (but not Bs), Bs (but not As), or both As and Bs. Hence:

$$\frac{2}{3}x = 6 + 7 + 9$$
$$\frac{2}{3}x = 22$$
$$x = 22 \times \frac{3}{2} = 33$$

16. **(A)** If $g(x) = x \cdot 2^x$, then:
   - $g(a + 1) = (a + 1) \cdot 2^{a+1}$
   - $g(a) = a \cdot 2^a$
   - $g(a + 1) - g(a) = (a + 1) \cdot 2^{a+1} - a \cdot 2^a$

$$= 2^a[(a + 1)2 - a]$$
$$= 2^a[2a + 2 - a]$$
$$= (a + 2)2^a$$

Appendix

Quick Review of Key Math Facts

Number Properties: Lessons 3-1 to 3-4, 7-5

INTEGERS AND DIVISIBILITY

- Integers include the counting numbers (positive integers), their opposites (negative integers), and 0:

$$\ldots, -4, -3, -2, -1, 0, 1, 2, 3, 4, \ldots$$

- If $a \div b$ has a 0 remainder, then a is evenly divisible by b. Thus, 6 is divisible by 3 since $6 \div 3 = 2$ remainder 0, but 6 is not divisible by 4 since $6 \div 4 = 1$ remainder 2.
- Even integers are integers that are divisible by 2:

$$\ldots, -6, -4, -2, 0, 2, 4, 6, \ldots$$

- Odd integers are integers that are not divisible by 2:

$$\ldots, -5, -3, -1, 1, 3, 5, \ldots$$

- A *prime number* is an integer greater than 1 that is divisible only by itself and 1:

$$2, 3, 5, 7, 11, 13, 17, \ldots$$

The only even integer that is prime is 2.

GENERAL INTEGERS

- If n represents an integer, then four consecutive integers are

$$n, n + 1, n + 2, \text{ and } n + 3$$

- If n represents an *even* integer, then four consecutive even integers are:

$$n, n + 2, n + 4, \text{ and } n + 6$$

- If n represents an *odd* integer, then four consecutive odd integers are:

$$n, n + 2, n + 4, \text{ and } n + 6$$

EXPONENTS

- An *exponent* indicates the number of times a number is to be used as a factor in a product. In 2^3, the exponent 3 means that 2, called the *base,* should be used as a factor 3 times, as in $2 \times 2 \times 2 = 8$.
- Anything raised to a zero exponent is 1. Thus, $5^0 = 1$.
- Anything raised to a negative exponent can be written with a positive exponent by using the rule:

$$x^{-n} = \frac{1}{x^n} (x \neq 0)$$

- The laws of exponents for multiplication and division work only when the bases are the same:

1. Product Law: $y^a \times y^b = y^{a+b}$

EXAMPLE: $2^3 \times 2^4 = 2^{3+4} = 2^7$

EXAMPLE: $2^5 \times 3^4 \neq 6^{5+4}$

EXAMPLE: $3x^2 \cdot 2x^5 = (3 \cdot 2)(x^2 \cdot x^5) = 6x^7$

2. Quotient Law: $\dfrac{y^a}{y^b} = y^{a-b}$

EXAMPLE: $\dfrac{2^7}{2^3} = 2^{7-3} = 2^4$

EXAMPLE: $\frac{8xy^3}{2y} = \frac{8x}{2} \cdot \frac{y^3}{y^1} = 4xy^2$

3. Power Law: $(y^a)^b = y^{a \times b}$

EXAMPLE: $(2^3)^4 = 2^{3 \times 4} = 2^{12}$

FACTORS AND FACTORING

- The *factors* of a whole number are those numbers that divide evenly into that number. For example, the factors of 6 are 1, 2, 3, and 6.
- Factoring breaks down a number into the product of two or more numbers. For example, $30 = 5 \times 6$ where 5 and 6 are factors of 30.
- The *prime factorization* of a number factors a number so that each factor is a prime number.

EXAMPLE: The prime factorization of 30 is

$$2 \times 3 \times 5$$

EXAMPLE: The prime factorization of 72 is

$$2^3 \times 3^2$$

INEQUALITY SYMBOLS

- $<$ means "is less than" as in $2 < 5$.
- $>$ means "is greater than" as in $7 > 3$.
- $\leq$ means "is less than *or* equal to."
- $\geq$ means "is greater than *or* equal to."

UNFAMILIAR OPERATION SYMBOLS

A new arithmetic operation can be created using one or more familiar arithmetic operations. The rule that tells how the new operation works, together with its symbol, may be given as an equation or an algebraic expression.

EXAMPLE: If $x \blacktriangle y = y^x + xy$, then $3 \blacktriangle 2 = 2^3 + (2)(3) = 8 + 6 = 14$.

EXAMPLE: Let $\uparrow x$ be defined as $x - \frac{2}{x}$ for all nonzero integers. If $\uparrow x = z$, where z is an integer, then a possible value of z is 1: when $x = 2$, $\uparrow x = z = 2 - \frac{2}{2} = 2 - 1 = 1$.

ARITHMETIC SEQUENCE

An ordered list of numbers forms an *arithmetic sequence* when each number after the first is produced by adding the same number, d, to the number that comes before it. The number d represents the *common difference* between any two consecutive terms. For an arithmetic sequence that contains n terms,

$$n\text{th term} = \text{First term} + (n - 1) \times d$$

EXAMPLE: In the sequence 5, 11, 17, . . . , the first term is 5, and each term after it is 6 more than the previous term. Since $d = 6$, the 16th term of this sequence is

$$5 + (16 - 1) \times 6 = 5 + (15 \times 6) = 5 + 90 = 95$$

Arithmetic: Lessons 3-5 to 3-8

FRACTIONS

- For the fraction $\frac{a}{b}$:

  1. b is never allowed to be 0.

  2. To change into a decimal, divide b into a since $\frac{a}{b}$ means $a \div b$.

  3. If $a < b$, the fraction is less than 1; if $a = b$, the fraction is equal to 1; and if $a > b$, the fraction is greater than 1.

- Combine fractions using the rule:

$$\frac{a}{b} + \frac{c}{d} = \frac{ad + bc}{bd}$$

EXAMPLE:

$$\frac{2}{5} + \frac{1}{3} = \frac{(2 \times 3) + (5 \times 1)}{15} = \frac{11}{15}$$

- Multiply fractions using the rule:

$$\frac{a}{b} \times \frac{c}{d} = \frac{a \times c}{b \times d}$$

EXAMPLE: $\frac{2}{3} \times \frac{5}{7} = \frac{10}{21}$

- Divide fractions by changing to multiplication using the rule:

$$\frac{a}{b} \div \frac{c}{d} = \frac{a}{b} \times \frac{d}{c} = \frac{a \times d}{b \times c}$$

EXAMPLE:

$$\frac{3}{7} \div \frac{7}{8} = \frac{3}{\underset{1}{\not{4}}} \times \frac{\overset{2}{\not{8}}}{7} = \frac{3 \times 2}{1 \times 7} = \frac{6}{7}$$

- Write a fraction in lowest terms by dividing out the greatest common factor of both the numerator and denominator.

EXAMPLE: To write $\frac{16}{24}$ in lowest terms, divide the numerator and denominator by the greatest whole number that divides 16 and 24 evenly:

$$\frac{16 \div 8}{24 \div 8} = \frac{2}{3}$$

PERCENT

- Percent means parts out of 100. For example, 60% means $\frac{60}{100}$ or 0.60 or $\frac{3}{5}$.
- To find the percent of a number, multiply the number by the percent in decimal form.

EXAMPLE: If the rate of sales tax is 6%, then the amount of sales tax on a \$40 item is 6% of \$40 or $0.06 \times \$40 = \$2.40$.

- To find what percent one number is of another, write the "is" number over the "of" number and multiply by 100%.

EXAMPLE: If the amount of sales tax on a \$65 item is \$5.20, then the tax rate is obtained by finding what percent 5.20 is of 65. Since

$$\frac{5.20 \text{ (number before } is)}{65 \text{ (number after } of)} \times 100\% = \frac{520}{65}\% = 8\%$$

the tax rate is 8%.

- To find a number when a percent of it is given, divide the given amount by the percent.

EXAMPLE: If 30% of some number n is 12, then

$$n = \frac{12}{0.30} = 40$$

- To find the percent of change (increase or decrease) use the formula

$$\%\text{ change} = \frac{\text{Amount of change}}{\text{Original amount}} \times 100\%$$

EXAMPLE: If the price of a stock drops from \$10 a share to \$8 a share, then

$$\%\text{ decrease} = \frac{10 - 8}{10} \times 100\%$$

$$= \frac{1}{5} \times 100\%$$

$$= 20\%$$

SQUARE ROOT

- The square root notation $\sqrt{n}$ means one of two equal nonnegative numbers whose product is n.

EXAMPLE: $\sqrt{16} = 4$ since $4 \times 4 = 16$

- If n is a number greater than 1, then:

1. $n^2 > n$.

EXAMPLE: If $n = 4$, then $4^2 > 4$ since $4^2 = 16$.

2. $n > \sqrt{n}$

EXAMPLE: If $n = 9$, $9 > \sqrt{9}$ since $\sqrt{9} = 3$.

3. $n > \frac{1}{n}$

EXAMPLE: If $n = 10$, $10 > \frac{1}{10}$.

- If n is a number between 0 and 1, then:

1. $n^2 < n$

EXAMPLE: If $n = 0.1$, then $(0.1)^2 < 0.1$.

2. $n < \sqrt{n}$

EXAMPLE: If $n = 0.49$, then $0.49 < \sqrt{0.49}$ since $\sqrt{0.49} = 0.7$.

3. $n < \frac{1}{n}$

EXAMPLE: If $n = 0.5$, then $0.5 < \frac{1}{0.5}$ since $\frac{1}{0.5} = 2$.

Algebra: Lessons 4-1 to 5-5

SIGNED NUMBERS

- Each number on a horizontal number line is greater than any number to its left.

  For example, $-\frac{1}{2} > -2$.

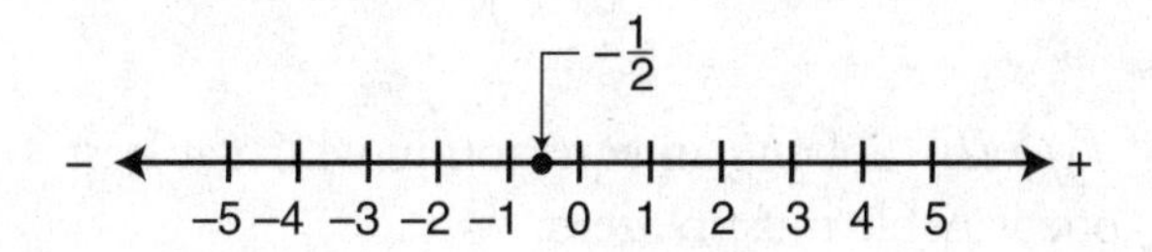

- The absolute value of a number n is indicated by writing $|n|$ and is equal to the number without its sign. Thus, $|-3| = 3$.
- To multiply or divide numbers that have the *same* sign, perform the operation and make the sign of the answer positive.

  EXAMPLE: $(-3) \times (-5) = +15$
- To multiply or divide numbers that have *different* signs, perform the operation and make the sign of the answer negative.

  EXAMPLE: $\frac{+15}{-3} = -5$
- To add numbers that have the *same* sign, add the numbers and attach the common sign to the sum.

  EXAMPLE: $(-3) + (-4) = -7$
- To add numbers that have *different* signs, subtract the absolute values and attach the sign of the number with the larger absolute value to the answer.

  EXAMPLE: $(-5) + (+2) = -(5 - 2)$
  $= -3$
- To subtract signed numbers, change to an equivalent addition example.

  EXAMPLE: $(-7) - (-2) = (-7) + (+2)$
  $= -5$

LINEAR EQUATIONS

- Solve a first-degree equation in one variable (letter) by isolating the variable.

  EXAMPLE: Solve: $3(x + 1) - x = 13$
  Remove parentheses: $3x + 3 - x = 13$
  Collect like terms: $2x = 13 - 3$
  Solve for x: $x = \frac{10}{2} = 5$

RATIO AND PROPORTION

- A *ratio* compares two quantities by division. The ratio of x to y is written as $x : y$ or as $\frac{x}{y}$. If x is 3 times as great as y, then $\frac{x}{y} = \frac{3}{1}$ or $3 : 1$.
- If $x : y$ and $y : z$, then $x : z$. Suppose the ratio of x to y is $2 : 3$ and the ratio of y to z is $4 : 7$. The ratio of x to z can be determined by changing the two ratios so that the second term of the first ratio is the same as the first term of the second ratio. Since $2 : 3$ is equivalent to $8 : \underline{12}$ and $4 : 7$ is equivalent to $\underline{12} : 21$, the ratio of x to z is $8 : 21$.
- A *proportion* is an equation that states that two ratios are numerically equal. To change the proportion $\frac{a}{b} = \frac{c}{d}$ into an equation without fractions, cross-multiply:

  $$\frac{a}{b} = \frac{c}{d} \text{ becomes } b \times c = a \times d$$

  When a proportion involves an unknown quantity, x, solve the equation obtained by cross-multiplying.

  EXAMPLE: Solve: $\frac{x}{x + 4} = \frac{3}{5}$

  Cross-multiply:

  $5x = 3(x + 4)$
  $5x = 3x + 12$

  Collect like terms:

  $2x = 12$

  Solve for x: $x = \frac{12}{2} = 6$

VARIATION

- Two variable quantities form a *direct variation* when as one quantity changes the other also changes so that their ratio does not change.

EXAMPLE: If 3 apples cost \$0.51, the cost of 5 apples can be determined using the fact that the number of apples and their cost form a direct variation. If x represents the cost of 5 apples, then:

$$\frac{\text{Number}}{\text{Cost}} = \frac{3}{\$0.51} = \frac{5}{x}$$

$$3x = 5 \times \$0.51 = \$2.55$$

$$x = \frac{\$2.55}{3} = \$0.85$$

- Two variable quantities form an *inverse variation* when as one quantity increases the other quantity decreases so that their product does not change.

EXAMPLE: Two people can paint a room in 6 hours. How long would it take three people to paint the same room? The number of painters and the hours required for the whole job form an inverse variation. If h is the number of hours it would take three people to paint the same room, then:

$$3 \times h = 2 \times 6$$

$$h = \frac{12}{3} = 4 \text{ hours}$$

SPECIAL TYPES OF EQUATIONS

- *Type I:* Solving an equation for a combination of variables.

EXAMPLE: If $3x + 3y = 24$, then you can solve for the value of $x + y$ by dividing each term of the equation by 3, which makes $x + y = 8$.

- *Type II:* Solving a system of two equations for a combination of variables.

EXAMPLE: If $x - 2y = 7$ and $2x - y = 20$, then you can solve for the value of $x - y$ by adding corresponding sides of the two equations. This makes $3x - 3y = 27$, so

$$x - y = \frac{27}{3} = 9$$

- *Type III:* Solving an equation in which at least one variable must be zero.

EXAMPLE: If $3x^2 + y^2 = 0$, then you can reason that two nonnegative numbers, $3x^2$ and y^2, can add up to zero only if each number is 0. This means that x and y are both 0.

- *Type IV:* Solving an equation for one letter in terms of another letter.

EXAMPLE: To solve $x + y = 7y - x$ for x, rearrange the terms of the equation so that like letters appear on the same side of the equation:

Add x to each side: $2x + y = 7y$

Subtract $3x$ from each side: $2x = 6y$

Divide each side by 2: $x = 3y$

EXAMPLE: If $5x - y = 3(x + y)$, the ratio of x to y can be determined by first solving for x in terms of y and then dividing both sides of the equation by y:

Remove the parentheses: $5x - y = 3x + 3y$

Add y to each side: $5x = 3x + 4y$

Subtract $3x$ from each side: $2x = 4y$

Divide each side by 2: $x = 2y$

Divide each side by $\frac{x}{y} = 2 = \frac{2}{1}$

Hence, the ratio of x to y is 2 : 1.

Translating Word Phrases

The accompanying table summarizes commonly used phrases and how they are translated into mathematical language. Remember that the order in which two numbers are added or multiplied does *not* matter, but the order in which two numbers are subtracted or divided *does* matter.

Arithmetic Operation	Related English Phrases	Examples
Addition (+)	Sum (the answer in addition), plus, increase in, increased by, more than	• The sum of x and y: $x + y$ • x increased by 2: $x + 2$ • 3 more than x: $x + 3$
Subtraction (−)	Difference (the answer in subtraction), take away, minus, decrease of, decreased by, diminished by, less, less than (reverses the order of the given numbers)	• x decreased by y: $x - y$ • x take away 2: $x - 2$ • h diminished by k: $h - k$ • x less y: $x - y$ • x less than y: $y - x$
Multiplication (×)	Product (the answer in multiplication), times, double (multiply by 2), triple (multiply by 3)	• x times y: xy or $x \cdot y$ or $(x)(y)$. • Product of 3 and p less than q: $3(q - p)$
Division (÷)	Quotient (the answer in division), divided by, ratio	• x divided by the sum of y and z: $\frac{x}{y + z}$ • The ratio of x to y: $\frac{x}{y}$

SOLVING AN INEQUALITY

Solve a first-degree inequality in the same way you solve a first-degree equation, with two exceptions:

- When multiplying or dividing both sides of an inequality by the same *negative* number, you must reverse the inequality.

EXAMPLE: Multiplying both sides of $5 > -3$ by -2 changes the inequality from $>$ to $<$:

$$(5) \times (-2) > (-3) \times (-2)$$

$$-10 < +6$$

EXAMPLE: To solve $1 - 2x > 9$ for x, first subtract 1 from each side, which makes $-2x > 8$. Then divide each side by -2 *and* reverse the inequality:

$$\frac{-2x}{-2} < \frac{8}{-2}$$

$$x < -4$$

- Solve a combined inequality by isolating the *variable* in the *middle* part of the inequality:

EXAMPLE: To solve $-3 \leq 2x - 1 \leq 9$ for x:

Add 1 to each member: $-2 \leq 2x \leq 10$

Divide each member by 2: $\frac{-2}{-2} \leq \frac{2x}{2} \leq \frac{10}{2}$

Simplify: $-1 \leq x \leq 5$

FACTORING AND QUADRATIC EQUATIONS

- Factoring out a common term from the sum or difference of two terms reverses the distributive law.

EXAMPLE: $x^2 + 3x = x(x + 3)$

- The difference of the squares of two quantities can be factored as the product of the sum and difference of the terms being squared.

EXAMPLE: $x^2 - 4 = (x)^2 - (2)^2$
$= (x + 2)(x - 2)$

EXAMPLE: $2y^3 - 18y = 2y(y^2 - 9)$
$= 2y(y + 3)(y - 3)$

- Binomials can be multiplied using FOIL.

EXAMPLE:

$$(x+5)(x-3) = \underbrace{(x)(x)}_{\text{First}} + \underbrace{(-3)(x)}_{\text{Outer}} + \underbrace{(5)(x)}_{\text{Inner}} + \underbrace{(5)(-3)}_{\text{Last}}$$
$$= x^2 + 2x - 15$$

- Shortcut formula for finding the product of the sum/difference of the same two terms:

$$(x + y)(x - y) - x^2 - y^2$$

EXAMPLE:

$$\underbrace{(2a + 3b)(2a - 3b)}_{\text{Think of } 2a \text{ as } x \text{ and } 3b \text{ as } y.} = (2a)^2 - (3b)^2$$
$$= 4a^2 - 9b^2$$

- Expansion formulas for squaring a binomial:

$$(x + y)^2 = (x + y)(x + y) = x^2 + 2xy + y^2$$

$$(x - y)^2 = (x - y)(x - y) = x^2 - 2xy + y^2$$

EXAMPLE: Given $x^2 + y^2 = 25$ and $xy = -3$. To find $(x + y)^2$, substitute into its expansion formula:

$$(x + y)^2 = x^2 + 2\overbrace{xy}^{-3} + y^2$$
$$= \underbrace{(x^2 + y^2)} + 2(-3)$$
$$= 25 + (-6)$$
$$= 19$$

- Factorable quadratic trinomials can be factored using the reverse of FOIL.

EXAMPLE: $x^2 + 2x - 15 = (x + a)(x + b)$ where a and b are picked so that $a \times b = -15$ and, at the same time, $a + b = +2$. Since $(+5) \times (-3) = -15$ and $(+5) + (-3) = +2$:
$x^2 + 2x - 15 = (x + 5)(x - 3)$

- To solve a factorable quadratic equation in which 0 appears alone on one side, factor the other side of the equation. Set each factor equal to 0 and solve the resulting equations.

EXAMPLE: If $x^2 + 3x = 0$, then $x(x + 3) = 0$ so $x = 0$ or $x + 3 = 0$, which makes $x = 0$ or $x = -3$.

EXAMPLE: If $x^2 - 4x + 4 = 0$, then $(x - 2)(x - 2) = 0$ so $x - 2 = 0$, which makes $x = 2$.

EXAMPLE: If $x^2 + 2x - 15 = 0$, then $(x + 5)(x - 3) = 0$ so $x + 5 = 0$ or $x - 3 = 0$, which makes $x = -5$ or $x = 3$.

Geometry: Lessons 6-1 to 6-4

ANGLES AND LINES

- *Vertical angles,* formed when two lines intersect, are equal in degree measure.

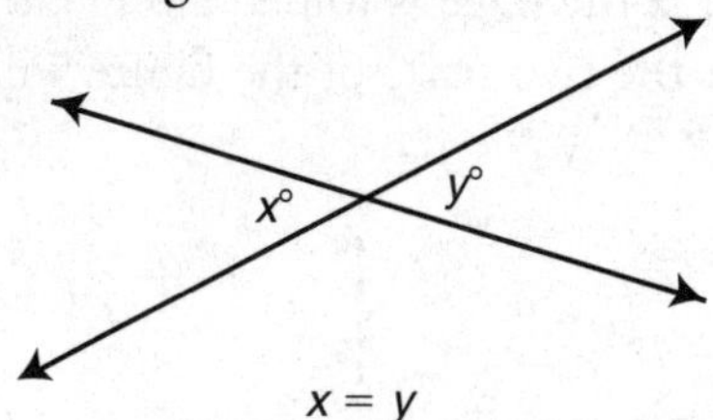

$x = y$

- When the noncommon sides of two adjacent angles form a line, the two angles add up to 180°.

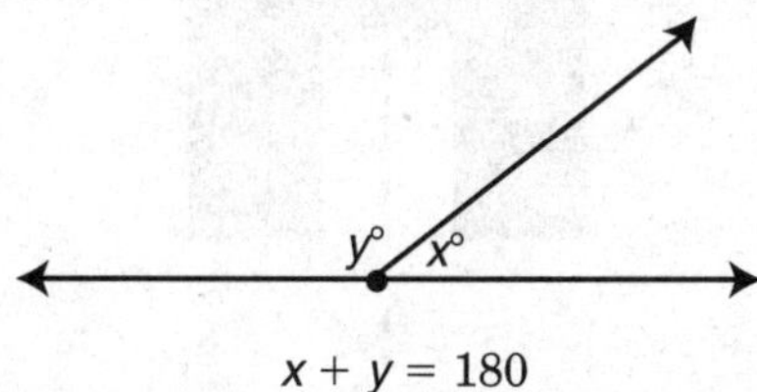

$x + y = 180$

- The sum of the nonoverlapping angles drawn about a point is 360°.

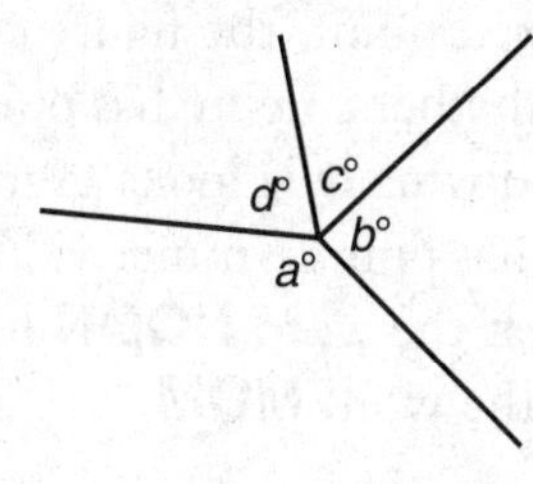

$a + b + c + d = 360$

- On the SAT, pairs of angles formed by parallel lines that look equal are equal, and pairs of angles that look unequal add up to 180°.

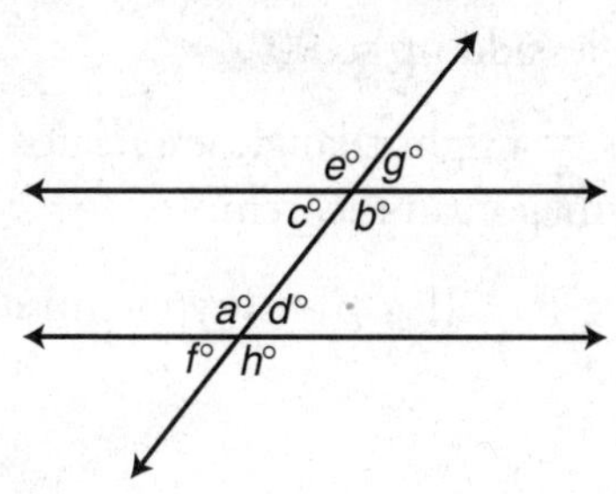

For example:

$a = b \quad a = e \quad a + g = 180$

$c = d \quad d = g \quad e + f = 180$

TRIANGLES AND POLYGONS

- The three interior angles of a triangle add up to 180°.
- An exterior angle of a triangle is equal to the sum of the two nonadjacent interior angles.

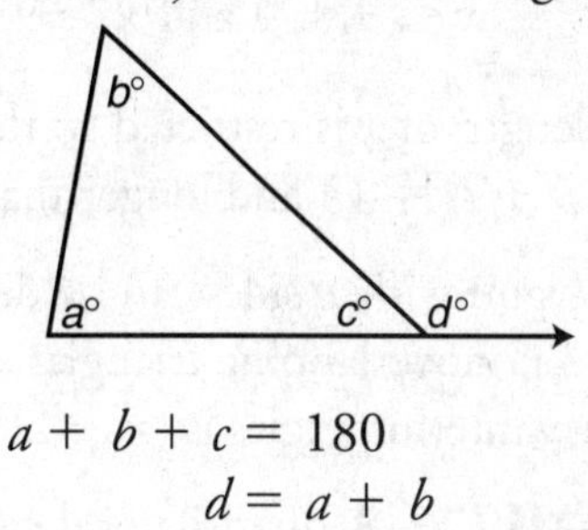

$a + b + c = 180$

$d = a + b$

- If a triangle has two equal sides (angles), then the angles (sides) opposite them are also equal.

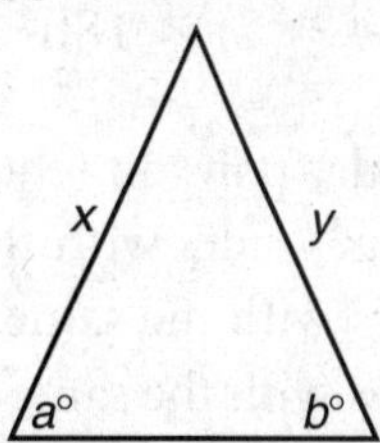

If $x = y$, then $a = b$.
If $a = b$, then $x = y$.

- If a triangle has three equal sides, then each angle measures 60°.

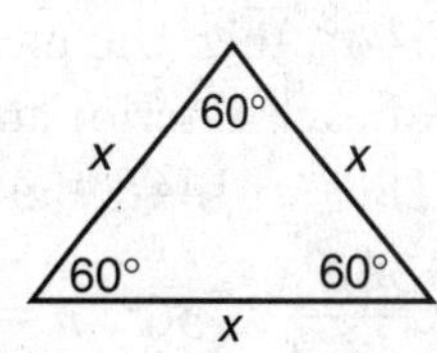

- Pairs of unequal angles of a triangle are opposite unequal sides with the larger angle facing the longer side.

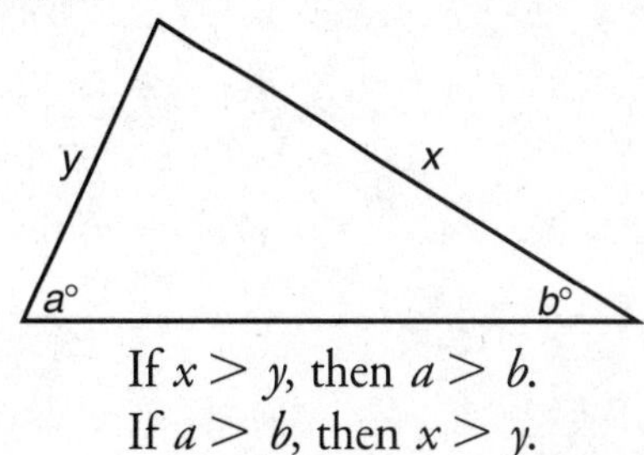

If $x > y$, then $a > b$.
If $a > b$, then $x > y$.

- Each side of a triangle must be shorter than the sum of the lengths of the other two sides *and* longer than their difference.

  EXAMPLE:

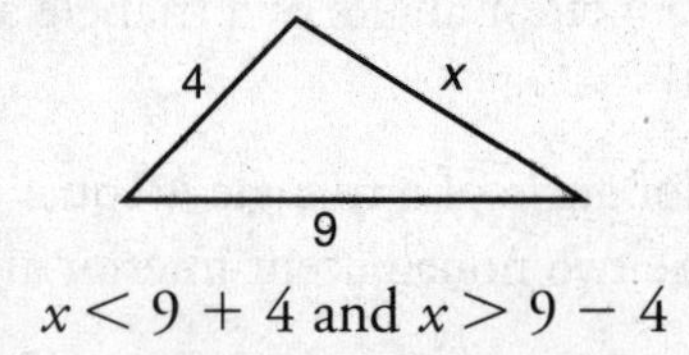

$$x < 9 + 4 \text{ and } x > 9 - 4$$

The length of x is restricted so that it is shorter than $9 + 4 = 13$ and longer than $9 - 4 = 5$.

- A polygon with n sides can be divided into $n - 2$ nonoverlapping triangles so that the sum of its n interior angles is $(n - 2) \times 180°$.

  EXAMPLE: A quadrilateral can be divided into 2 triangles so that its four angles add up to $(4 - 2) \times 180° = 360°$.

- An n-sided regular polygon (equilateral and equiangular) has n sides with the same length, n interior angles with the same measure, and n exterior angles with the same measure. The measure of each exterior angle is $\frac{360°}{n}$, and the interior angle and exterior angles at each vertice are supplementary.

  EXAMPLE: If the measure of each interior angle of a regular polygon is 144°, then the measure of each exterior angle is $180° - 144° = 36°$. Since $\frac{360°}{n} = 36°$, $n = 10$, so the regular polygon has 10 sides.

SYMMETRY

A figure has a *line of symmetry* if a line can be drawn that divides the figure into two mirror image parts. The letter **H** in the accompanying figure has both a vertical line of symmetry and a horizontal line of symmetry. If the page is folded along either line of symmetry, the two parts of the figure will coincide exactly.

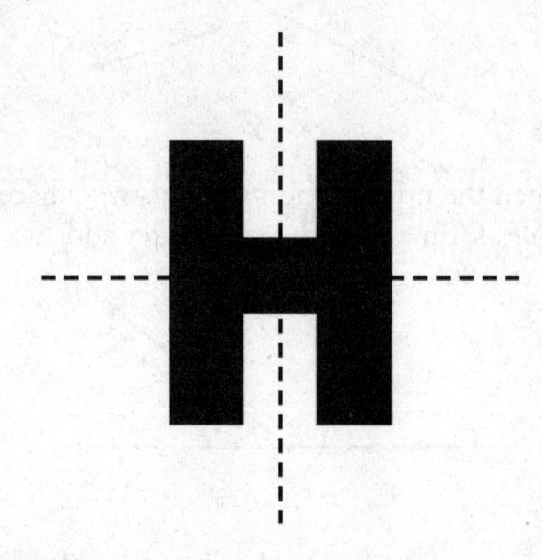

A figure has *point symmetry* if after a 180° rotation about a fixed point, the figure coincides with itself. To test whether a figure has point symmetry, turn it upside down. If it looks exactly the same, then the figure has point symmetry. You can verify, for example, that the word **NOON** has point symmetry but not the word **MOM**.

RIGHT TRIANGLES

- In a right triangle, the longest side is opposite the right (90°) angle and is called the *hypotenuse*. The other two sides are called *legs*. The two acute angles add up to 90°.
- The sides of a right triangle are related according to the Pythagorean theorem:

$$(\text{leg}_1)^2 + (\text{leg}_2)^2 = (\text{hypotenuse})^2$$

EXAMPLE:

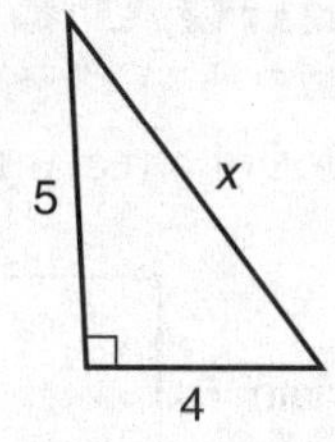

$$4^2 + 5^2 = x^2$$
$$16 + 25 = x^2$$
$$41 = x^2$$
$$x = \sqrt{41}$$

A set of three positive integers that satisfies the Pythagorean theorem is called a *Pythagorean triple.* Multiplying each of the numbers of a Pythagorean triple by the same positive integer produces another Pythagorean triple. Since 3–4–5 is a Pythagorean triple, so are 6–8–10, 9–12–15, 12–16–20, and so forth. Other basic Pythagorean triples include 5–12–13, 8–15–17, and 7–24–25.

Pythagorean Triples Worth Memorizing

$3n$–$4n$–$5n$

$5n$–$12n$–$13n$

$8n$–$15n$–$17n$

$7n$–$24n$–$25n$

where n can be any positive integer.

EXAMPLE: If the length and width of a rectangle are 6 and 8, then you don't need the Pythagorean theorem to find the diagonal (hypotenuse). Since 3 : 4 : *5* = 6 : 8 : *10* (where $n = 2$), the length of the diagonal is 10.

Area: Lessons 6-5 and 6-6

AREA OF A PARALLELOGRAM

- A parallelogram has these properties:
  1. Opposite sides are parallel and equal.
  2. Opposite angles are equal.
  3. Diagonals cut each other in half.

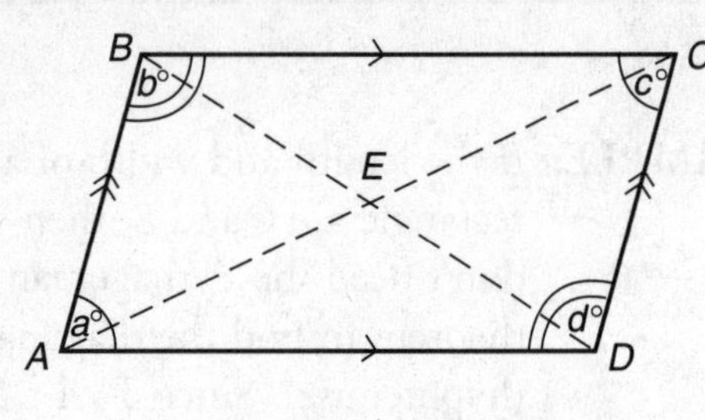

$$\overline{AD} \parallel \overline{BC} \quad AD = BC$$
$$\overline{AB} \parallel \overline{DC} \quad AB = CD$$
$$a = c \quad BE = DE$$
$$b = d \quad AE = CE$$

- The area A of a parallelogram is base times height.

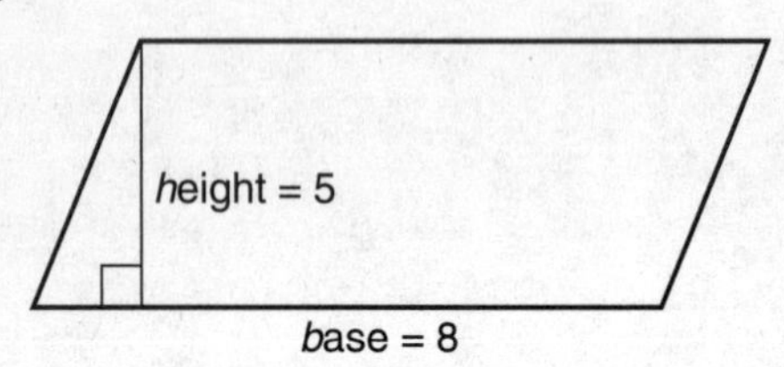

$$\begin{aligned} A &= base \times height \\ &= 8 \times 5 \\ &= 40 \end{aligned}$$

AREA OF A RECTANGLE

- A rectangle has these properties:
  1. All properties of a parallelogram.
  2. Four 90° angles.
  3. Equal diagonals.

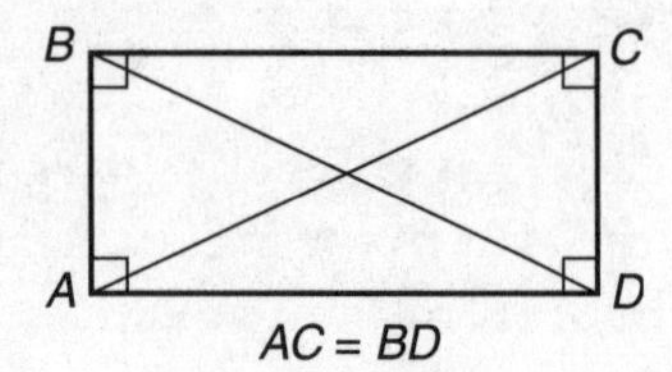

- The area A of a rectangle is length times width.

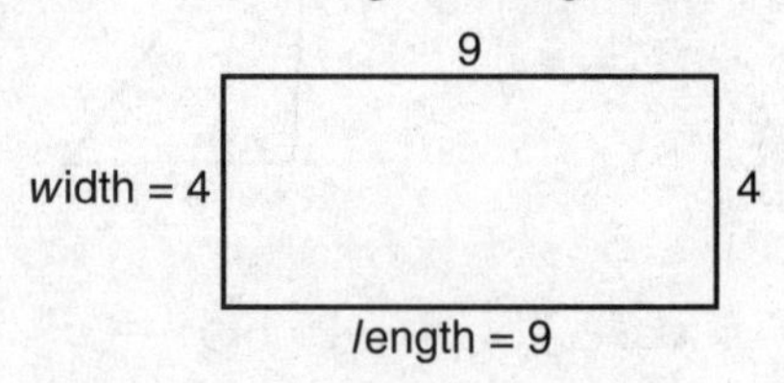

$$\begin{aligned} A &= length \times width \\ &= 9 \times 4 \\ &= 36 \end{aligned}$$

AREA OF A SQUARE

- A square has these properties:
  1. All properties of a rectangle.
  2. Four equal sides.
  3. Diagonals that intersect at 90° angles and that cut opposite angles of the square in half.
- The length of a diagonal is $\sqrt{2}$ times the length of a side.
- Each side length is $\frac{1}{2} \times \sqrt{2} \times$ diagonal.

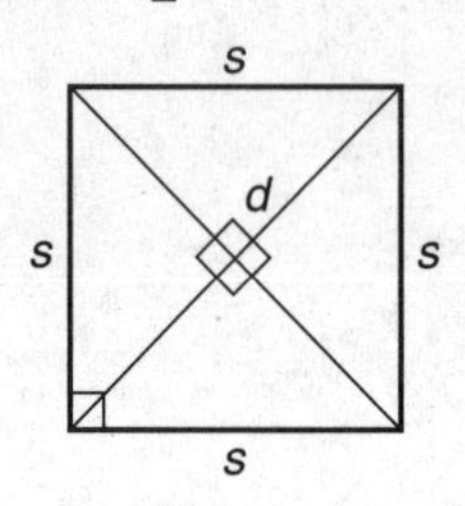

$$d = \sqrt{2} \times s$$
$$s = \frac{1}{2} \times \sqrt{2} \times d$$

- The area A of a square is side times side.

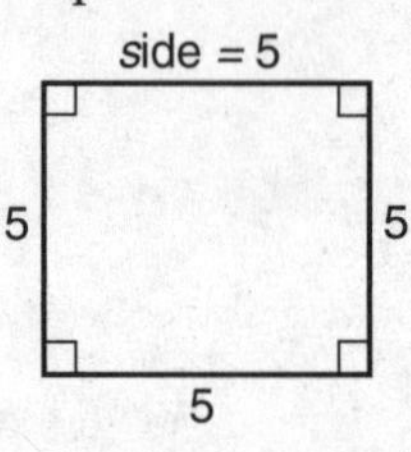

$$\begin{aligned} A &= side \times side \\ &= 5 \times 5 \\ &= 25 \end{aligned}$$

- If the diagonal of a square is known, then the area is one-half times the square of the diagonal.

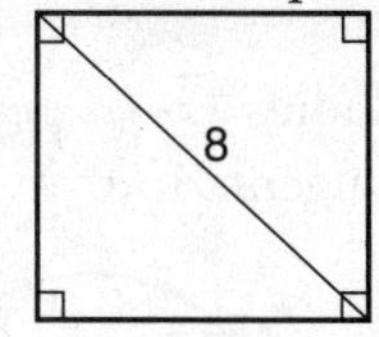

$$A = \frac{1}{2} \times (diagonal)^2$$
$$= \frac{1}{2} \times (8)^2$$
$$= 32$$

AREA OF A TRIANGLE

- The area A of a triangle is one-half times the base times the height.

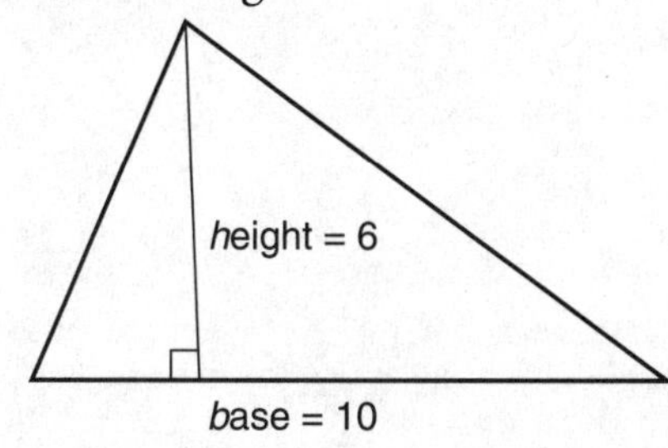

$$A = \frac{1}{2} \times base \times height$$
$$= \frac{1}{2} \times 10 \times 6$$
$$= 30$$

AREA OF A TRAPEZOID

- A trapezoid has exactly one pair of sides parallel, called the *bases.*
- The area A of a trapezoid is one-half times the height times the sum of the two bases.

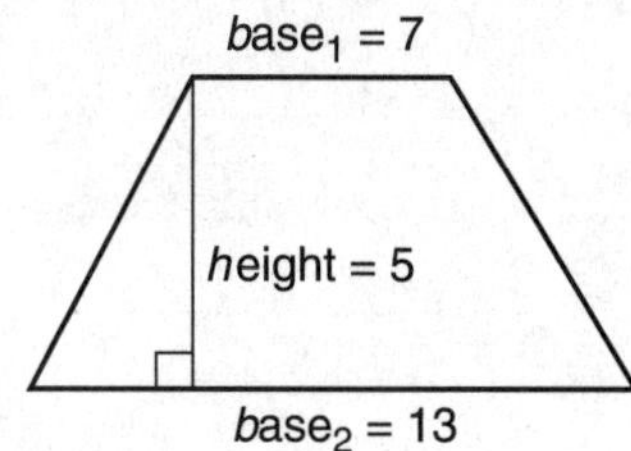

$$A = \frac{1}{2} \times height \times (base_1 + base_2)$$
$$= \frac{1}{2} \times 5 \times (7 + 13)$$
$$= \frac{1}{2} \times 5 \times 20$$
$$= \frac{1}{2} \times 100$$
$$= 50$$

AREA AND CIRCUMFERENCE OF A CIRCLE

- The perimeter of a circle is called the *circumference* where:

$$\text{Circumference} = \pi \times diameter$$
$$= 2 \times \pi \times radius$$

- The area A of a circle is

$$A = \pi \times (radius)^2$$

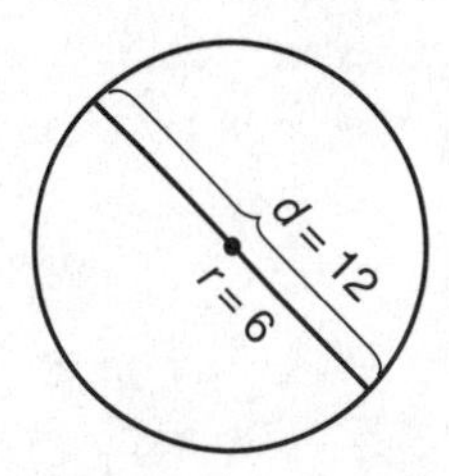

$$\text{Circumference} = \pi d$$
$$= 12\pi$$
$$\text{Circumference} = 2\pi r$$
$$= 2 \times \pi \times 6$$
$$= 12\pi$$
$$A = \pi \times (6)^2$$
$$= 36\pi$$

AREA OF A SHADED REGION

- To find the area of a shaded region, try subtracting the areas of the figures that overlap to form that region.

EXAMPLE: Suppose each circle below has an area of 25π.

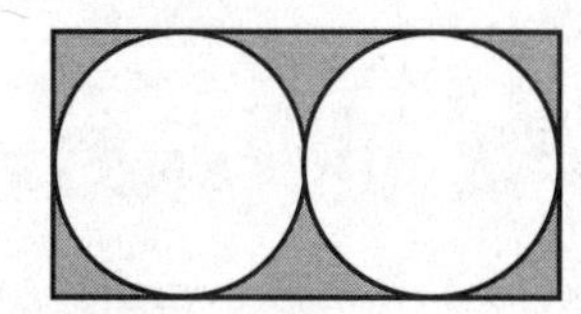

Since $25\pi = \pi r^2 = \pi(5^2)$, the radius of each circle is 5, so the diameter of each circle is 10. Hence, the width of the rectangle is one diameter or 10 and the length of the rectangle is two diameters or 20. The area of the rectangle is $20 \times 10 = 200$. The area of the shaded region is equal to the difference between the area of the rectangle and the sum of the areas of the two circles, which is $200 - 50\pi$.

Radius ⊥ Tangent

- If a line is tangent to a circle, then a radius drawn to the point of contact is perpendicular to the tangent at that point.

EXAMPLE: Radius $\overline{OA}$ is perpendicular to tangent $\overline{AB}$.

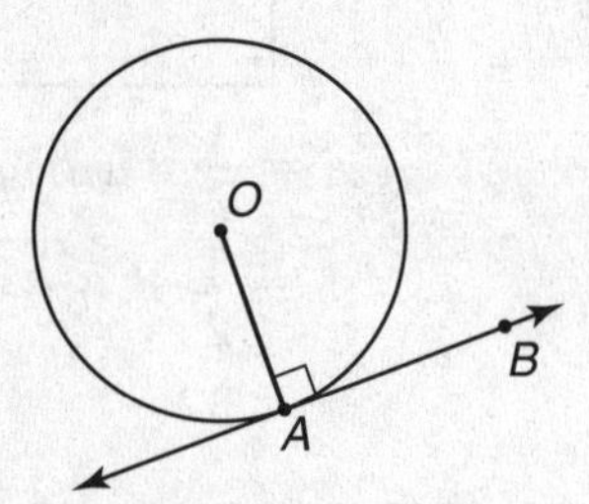

Solids: Lesson 6-7

CUBE AND RECTANGULAR BOX

- A cube has 6 equal square sides called faces. The length of each face is called the *edge length.*
- The surface area A of a cube is 6 times the *e*dge length times the *e*dge length:

$$A = 6 \times e \times e$$

- The volume V of a cube is *e*dge length times *e*dge length times *e*dge length:

$$V = e \times e \times e$$

EXAMPLE: If length of an edge of a cube is 4, then the surface area is $6 \times 4 \times 4 = 96$ and the volume of the cube is $4 \times 4 \times 4 = 64$.

- The volume V of a rectangular box is *l*ength times *w*idth times *h*eight:

$$V = l \times w \times h$$

- The length of a diagonal, d, drawn in a rectangular box with dimensions $l \times w \times h$ is $\sqrt{l^2 + w^2 + h^2}$:

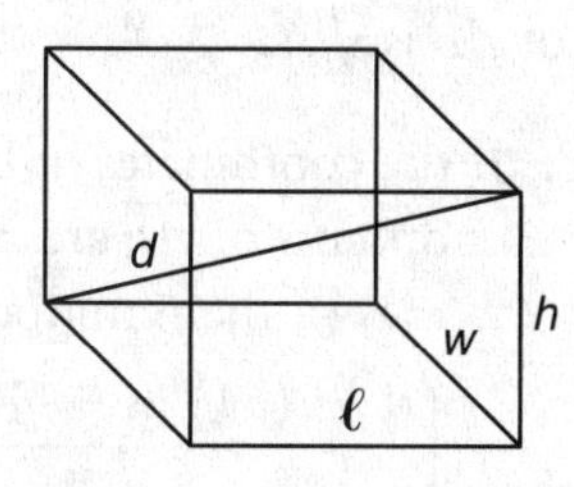

$$d = \sqrt{\ell^2 + w^2 + h^2}$$

CIRCULAR CYLINDER

- The volume V of a circular cylinder is the area of its circular base times the height of the cylinder:

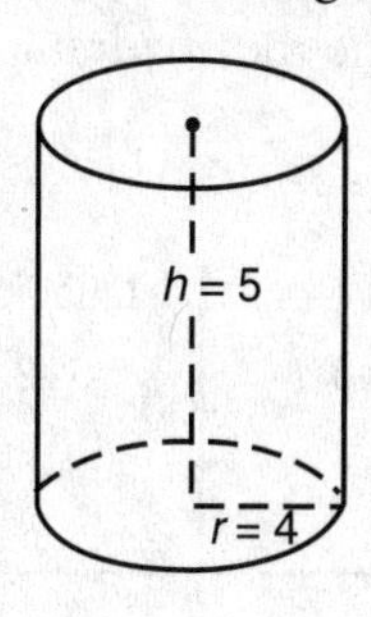

$$V = \overbrace{\pi r^2}^{\text{Area of base}} \times h$$
$$= \pi \times 4^2 \times 5$$
$$= 80\pi$$

Coordinate Geometry: Lesson 6-8

LOCATING POINTS IN QUADRANTS

- The signs of the *x*- and *y*-coordinates of point $P(x, y)$ determine the quadrant in which *P* is located.

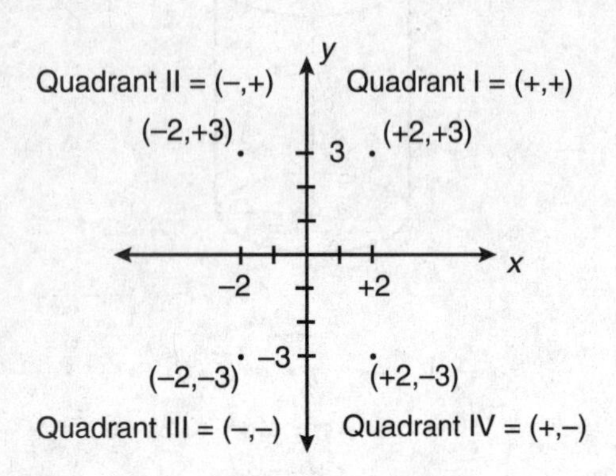

SLOPE

- The slope of a line is a number that represents the steepness of the line.
- The slope of a nonvertical line that passes through $A(x_A, y_A)$ and $B(x_B, y_B)$ is:

$$\text{Slope} = \frac{\text{Vertical change (difference) in } y}{\text{Horizontal change (difference) in } x}$$

$$= \frac{y_B - y_A}{x_B - x_A}$$

EXAMPLE: The slope of the line that passes through $A(-1,2)$ and $B(1,8)$ is

$$\frac{\text{difference in } y}{\text{difference in } x} = \frac{8-2}{1-(-1)} = \frac{6}{1+1} = 3$$

- Facts about slopes:
  1. The slope of a horizontal line is 0 and the slope of a vertical line is not defined.
  2. A line that rises from left to right has a positive slope.
  3. A line that falls from left to right has a negative slope.
  4. Parallel lines have the same slope.
  5. Perpendicular lines have slopes that are negative reciprocals.
  6. If a line that passes through the origin contains the point (a, b), then its slope is $\frac{b}{a}$.

MIDPOINT AND DISTANCE

- If the coordinates of $\overline{AB}$ are $A(x_A, y_A)$ and $B(x_B, y_B)$, then
  1. The coordinates of the midpoint of $\overline{AB}$ are the averages of the corresponding coordinates of the endpoints: $\left(\frac{x_A + x_B}{2}, \frac{y_A + y_B}{2}\right)$
  2. The length of $\overline{AB}$ is $\sqrt{(x_B - x_A)^2 + (y_B - y_A)^2}$

EXAMPLE: If the coordinates of the endpoints of $\overline{AB}$ are $A(1,8)$ and $B(-1,4)$, then: midpoint

$$= \left(\frac{1+(-1)}{2}, \frac{8+4}{2}\right) = (0,6)$$

$$AB = \sqrt{(-1-1)^2 + (4-8)^2}$$

$$= \sqrt{(-2)^2 + (-4)^2}$$

$$= \sqrt{4+16}$$

$$= \sqrt{20}$$

$$= \sqrt{4} \times \sqrt{5}$$

$$= 2\sqrt{5}$$

EQUATION OF A LINE

- The set of all points that lie on a nonvertical line can be described by an equation that has the form $y = mx + b$, where m is the slope of the line and b is the y-coordinate of the point at which the line crosses the y-axis.

EXAMPLE: The equation of line p is $y = 3x - 2$:

1. The slope of line p is 3 and the y-intercept is -2.
2. The slope of a line that is *parallel* to line p is 3, since parallel lines have the same slope.
3. The slope of a line that is *perpendicular* to line p is $-\frac{1}{3}$, since perpendicular lines have slopes that are negative reciprocals.

EXAMPLE: To write an equation of the line that contains the points (3,1) and (0,7), find the values of m and b in $y = mx + b$:

1. $m = \dfrac{7 - 1}{0 - 3} = -\dfrac{6}{3} = -2$
2. $b = 7$, since b is the value of y when $x = 0$, which is given by (0,7).
3. $y = -2x + 7$

EXAMPLE:

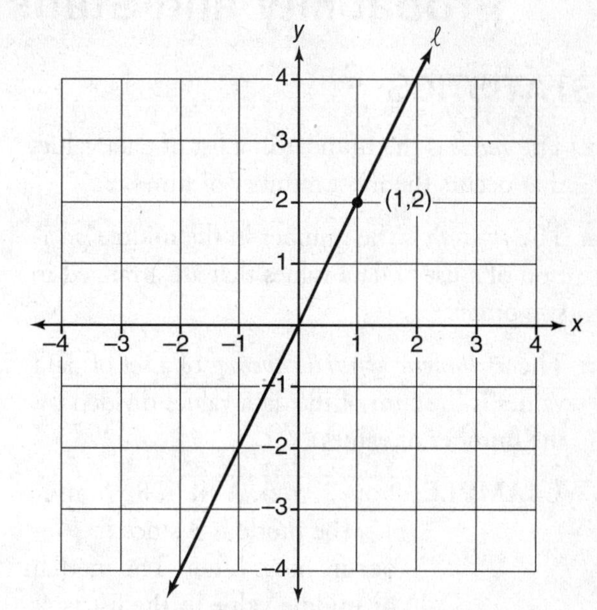

To write an equation of the line perpendicular to line ℓ at (1,2), first note that the slope of line ℓ is $\frac{2}{1} = 2$. The slope of the perpendicular is $-\frac{1}{2}$, so its equation has the form $y = -\frac{1}{2}x + b$. Because the coordinates of (1,2) must satisfy this equation, $2 = -\frac{1}{2}(1) + b$ and $b = \frac{5}{2}$. Hence, an equation of the perpendicular to line ℓ at (1,2) is $y = -\frac{1}{2}x + \frac{5}{2}$.

Probability and Statistics: Lessons 7-1 to 7-3

STATISTICS

- The *mode* is the number in a list of data values that occurs the most number of times.
- The *median* is the number in the middle position of a list of data values that are arranged in size order.
- The *arithmetic mean* or *average* of a set of data values is the sum of the data values divided by the number of values.

EXAMPLE: For 2, 3, 3, 3, 4, 5, 8, 9, and 17, the mode is 3 since it occurs most often. The median or middle value in the list is 4 since there are the same number of values below and above 4. The average is

$$\frac{2+3+3+3+4+5+8+9+17}{9} = \frac{54}{9} = 6$$

- If a data set contains an even number of values, the median is the average (arithmetic mean) of the two middle numbers.

EXAMPLE: For 2, 4, 6, 8 the median is

$$\frac{4+6}{2} = 5$$

- The average of a set of values times the number of values in the set gives the sum of the values.

EXAMPLE: If the average of 5 numbers is 16, then the sum of the 5 numbers is $16 \times 5 = 80$.

- A *weighted average* counts some data values more times than other data values.

EXAMPLE: John scored 80, 85, and 75 on three class exams and received a 90 on the final exam. If the final exam grade is weighted *three* times as much as a class test, John's weighted exam average is:

$$\frac{(80+85+75)+(3\times 90)}{6} = \frac{510}{6} = 85$$

PROBABILITY

- The probability that a future event will happen is expressed as a fraction with a value from 0 to 1. If an event can happen in r out of n equally likely ways, then the probability that it will happen is the fraction $\frac{r}{n}$.
- If an event is certain to happen, its probability is 1.
- If an event is impossible, its probability is 0.
- If the probability that an event will happen is p, then the probability that the event will *not* happen is $1 - p$.

EXAMPLE: The probability of picking a red marble from a jar that contains only 3 red marbles and 4 green marbles is $\frac{3}{3+4}$ or $\frac{3}{7}$. The probability that a red marble will *not* be picked is $1 - \frac{3}{7} = \frac{4}{7}$.

GEOMETRIC PROBABILITY

- Finding the probability of an event may require calculating the ratio of the areas of two regions.

  EXAMPLE: Suppose a point is picked at random from the interior of circle *P* shown in the accompanying figure.

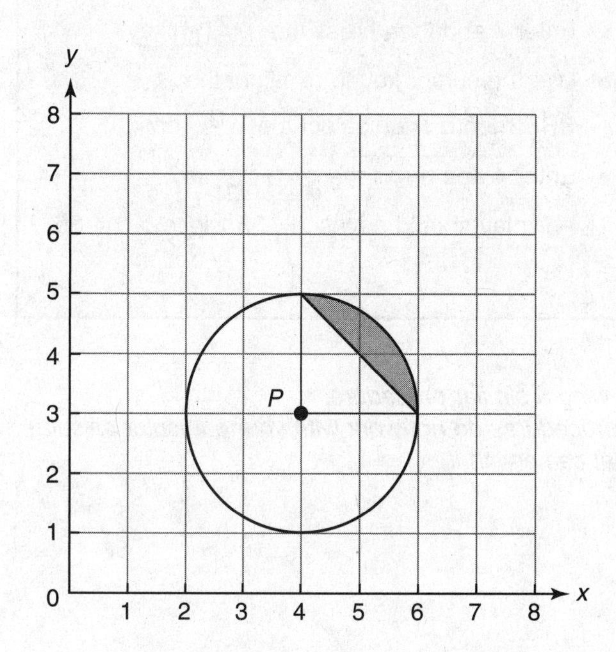

The probability that the point will lie in the shaded region equals the fraction:

$$\frac{\text{area of shaded region}}{\text{area of circle}}$$

1. Since the radius of the circle is 2, the area of the circle is $\pi \times 2^2 = 4\pi$.
2. The area of the right triangle is $\frac{1}{2} \times 2 \times 2 = 2$. The region bounded by the radii and the circle whose chord is the hypotenuse of the right triangle is a quarter of a circle, so its area is one-fourth of the area of the circle: $\frac{1}{4} \times 4\pi = \pi$. The area of the shaded region is $\pi - 2$.
3. The probability that the point will lie in the shaded region is $\frac{\pi - 2}{4\pi}$.

FACTORIAL NOTATION

- If n is a positive integer, then the notation $n!$ represents the product of all integers from n to 1, as in

$$5! = 5 \times 4 \times 3 \times 2 \times 1 = 120$$

COUNTING FORMULAS

- The number of different arrangements in which order matters of n objects is $n!$. For example, four boys can be arranged in a line in $4! = 4 \times 3 \times 2 \times 1 = 24$ different ways.
- The number of different ways in which r objects can be selected from a larger set of n objects when order does not matter is represented by the notation ${}_nC_r$ where

$${}_nC_r = \frac{n!}{r! \times (n-r)!}$$

For example, from a group of 7 students, a committee of 3 students can be selected in ${}_nC_r$ different ways, where $n = 7$ and $r = 3$:

$$\begin{aligned} {}_7C_3 &= \frac{7!}{3! \times (7-3)!} \\ &= \frac{7 \times 6 \times 5 \times \cancel{4}!}{3! \times \cancel{4}!} \\ &= \frac{7 \times \cancel{6} \times 5}{\cancel{3 \times 2} \times 1} \\ &= 35 \end{aligned}$$

Rather than memorize this formula, learn how your calculator can be used to do the work.

Using a Calculator to Evaluate ${}_nC_r$

Scientific Calculator

With a scientific calculator, you can use a "second" function key to evaluate combinations. To evaluate ${}_7C_4$:

- Enter 7.
- Press the [INV] or [2nd] or [SHIFT] key (depending on your calculator) and then press the key that has ${}_nC_r$ labeled above it.
- Enter 4 and press the [=] key.

The display should now show 35, since ${}_7C_4 = 35$.

Graphing Calculator

Using a graphing calculator, such as the TI 83/84, you can evaluate combinations from the **PRB** (**PR**o**B**ability) menu. To evaluate ${}_7C_4$:

- Enter 7 and then press the [MATH] key.
- Use the cursor key to highlight the **PRB** menu. Then select the ${}_nC_r$ option.
- Enter 4 and press the [ENTER] key.

The display should now show 35, since ${}_7C_4 = 35$.

NOTE:

- *The expressions ${}_nP_r$ and $n!$ can be evaluated following a similar procedure.*
- *Not all calculators work in the same way. If these procedures do not work with your particular calculator, you will need to read the instruction manual that came with it.*

Algebra II: Lessons 8-1 to 8-6

EXPONENT LAWS

- $x^0 = 1$

  EXAMPLE: $7^0 = 1$

- $x^{-n} = \frac{1}{x^n}$

  EXAMPLE: $5^{-2} = \frac{1}{5^2} = \frac{1}{25}$ and $\frac{1}{2^{-3}}$

  $= 2^3 = 8$

- $\left(\frac{a}{b}\right)^{-n} = \left(\frac{b}{a}\right)^n$

  EXAMPLE: $\left(\frac{4}{3}\right)^{-2} = \left(\frac{3}{4}\right)^2 = \frac{9}{16}$

- $x^{\frac{1}{n}} = \sqrt[n]{x}$ when n is a positive integer greater than 2.

  EXAMPLE: $8^{\frac{1}{3}} = \sqrt[3]{8} = 2$

- $x^{\frac{k}{n}} = \sqrt[n]{x^k} = \left(\sqrt[n]{x}\right)^k$ where k and n are positive integers, $k \geq 1$, $n \geq 2$, and $\frac{k}{n}$ is in lowest terms.

  EXAMPLE: $64^{\frac{2}{3}} = \left(\sqrt[3]{64}\right)^2 = (4)^2 = 16$

ABSOLUTE VALUE

- To remove an absolute value sign, use these rules, where d is a positive number:

  1. If $|ax \pm b| = d$, then $ax \pm b = d$ or $ax \pm b = -d$.

  2. If $|ax \pm b| \leq d$, then $\pm d \leq ax \pm b \leq d$.

  3. If $|ax \pm b| \geq d$, then $ax \pm b \leq -d$ or $ax \pm b \geq d$.

  EXAMPLE: If $|x - 3| = 5$, then $x - 3 = 5$ or $x - 3 = -5$

RADICAL EQUATIONS

- To solve a radical equation, isolate the radical and then raise both sides of the equation to the power that eliminates the radical.

  EXAMPLE: If $\sqrt{2x + 1} - 3 = 4$, then:

$$\sqrt{2x + 1} = 7$$
$$\left(\sqrt{2x + 1}\right)^2 = 7^2$$
$$2x + 1 = 49$$
$$2x = 48$$
$$x = \frac{48}{2} = 24$$

EXPONENTIAL EQUATIONS

- To solve an equation in which the variable is in an exponent, rewrite each side of the equation as a power of the same base. Then solve the equation obtained by setting the exponents equal to each other.

  EXAMPLE: If $16^{x+4} = 32^{2x-10}$, then

$$16^{x+4} = 32^{2x-10}$$
$$(2^4)^{x+4} = (2^5)^{2x-10}$$
$$2^{4x+16} = 2^{10x-50}$$
$$4x + 16 = 10x - 50$$
$$-6x = -66$$
$$x = \frac{-66}{-6} = 11$$

FUNCTIONS

- A *function* is a set of ordered pairs in which no two ordered pairs of the form (x,y) have the same x-value but have different y-values. If $f(x) = 3^x - 1$, then $f(2)$ is the value of y when x is replaced by 2 in the equation $y = 3^x - 1$. Thus:
  $$y = f(2) = 3^2 - 1 = 9 - 1 = 8,$$
  since $f(2) = 8$, $(2,8)$ is a point on the graph of $f(x) = 3^x - 1$.
- For a function represented as a graph, the *domain* is the set of the x-coordinates of all points, and only those points, that lie on the graph. The corresponding set of y-coordinates of those points is the *range* of the function. Unless otherwise indicated, you can assume that the *domain* of any function f is the set of real numbers x for which $f(x)$ is a real number. The *range* is the set of values that y takes on as x runs through each of its possible values. For the function $f(x) = \sqrt{x - 3}$, only values of x greater than or equal to 3 are allowed. Hence, the domain is the set of all numbers $x \geq 3$. Since the square root of a nonnegative number is also nonnegative, the range is the set of all numbers $y \geq 0$.

Vertical Line Test

- A graph may or may not represent a function. A graph is a function if any vertical line intersects it in *at most* one point.

LINEAR FUNCTION

- A *linear function* is a function that can be written in the form $y = mx + b$, where $m \neq 0$. The graph of a linear function is a line whose slope represents the constant rate at which y changes for each unit change in x.

ABSOLUTE VALUE FUNCTION

- An *absolute value function* is a function that can be written in the form $y = |x|$. Here is the graph of $y = |x|$:

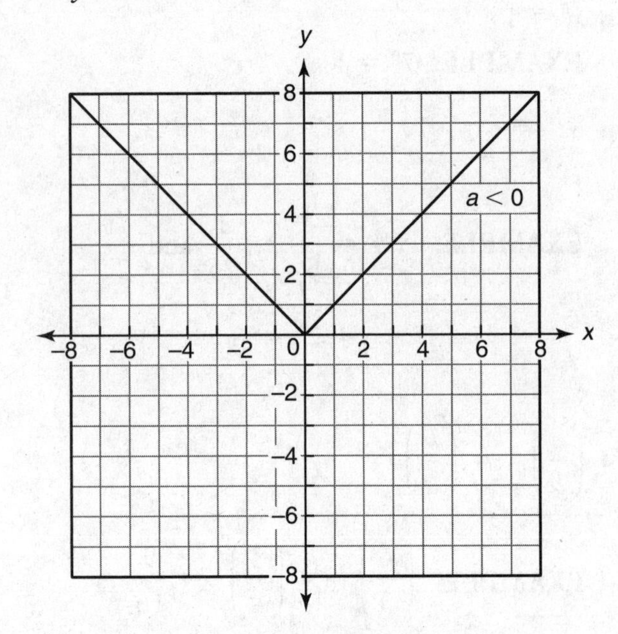

QUADRATIC FUNCTION

- A *quadratic function* is a function that can be written in the form $y = ax^2 + bx + c$, where a, b, and c stand for numbers and $a \neq 0$. The graph of a quadratic function in x is a "**U**-shaped" curve called a *parabola* with a vertical axis of symmetry.
- When $a > 0$ the parabola has a *minimum* turning point, as shown in the accompanying graph:

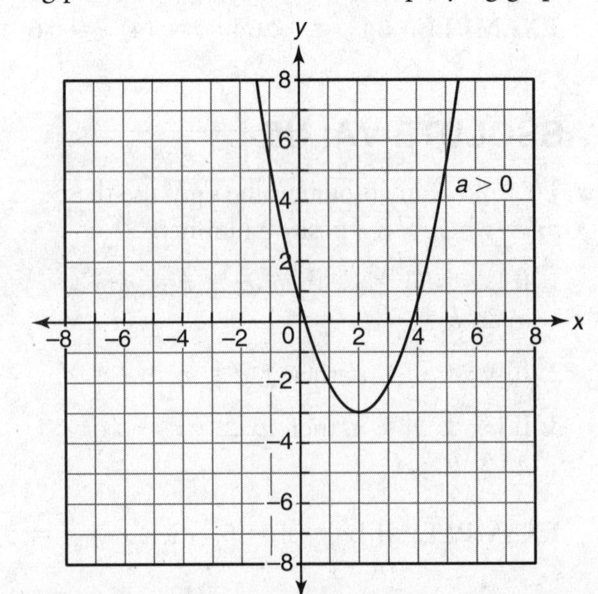

- When $a < 0$ the parabola has a *maximum* turning point, as shown in the accompanying graph:

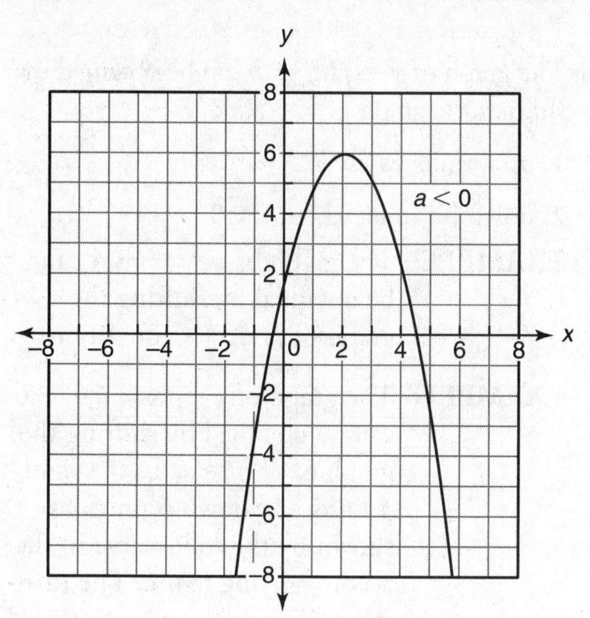

EXPONENTIAL FUNCTION

- An *exponential function* is a function that can be written in the form $y = b^x$, where b stands for a positive number that is not equal to 1.

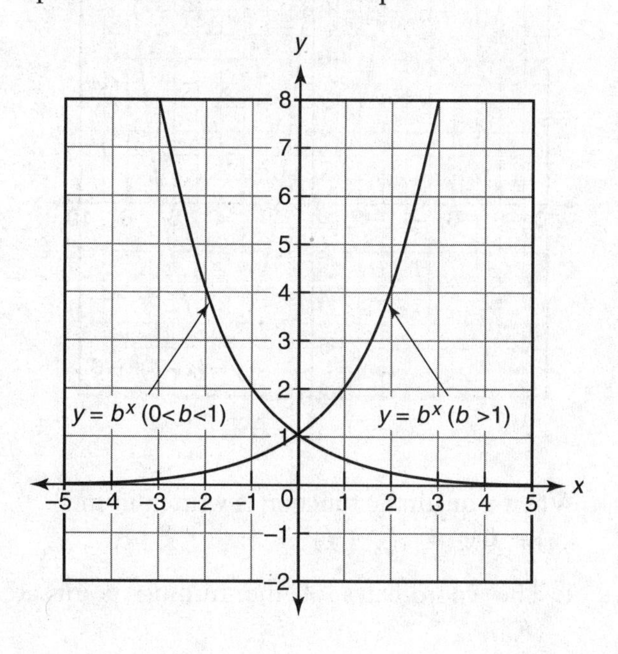

TRANSFORMATIONS

- A *transformation* of the graph of a function "moves" each point of the graph according to a given rule. If two functions are related to each other by a simple transformation, then knowing what the graph of one function looks like enables you to easily determine what the graph of the related function looks like.

EXAMPLE: Replacing $f(x)$ with $-f(x)$ reflects the graph of the function f in the x-axis. For example, $y = f(x) = x^2$ is a parabola, as shown by the solid curve in the accompanying figure. The graph of $y = -x^2$ can be obtained by reflecting ("flipping") the graph of $y = x^2$ over the x-axis, as shown by the broken curve in the accompanying figure.

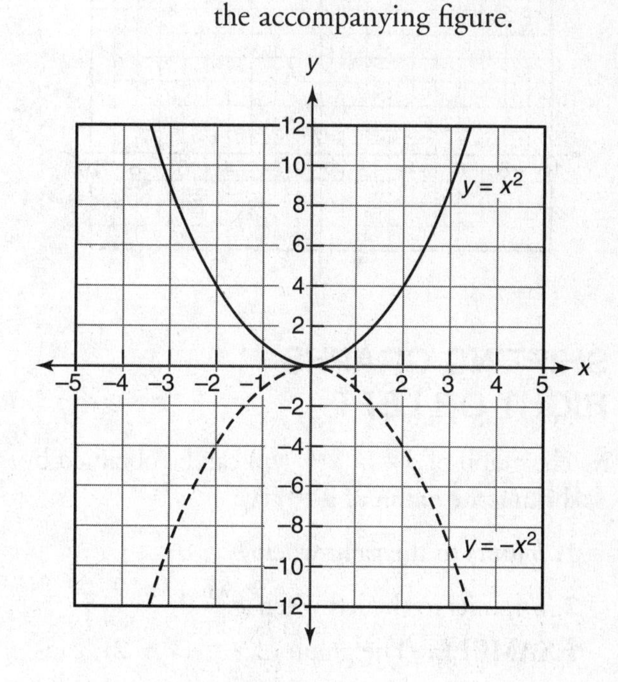

FLIPPING GRAPHS OVER AXES

- The graph of $y = -f(x)$ is the reflection of the graph of $y = f(x)$ over the x-axis.
- The graph of $y = f(-x)$ is the reflection of the graph of $y = f(x)$ over the y-axis.

EXAMPLE: The graphs of $y = f(x) = \sqrt{x}$ and $y = f(-x) = \sqrt{-x}$ are reflections of each other over the y-axis:

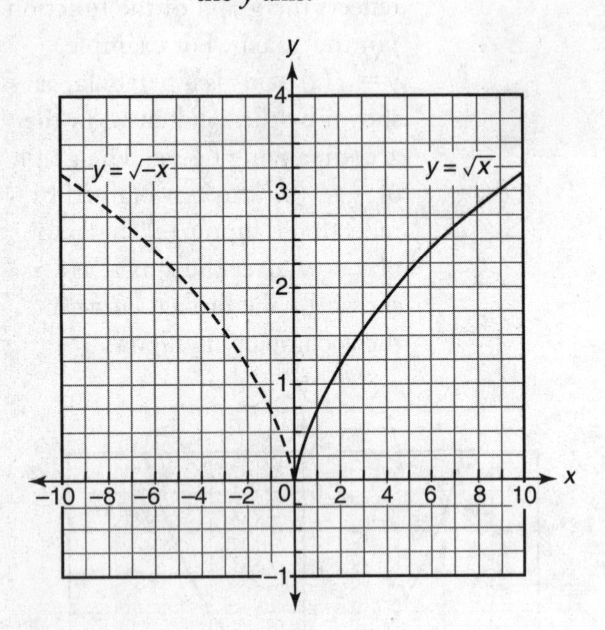

SHIFTING GRAPHS RIGHT OR LEFT

- The graph of $y = f(x - (h))$ can be obtained by shifting the graph of $y = f(x)$:

1. h units to the right when $h > 0$
2. $|h|$ units to the left when $h < 0$

EXAMPLE: The graph of $y = (x - 2)^2$ can be obtained by shifting the graph of $f(x) = x^2$ to the right 2 units since $h = +2$.

EXAMPLE: The graph of $y = (x + 3)^2$ can be obtained by shifting the graph of $f(x) = x^2$ to the left 3 units, since $y = (x + 3)^2 = (x - (-3))^2$, so $h = -3$.

SHIFTING GRAPHS UP OR DOWN

- The graph of $y = f(x) + h$ can be obtained by shifting the graph of $y = f(x)$:

1. up h units when $h > 0$
2. down $|h|$ units when $h < 0$

EXAMPLE: The graph of $y = x^2 + 4$ can be obtained by shifting the graph of $f(x) = x^2$ up 4 units.

EXAMPLE: The graph of $y = (x - 4)^2 - 6$ can be obtained by shifting the graph of $f(x) = x^2$ to the right 4 units and down 6 units, as shown by the solid curve in the accompanying figure. The turning point of the new graph is (4,6).

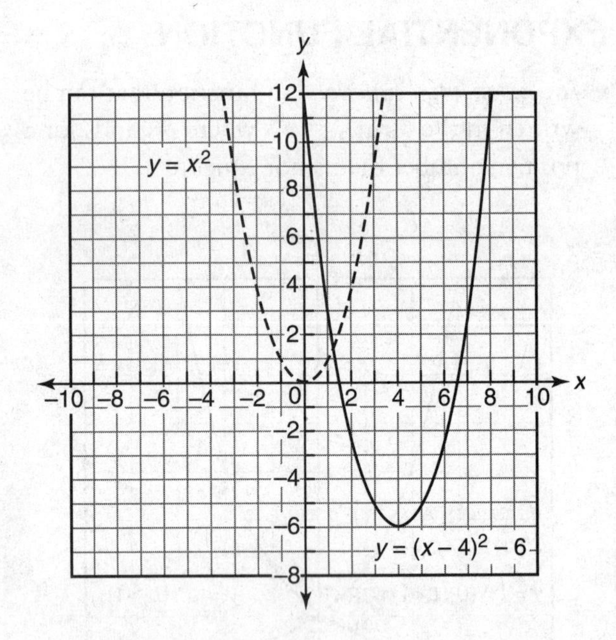

- When a quadratic function is written in the form $f(x) = a(x - h)^2 + k$:

1. The coordinates of the turning point are (h,k).
2. If $a > 0$, the turning point is the lowest point on the graph.
3. If $a < 0$, the turning point is the highest point on the graph.